INTRODUCTION TO THE THEORY OF COMPLEX SYSTEMS

Introduction to the Theory of Complex Systems

Stefan Thurner, Rudolf Hanel, and Peter Klimek

OXFORD
UNIVERSITY PRESS

OXFORD
UNIVERSITY PRESS

Great Clarendon Street, Oxford, OX2 6DP,
United Kingdom

Oxford University Press is a department of the University of Oxford.
It furthers the University's objective of excellence in research, scholarship,
and education by publishing worldwide. Oxford is a registered trade mark of
Oxford University Press in the UK and in certain other countries

Published in the United States of America by Oxford University Press
198 Madison Avenue, New York, NY 10016, United States of America

British Library Cataloguing in Publication Data
Data available

Library of Congress Control Number: 2018947065

ISBN 978–0–19–882193–9
DOI: 10.1093/oso/9780198821939.001.0001

Printed and bound by
CPI Group (UK) Ltd, Croydon, CR0 4YY

Links to third party websites are provided by Oxford in good faith and
for information only. Oxford disclaims any responsibility for the materials
contained in any third party website referenced in this work.

Preface

This book is for people who are interested in the science of complex adaptive systems and wish to have more than just a casual understanding of it. As with all the sciences, understanding of complex adaptive systems is reached solely in a quantitative, predictive, and ultimately experimentally testable manner. Complex adaptive systems are dynamical systems that are able to change their structure, their interactions, and, consequently, their dynamics as they evolve in time. This is not a book about complicated systems, even though most complex systems are complicated. Indeed, over the last 300 years, scientists have usually dealt with complicated systems that are neither complex nor adaptive.

The theory of complex systems is the theory of generalized time-varying interactions between elements that are characterized by states. Interactions typically take place on networks that connect those elements. The interactions involved may cause the states of the elements themselves to alter over time. The essence of a complex system is that the interaction networks may change and rearrange as a consequence of changes in the states of the elements. Thus, complex systems are systems whose states change as a result of interactions and whose interactions change concurrently as a result of states. Due to this chicken–egg-type problem, complex systems show an extremely rich spectrum of behaviour: they are adaptive and co-evolutionary; they show path-dependence, emergence, power laws; they have rich phase diagrams; they produce and destroy diversity; they are inherently prone to collapse; they are resilient, and so on. The theory of complex systems tries to understand these properties based on its building blocks and on the interactions between those building blocks that take place on networks. It combines mathematical and physical principles with concepts borrowed from biology and the social sciences; it uses new computational techniques and, with the advent of comprehensive large-scale data sets, is becoming experimentally testable. The goal of the theory of complex systems is to understand the dynamical *systemic* outcomes of interconnected systems, and its ultimate goal is to eventually control and design systemic properties of systems such as the economy, the financial system, social processes, cities, the climate, and ecology. The theory of complex systems builds partly on previous attempts to understand systems that interact in non-trivial ways, such as game theory, cybernetics, or systems theory. However, in its current state, the science of complex systems goes well beyond these earlier developments, in so many ways, in fact, that it can be regarded as an independent scientific branch, which—due to its quantitative, predictive, and testable nature—is a natural science.

Even though it is fair to say that the theory of complex systems is not yet complete, in recent years, it has become quite clear just what the theory is going to look like. Its elements and structure are emerging. The current state of the theory of complex systems is comparable perhaps to the state of quantum mechanics in the 1920s, before the famous Copenhagen meetings and Werner Heisenberg's book. At that time, quantum mechanics was a collection of experimental and theoretical bits and pieces, which had not yet been seen within a fully comprehensive framework. Nevertheless, it was clear that, one day soon, such a framework would exist. The present situation can be compared to an archaeological project, where a mosaic floor has been discovered and is being excavated. While the mosaic is only partly visible and the full picture is still missing, several facts are becoming clear: the mosaic exists; it shows identifiable elements (for instance, people and animals engaged in recognizable activities); there are large patches missing or still invisible, but experts can already tell that the mosaic represents a scene from, say, Homer's *Odyssey*. Similarly, for dynamical complex adaptive systems, it is clear that a theory exists that, eventually, can be fully developed. There are those who say that complex systems will never be understood or that, by their very nature, they are incomprehensible. This book will demonstrate that such statements are incorrect. The elements of a theory of complex systems are becoming clear: dynamical multilayer networks, scaling, statistical mechanics of algorithmic dynamics, evolution and co-evolution, and information theory. The essence of this book is to focus on these components, clarify their meaning in the context of complex systems, and endow the reader with a mathematical skill set to be applied on concrete problems in the world of complex systems.

The book is written in mathematical language because this is the only way to express facts in a quantitative and predictive manner and to make statements that are unambiguous. We aim for consistency. The book should be comprehensible so that no-one with an understanding of basic calculus, linear algebra, and statistics need refer to other works. The book is particularly designed for graduate students in physics or mathematics. We try to avoid ambiguous statements while, at the same time, being as general as possible. The hope is that this work will serve as a textbook and as a starting point for journeys into new and unexplored territory.

Many complex systems are often sensitive to details in their internal setup, to initial and to boundary conditions. Concepts that proved to be extremely robust and effective in non-complex systems, such as the central limit theorem, classical statistical mechanics, or information theory, lose their predictive power when confronted with complex systems. Extreme care is thus needed in any attempt to apply these otherwise distinguished concepts to complex systems: doing so could end in confusion and nonsensical results. In several concrete examples, we will demonstrate the importance of understanding what these methods mean in the context of complex systems and whether they can or cannot be applied. We will discuss how some of these classical concepts can be generalized to become useful for understanding complex systems.

The book is also a statement about our belief that the exact sciences may be entering a phase of transition from a traditional analytical description of nature, as used with tremendous success since Galileo and Newton, towards an *algorithmic* description. Whereas the analytical description of nature is, conceptually, based largely on differential equations and analytical equations of motion, the algorithmic view takes into account evolutionary and co-evolutionary aspects of dynamics. It provides a framework for systems that can endogenously change their internal interaction networks, rules of functioning, dynamics, and even environment, as they evolve in time. Algorithmic dynamics, which is characteristic of complex dynamical systems, may be a key to the quantitative and predictive understanding of many natural and man-made systems. In contrast to physical systems, which typically evolve analytically, algorithmic dynamics describe certainly how living, social, environmental, and economic systems unfold. This algorithmic view is not new but has been advocated by authors like Joseph A. Schumpeter, Stuart Kauffman, and Brian Arthur. However, it has not, to date, been picked up by mainstream science, and it has never been presented in the context of the theory of complex systems.

This book is based on a two-semester course, that has been held at the Medical University of Vienna since 2011. We are grateful to our students and to Kathryn Platzer and Anita Wanjek for helping us with the manuscript.

ST Vienna January 2018

Contents

1

Introduction to Complex Systems

1.1 Physics, biology, or social science?

The science of complex systems is not an offspring of physics, biology, or the social sciences, but a unique mix of all three. Before we discuss what the science of complex systems is or is not, we focus on the sciences from which it has emerged. By recalling what physics, biology, and the social sciences are, we will develop an intuitive feel for complex systems and how this science differs from other disciplines. This chapter thus aims to show that the science of complex systems combines physics, biology, and the social sciences in a unique blend that is a new discipline in its own right. The chapter will also clarify the structure of the book.

1.2 Components from physics

Physics makes quantitative statements about natural phenomena. Quantitative statements can be formulated less ambiguously than qualitative descriptions, which are based on words. Statements can be expressed in the form of predictions in the sense that the trajectory of a particle or the outcome of a process can be anticipated. If an experiment can be designed to test this prediction unambiguously, we say that the statement is experimentally testable. Quantitative statements are validated or falsified using quantitative measurements and experiments.

> Physics is the experimental, quantitative, and predictive science of matter and its interactions.

Pictorially, physics progresses by putting specific questions to nature in the form of experiments; surprisingly, if the questions are well posed, they result in concrete answers that are robust and repeatable for an arbitrary number of times by anyone who can do the same experiment. This method of generating knowledge about nature, by using experiments to ask questions of it, is unique in the history of humankind and is called the *scientific method*. The scientific method has been at the core of all technological progress since the time of the Enlightenment.

Introduction to the Theory of Complex Systems. Stefan Thurner, Rudolf Hanel, and Peter Klimek,
Oxford University Press (2018). © Stefan Thurner, Rudolf Hanel, and Peter Klimek.
DOI: 10.1093/oso/9780198821939.001.0001

Physics deals with matter at various scales and levels of granularity, ranging from macroscopic matter like galaxies, stars, planets, stones, and projectiles, to the scale of molecules, atoms, hadrons, quarks, and gauge bosons. There are four fundamental forces at the core of all interactions between all forms of matter: gravity, electromagnetism and two types of nuclear force: the weak and the strong. According to quantum field theory, all interactions in the physical world are mediated by the exchange of gauge bosons. The graviton, the boson for gravity, has not yet been confirmed experimentally.

1.2.1   The nature of the fundamental forces

The four fundamental forces are very different in nature and strength. They are characterized by a number of properties that are crucial for understanding how and why it was possible to develop physics without computers. These properties are set out here.

Usually, the four fundamental forces are homogeneous and isotropic in space (and time). Forces that are homogeneous act in the same way everywhere in space; forces that are isotropic are the same, regardless of the direction in which they act. These two properties drastically simplify the mathematical treatment of interactions in physics. In particular, forces can be written as derivatives of potentials, two-body problems can effectively be treated as one-body problems, and the so-called mean field approach can be used for many-body systems. The mean field approach is the assumption that a particle reacts to the single field generated by the many particles around it. Often, such systems can be fully understood and solved even without computers. There are important exceptions, however; one being that the strong force acts as if interactions were limited to a 'string', where flux-tubes are formed between interacting quarks, similar to type II superconductivity.

The physical forces differ greatly in strength. Compared to the strong force, the electromagnetic force is about a thousand times weaker, the weak force is about 10^{16} times weaker, and the gravitational force is only 10^{-41} of the strength of the strong force [405]. When any physical phenomenon is being dealt with, usually only a single force has to be considered. All the others are small enough to be safely neglected. Effectively, the superposition of four forces does not matter; for any phenomenon, only one force

Matter	Interaction types	Characteristic length scale
macroscopic matter	gravity, electromagnetism	all ranges
molecules	electromagnetism	all ranges
atoms	electromagnetism, weak force	$\sim 10^{-18}$ m
hadrons and leptons	electromagnetism, weak and strong force	$10^{-18} - 10^{-15}$ m
quarks and gauge bosons	electromagnetism, weak and strong force	$10^{-18} - 10^{-15}$ m

is relevant. We will see that this is drastically different in complex systems, where a multitude of different interaction types of similar strength often have to be taken into account simultaneously.

Typically, physics does not specify which particles interact with each other, as they interact in identical ways. The interaction strength depends only on the relevant interaction type, the form of the potential, and the relative distance between particles. In complex systems, interactions are often *specific*. Not all elements, only certain pairs or groups of elements, interact with each other. Networks are used to keep track of which elements interact with others in a complex system.

1.2.2   What does predictive mean?

Physics is an experimental and a predictive science. Let us assume that you perform an experiment repeatedly; for example, you drop a stone and record its trajectory over time. The predictive or theoretical task is to predict this trajectory based on an understanding of the phenomenon. Since Newton's time, understanding a phenomenon in physics has often meant being able to describe it with differential equations. A phenomenon is understood dynamically if its essence can be captured in a differential equation. Typically, the following three-step process is then followed:

1. Find the differential equations to encode your understanding of a dynamical system. In the example of our stone-dropping experiment, we would perhaps apply Newton's equation,

$$m\frac{d^2x}{dt^2} = F(x),$$

   where t is time, $x(t)$ is the trajectory, m is mass of the stone, and F is force on the stone. In our case, we would hope to identify the force with gravity, meaning that $F = gm$.

2. Once the equation is specified, try to solve it. The equation can be solved using elementary calculus, and we get, $x(t) = x_0 + v_0 t + \frac{1}{2}gt^2$. To make a testable prediction we have to fix the boundary or initial conditions; in our case we have to specify what the initial position x_0 and initial velocity v_0 are in our experiment. Once this is done, we have a prediction for the trajectory of the stone, $x(t)$.

3. Compare the result with your experiments. Does the stone really follow this predicted path $x(t)$? If it does, you might claim that you have understood something on a quantitative, predictive, and experimental basis. If the stone (repeatedly) follows another trajectory, you have to try harder to find a better prediction.

Fixing initial or boundary conditions means simply taking the system out of its context, separating it from the rest of the universe. There are no factors, other than the boundary conditions, that influence the motion of the system from the outside. That such a separation of systems from their context is indeed possible is one reason why physics has been so successful, even before computing devices became available. For many complex systems, it is impossible to separate the dynamics from the context in a clear way. This means that many outside influences that are not under experimental control will simultaneously determine their dynamics.

In principle, the same thinking used to describe physical phenomena holds for arbitrarily complicated systems. Assume that a vector $X(t)$ represents the state of a system at a given time (e.g. all positions and momenta of its elements), we then get a set of equations of motion in the form,

$$\frac{d^2 X(t)}{dt^2} = G(X(t)),$$

where G is a high-dimensional function. Predictive means that, in principle, these equations can be solved. Pierre-Simon Laplace was following this principle when he introduced a hypothetical daemon familiar with the Newtonian equations of motion and all the initial conditions of all the elements of a large system (the universe) and thus able to solve all equations. This daemon could then predict everything. The problem, however, is that such a daemon is hard to find. In fact, these equations can be difficult, even impossible, to solve. Already for three bodies that exert a gravitational force on each other, the famous three-body problem (e.g. Sun, Earth, Moon), there is no general analytical solution provided by algebraic and transcendental functions. This was first demonstrated by Henri Poincaré and paved the way for what is today called chaos theory. In fact, the strict Newton–Laplace program of a predictable world in terms of unambiguously computable trajectories is completely useless for most systems composed of many particles. Are these large systems not then predictable? What about systems with an extremely large number of elements, such as gases, which contain of the order of $\mathcal{O}(10^{23})$ molecules?

Imagine that we perform the following experiment over and over again: we heat and cool water. We gain the insight that if we cool water to 0°C and below, it will freeze, that if we heat it to 100°C it will start to boil and, under standard conditions, ultimately evaporate. These phase transitions will happen with certainty. In that sense, they are predictable. We cannot predict from the equations of motion which molecule will be the first to leave the liquid. Given appropriate instrumentation, we can perhaps measure the velocity of a few single gas molecules at a point in time, but certainly not all 10^{23}. What can be measured is the probability distribution that a gas molecule is observed with a specific velocity v,

$$p(v) \sim v^2 \exp\left(-\frac{mv^2}{2kT}\right),$$

where T is temperature, and k is Boltzmann's constant. Given this probability distribution, it is possible to derive a number of properties of gases that perfectly describe their *macroscopic* behaviour and make them predictable on a macroscopic (or systemic) level.

For non-interacting particles, these predictions can be extremely precise. The predictions immediately start to degenerate as soon as there are strong interactions between the particles or if the number of particles is not large enough. Note that the term prediction now has a much weaker meaning than in the Newton–Laplace program. The meaning has shifted from being a description based on the exact knowledge of each component of a system to one based on a probabilistic knowledge of the system. Even though one can still make extremely precise predictions about multiparticle systems in a probabilistic framework, the concept of determinism is now diluted. The framework for predictions on a macroscopic level about systems composed of many particles on a probabilistic basis is called statistical mechanics.

1.2.3 Statistical mechanics—predictability on stochastic grounds

The aim of statistical mechanics is to understand the macroscopic properties of a system on the basis of a statistical description of its microscopic components. The idea behind it is to link the microscopic world of components with the macroscopic properties of the aggregate system. An essential concept that makes this link possible is Boltzmann–Gibbs entropy.

A system is often prepared in a macrostate, which means that aggregate properties like the temperature or pressure of a gas are known. There are typically many possible microstates that are associated with that macrostate. A microstate is a possible microscopic configuration of a system. For example, a particular microstate is one for which all positions and velocities of gas molecules in a container are known. There are usually many microstates that can lead to one and the same macrostate; for example, the temperature and pressure in the container. In statistical mechanics, the main task is to compute the probabilities for the many microstates that lead to that single macrostate. In physics, the macroscopic description is often relatively simple. Macroscopic properties are often strongly determined by the phase in which the system is. Physical systems often have very few phases—typically solid, gaseous, or liquid.

Within the Newton–Laplace framework, traditional physics works with extreme precision for very few particles or for extremely many non-interacting particles, where the statistical mechanics of Boltzmann–Gibbs applies. In other words, the class of systems that can be understood with traditional physics is not that big. Most systems are composed of many strongly interacting particles. Often, the interactions are of multiple types, are non-linear, and vary over time. Very often, such systems are complex systems.

1.2.4 The evolution of the concept of predictability in physics

The concept of prediction and predictability has changed in significant ways over the past three centuries. Prediction in the eighteenth century was quite different from the concept of prediction in the twenty-first. The concept of determinism has undergone at least three transitions [300].

In the *classical mechanics* of the eighteenth and nineteenth centuries, prediction meant the exact prediction of trajectories. Equations of motion would make exact statements about the future evolution of simple dynamical systems. The extension to more than two bodies has been causing problems since the very beginning of Newtonian physics; see, for example, the famous conflict between Isaac Newton and John Flamsteed on the predictability of the orbit of the Moon. By about 1900, when interest in understanding many-body systems arose, the problem became apparent. The theory of Ludwig Boltzmann, referred to nowadays as statistical mechanics, was effectively based on the then speculative existence of atoms and molecules, and it drastically changed the classical concept of predictability.

In *statistical mechanics*, based on the assumption that atoms and molecules follow Newtonian trajectories, the law of large numbers allows stochastic predictions to be made about the macroscopic behaviour of gases. Statistical mechanics is a theory of the macroscopic or collective behaviour of non-interacting particles. The concepts of predictability and determinism were subject to further change in the 1920s with the emergence of quantum mechanics and non-linear dynamics.

In *quantum mechanics*, the concept of determinism disappears altogether due to the fundamental simultaneous unpredictability of the position and momentum of the (sub-)atomic components of a system. However, quantum mechanics still allows us to make extremely high-quality predictions on a collective basis. Collective phenomena remain predictable to a large extent on a macro- or systemic level.

In *non-linear systems*, it became clear that even in systems for which the equations of motion can be solved in principle, the sensitivity to initial conditions can be so enormous that the concept of predictability must, for all practical purposes, be abandoned. A further crisis in terms of predictability arose in the 1990s, when interest in more general forms of interactions began to appear.

In *complex systems*, the situation is even more difficult than in quantum mechanics, where there is uncertainty about the components, but not about its interactions. For many complex systems, not only can components be unpredictable, but the interactions between components can also become specific, time-dependent, non-linear, and unpredictable. However, there is still hope that probabilistic predictions about the dynamics and the collective properties of complex systems are possible. Progress in the science of complex systems will, however, be impossible without a detailed understanding of the dynamics of how elements specifically interact with each other. This is, of course, only possible with massive computational effort and comprehensive data.

1.2.5 Physics is analytic, complex systems are algorithmic

Physics largely follows an *analytical* paradigm. Knowledge of phenomena is expressed in analytical equations that allow us to make predictions. This is possible because interactions are homogeneous, isotropic, and of a single type. Interactions in physics typically do not change over time. They are usually given and fixed. The task is to work out specific solutions regarding the evolution of the system for a given set of initial and boundary conditions.

This is radically different for complex systems, where interactions themselves can change over time as a consequence of the dynamics of the system. In that sense, complex systems change their internal interaction structure as they evolve. Systems that change their internal structure dynamically can be viewed as *machines* that change their internal structure as they operate. However, a description of the operation of a machine using analytical equations would not be efficient. Indeed, to describe a steam engine by seeking the corresponding equations of motion for all its parts would be highly inefficient. Machines are best described as *algorithms*—a list of rules regarding how the dynamics of the system updates its states and future interactions, which then lead to new constraints on the dynamics at the next time step. First, pressure builds up here, then a valve opens there, vapour pushes this piston, then this valve closes and opens another one, driving the piston back, and so on.

Algorithmic descriptions describe not only the evolution of the states of the components of a system, but also the evolution of its internal states (interactions) that will determine the next update of the states at the next time step. Many complex systems work in this way: states of components and the interactions between them are simultaneously updated, which can lead to the tremendous mathematical difficulties that make complex systems so hard to understand. These difficulties in their various forms will be addressed time and again in this book. Whenever it is possible to ignore the changes in the interactions in a dynamical system, analytic descriptions become meaningful.

> Physics is generally analytic, complex systems are algorithmic. Quantitative predictions that can be tested experimentally can be made within the analytic or the algorithmic paradigm.

1.2.6   What are complex systems from a physics point of view?

From a physics point of view, one could try to characterize complex systems by the following extensions to physics.

- Complex systems are composed of many elements, components, or particles. These elements are typically described by their state, such as velocity, position, age, spin, colour, wealth, mass, shape, and so on. Elements may have stochastic components.

- Elements are not limited to physical forms of matter; anything that can interact and be described by states can be seen as generalized matter.

- Interactions between elements may be specific. Who interacts with whom, when, in what form, and how strong is described by interaction networks.

- Interactions are not limited to the four fundamental forces, but can be of a more complicated type. Generalized interactions are not limited to the exchange of gauge bosons, but can be mediated through exchange of messages, objects, gifts, information, even bullets, and so on.

continued

- Complex systems may involve superpositions of interactions of similar strengths.
- Complex systems are often chaotic in the sense that they depend strongly on the initial conditions and details of the system. Update equations that algorithmically describe the dynamics are often non-linear.
- Complex systems are often driven systems. Some obey conservation laws, some do not.
- Complex systems can exhibit a rich phase structure and have a huge variety of macrostates that often cannot be inferred from the properties of the elements. This is sometimes referred to as *emergence*. Simple forms of emergence are, of course, already present in physics. The spectrum of the hydrogen atom or the liquid phase of water are emergent properties of the involved particles and their interactions.

With these extensions, we can derive a physics-based definition for what the theory of complex systems is.

The theory of complex systems is the quantitative, predictive and experimentally testable science of generalized matter interacting through generalized interactions.

Generalized interactions are described by the interaction type and who interacts with whom at what time and at what strength. If there are more than two interacting elements involved, interactions can be conveniently described by time-dependent networks,

$$M_{ij}^{\alpha}(t),$$

where i and j label the elements in the system, and α denotes the interaction type. $M_{ij}^{\alpha}(t)$ are matrix elements of a structure with three indices. The value $M_{ij}^{\alpha}(t)$ indicates the strength of the interaction of type α between element i and j at time t. $M_{ij}^{\alpha}(t)=0$ means no interaction of that type. Interactions in complex systems remain based on the concept of exchange; however, they are not limited to the exchange of gauge bosons. In complex systems, interactions can happen through communication, where messages are exchanged, through trade where goods and services are exchanged, through friendships, where bottles of wine are exchanged, and through hostility, where insults and bullets are exchanged.

Because of more specific and time-varying interactions and the increased variety of types of interaction, the variety of macroscopic states and systemic properties increases drastically in complex systems. This diversity increase of macrostates and phenomena emerges from the properties both of the system's components and its interactions. The phenomenon of collective properties arising that are, a priori, unexpected from the elements alone is sometimes called *emergence*. This is mainly a consequence of the presence of generalized interactions. Systems with time-varying generalized interactions can exhibit an extremely rich phase structure and may be adaptive. Phases may co-exist in particular complex systems. The plurality of macrostates in a system leads to new types of questions that can be addressed, such as: what is the number of macrostates? What are their co-occurrence rates? What are the typical sequences of occurrence? What are the life-times of macrostates? What are the probabilities of transition between macrostates? As yet, there are no general answers to these questions, and they remain a challenge for the theory of complex systems. For many complex systems, the framework of physics is incomplete. Some of the missing concepts are those of non-equilibrium, evolution, and co-evolution. These concepts will be illustrated in the sections that follow.

1.2.7   A note on chemistry—the science of equilibria

In chemistry, interactions between atoms and molecules are specific in the sense that not every molecule binds to (interacts with) any other molecule. So why is chemistry usually not considered to be a candidate for a theory of complex systems? To a large extent, chemistry is based on the law of mass action. Many particles interact in ways that lead to equilibrium states. For example, consider two substances A and B that undergo a reaction to form substances S and T,

$$\alpha A + \beta B \rightleftharpoons \sigma S + \tau T,$$

where $\alpha, \beta, \sigma, \tau$ are the stoichiometric constants, and k_+ and k_- are the forward and backward reaction rates, respectively. The forward reaction happens at a rate that is proportional to $k_+\{A\}^\alpha\{B\}^\beta$, the backward reaction is proportional to $k_-\{S\}^\sigma\{T\}^\tau$. The brackets indicate the active (reacting) masses of the substances. Equilibrium is attained if the ratio of the reaction rates equals a constant K,

$$K = \frac{k_+}{k_-} = \frac{\{S\}^\sigma\{T\}^\tau}{\{A\}^\alpha\{B\}^\beta}.$$

Note that the solution to this equation gives the stationary concentrations of the various substances. Technically, these equations are fixed point equations. In contrast to chemical reactions and statistical mechanics, many complex systems are characterized by being out-of-equilibrium. Complex systems are often so-called driven systems, where the system is (exogenously) driven away from its equilibrium states. If there is no equilibrium, there is no way of using fixed-point-type equations to solve the problems. The mathematical difficulties in dealing with out-of-equilibrium or non-equilibrium systems are tremendous and generally beyond analytical reach. One way that offers a handle on understanding driven out-of-equilibrium systems is the concept of self-organized criticality, which allows essential elements of the statistics of complex systems to be understood; in particular, the omnipresence of power laws.

Many complex systems are driven systems and are out-of-equilibrium.

By comparing the nature of complex systems and basic equilibrium chemistry, we learn that the mere presence of specific interactions does not automatically lead us to complex systems. However, cyclical catalytic chemical reactions [22, 113, 205], are classic prototypes of complex systems.

1.3 Components from the life sciences

We now present several key features of complex systems that have been adopted from biology. In particular, we discuss the concepts of evolution, adaptation, self-organization, and, again, networks.

The life sciences describe the experimental science of living matter. What is living matter? A reasonable minimal answer has been attempted by the following three statements [223]:

- Living matter must be self-replicating.
- It must run through at least one Carnot cycle.
- It must be localized.

Life without self-replication is not sustainable. It is, of course, conceivable that non-self-replicating organisms can be created that live for a time and then vanish and have to be recreated. However, this is not how we experience life on the planet, which is basically a single, continuous, living germ line that originated about 3.5 billion years ago, and has existed ever since. A Carnot cycle is a thermodynamic cyclical process that converts thermal energy into work, or vice versa. Starting from an initial state, after the cycle is completed, the system returns to the same initial state. The notion that living matter must perform at least one Carnot cycle is motivated by the fact that all living organisms use energy gradients (usually thermal) to perform work of some kind. For example, this work could be used for moving or copying DNA molecules. This view also pays tribute to the fact that all living objects are out-of-equilibrium and constantly driven by energy gradients. If, after performing work, a system were not able to reach its previous states, it would be hard to call it a living system. Both self-replication and Carnot cycles require some sort of localization. On this planet, this localization typically happens at the level of cells.

Living matter uses energy and performs work on short timescales without significantly transforming itself. It is constantly driven by energy gradients and is out-of-equilibrium. Self-replication and Carnot cycles require localization.

1.3.1 Chemistry of small systems

Living matter, as we know it on this planet, is a self-sustained sequence of genetic activity over time. By genetic activity we mean that genes (locations on the DNA) can be turned

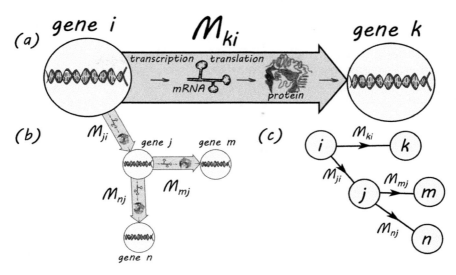

Figure 1.1 *Schematic view of genetic activity and what a link M_{ki} means in a genetic regulatory network. (a) Gene i activates gene k if something like the following process takes place: the activity of gene i means that a specific sub-sequence of the deoxyribonucleic acid (DNA) (gene) is copied into a complementary structure, an mRNA molecule. This mRNA molecule from gene i, might get 'translated' (copied again) into a protein of type i. This protein can bind with other proteins to form a cluster of proteins, a 'complex'. Such complexes can bind to other regions of the DNA, say, the region that is associated with gene k, and thereby cause the activation of gene k. (b) Gene i causes gene j to become active, which activates genes m and n. (c) The process, where the activity of gene i triggers the activity of other genes, can be represented as a directed genetic regulatory network. Complexes can also deactivate genes. If gene j is active, a complex might deactivate it.*

'on' and 'off'. If a gene is on, it triggers the production of molecular material, such as ribonucleic acid (RNA) that can later be translated into proteins. A gene is typically turned on by a cluster of proteins that bind to each other to form a so-called 'complex'. If such a cluster binds to a specific location on the DNA, this could cause a copying process to be activated at this position; the gene is then active or 'on'; see Figure 1.1.

Genetic activity is based on chemical reactions that take place locally, usually within cells or their nuclei. However, these chemical reactions are special in the sense that only a few molecules are involved [341]. In traditional chemistry, reactions usually involve billions of atoms or molecules. What happens within a cell is chemistry with a *few* molecules. This immediately leads to a number of problems:

- It can no longer be assumed that molecules meet by chance to react.
- With only a few molecules present that might never meet to react, the concept of equilibrium becomes useless.
- Without equilibrium, there is no law of mass action.

If there is no law of mass action, how can chemistry be done? Classical equilibrium chemistry is inadequate for dealing with molecular mechanisms in living matter. In cells, molecules are often actively transported from the site of production (typically, the nucleus, for organisms that have one) to where they are needed in the cell. This means that diffusion of molecules no longer follows the classical diffusion equation. Instead, molecular transport is often describable by an anomalous diffusion equation of the form,

$$\frac{d}{dt}p(x,t) = D\frac{d^{2+v}}{dx^{2+v}}p(x,t)^{\mu},$$

where $p(x,t)$ is the probability of finding a molecule at position x at time t, D is the diffusion constant, and μ and v are exponents that make the diffusion equation non-linear.

Chemical binding often depends on the three-dimensional structure of the molecules involved. This structure can depend on the 'state' of the molecules. For example, a molecule can be in a normal or a phosphorylated state. Phosphorylation happens through the addition of a phosphoryl group (PO_3^{2-}) to a molecule, which may change its entire structure. This means that for a particular state of a molecule it binds to others, but does not bind if it is in the other state. A further complication in the chemistry of a few particles arises with the reaction rates. By definition, the term reaction rate only makes sense for sufficiently large systems. The speed of reactions depends crucially on the statistical mechanics of the underlying small system and fluctuation theorems may now become important [122].

1.3.2 Biological interactions happen on networks—almost exclusively

Genetic regulation governs the temporal sequence of the abundance of proteins, nucleic material, and metabolites within any living organism. To a large extent, genetic regulation can be viewed as a discrete interaction: a gene is active or inactive; a protein binds to another or it does not; a molecule is phosphorylated or not. Discrete interactions are well-described by networks. In the context of the life sciences, three well-known networks are the metabolic network, the protein–protein binding network, and the Boolean gene-regulatory network. The metabolic network[1] is the set of linked chemical reactions occurring within a cell that determine the cell's physiological and biochemical properties. The metabolic network is often represented in networks of chemical reactions, where nodes represent substances and directed links (arrows) correspond to reactions or catalytic influences. The protein–protein networks represent empirical findings about protein–protein interactions (binding) in network representations [102]. Nodes are proteins, and links specify the interaction type between them. Different interaction types include stable, transient, and homo- or hetero-oligomer interactions.

---

[1] For an example of what metabolic networks look like, see http://biochemical-pathways.com/#/map/1

1.3.3 What is evolution?

'Nothing in biology makes sense except in the light of evolution'. Theodosius Dobzhansky

Evolution is a natural phenomenon. It is a process that increases and destroys diversity, and it looks like both a 'creative' and a 'destructive' process. Evolution appears in biological, technological, economical, financial, historical, and other contexts. In that sense, evolutionary dynamics is universal. Evolutionary systems follow characteristic dynamical and statistical patterns, regardless of the context. These patterns are surprisingly robust and, as a natural phenomenon, they deserve a quantitative and predictive scientific explanation.

What is evolution? Genetic material and the process of replication involve several stochastic components that may lead to variations in the offspring. Replication and variation are two of the three main ingredients of evolutionary processes. What evolution means in a biological context is captured by the classic Darwinian narrative. Consider a population of some kind that is able to produce offspring. This offspring has some random variations (e.g. mutations). Individuals with the optimal variations with respect to a given environment have a selection advantage (i.e. higher fitness). Fitness manifests itself by higher reproductive success. Individuals with optimal variations will have more offspring and will thus pass their particular variations on to a new generation. In this way 'optimal' variations are selected over time. This is certainly a convincing description of what is going on; however, in this form it may not be useful for predictive science. How can we predict the fitness of individuals in future generations, given that life in future environments will look very different from what it is today? Except over very short time periods, this is a truly challenging task that is far from understood. There is a good prospect, however, of the *statistics* of evolutionary systems being understood. The Darwinian scenario fails to explain essential features about evolutionary systems, such as the existence of boom and crash phases, where the diversity of systems radically changes within short periods of time. An example is the massive diversification (explosion) of species and genera about 500 million years ago in the Cambrian era. It will almost certainly never be possible to predict what species will live on Earth even 500 years from now, but it may be perfectly possible to understand the statistics of evolutionary events and the factors that determine the statistics. In particular, statistical statements about expected diversity, diversification rates, robustness, resilience, and adaptability are coming within reach. In Chapter 5 we will discuss approaches to formulating evolutionary dynamics in ways that make them accessible both combinatorially and statistically.

The concept of evolution is not limited to biology. In the economy, the equivalent of biological evolution is innovation, where new goods and services are constantly being produced by combination of existing goods and services. Some new goods will be selected in markets, while the majority of novelties will not be viable and will vanish. The industrial revolution can be seen as one result of evolutionary dynamics, leading, as it did, to an ongoing explosion of diversification of goods, services, and innovations.

Another example of evolutionary dynamics outside biology is the sequence of invention and discovery of chemical compounds. The history of humankind itself is an example of evolutionary dynamics. Evolutionary dynamics can take place simultaneously at various scales. In biological settings, it works at the level of molecules, cells, organisms, and populations; in economic settings, it can work at product, firm, corporation, and country level. A famous application of evolutionary dynamics in computer science are so-called genetic algorithms [194]. These algorithms mimic natural selection by iteratively producing copies of computer code with slight variations. Those copies that perform best for a given problem (usually an optimization task) are iteratively selected and are passed onto the next 'generation' of codes.

1.3.3.1   *Evolution is not physics*

To illustrate that evolution is not a process that can be described with traditional physics, we define an evolutionary process as a three-step process:

1. A new thing comes into existence within a given *environment.*
2. The new thing has the chance to interact with its environment. The result of this interaction is that it gets 'selected' (survives) or is destroyed.
3. If the new thing gets selected in the environment, it becomes part of this environment (boundary) and thus transforms the old environment into a new one. New and arriving things in the future will experience the new environment. In that sense, evolution is an algorithmic process that co-evolves its boundaries.

If we try to interpret this three-step process in terms of physics, we immediately see that even if we were able to write down the dynamics of the system in the form of equations of motion, we would not be able to fix the system's boundary conditions. Obviously, the environment plays the role of the boundary conditions within which the interactions happen. The boundary conditions evolve as a consequence of the dynamics of the system and change at every instant. The dynamics of the boundary conditions is dynamically coupled with the equations of motion. Consequently, as the boundary conditions cannot be fixed, this set of equations cannot, in general, be solved and the Newtonian method breaks down. A system of dynamical equations that are coupled dynamically to their boundary conditions is a mathematical monster. That is why an algorithmic process like evolution is hard to solve using analytical approaches.[2]

The second problem associated with evolutionary dynamics, from a physics point of view, is that the phasespace is not well-defined. As new elements may arrive at any point in time, it is impossible to prestate what the phasespace of such systems will be. Obviously, this poses problems in terms of producing statistics with these systems. The situation could be compared to trying to produce statistics by rolling a dice, whose number of faces changes from one throw to the next.

---

[2] Such systems can be treated analytically whenever the characteristic timescales of the processes involved are different. In our example, this would be the case if the dynamics of the interactions of the 'new thing' with the environment happens on a fast timescale, while changes in the environment happen slowly.

Evolutionary dynamics is radically different from physics for two main reasons:

- In evolutionary systems, boundary conditions cannot be fixed.
- In evolutionary systems, the phasespace is not well defined—it changes over time. New elements may emerge that change the environment and therefore also the dynamics for all the existing elements of the system.

Evolutionary aspects are essential for many complex systems and cannot be ignored. A great challenge in the theory of complex systems is to develop a consistent framework that is nevertheless able to deal with evolutionary processes in quantitative and predictive terms. We will see how a number of recently developed mathematical methods can be used to address and deal with these two fundamental problems. In particular, in Chapter 5, we will discuss combinatorial evolution models. These models are a good example of how algorithmic approaches lead to quantitative and testable predictions.

1.3.3.2 *The concept of the adjacent possible*

A helpful steppingstone in addressing the problem of dynamically changing phasespaces is the concept of the *adjacent possible*, proposed by Stuart Kauffman [223]. The adjacent possible is the set of all possible states of the world that *could* potentially exist in the next time step, given the present state of the world. It drastically reduces the size of phasespace from all possible states to a set of possibilities that are conditional on the present. Obviously, not everything can be produced within the next time step. There are many states that are impossible to imagine, as the components required to make them do not yet exist. In other words, the adjacent possible is the subset of all possible worlds that are *reachable* within the next time step and depends strongly on the present state of the world. In this view, evolution is a process that continuously 'fills' its adjacent possible. The concrete realization of the adjacent possible at one time step determines the adjacent possible at the next time step.

Thus, in the context of biological evolution or technological innovation, the adjacent possible is a huge set, in which the present state of the world determines the potential realization of a vast number of possibilities in the next time step. Typically, the future states are not known. In contrast, in physics, a given state often determines the next state with high precision. This means that the adjacent possible is a very small set. For example, the adjacent possible of a falling stone is given by the next position (point) on its parabolic trajectory. In comparison, the adjacent possible of an ecosystem consists of all organisms that can be born within the next time step, with all possible mutations and variations that can possibly happen—a large set of possibilities indeed. The concept of the adjacent possible introduces path-dependence in the stochastic dynamics of phasespace. We will discuss the statistics of path-dependent evolutionary processes in Chapters 5 and 6.

1.3.3.3 *Summary evolutionary processes*

Evolutionary processes are relevant to the treatment of complex systems for the following reasons.

- For evolutionary systems, boundary conditions cannot usually be fixed. This means that it is impossible to take the system apart and separate it from its context without massively altering and perhaps even destroying it. The concept of reductionism is inadequate for describing evolutionary processes.

- Evolutionary complex systems change their boundary conditions as they unfold in time. They co-evolve with their boundary conditions. Frequently, situations are difficult or impossible to solve analytically.

- For complex systems, the adjacent possible is a large set of possibilities. For physics, it is typically a very small set.

- The adjacent possible itself evolves.

- In many physical systems, the realization of the adjacent possible does not influence the next adjacent possible; in evolutionary systems, it does.

1.3.4   Adaptive and robust—the concept of the edge of chaos

Many complex systems are robust and adaptive at the same time. The ability to adapt to changing environments and to be robust against changing environments seem to be mutually exclusive. However, most living systems are clearly adaptive and robust at the same time. As an explanation for how these seemingly contradictory features could co-exist, the following view of the *edge of chaos* was proposed [246]. Every dynamical system has a maximal Lyapunov exponent, which measures how fast two initially infinitesimally close trajectories diverge over time. The exponential rate of divergence is the Lyapunov exponent, λ,

$$|\delta X(t)| \sim e^{\lambda t} |\delta X(0)|,$$

where $|\delta X(t)|$ is the distance between the trajectories at time t and $|\delta X(0)|$ is the initial separation. If λ is positive, the system is called *chaotic* or strongly mixing. If λ is negative, the system approaches an attractor, meaning that two initially infinitesimally separated trajectories converge. This attractor can be a trivial point (fixed point), a line (limit cycle), or a fractal object. An interesting case arises when the exponent λ is exactly zero. The system is then called quasi-periodic or at the 'edge of chaos'. There are many low-dimensional examples where systems exhibit all three possibilities—they can be chaotic, periodic, or at the edge of chaos, depending on their control parameters. The simplest of these is the logistic map.

The intuitive understanding of how a system can be adaptive and robust at the same time if it operates at the edge of chaos, is given by the following. If λ is close to zero, it takes only tiny changes in the system to move it from a stable and periodic mode (λ slightly negative) to the chaotic phase (λ positive). In the periodic mode, the system is stable and robust; it returns to the attractor when perturbed. When it transits into the chaotic phase, say, through a strong perturbation in the environment, it will sample large regions

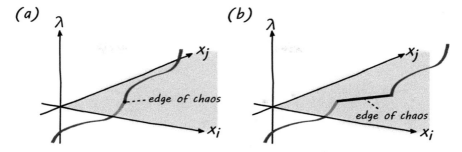

Figure 1.2 *Schematic view of the edge of chaos. (a) Shows the largest Lyapunov exponent in the parameter space (indicated by x_i and x_j) of a dynamical system. The largest exponent dominates the dynamics of the system. For every point in parameter space, the Lyapunov exponent is either positive (chaotic), negative (periodic), or exactly zero—which means at the 'edge of chaos'. The edge of chaos offers a pictorial understanding of how systems can be adaptive and robust at the same time. Systems at the edge of chaos can be thought of as being usually just inside the regular regime. They can transit to the chaotic region very quickly. Once in the chaotic regime, vast volumes of phasespace can be explored which possibly contain regions that are 'better' for the system. Once such a region is found, the system can transit back to the periodic region. Note that the region of parameter space where the Lyapunov exponent is zero is tiny. Some evolutionary complex systems show that this region can be drastically inflated. This will be discussed in Chapter 5. (b) Shows a situation for an inflated edge of chaos.*

of phasespace very quickly. By sampling large volumes of phasespace, the system has the chance to find new 'solutions' that are compatible with the perturbed environment and can then again settle into a periodic phase. The system adapted by sampling other optima. This situation is comparable to a simulated annealing procedure in a computer code.

1.3.4.1 How does nature find the edge of chaos?

For many non-linear dynamical systems, the set of points at which the Lyapunov exponents are exactly zero is very limited and often of measure zero; see Figure 1.2a. If living systems operate at the edge of chaos, how did evolution find and select these points? How can a mechanism of evolution detect something that is of measure zero? One possible explanation is that the regions where the Lyapunov exponent is zero is not of measure zero, but are extended areas in parameter space. Indeed, simple evolutionary models do show inflated regions of the edge of chaos [356]; see Figure 1.2b. We discuss situations where this possibility seems to be realized in Chapter 5. Another explanation is self-organized criticality. This is a mechanism that allows systems to endogenously organize themselves to operate at a critical point that separates the chaotic from the regular regime [24, 226]. A critical point in physics is a point in parameter space where a system is at a phase transition point. These points are reached at conditions (temperature, pressure, slope in a sand pile, etc.) where characteristic quantities (e.g. correlation length) become divergent. Think, for example, of the critical temperature at which a magnetization transition occurs, the Curie temperature T_c. Above that temperature there is no magnetization. Approaching the Curie temperature shows that, at the critical

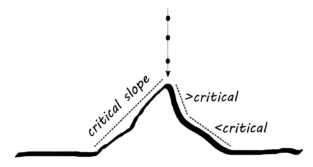

Figure 1.3 *Sand pile models are models for self-organized criticality. These systems self-organize towards a critical state, which is characterized by the fact that the system develops dynamics that are felt across the entire system; in other words, it develops diverging correlation lengths. In sand piles, the system organizes towards a critical slope, above which avalanches of sand will occur to reduce the slope and below which sand is deposited close to the sites where the sand hits, which makes the slope steeper. The size distribution of avalanches is a power law, meaning that it covers a large spectrum of sizes. The frequency distribution of the occurrence of avalanches is also a power law. Besides being a playground for understanding self-organized criticality, sand pile models have practical applications in the science of earthquakes and collapse.*

point, quantities, such as the magnetic susceptibility χ start to diverge as a power law, $\chi = (T - T_c)^{-\gamma}$, where γ is a critical exponent. In many physical systems, these transition points are unique and parameters have to be fine-tuned to these critical points to find the power law behaviour. Self-organized critical systems manage to find these critical points endogenously, without any fine-tuning. Self-organized criticality seems to be realized in a wide range of systems, including sand piles, precipitation, heartbeats, avalanches, forest fires, earthquakes, financial markets, combinatorial evolution, and so on. It often occurs in slowly driven systems, where driven means that they are driven away from equilibrium.

Self-organized critical systems are dynamical, out-of-equilibrium systems that have a critical point as an attractor. These systems are characterized by (approximate) scale invariance or 'scaling'. Scale invariance means the absence of characteristic scales, and it often manifests itself in the form of power laws in the associated probability distribution functions. Self-organized criticality is one of the classic ways of understanding the origin of power laws, which are omnipresent in complex systems. Other ways of understanding power laws include criticality, multiplicative processes with constraints, preferential dynamics, extremal optimization methods, and sample space reducing processes. We will discuss the mechanisms for understanding the origin of power laws in Chapter 3.

1.3.4.2 Intuition behind self-organized critical systems—sand pile models

Imagine sand is dropped onto a table as shown in Figure 1.3. A pile gradually builds up. If you consider the slopes of the pile, you will observe that they are not constant but that they vary in terms of their slope angles. If the slope becomes too steep, avalanches go off and the slope becomes flatter again. If the slope is flat, sand becomes deposited and the slope becomes steeper. In other words, the pile *self-organizes* itself towards a critical slope. The system is robust and adaptive.

1.3.5 Components taken from the life sciences

Let us now put together the components we need for a description of complex systems adapted from the life sciences.

- Interactions between elements happen specifically. They take place on networks. Often, interactions are discrete; they either happen or not. Often, systems operate algorithmically.
- Complex systems are out-of-equilibrium, which makes them hard to deal with analytically. If they are self-organized critical, the statistics behind them can be understood.
- Many complex systems follow evolutionary dynamics. As they progress, they change their own environment, their context, or their boundary conditions.
- Many complex systems are adaptive and robust at the same time. They operate at the 'edge of chaos'. Self-organized criticality is a mechanism that regulates systems towards their edge of chaos.
- Most evolutionary complex systems are path-dependent and have memory. They are therefore non-ergodic and non-Markovian. The adjacent possible is a way of conceptualizing the evolution of the 'reachable' phasespace.

1.4 Components from the social sciences

Social science is the science of social interactions and their implications for society.

Usually, social science is neither very quantitative or predictive, nor does it produce experimentally testable predictions. It is largely qualitative and descriptive. This is because, until recently, there was a tremendous shortage of data that are both time-resolved (longitudinal) and multidimensional. The situation is changing fast with the new tendency of homo sapiens to leave electronic fingerprints in practically all dimensions of life. The centuries-old data problem of the social sciences is rapidly disappearing. Another fundamental problem in the social sciences is the lack of reproducibility or repeatability. On many occasions, an event takes place once in history and no repeats are possible.

As in biology, social processes are hard to understand mathematically because they are evolutionary, path-dependent, out-of-equilibrium, and context-dependent. They are high-dimensional and involve interactions on multiple levels. The methodological tools used by traditional social scientists have too many shortcomings to address these issues appropriately. Unfortunately, the tools of social science practised today rarely extend much beyond linear regression models, basic statistics, and game theory. In some branches of social science, such as economics, there is a tradition of ignoring the scientific principle in the sense that data, even if they are available, are not taken seriously. There are influential 'theories' that are in plain contrast to experimental evidence, including the capital asset pricing model, the efficient market hypothesis, or the Markowitz portfolio theory. These concepts earned their inventors Nobel Prizes. What did also not occur to date in the social sciences are the massive joint and coordinated efforts between scientists that have been taking place in physics (CERN), biology (the genome project), and climate science. However, two important components have been developed in the social sciences and play a crucial role in the theory of complex systems:

- *Multilayer interaction networks.* In social systems, interactions happen simultaneously at more or less the same strength scale on a multitude of superimposed interaction networks. Social scientists, in particular sociologists, have recognized the importance of social networks[3] since the 1970s [156, 398].

- *Game theory.* Another contribution from the social sciences, game theory, is a concept that allows us to determine the outcome of rational interactions between agents trying to optimize their payoff or utility [393]. Each agent is aware that the other agent is rational and that he/she also knows that the other agent is rational. Before computers arrived on the scene, game theory was one of the very few methods of dealing with complex systems mathematically. Game theory can easily be transferred to dynamical situations, and it was believed for a long time that iterative game-theoretic interactions were a way of understanding the origin of cooperation in societies. This view is now severely challenged by the discovery of so-called zero-determinant games [316]. Game theory was first developed and used in economics but has penetrated other fields of the social and behavioural sciences.

1.4.1   Social systems are continuously restructuring networks

Social systems can be thought of as time-varying multilayer (multiplex) networks. Nodes are individuals or institutions, and links are interactions of different types. Interactions

---

[3] Interestingly, until very recently, networks were not recognized as relevant by the 'queen' of the social sciences, economics, even though networks clearly dominate practically every domain of the economy. Mainstream economics has successfully ignored networks—production networks, distribution networks, trading and consumption networks, ownership networks, information networks, and financial networks—for more than a century in favour of a rather unrealistic equilibrium view of the economy.

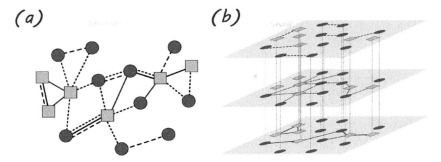

Figure 1.4 *Two schematic representations of the same multilayer network. Nodes are characterized by a two-dimensional state vector. The first component is given by colours (light- and dark-grey) the second by shapes (circles, squares). Nodes interact through three types of interaction that are represented by (full, broken, and dotted) lines. (a) Shows the projection of the multilayer network to a single layer, whereas in (b) each type of link is shown in a different layer. The system is complex if states simultaneously change as a function of the interaction network and if interactions change as a function of the states; see Equation 1.1. The multilayer network could represent a network of banks at a given moment, where shapes represent the wealth of the bank and the links could represent financial assets that connect banks with each other. A full line could mean a credit relation; a broken line could represent derivatives trading, and dotted lines indicate if one bank owns shares in another bank. Depending on that network, the banks will make a profit or loss in the next time step.*

change over time. The types of link can be friendship, family ties, exchange of goods, payments, trust, communication, enmity, and so on. Every type of link is represented by a separate network layer; see, for example, Figure 1.4. Individuals interact through a superposition of these different interaction types (multilayer network), which happen simultaneously and are often of the same order of magnitude in 'strength'. Often, networks at one level interact with networks at other levels. Networks that characterize social systems show a rich spectrum of growth patterns and a high level of plasticity. This plasticity of networks arises from restructuring processes through link creation, relinking, and link removal. Understanding and describing the underlying restructuring dynamics can be very challenging; however, there are a few typical and recurring dynamical patterns that allow for scientific progress. We will discuss these in Chapter 4.

Individuals are characterized by states, such as their wealth, gender, education level, political opinion, age, and so on. Some of these states can dynamically change over time. States typically have an influence on the linking dynamics of the corresponding individual (node). If that is the case, a tight connection exists between network structure and node states. The joint dynamics of network restructuring and changes of states by individuals is a classic example of co-evolution.

1.5 What are Complex Systems?

We present a one-sentence summary of what complex systems are, which covers most of the features discussed in the previous sections:

Complex systems are co-evolving multilayer networks.

This statement summarizes the following ten facts about complex systems and provides an intuitive picture of the essence of complex systems.

1. Complex systems are composed of many elements. These are labelled with latin indices, i.

2. These elements interact with each other through one or more interaction types, labelled with greek indices, α. Interactions are often specific between elements. To keep track of which elements interact, we use networks. Interactions are represented as links in the interaction networks. The interacting elements are the nodes in these networks. Every interaction type can be seen as one network layer in a multilayer network; see Figure 1.4. A multilayer network is a collection of networks linking the same set of nodes. If these networks evolve independently, multilayer networks are just superpositions of networks. However, there are often interactions between interaction layers.

3. Interactions are not static but change over time. We use the following notation to keep track of interactions in the system. The strength of an interaction of type α between two elements i and j at time t is denoted by,

$$M_{ij}^{\alpha}(t).$$

Interactions can be physical, chemical, social, or symbolic. Most interactions are mediated through some sort of exchange process between nodes. In that sense, interaction strength is often related to the quantity of objects exchanged (gauge bosons for physical interactions, electrons for chemical interactions, financial assets for economical interactions, bottles of wine for positive social interactions, and bullets for aggressive ones, etc.). Interactions can be deterministic or stochastic.

4. Elements are characterized by states. States can be scalar; if an element has various independent states, it will be described by a state vector or a state tensor. States are not static but evolve with time. We denote state vectors by,

$$\sigma_i(t).$$

States can be the velocity of a planet, spin of an electron, state of phosphorylation of a protein capitalization of a bank, or the political preference of a person. State changes can be deterministic or stochastic. They can be the result of an endogenous dynamics or of external driving.

5. Complex systems are characterized by the fact that states and interactions are often not independent but evolve together by mutually influencing each other; states and interactions *co-evolve*. The way in which states and interactions are coupled can be deterministic or stochastic.

6. The dynamics of co-evolving multilayer networks is usually highly non-linear.

7. Complex systems are context-dependent. Multilayer networks provide that context and thus offer the possibility of a self-consistent description of complex systems. To be more precise, for any dynamic process happening on a given network layer, the other layers represent the 'context' in the sense that they provide the only other ways in which elements in the initial layer can be influenced. Multilayer networks sometimes allow complex systems to be interpreted as 'closed systems'. Of course, they can be externally driven and usually are dissipative and non-Hamiltonian. In that sense, complex systems are hard to describe analytically.

8. Complex systems are *algorithmic*. Their algorithmic nature is a direct consequence of the discrete interactions between interaction networks and states.

9. Complex systems are path-dependent and consequently often non-ergodic. Given that the network dynamics is sufficiently slow, the networks in the various layers can be seen as a 'memory' that stores and records the recent dynamical past of the system.

10. Complex systems often have memory. Information about the past can be stored in nodes, if they have a memory, or in the network structure of the various layers.

In this book, we assume that a co-evolving multilayer network structure is the fundamental dynamical backbone of complex systems. This assumption not only provides a simple conceptual framework, it also allows us to explain several of the essential properties of complex systems. These include:

- the emergence and origin of power laws,
- self-organized criticality,
- collapse and boom phases in evolutionary dynamics,
- the nature of phase transitions,
- the origin of the edge of chaos,
- statistics of path-dependent processes.

A snapshot of a co-evolving multilayer network is shown in Figure 1.4. Nodes are represented by a state vector with two components, colour (light- and dark-grey) and shape (circles and boxes). Nodes interact through three types of interaction (full, broken, and dotted lines). The system is a complex system if states change as a function (deterministic or stochastic) of the interaction network and, simultaneously, interactions (the networks) change as a function of the states. See also Equations (1.1). For example, the multilayer network shown could represent a network of banks on a given day, where the shape of the nodes represents the wealth of the bank (circle indicates rich, box indicates poor) and the colour represents the gender of the CEO of the bank (light-grey is female, dark-grey is male). The links represent financial contracts (assets) between banks; a full line could mean a credit relation, a broken line could represent derivatives trading, and dotted lines could indicate that one bank owns shares in another. The set of links associated with a bank can be seen as its portfolio. Clearly, the wealth state of a bank will influence the network structure. If a bank is poor, it is not allowed to issue new credits. At the same time, the network structure of assets (the portfolio) has a huge effect on the future wealth of the banks. On the other hand, the gender of the CEO may have little effect on the interbank network structure, and the networks will certainly not have any effect on the gender of the CEO. While the wealth-asset network is a complex system, the gender-network system is not.

1.5.1   What is co-evolution?

To provide some insights into what we mean by co-evolution, we formulate it in a slightly more formal way. In general, interactions can change the states of the elements. In physics, gravitational interaction changes the momentum of massive objects; electromagnetic interactions lead to spin flips; chemical interactions may change the binding state of proteins; economic interactions change the portfolios of traders; and social interactions may change sympathy levels.

The interaction partners of a node in a network (or multilayer network) can be seen as the local 'environment' of that node. The environment often determines the future state of the node. In complex systems, interactions can change over time. For example, people establish new friendships or economic relations; countries terminate diplomatic relations. The state of nodes determines (fully or in part) the future state of the link, whether it exists in the future or not and, if it exists, the strength and the direction that it will have.

The essence of co-evolution can be encompassed in the statement:

- The state of the network (topology and weights) determines the future states of the nodes.
- The state of the nodes determines the future state of the links of the network.

More formally, co-evolving multiplex networks can be written as,

$$\frac{d}{dt}\sigma_i(t) \sim F\left(M_{ij}^{\alpha}(t), \sigma_j(t)\right)$$
$$\frac{d}{dt}M_{ij}^{\alpha}(t) \sim G\left(M_{ij}^{\beta}(t), \sigma_j(t)\right). \tag{1.1}$$

Here, the derivatives mean 'change within the next time step' and should not be confused with real derivatives. The first equation means that the states of element i change as a 'function', F, that depends on the present states of σ_i and the present multilayer network states, $M_{ij}^{\alpha}(t)$. The function F can be deterministic or stochastic and contains all summations over greek indices and all j. The first equation depicts the analytical nature of physics that has characterized the past 300 years. Once one specifies F and the initial conditions, say, $\sigma_i(t=0)$, the solution of the equation provides us with the trajectories of the elements of the system. Note that in physics the interaction matrix $M_{ij}^{\alpha}(t)$ represents the four forces. Usually, it only involves a single interaction type α, is static, and only depends on the distance between i and j. Typically, systems that can be described with the first equation alone are not called complex, however complicated they may be.

The second equation specifies how the interactions evolve over time as a function G that depends on the same inputs, states of elements and interaction networks. G can be deterministic or stochastic. Now interactions evolve in time. In physics this is very rarely the case. The combination of both equations makes the system a co-evolving complex system. Co-evolving systems of this type are, in general, no longer analytically solvable.[4] One cannot solve these systems using the rationale of physics because the environment— or the boundary conditions—specified by M change as the system evolves. From a practical point of view, Equations 1.1 are useless until the functions G and F are specified. Much of the science of complex systems is related to identifying and giving meaning to these functions for a concrete system at hand. This can be done in an analytical or algorithmic way, meaning that F and G can be given by analytical expressions or algorithmic 'update rules'. Both can be deterministic or stochastic.

More and more data sets containing full information about a system are becoming available, meaning that all state changes and all interactions between the elements are recorded. It is becoming technically and computationally possible to monitor cell-phone communication networks on a national scale, to monitor all genetic molecular activities within a cell, or all legal financial transactions on the planet. Data that contain time-resolved information on states and interactions can be used to actually visualize Equations (1.1); all the necessary components are listed in the data at any point in time: the interaction networks M_{ij}^{α}, the states of the elements σ_i, and all the changes $\frac{d}{dt}\sigma_i$ and $\frac{d}{dt}M_{ij}^{\alpha}$ over time. Even though Equations (1.1) might not be analytically solvable, it is becoming possible for more and more situations to 'watch' them. It is often possible to formulate agent-based models of complex systems in the exact form of Equations (1.1), and this allows us to make quantitative and testable predictions.

The structure of Equations (1.1) is, of course, not the most general possible. Immediate generalizations would be to endow the multilayer networks with a second greek index, $M_{ij}^{\alpha\beta}$, which would allow us to capture cross-layer interactions between elements. It is conceivable that elements and interactions are embedded in space and time; indices labelling the elements and interactions could carry such additional information, $i(x,t,\ldots)$ or $\{ij\}^{\alpha\beta}(x,t,\ldots)$. Finally, one could introduce memory to the elements and interactions of the system.

1.5.2 The role of the computer

The science of complex systems is unthinkable without computers. The analytical tools available until the 1980s were good enough to address problems based on differential

---

[4] Except for simple examples or situations, where the timescale of the dynamics of the states is clearly different from the dynamics of the interaction networks.

equations, for systems in equilibrium, for linear (or sufficiently linearizable) systems, and for stochastic systems with weak interactions. The problems associated with evolutionary processes, non-ergodicity, out-of-equilibrium systems, self-organization, path-dependence, and so on, were practically beyond scientific reach, mainly because of computational limits. The computational power to address these issues has only become available in the past decades.

Often, computer simulations are the only way of studying and developing insights into the dynamics of complex systems. Simulations are often referred to by agent-based models, where elements and their interactions are modelled and simulated dynamically. Agent-based models allow us to study the collective outcomes of systems comprising elements with specific properties and interactions. The algorithmic description of systems in terms of update rules for states and interactions is fully compatible with the way computer simulations are done. In many real-world situations there is only a single history and this cannot be repeated. This occurs in social systems, in the economy, or in biological evolution. Computer simulations allow us to create artificial histories that are statistically equivalent copies. Ensembles of histories can be created that help us understand the systemic properties of the systems that lead to them. Without simulations, predictive statements about systemic properties like robustness, resilience, efficiency, likelihood of collapse, and so on, would never be possible. The possibility of creating artificial histories solves the problem of repeatability.

With the current availability of computer power and high-dimensional and time-resolved data, computational limits or data issues no longer pose fundamental bottlenecks for understanding complex systems. The bottleneck for progress is the theory of complex systems, a mathematically consistent framework in which data can be systematically transformed into useful knowledge.

> The computer has fundamentally changed the evolution of science. It has finally opened and paved the way to understanding complex adaptive systems as a natural science on a quantitative, predictive, and testable basis.

1.6   The structure of the book

Complex systems span an immense universe of phenomena, systems, and processes. We believe that most of these can be mapped, in one way or another, into the framework of the stochastic, co-evolving, multilayer networks that we sketched in Equations (1.1). To be able to treat those systems in quantitative and predictive ways, we need tools and concepts for random processes, networks, and evolutionary dynamics. These needs define the structure of the book.

Complex systems involve many different sources of randomness in their components, interactions, processes, time series, structures, and so on. In Chapter 2 we review the basic notions of randomness and statistics that will be needed in various parts of the book. In Chapter 3 we discuss the notion of scaling and learn why it is helpful. We review the classic routes to understanding the origin of power laws, which are omnipresent in the statistical description of complex systems. One of the central backbones of complex systems are dynamical networks, which tell us how the building blocks interact with each other. Dynamics can happen on networks (e.g. diffusion on networks), or the networks dynamically rearrange themselves. The notions and basic concepts of networks, their structures, characteristics, functions, and ultimately their dynamics will be developed in Chapter 4.

Evolutionary dynamics is central to many complex adaptive systems. After reviewing classic ways of understanding the dynamics of evolutionary systems, in Chapter 5 we show how a general model of evolutionary systems can be related to co-evolving network structures. We will learn that this approach is an example of an algorithmic description of systems. We will further see that evolutionary systems have familiar phase diagrams and can be naturally associated to self-organized critical systems.

Finally, in Chapter 6 we probe how far methods from statistical mechanics, information theory, and statistical inference methods can be used in the context of stochastic complex systems. These systems are typically path-dependent, evolutionary, non-Markovian, and non-ergodic; thus, the methods that we have learned in physics, statistics, and information theory should be inapplicable. The chapter is designed to show that a careful generalization of the classic concepts of entropy, information production, and statistical inference methods also allows us to use these concepts for (simple) complex systems and for path-dependent processes in particular.

All chapters start in a relatively simple fashion and become more difficult towards the end. We conclude most chapters with extensive examples that should provide an impression of the status of actual research.

1.6.1   What has complexity science contributed to the history of science?

We conclude this chapter with a somewhat incomplete reading list containing several classics and a few more recent contributions from the science of complex systems. Some of them have already changed our world view:

- network theory [291],
- genetic regulatory networks [221, 222],
- Boolean networks [225],
- self-organized criticality [24, 226],
- genetic algorithms [194],
- auto-catalytic networks [22, 113, 205, 206],
- econophysics [66, 124, 261],
- theory of increasing returns [13],

- origin and statistics of power laws [93, 137, 290, 346],
- mosaic vaccines [31],
- statistical mechanics of complex systems [175, 178, 282, 378],
- networks models in epidemiology [88, 301],
- complexity economics [94, 189, 248, 124],
- systemic risk in financial markets [35, 311],
- allometric scaling in biology [71, 407],
- science of cities [36, 49].

In this book, we want to take a few steps in new directions. In particular, we want to clarify the origin of power laws, especially in the context of driven non-equilibrium systems. We aim to derive a framework for the statistics of driven systems. We try to categorize probabilistic complex systems into equivalence classes that characterize their statistical properties. We present a generalization of statistical mechanics and information theory, so that they finally become applicable and useful for complex systems. In particular, we derive an entropy concept for complex systems and carefully discuss its meaning. Finally, we make an attempt to unify the many classical approaches to evolution and co-evolution into a single mathematical, algorithmic framework. The overarching theme of the book is to contribute to methodology for understanding the co-evolutionary dynamics of states and interactions.

2

Probability and Random Processes

2.1 Overview

Phenomena, systems, and processes are rarely deterministic, but contain stochastic, probabilistic, or random components. Even if nature were to follow purely deterministic laws, which at a quantum level it does not, our scientific notion of the world would still have to remain probabilistic to a large extent. We do not have the means to observe and record deterministic trajectories with the infinite precision required to deterministically describe a world full of non-linear interactions.[1] To some extent we can increase experimental precision, storage capacity, and computing power. Indeed, what we can do today would have been unthinkable only a few decades ago. However, the kind of precision needed to conduct science of non-linear systems at a deterministic level is not only temporarily out-of-reach, but is actually outside our world, especially when biological, social, and economical phenomena are being considered. For that fundamental reason, a probabilistic description of most phenomena in this world is necessary. We will use the terms stochastic, probabilistic, and random interchangeably.

Systems can be intrinsically probabilistic, in which case they typically contain one or more 'sources of randomness'. If systems are deterministic, but we lack detailed information about their states and trajectories, randomness is introduced as a consequence of the coarse-grained description. In both situations, if we want to scientifically understand these systems, it is necessary to understand and quantify the ignorance and randomness involved. Probability theory provides us with the tools for this task. Here, we will not present a proper course in probability theory, which is far beyond the scope of this book, but we will provide a crash course on the most important notions of probability and random processes.[2] In particular, we will try to attach a precise meaning to words like odds, probability, expectation, variance, and so on. We will begin by describing the most elementary stochastic event—the trial—and develop the notion of urn models, which will

---

[1] Non-linear or chaotic deterministic systems can often effectively be seen as random processes. Uncertainties about their initial conditions grow exponentially as the system evolves. After some time even small uncertainties increase to a point where it becomes impossible to predict any future states or trajectories based on the underlying deterministic equations.

[2] Readers familiar with probability and random processes may skip the chapter.

Introduction to the Theory of Complex Systems. Stefan Thurner, Rudolf Hanel, and Peter Klimek,
Oxford University Press (2018). © Stefan Thurner, Rudolf Hanel, and Peter Klimek.
DOI: 10.1093/oso/9780198821939.001.0001

play an important role throughout the book. We will discuss basic facts about random variables and the elementary operations that can be performed. Such operations include sums and products and how to compare random variables with each other. We will then learn how to compose simple stochastic processes from elementary stochastic events, such as *independent identical distributed* (i.i.d.) trials and how to use *multinomial statistics*. We will discuss random processes as temporal sequences of trials. If the variables in these sequences are identical and independent, we call them Bernoulli processes. If trials depend only on the outcome of the previously observed variable, they are Markov processes, and if the evolution of the processes depends on their history, we call them path-dependent. We will discuss processes that evolve according to update rules that include random components.

We will learn how the frequency of observed events can be used to estimate probabilities of their occurrence and how frequentist statistics is built upon this notion. We will touch upon the basic logic of Bayesian reasoning, which is important for testing the likelihood of different hypotheses when we are confronted with specific data. We present a number of classical distribution functions, including power laws and other *fat*- or *heavy*-tailed distributions. These are important for situations where *extreme events* occur frequently. A recurring theme in the context of complex systems is that frequent and large outliers are the rule rather than the exception. Understanding the origin of power law distributions, or other fat-tailed distributions is essential for the realistic assessment of risks, systemic risk in particular, which—despite its importance for society—is still widely based on Gaussian statistics. Gaussian statistics drastically underestimates the frequency of extreme events.

In probability theory, we are dealing with 'known unknowns'. This means that we know the stochastic system and what types of event it is possible for us to observe. What we do not know are the *exact* future outcomes (the unknown), which we cannot predict because of the stochastic nature of the system. However, we do know the *range of possible outcomes* (the known). Take the throwing of a dice, for instance. We do not know the outcome of the next throw, but we do know all the possible outcomes, namely, $1, 2, \cdots, 6$, and we might know the associated probabilities. The range of possible outcomes is called the *sample space* of the considered random variable. Statistics and probability theory provide us with a set of tools that allows us to deal with these known unknowns. Many complex systems, however, are systems that involve 'unknown unknowns'. This happens, for example, if it is not possible a priori to specify all the possible outcomes that one might observe in a process. This situation is present in the context of evolution, co-evolution, and innovation, where processes 'invent' new possibilities. It is usually impossible to prestate the future sample space of open-ended evolutionary processes. However, for some driven dissipative systems, for which the dynamics of the sample space is known, there are fascinating new possibilities for understanding the statistics of driven systems on the basis of the dynamics of the sample space.

In this chapter, we will not deal with unknown unknowns. This will partly be covered in Chapter 5. We merely touch upon this tricky subject with a few remarks on so-called urn processes and dynamic sample spaces in Section 2.1.1.5. Here, we mainly focus on situations with static and simple examples of dynamically evolving sample spaces.

The latter will occur in the context of reinforcement processes and driven dissipative systems. Systems of this nature will be discussed in more detail in Chapters 3 and 6.

2.1.1   Basic concepts and notions

2.1.1.1   *Trials, odds, chances, and probability*

In everyday language the notions of *odds*, *chances*, and *probability* are often used synonymously. In probability theory they carry a distinct meaning, which we should be aware of. We start by defining the most elementary process, the *trial*.

> Any process that selects or chooses a particular event from a set of available and possible events is called a trial. The set of all possible events is called the *sample space* of that trial.

In a probabilistic trial, the choice is made randomly. As a consequence, its outcome cannot be predicted with certainty from previously observed trials. For example, in a dice game such as Yahtzee, each throw of the dice is a trial that randomly selects one of six possible outcomes. What is needed next is a possibility of characterizing the *quality* of possible outcomes. For that reason we introduce the notion of *favourable* and *unfavourable* events, and for this we need the concept of a bet.

> A *bet* divides all *possible events* into *favourable* and *unfavourable events*. We say a bet is won if a favourable event is selected by a trial. If an unfavourable event occurs, the bet is lost.

The words *chances* and *odds* characterize two different but related ways of characterizing the possible outcomes of a bet. They assign a numerical value to a bet, which allows us to express a 'degree of certainty' about the outcome of a trial.

> Assume that a trial can select from W distinct but otherwise equivalent possible events and that a bet divides those events into n_f favourable, and $n_u = W - n_f$ unfavourable events. Then we define:
>
> - the *odds* of winning the bet are $o_{\text{of}} = n_f : n_u$
> - the *odds* against winning the bet are $o_{\text{against}} = n_u : n_f$
> - the *chances* of winning the bet are $q_{\text{of}} = n_f : W$
> - the *chances* against winning the bet are $q_{\text{against}} = n_u : W$
>
> The symbol '$x : y$' stands for the ratio of x to y.

For example, the odds of throwing a six with a fair dice with six faces are $1 : 5$; the odds against throwing a six are $5 : 1$. The chances of throwing a six are $1 : 6$, while the chances of throwing no six are $5 : 6$. While chances add up to one, $q_{\text{of}} + q_{\text{against}} = 1$, for odds we have $o_{\text{of}} o_{\text{against}} = 1$. Moreover, the following relations hold,

$$q_{\text{of}} = \frac{o_{\text{of}}}{1 + o_{\text{of}}} \quad \text{and} \quad o_{\text{of}} = \frac{q_{\text{of}}}{1 - q_{\text{of}}}. \tag{2.1}$$

Odds have a practical interpretation. They characterize the expected number of times of winning identical independent bets repeatedly in a row (see Exercise 2.1). Sometimes, *chance* is immediately taken for *probability*. However, it is helpful to distinguish the empirical notion of chance from the more abstract notion of a probability measure. The difference is similar to the way one would distinguish the empirical notion of a distance from the abstract notion of a metric. Later, in Section 2.1.1.5, we will use the notion of chance to define *probabilities*.

2.1.1.2   *Experiments, random variables, and sample space*

To discuss the fundamental concept of a random variable we must first clarify what an experiment is and what a measurement is.

Experiment, measurement, observation. An experiment is a process that retrieves information about another (random) process. It defines which properties will be observed (measured) and how observations are obtained and recorded. An *experiment* associates *measurable* quantities, called *observables*, with a process and assigns a random variable X to each observable. The simplest experiment is a *trial*, which assigns a definite observed value x to the random variable X as the outcome of the trial. By performing a trial we measure (or sample) a random variable. Experiments can consist of many trials on various variables. We use the terms *to measure, to observe,* and *to sample* synonymously.

Thus, any property of a process that can be experimentally observed and recorded in a trial can become a *variable* X of the process. The experiment determines which of the potentially observable variables of a process will actually be sampled in the experiment. An experiment A may sample and record property X of a process, while a different experiment B may sample and record both X and Y. For example, X could be the face value of a dice that is thrown; the other observable Y of the same dice could be the position at which the dice comes to a rest on the table. We can now specify what we mean by a random variable.

A *random variable* X is a variable whose possible outcomes can be drawn from a range of possible values. The collection of those values is called the *sample space* of the random variable. Random variables X can be sampled in trials with known or unknown odds of observing particular values. The value of the variable is uncertain before measurement and cannot be inferred with certainty from previous observations of the same system or process. One can only bet on its outcome (compare [355]). Variables of a process that can be predicted with certainty are called *deterministic*.

2.1.1.3 Systems and processes

We will use the notions of *system* and *process* almost synonymously. The term *system* highlights the systemic aspect of a process and its set of observables that can be observed together. The term *process* highlights the temporal aspect of a system and how a system and its variables evolve over time in successive observations. Let us think of random variables that are independent in the same sense that the outcome of one throw of a dice does not influence the outcome of a later trial. It then does not really matter if we think of N dice as a system of N dice that are all thrown at once in a single trial or as a process in which we throw one dice N times in a row in N successive trials. Here the ideas of system and process are indistinguishable.[3] For complex processes with interdependent variables, systemic and temporal aspects need to be very carefully distinguished. In any case, it is useful to characterize what we mean by a random process.

A *random process* is a process that can be studied in an experiment and involves the repeated observation of random variables. Mathematically, we can think of a process as a map, $X : n \rightarrow X(n)$, that associates a time point n (of a discrete or continuous time line) with a random variable $X(n)$ that is sampled at time n. In simple processes the random variable may consist of a single value, but it can also be multivalued; in other words, a vector of properties is sampled simultaneously. Famous random processes are Bernoulli processes, Poisson processes, and Brownian or Wiener processes. Random processes can be arbitrarily complicated in terms of the number of stochastic observables and their interrelations, such as path-dependent processes, where the chances of observing the next event depend on the history of the process.

2.1.1.4 Urns

Probability theory is a collection of tools that allows us to understand random processes based on a few elementary processes. Complicated processes are often composed of, or can be segmented into, simple elementary processes. For example, elementary processes include the tossing of coins or the throwing of dice. Another useful class of elementary random processes are so-called *urn models*.

Urns are containers filled with balls (or any other type of object) that are indistinguishable from each other except for one property, which may be the colour or a number

---

[3] That is why for such simple processes, so-called ensemble averages and time averages of observables yield identical results.

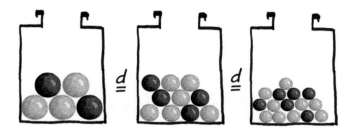

Figure 2.1 *Urn model. The three urns are equivalent in the sense that the chances of drawing a grey ball q_G are 3 : 5. The probability is determined by the ratio given by the chances, $P(G) = 0.6$. Even though the absolute numbers that determine the chances in the three urns are different, the three urns are equivalent in terms of the probability of drawing a black or grey ball.*

written on the ball. Balls are then drawn blindly from that urn; thus a specific colour can only be drawn by chance. When we draw a ball we call that an *urn trial*. Urns allow us to set up and prepare the chances of observing specific outcomes in an experiment. Compare, for example a two-sided coin. A fair coin will always have the chances of 1 : 2 possibilities. It would be difficult to design a biased coin that gives 'heads' in 7 out of 61 possibilities. However, all rational chances, such as 7 : 61, can easily be realized using an urn. For example, place 7 white and $54 = 61 - 7$ black balls into an urn. You have prepared the urn in such a way that the chances of drawing white are 7 : 61 and the chances of drawing black are 54 : 61. Of course, you have the same chances if you use multiples of those numbers, for instance, 21 white balls and 162 black balls; see also Figure 2.1. What is important is not the absolute number of balls but their ratios q, which we will call *probability density* or *probability distribution*; see Section 2.1.1.5. Real-valued probabilities can always be thought of as limits of rational-valued probabilities.

Once a ball is drawn in an urn trial, we can decide what to do next. The decision can depend on the outcome of previous urn trials. For example, the ball chosen can be placed back into the urn (drawing with replacement) or not (drawing without replacement). This gives us the opportunity to use urns to model probabilistic processes that can even involve networks of interdependent urns 'connected' through *update rules*. Update rules specify how the outcome of one trial with one urn determines how the content of another urn needs to be prepared for the next trial. Update rules are often presented as *algorithms* that tell us how the sample space and the chances of drawing elements evolve over time.

Processes, where a random variable X is sampled in independent (memoryless) trials over a discrete sample space that does not change over time, are called *Bernoulli processes*. We use the term 'Bernoulli process' for situations in which trials sample from two or more (finitely many) elements.[4] An example of a Bernoulli process is the algorithm 'draw a ball—record its colour—put it back in the urn—repeat', which is called *drawing with replacement*. It keeps the chances of drawing particular colours from the urn unchanged

---

[4] In the literature a Bernoulli process is sometimes specifically defined as a two-state process (binary process). Processes with more than two states are called *multivalued* Bernoulli processes or *categorical* processes.

and follows a *multinomial* distribution; see Section 2.5.1.3. Another example of an algorithm is drawing without replacement, which is a simple example of a history-dependent process as, whenever a ball is drawn, the contents of the urn change, and consequently, so do the odds for sampling balls in the next trial. This process type follows a *hypergeometric* distribution.

Urns are especially helpful for understanding self-reinforcing complex systems. In this case, we will not just 'draw and replace' a ball but 'draw, replace, and reinforce' by replacing a drawn ball of a certain type (say, black) with $\delta > 1$ balls of the same type (black). These urn processes are called Pólya urns, or Pólya urn processes, which show the *rich-get-richer* or the *winner-takes-all* phenomenon. We will discuss these processes in detail in Chapter 6.

It is possible to design arbitrarily complicated algorithms, that is, sets of rules that specify how to prepare and modify the sample space for the next *experiment* based on what has happened in the past. These rules can be deterministic or stochastic. For example, one could think of an algorithm like, 'whenever you draw a red ball, add three balls of a colour not previously used in the urn process', or 'if you draw green, draw another ball and remove all balls of this colour from the urn'. One may even think of networks of urn processes, where an event sampled in one urn may alter the content of adjacent urns in the network. This flexibility makes urn processes an incredibly useful mathematical vehicle for exploring the properties of many path-dependent random processes.

2.1.1.5   *Probability and distribution functions*

Since Kolmogorov, the mathematical foundations of probability theory have been based on set theory and measure theory. For example, the different colours that can be drawn from an urn can be mapped into an index-set Ω, which is simply the (discrete) sample space of the random variable. An urn containing otherwise identical balls of W different colours will therefore have a sample space of the form $\Omega = \{1, 2, 3, \cdots, W\}$. If we draw colour $i \in \Omega$ in an experiment, we say that we have *observed, measured,* or *sampled i.*

The chances of drawing the possible events $i \in \Omega$ in an urn trial depend on the urn's contents before the trial. The urn experiment can be prepared by loading the empty urn with a_i balls for each distinct colour i. We collect those numbers in a vector $a = (a_1, a_2, \cdots, a_W)$. We call a_i the *multiplicity* of events or states i. By placing all $|a| = \sum_i a_i$ balls in the urn, the chances of observing i in the next experiment are given by a_i from $|a|$ possibilities. In complex systems, often-severe conceptual and practical problems arise if Ω is not static but evolves in time, or worse, if Ω cannot be prestated at all.

We identify the numbers q_i that are defined by $q_i : 1 = a_i : |a|$ with the *probability P* of observing i, and write,

$$P(\text{to draw } i \text{ from the sample space } \Omega | \text{given } q) = q_i. \qquad (2.2)$$

By defining the numbers $q_i = a_i/|a|$, the chances $a_i : |a|$ of observing a particular trial outcome $X = i$ can be identified with the *probability* P of sampling i.

> A function $q : i \rightarrow q_i$ that maps each possible event $i \in \Omega$ to its *probability* q_i is called the *probability distribution function* of X over Ω. The distribution function is identical to the vector $q = (q_1, \cdots, q_W)$. Note that $|q| = \sum_{i=1}^{W} q_i = 1$ and $q_i \geq 0$ for all i.

The notion of distribution functions is more complicated for continuous sample spaces. For continuous sample spaces such as $\Omega = [0, 1]$, a random variable X over Ω has a probability of $P(y \leq X \leq x)$ of being observed in the interval $[y, x]$. This means that the probability of X exactly taking the value x is vanishing, $P(x \leq X \leq x) = 0$. However, the limit $\lim_{y \rightarrow x} P(y < X \leq x)/|x - y| = \rho(x)$ defines the so-called *probability density function* (pdf) ρ for the random variable X and $P(y \leq X \leq x) = \int_{y}^{x} dx \rho(x)$. For further reading see, for instance [319] (p. 25).

Suppose that we perform an urn experiment, where we repeatedly sample with replacement in N subsequent trials. After the experiment we observe that we sampled colour i in k_i out of N trials and $\sum_{i=1}^{W} k_i = N$. We call $k = (k_1, \cdots, k_W)$ the *histogram* of the experiment and $p_i = k_i/N$ the *relative frequencies* of observing i. Sometimes, $p = (p_1, \cdots, p_W)$ is also called *sample distribution* or *empirical distribution*. Just like probabilities q, the relative frequencies are normalized, $\sum_{i=1}^{W} p_i = 1$. In identical random experiments (with i.i.d. variables) the probabilities q are fixed, while the relative frequencies p may vary from experiment to experiment.

> A function $k : i \rightarrow k_i$ that maps each possible event $i \in \Omega$ to its empirically observed *frequencies* k_i, obtained from an experiment sampling a random variable X over Ω for N times, is called the *histogram* of the experiment. The normalized frequencies $p = k/N$ are called the *relative frequency distribution function*.

2.1.2   Probability and information

The notions of *probability* and *information* are tightly related. In the context of probability, information appears in at least four different ways that will be used in various sections of the book. We mention these briefly without any claim of being complete.

Prior information is the knowledge that we have about some observable *prior* to a trial. We can use this information to make a prediction. If we want to predict the outcome of sampling the observable, then the natural choice is to place our bet on the outcome with the highest chances of being sampled. The random variable representing the observable may depend on what we know about the observable before the trial. Consequently, the probabilities also depend on what we know before the trial.[5]

---

[5] To understand that probabilities can depend on prior knowledge, consider this example: a friend A meets you in a bar. A is accompanied by another person B, that you have never met and know absolutely nothing

Posterior information Typically, when dealing with a new process, we know little about it; our prior information about the process is limited. We may know the type of process we have before us and thus the parameters that we need to determine in order to fully specify the process. Typically, we do not know the exact value of those parameters. We obtain this information *after* experiments have been performed (a posteriori). This information is then available to us in observation records or histograms. With this empirical data we can infer the parameters of the random process with more certainty than prior to the experiment. Posterior estimates of a priori unknown parameters of a process, based on empirical data, can be made by using *Bayesian inference*, which we will discuss in Section 2.2.5.

Information-theoretic information Frequently, we wish to communicate important information to friends on our cell phone, such as, 'I'm calling you with my cellphone'. In this context the concept of information becomes exactly quantifiable in terms of the code length that is minimally required to transmit the information through the information channel that consists of cell phones and transmission towers. The information content of a message is typically measured by the frequencies of letters (typically from a binary alphabet consisting of 0 and 1) required to communicate that message. We will deal with this code-based notion of information in more detail in Chapter 6.

Algorithmic information There is another notion of information based on the minimal length of codes. Here, there is no sender involved wanting to send a message to a receiver; however, there is a computer running computer codes that algorithmically predict the behaviour of a system or a process. The minimum number of bits required to write a code that, when run on a computer, emulates a specific (random) process, is called the *algorithmic complexity* or the *Kolmogorov–Chaitin–Solomonoff complexity* of that process [77, 236, 344]. The Kolmogorov–Chaitin–Solomonoff complexity is not identical to the information-theoretic information rates. However, the two notions are closely related.[6] In this book, we will not discuss algorithmic

---

about. The conversation touches upon the topic of tattoos, and A proposes the following wager to you: if you can guess correctly whether or not B has a tattoo, then the next drinks are on A, otherwise you are the one who pays. Suppose B does not have a tattoo. While B knows exactly that she has no tattoo, *you* do not have that knowledge. Thus, the question, tattoo or no tattoo, can be decided with certainty by B (as long as she is not blind and does not suffer from amnesia) while for you both possibilities exist. The random experiment therefore is not 'tattoo-or-no-tattoo', which would obviously lead to ambiguous results, depending on who you ask. A reasonable random experiment is '*you predicting whether B has a tattoo or not*' while the process '*B predicting whether B has a tattoo or not*' is another (deterministic) experiment. If we write $P(B$ has a tattoo$|q)$ then the probabilities q stand for the information available to the predictor (the person, machine, or algorithm that predicts). This information conditions the probability. q refers to the chances of the person making the prediction, and the probabilities $P(B$ has a tattoo$|q_{you})$ and $P(B$ has a tattoo$|q_B)$ are distinct probabilities and will have distinct values.

[6] In a nutshell, if we know that Kolmogorov complexity of a system (symbol string) A is given by the length of the code of the shortest program B that prints A, then it is plausible that the program B can be interpreted as a string that results from a loss-free compression algorithm applied to string A. Here, the information-theoretic aspect enters. A can again be obtained from B by re-expanding the compressed sequence. However, as B is the program printing A, B needs to be a little longer than the compressed version of A itself, as we have to add the code for expansion.

measures of complexity. The topic has been discussed extensively, and there is excellent classical work to which we refer the interested reader [146, 147, 255].

2.1.2.1 Why are random processes important?

Perhaps the most important practical reason for studying probability theory is that the uncertainties associated with individual trials sometimes translate into quite predictable phenomena on an aggregate scale. It may be possible to link elementary stochastic processes (micro description) to aggregate systemic properties of systems composed of such elementary processes (macro description). In particular, the *frequencies* of events observed in populations of trials (distribution functions) often follow regularities or 'laws' that tell us how these frequencies are related to other frequencies or parameters that characterize a process. Sometimes such laws allow us to obtain predictability of stochastic systems at a coarse-grained macroscopic level of description. If we can understand distribution functions at the macro level on the basis of the underlying stochastic dynamics, there is frequently much to be gained—for example, if we understand how to reduce the frequency of occurrence of a particular disease at the population level by randomly vaccinating at the individual level [301]. These laws clarify why we observe specific distribution functions and why certain processes produce them. They allow us to understand how distribution functions change over time and how to decompose processes into deterministic and simple random components. Sometimes even physical laws can be derived from laws that govern random variables. *Statistical physics* is perhaps the most famous example. Here, the *law of large numbers* is a probabilistic law of this kind, governing how velocities of gas molecules are distributed. This then allows us to deduce the relations between pressure, temperature, and volume of gases. Since Boltzmann's times [59, 60] the ideal gas law has become a probabilistic law based on *average* numbers of molecules bouncing into the walls of a container. In that sense, thermometers are physical tools for measuring statistical properties of gas molecules at the macro level. One of today's greatest challenges is to discover and develop the analogue of 'thermometers' for complex non-equilibrium systems, such as ecosystems, financial systems, or societies. In this book, we encounter probability theory in the context of network theory, data analytics and time series analysis, information theory, risk assessment, prediction and forecasts, epidemiology, classification problems, parameter estimation, and hypothesis testing. Before we dive into the heart of the chapter, we summarize the main points so far.

- Probability theory deals with *possibilities* and the *uncertainty* of their realization. One can *bet* on the realization of different possibilities.
- The prospects of winning a bet are characterized in terms of *odds* or *chances* for or against the realization of an event. The notion of *probability* is derived from, and closely related to, the notion of chances.
- Random processes are characterized as independent or interdependent *trials*.

- The collection of all the possible outcomes of a trial is called *sample space*.
- In an experiment a process is characterized by observable *variables*, which are *deterministic* if their value can be predicted with *certainty*, and *stochastic* otherwise.
- The function that maps the elements of the sample space to their probability of occurrence is called the *probability distribution function*.
- *Urn processes* allow us to conceptualize and model random processes by controlling the evolution of the sample space of the urn (contents of the urn). They can be seen as elementary 'programmable' random generators, which can be used to implement algorithms that model complicated random processes.
- Distinct ways of drawing from an urn correspond to distinct random processes that are often associated with distinct distribution functions. For example, *drawing with replacement* and *drawing without replacement* lead to the *multinomial* and the *hypergeometric* distribution, respectively.
- Probability theory provides a means of dealing with insufficient information and uncertainty under given conditions. It allows us to predict the occurrence frequencies of (typically independent) events in large samples, large systems, or long time series.
- In terms of notation, in the remainder of the book, we will use q for denoting chances (probabilities of elementary events), p for relative frequencies, k for histograms, W for the number of states $i = 1, \cdots, W$, and $n = 1, \cdots, N$ (or $t = 1, \cdots, T$) for numbering the elements (or time steps).

2.2   Probability

So far we have discussed basic notions of probability theory derived from the concept of odds and chances in experiments performed with a random variable X. We write $\Omega(X)$ to denote the sample space of variable X. We have identified the chances q_i in urn models of drawing colour i with the probability $P(X = i|q) = q_i$. The distribution function q determines the conditions under which the random variable X is drawn. We have also identified the empirical distribution functions $p_i = k_i/N$ as the normalized histogram k_i, that is, the relative frequency of occurrence of i. In the following, we will see how probability can be characterized as a property of sample spaces.

2.2.1   Basic probability measures and the Kolmogorov axioms

Perhaps the most influential formalization of *probabilities* is Kolmogorov's set theoretic approach [237], which has been tremendously successful. It has even been noted that

'One might come to believe that probability theory began with Kolmogorov' [209]. Kolmogorov's axioms can be formulated as follows:

Kolmogorov axioms. Any function P fulfilling the following three axioms is called a *probability measure* over the sample space Ω.

KA1 (positivity) for any subset $A \subset \Omega$, $P(A)$ is a real number and $P(A) \geq 0$.

KA2 (unitarity) $P(\Omega) = 1$.

KA3 (σ-additivity) for any countable sequence of mutually disjoint subsets $(A_1, A_2, \cdots)$ of Ω, $P\left(\bigcup_{i=1}^{\infty} A_i\right) = \sum_{i=1}^{\infty} P(A_i)$.

Axiom KA3 is formulated in this way in order to deal with sample spaces that contain a continuum of infinitely many elements, such as the interval of real numbers between zero and one, $[0, 1]$. However, if sample spaces are discrete and finite $\Omega = \{1, 2, \cdots, W\}$, axiom KA3 simply states that for any subset $A \subset \Omega$,

$$P(A) = \sum_{i \in A} P(i). \tag{2.3}$$

Suppose the chances of drawing different colours $i \in \Omega$ in an urn experiment are given by $q_i = a_i : |a|$. What are the chances of drawing a particular colour i *or* another particular colour j? The chances of drawing i or j are given by $a_i + a_j$ out of $|a|$ possibilities and $P(i \text{ or } j|q) = q_i + q_j$. As a result, the probability of drawing any colour contained in the urn $P(\text{draw any } i \in \Omega|q) = 1$, which exactly fulfils KA2. In other words, we identify certainty with a probability value 1.

We sometimes write $P(X = i)$ instead of $P(i)$ or $P(X \in A)$ instead of $P(A)$, to explicitly indicate that it is the random variable X that assumes the value i, or one of the values in A. Sometimes distinct probability measures P need to be distinguished from distinct random variables $X(n)$, $n = 1, 2, \cdots$, which are considered in the same context. This can be done by writing $P(X(n) \in A)$, or alternatively, if $q(n)$ is the distribution function of $X(n)$, by writing $P(A|q(n))$. Moreover, if we are interested in particular ranges of values of A, we can define those by using statements of the form: expression$(X) = $ true. For example, $P(x_{\min} \leq X \leq x_{\max})$ denotes the probability that we observe that X has a value larger than $x_{\min}$ and smaller than $x_{\max}$.

For a random number X, the function,

$$Q(x) = P(X \leq x), \tag{2.4}$$

is called the *cumulative distribution function* (cdf) of X. The probability distribution function (pdf) $q(x)$ can be recovered using differences $q(i) = Q(i) - Q(i-1)$ (for discrete variables) or derivatives (for continuous variables), $q(x) = \frac{d}{dx} Q(x)$. For continuous variables $q(x)$ is also called the *probability density function*. Discrete variables $q(i)$ are often denoted by q_i and are called the *weights* of i.

2.2.2 Histograms and relative frequencies

While chances (or odds) tell us something about the probability of a random variable X taking a particular value, repeated observations of a random process, $n \to X(n)$, $n = 1, 2, \ldots, N$, provide us with collections of definite values $x(N) = (x_1, x_2, \cdots, x_N)$. These are called the *sample*. This is one particular realization of the process $(X(1), X(2) \cdots, X(N))$. Each random variable $X(n)$ belongs to a sample space $\Omega(X(n))$, and each sampled value x_n is drawn from $\Omega(X(n))$. We call $x(N)$ a *realization*, a *sequence of observations*, or a *sample* of the random process. Probability theory provides us with *statistical* tools to analyse samples $x(N)$. The simplest way of characterizing samples $x(N)$ is in terms of their *histograms* and *relative frequency distributions*.

Given a sequence of observations $x(N) = (x_1, x_2, \cdots, x_N)$ with $x_n \in \Omega = \{1, 2, \cdots, W\}$,

- the *histogram* $k = (k_1, \cdots, k_W)$ counts the number of events $i \in \Omega$ recorded in $x(N)$. $k_i(x)$ denotes the number of times i appears in the sequence $x(N)$;
- the function $p = (p_1, \cdots, p_W)$ that maps each possible event $i \in \Omega$ to the normalized histogram $p_i = k_i/N$ is the *empirical distribution function*. p are also called *relative frequencies*.

Relative frequency distributions are non-negative ($p_i \geq 0$) and normalized ($\sum_{i \in \Omega} p_i = 1$). Note that one should strictly distinguish between probabilities $P(i) = q_i$ and relative frequencies p_i, even though both fulfil Kolmogorov's probability axioms for discrete sample spaces. *Probability* is a property of the random variable; *relative frequency* is a property of the experiment. Note that the entire histogram is also a random variable. It is unknown prior to the sampling experiment. Note that questions such as 'how probable is it to find the relative frequencies p after N observations of a random process?' can be answered for several classes of random processes. In other words, one can compute how empirical distribution functions are distributed. We will discuss this in Section 2.5.1.3.

2.2.3 Mean, variance, and higher moments

The notions of *mean* (sometimes also called *expectation value*, *average*, or *first moment*) and *variance*, refer either to measures corresponding to a random variable X or to samples $x(N) = (x_1, \cdots, x_N)$. To distinguish the notions, the latter are sometimes called *sample mean* and *sample variance*; see Figure 2.2. We can define expectation values and sample means of functions f that map the sample space Ω into the real numbers.

The expectation value of a function f, $\langle f(X) \rangle$, with respect to random variable X over Ω is given by,

$$\langle f(X) \rangle = \sum_{i \in \Omega} P(X = i) f(i). \tag{2.5}$$

continued

Note that in this context $i \in \Omega$ need not have a numerical value. One sometimes finds the notation $E(f)$ instead of $\langle f \rangle$ to denote the expectation value.[a]

[a] For random variables X over continuous sample spaces Ω, the expectation value is computed as an integral over the probability density distribution ρ of X. The expectation value becomes $\langle X \rangle = \int_{x\in\Omega} dx \rho(x) x$.

Similarly, one defines the sample mean of a function:

The sample mean $\langle f(x) \rangle$ of a function f with respect to the sample $x(N) = (x_1, \cdots, x_N)$ over Ω, is given by,

$$\langle f(x) \rangle = \frac{1}{N} \sum_{n=1}^{N} f(x_n). \tag{2.6}$$

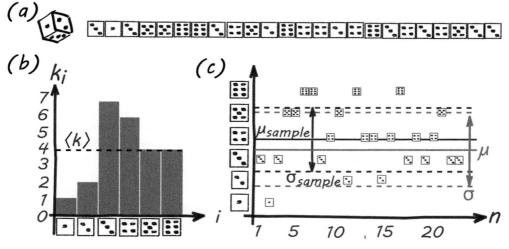

Figure 2.2 *Illustration of some basic notation. A fair dice with $W = 6$ faces has a sample space $\Omega = \{1,2,3,4,5,6\}$, where the chances of throwing a particular face are $1 : 6$. The probability of throwing face i is therefore $q_i = 1/6$. (a) We throw the dice $N = 24$ times and record the outcomes of those experiments in the sample $x(N) = (x_1, \cdots, x_{24})$. The histogram $k = (k_1, \cdots, k_6)$, that is, the relative frequency distribution $p_i = k_i/N$ of sequence $x(N)$ is shown in (b). (c) The sample mean, μ_{Sample} and sample variance, σ_{Sample} (black) are different from the expectation value of the random process X that generates the sequences $x(N)$, $\langle X \rangle = \mu = 3.5$ and its variance $\sigma^2(X) = 2.9167$ (grey).*

If the elements in sample space Ω are numbers, so-called *moments* can be computed. The expectation value and the variance can then be defined as moments.

Moments are the expectation values of the functions $f(x) = x^m$, $m = 0, 1, 2, \cdots$. The *m'th moment* of a random variable X is defined as,

$$\langle X^m \rangle = \sum_{i \in \Omega} P(X = i) i^m. \tag{2.7}$$

- The expectation value $\mu(X) = \langle X \rangle$ is the first moment of X.
- The variance of X is defined by $\sigma^2(X) = \langle (X - \langle X \rangle)^2 \rangle$.

Analogously, for samples the *m'th sample moment* of a sample $x(N)$ is given by,

$$\langle x^m \rangle = \frac{1}{N} \sum_{n=1}^{N} x_n^m. \tag{2.8}$$

- The sample mean $\mu(x) = \langle x \rangle$ is the first sample moment of x.
- The sample variance of x is defined by $\sigma^2(x) = \langle (x - \langle x \rangle)^2 \rangle$.

Sometimes in the literature the sample variance gets 'corrected',[7] by a factor $N/(N-1)$. The variance is a measure that informs us how strongly the random variable X (or the samples $x(N)$) is concentrated around its mean.

There are measures called *cumulants* that are closely related to moments, but might in some cases be more informative than moments. In particular, cumulants allow us to detect a fundamental property of random processes that distinguishes simple (memoryless) processes from more complicated (path-dependent) ones. One practical use of cumulants is that simply by evaluating the third cumulant of a process (compare Exercise 2.8) one can infer if a process is a Markov process. If it is, the third cumulant vanishes asymptotically. If the third cumulant does not vanish, the process is path-dependent [143]. Cumulants are the coefficients $\kappa_n(X)$ in the expansion of the so-called cumulant-generating function,

$$K(t) = \log \langle e^{tX} \rangle = \sum_{n=0}^{\infty} \kappa_n(X) \frac{t^n}{n!}. \tag{2.9}$$

Note that $\kappa_1(X) = \mu(X)$ and $\kappa_2(X) = \sigma^2(X)$.

---

[7] The corrected sample variance can be obtained as a Bayesian estimate of the true variance of the variable X that generates the sample.

2.2.4 More than one random variable

In path-dependent processes (e.g. Pólya urn processes), subsequent random variables $X(n)$, $n = 1, 2, \cdots$, will not be identical. How do we find out if random variables are identical? We need a formal way of stating that one random variable is equivalent to another. We also need formal tools to detect if observing one random variable changes the probability distribution of another random variable.

2.2.4.1 Comparing random variables

If we have two real numbers x and y, by writing $x = y$ we mean that x and y have the same value. This notion of equality is too strong for random variables X and Y, as they have no definite values but take values with a certain probability. If we throw two identical dice, then only in one out of six throws will both show the same face value. Nevertheless, we would call the two dice equivalent.

> If X and Y are two random variables over the same sample space Ω then
>
> (i) we call X and Y identical (in distribution) if X and Y have identical distribution functions. In this case, we write $X \overset{d}{=} Y$;
>
> (ii) we call X and Y of the same *type*, if there are two real numbers a and b such that,
>
> $$X \overset{d}{=} aY + b. \tag{2.10}$$
>
> The definition follows [293].

Note that for two random variables to be of the same type, the observable events have to be numbers; they cannot be colours. In that case, the elements of the sample space $x \in \Omega(X)$ can be mapped one-to-one into elements $y \in \Omega(Y)$ through an *affine transformation*. In other words, we can multiply and add constants a and b to the elements of the sample spaces $\Omega(X)$ and $\Omega(Y)$.

2.2.4.2 Joint and conditional probabilities

We have discussed the meaning of a probability $P(X \in A)$ of observing a single random variable X in the range (or set) A. However, a process may incorporate more than one random variable that we can observe at the same time. Suppose that we observe a random variable X_1 over sample space $\Omega(X_1)$ and a variable X_2 over $\Omega(X_2)$; then we can treat (X_1, X_2) as a joint random variable over $\Omega(X_1, X_2) = \Omega(X_1) \times \Omega(X_2)$ where,

$$\Omega(X_1) \times \Omega(X_2) = \{(x_1, x_2) | x_1 \in \Omega(X_1) \text{ and } | x_2 \in \Omega(X_2)\}, \tag{2.11}$$

is called the *Cartesian product* of the two sample spaces. The Cartesian product of two sets is simply the set of all possible ordered pairs (x_1, x_2) that we can generate from the two sets. We can now define the joint probability.

We call a measure $P(X_1 \in A_1, X_2 \in A_2) = P((X_1, X_2) \in A_1 \times A_2)$ the *joint probability* of the random variables X_1 and X_2. Joint probabilities have to respect the following consistency conditions,

$$P(X_1 \in A_1) = P(X_1 \in A_1, X_2 \in \Omega_2)$$
$$P(X_2 \in A_2) = P(X_1 \in \Omega_1, X_2 \in A_2). \tag{2.12}$$

These conditions state that if one of the two variables can take any possible value, then the joint probability reduces to a probability that measures the chances of finding the other variable within the specified range. These probability measures $P(X_1 \in A_1)$ and $P(X_2 \in A_2)$ are called *marginal probabilities*. The process of obtaining the marginal probabilities is called *marginalization*.

If two random variables X_1 and X_2 can be observed such that the chances of observing one variable do not depend on the realization of the other variable, then the two random variables are called *statistically independent*.

Two random variables are called *statistically independent* if,

$$P(X_1 \in A_1, X_2 \in A_2) = P(X_1 \in A_1)P(X_2 \in A_2), \tag{2.13}$$

for all $A_1 \subset \Omega_1$ and $A_2 \subset \Omega_2$. Here the symbol $\subset$ means 'subset of'.

An illustration of the concept of statistical independence is shown in Figures 2.3 and 2.4. Imagine an urn containing two black (B) and three grey (G) balls. The chances of drawing B are two out of five and of drawing G are three out of five. If we draw with replacement, that is, we draw a ball, record the colour, and then put it back into the urn, our experiment will not change the chances of drawing B or G in the next experiment. For processes with replacement, subsequent experiments, $X(n)$ and $X(n+1)$, $n = 1, 2, 3, \cdots$, are identical in distribution and statistically independent. However, if we do not replace the ball (say, B) after drawing it, there are only one B and three G balls left in the urn for the next trial. The chances of drawing B are now one in four. In drawing without replacement, subsequent variables, $X(n)$ and $X(n+1)$, are no longer identical in distribution and no longer statistically independent. The distribution of the second variable clearly depends on the outcome of the first variable. The same is true for Pólya urn processes, where additional balls of the same colour are added after every draw. Again, subsequent experiments are not independent.

If $X(n)$, $n = 1, 2, \cdots, N$ are statistically independent variables and identical in distribution those random variables are called *independent identically distributed*. In short, $X(n)$ are i.i.d. variables.

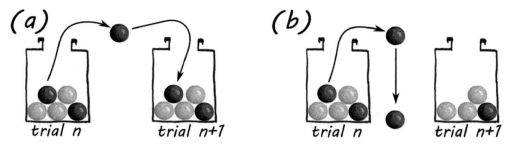

Figure 2.3 *(a) Illustration of statistical independence in an urn processes with replacement and (b) statistical dependence in the case of no replacement.*

We are now in a position to ask how probable it is to observe $X_1 \in A_1$, given that we observed $X_2 \in A_2$. This results in the definition of the conditional probability.

The probability of observing $X_1 \in A_1$, given that we know that $X_2 \in A_2$, is called the *conditional probability*. It is defined as,

$$P(X_1 \in A_1 | X_2 \in A_2) = \frac{P(X_1 \in A_1, X_2 \in A_2)}{P(X_2 \in A_2)}. \qquad (2.14)$$

As a consequence, if X_1 and X_2 are statistically independent random variables, the conditional probability reduces to the marginal probability, $P(X_1 \in A_1) = P(X_1 \in A_1 | X_2 \in A_2)$. This is generally true for Bernoulli processes, which we will discuss in Section 2.5.1.1.

If X_1 and X_2 are two independent random variables (imagine two distinct urns), what is the probability of jointly drawing $i \in \Omega(X_1)$ and $j \in \Omega(X_2)$? If the respective chances are $q_i = a_i : |a|$ and $q'_j = a'_j : |a'|$, the chances of observing the pair (i,j) are given by $a_i a'_j$ out of $|a||a'|$ possibilities. Thus, $P(X_1 = i$ and $X_2 = j|q \times q') = q_i q'_j$, or $P(X_1 = i$ and $X_2 = j|q \times q') = P(X_1 = i|q)P(X_2 = j|q')$.

Why does 'X_1 *and* X_2' mean to *multiply* two probabilities? Suppose we can draw from two events 0 and 1 in two identical independent random experiments with chances q_0 and q_1. Suppose $q_0 = 2/5$ and $q_1 = 3/5$. We can represent these chances by putting five balls into an urn, two black and three grey; see Figure 2.4. Let us replace the balls after each trial, so that the chances of picking balls remain the same. To pick black twice in two trials, we have two out of five possibilities for the first ball *and* then two out of five possibilities again. This means that for $N = 2$ trials we have to choose from $4 = 2 \times 2$ out of $25 = 5 \times 5$ possibilities to pick a black ball twice, that is, $P(\text{black and black}) = q_0^2$. Similarly, we can construct all other probabilities to draw any combination of black and grey balls.

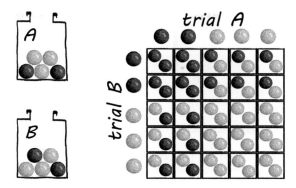

Figure 2.4 *Consider the urn experiment with replacement. The reason why the probability of drawing i and j independently is given by the product of P(i and j) = P(i)P(j) can be visualized by drawing a table of all possible outcomes of the two independent experiments. To draw two black balls there are four in twenty-five possibilities; to draw two grey balls there are nine possibilities. To draw one black and one grey ball there are 12 = 6 + 6 possibilities.*

The notion of *statistical independence* in Equation (2.13) is built on the product property of joint observations. The product captures the insight that two random experiments are statistically independent if the outcome of one experiment does not influence the chances for the outcomes of another variable, and vice versa.

2.2.5 A note on Bayesian reasoning

Bayesian reasoning provides us with a simple and effective tool that can be used to estimate the relative likelihood of two or more alternative hypotheses, or to estimate uncertain parameters that condition random processes. Computations in Bayesian arguments quickly turn out to be lengthy; however, the basic idea underlying all Bayesian reasoning is easily explained.

Consider two random variables X and Y with their respective sample spaces $\Omega(X)$ and $\Omega(Y)$ with elements $i \in \Omega(X)$ and $f \in \Omega(Y)$. Given their joint probability $P(X = i, Y = f)$ and the marginal probabilities $P(X = i)$ and $P(Y = f)$, defined in Equation (2.12), the conditional probabilities are given by $P(X = i | Y = f) = P(X = i, Y = f)/P(Y = f)$, and $P(Y = f | X = i) = P(X = i, Y = f)/P(X = i)$.

From this observation *Bayes' rule* follows,

$$P(X = i | Y = f) = P(Y = f | X = i)\frac{P(X = i)}{P(Y = f)}. \tag{2.15}$$

It allows us to flip the arguments in the conditional probabilities.

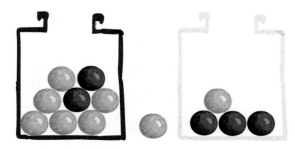

Figure 2.5 *Illustration of a typical Bayesian problem. There are two urns, black and white, each filled with black and grey balls. The chances of drawing each type of ball from both urns are known. Imagine that a blindfolded person draws a ball from one of the urns but does not know from which. The ball drawn is grey. What is the probability that the urn it was drawn from was the black one? In other words, we are testing the hypothesis that the 'urn was black' given the sample 'the ball is grey'. The prior information is that the urn was either black or white.*

What must appear as a mathematical triviality becomes a powerful tool for probabilistic reasoning if we read Bayes' rule in the following way. If we know the process (or a model of it) that maps uncertain initial conditions i (such as the parameters of the model) to the final observations f (data produced by the process), that is, if we know $P(Y=f|X=i)$, then we can infer the probability distribution of an unknown initial condition i, conditional on observed data f. In other words, Bayes allows us to ask questions like: given specific data, what are the best parameters of a model to explain it? Or what is the best model (if we have more than one) that describes the data? Bayes' principle further allows us to deal with cause and effect or other forms of statistical interdependencies of variables simultaneously. If new data become available, Bayes' rule provides a scheme to iteratively update our level of certainty by mapping prior probabilities (before new data have been considered) to posterior distribution functions (after new data have been considered).

The following example is a typical application of Bayesian inference. Assume that you have two urns, one black, the other white; both are filled with black and grey balls. You know the chances of drawing black (B) and grey (G) balls for both urns, $P(\text{ball} = G|\text{urn} = \text{black})$ and $P(\text{ball} = G|\text{urn} = \text{white})$. Imagine you close your eyes, and somebody hands you one of the two urns to draw from. You do not see which urn it is. After you open your eyes, you see that you have drawn a grey ball; see Figure 2.5. The question now is: what is the probability that the urn you drew from was black? In other words, we want to know the value of $P(\text{urn} = \text{black}|\text{ball} = G)$. The answer to that question is given by Bayes' rule in Equation (2.15),

$$
\begin{aligned}
P(\text{urn} = \text{black}|\text{ball} = G) &= \frac{P(\text{ball}=G|\text{urn}=\text{black})P(\text{urn}=\text{black})}{P(\text{ball}=G)} \\
&= \frac{P(\text{ball}=G|\text{urn}=\text{black})P(\text{urn}=\text{black})}{P(\text{ball}=G|\text{urn}=\text{black})P(\text{urn}=\text{black})+P(\text{ball}=G|\text{urn}=\text{white})P(\text{urn}=\text{white})} \\
&= \left(1 + \frac{P(\text{ball}=G|\text{urn}=\text{white})}{P(\text{ball}=G|\text{urn}=\text{black})}\frac{P(\text{urn}=\text{white})}{P(\text{urn}=\text{black})}\right)^{-1}.
\end{aligned}
$$

$$(2.16)$$

If we know that the conditional probabilities have the values $P(\text{ball} = G|\text{urn} = \text{black}) = \frac{6}{8}$ and $P(\text{ball} = G|\text{urn} = \text{white}) = \frac{1}{4}$, then we still face the problem of not knowing the value of the prior distributions $P(\text{urn} = \text{black})$. With no additional information available, our best guess is to assume a $50:50$ chance; $P(\text{urn} = \text{black}) = P(\text{urn} = \text{white}) = \frac{1}{2}$. This is the so-called maximum ignorance assumption for the prior distributions. From this, it follows that $P(\text{urn} = \text{black}|\text{ball} = G) = 1/(1 + (1/4)/(6/8)) = 1/(1 + (1/3)) = 3/4$; the probability we have drawn from the black urn is 0.75.

Questions of this type arise in countless situations. For example, in medicine, you can rephrase this example by identifying the sampled black ball with a patient with tumor X, of subtype Y, and genome Z. If you are treating this patient, you can ask for the probability that the box was white, which in the medical example could mean the probability that medication B works for that patient. In other words: to which class (box) of patients does this patient belong? To the class where medication B works or to the other where it does not?

Bayesian statistics is perfectly applicable to these types of question, and yields testable predictions. In other words, Bayesian inference is made for problems where you are given a sample $x(N) = (x_1, \cdots, x_N)$ that records the outcome of N typically identical and independent experiments. However, in the experiment we do not know the distribution function q_i of drawing particular values, $i \in \Omega$. Using Bayes' rule one can estimate the distribution function q (for instance, by predicting the most likely q_i) from data x and perhaps additional available information I.

Let us consider another example. Suppose you are drawing from an urn with replacement. You are given the information that the urn only contains $W = 2$ colours, grey (G) and black (B). You also know that there are only $M = 5$ balls in the urn. This information might be collected in the vector $I = (W, M)$. However, you do not know how many balls are G and how many are B. Let n_G be the number of G and $n_B = N - n_G$ the number of B balls. As you know there are two colours, n_B has to be one of the numbers $n_B \in \{1, 2, 3, 4\}$. Suppose you have drawn $N = 20$ times from the urn and obtained the sample $x(N) = (x_1, \cdots, x_{20})$, which contains $k_G = 13$ times G and $k_B = 7$ times B. What is the sample $x(N)$ telling us about the probability of finding n_G and n_B balls in the urn? In other words, we would like to know the probability[8] $P(q|x, I)$ of observing the distribution function q given the samples $x(N)$ and the information I. One may also interpret this as an example of *Bayesian hypothesis testing*. n_B determines q uniquely given the information I, and there are four possible hypotheses, $H_b = \{n_G = b\}$, $b = 1, \cdots, 4$, to choose from. Moreover $q_B(b) = b/5$ and $q_G(b) = 1 - b/5$. We therefore want to know $P(H_b|x, I)$, the probability of the hypotheses H_b being true. We can use Bayes' rule and the definition of conditional probability in Equation (2.14) to obtain,

$$
\begin{aligned}
P(H_b|x, I) &= P(H_b, x, I)/P(x, I) \\
&= P(x|H_b, I)P(H_b, I)/P(x, I) \\
&= P(x|H_b, I)P(H_b|I)/P(x|I).
\end{aligned} \tag{2.17}
$$

---

[8] We now drop the (N) in the notation of $x(N)$ for a moment.

Estimating $P(H_b|x,I)$ now becomes a matter of estimating $P(x|q(b),I) = P(x|H_b,I)$, $P(H_b|I)$ and $P(x|I)$; see Exercise 2.2. Be careful with the notation of conditional ($|$) and joint ($,$) distributions. We now take the following steps.

1. *The prior probabilities*: Given the information I, the best (maximal ignorance) estimate of $P(H_b|I) = 1/4$, for any of the four possible values $b \in \{1,2,3,4\}$. Note that this choice depends on the fact that we have no previous information about the urn, as this is our first sequence of N experiments.[9] $P(H_b|I)$ are called the *prior probabilities* of H_b, as they represent the probability of the predictor guessing H_b correctly before the experiment.

2. *The random process*: $P(x|q,I)$ is the probability of drawing a particular sample x containing k_B times B and k_G times G in $N = k_B + k_G$ independent trials. This probability is known to us from basic probabilistic and combinatorial facts,

$$P(x|q,I) = q_B^{k_B} q_G^{k_G}. \tag{2.18}$$

3. *Normalization*: We next compute $P(x|I)$. This can be done by marginalizing $P(x, H_b|I)$ with respect to H_b. This means summing $P(x, H_b|I)$ over all hypotheses b. Using the properties of conditional probabilities, we find $P(x, H_b|I) = P(x|H_b,I)P(H_b|I)$. Using $P(H_b|I) = 1/4$, it follows that,

$$P(x|I) = \sum_{b=1}^{4} P(x, H_b|I) = \sum_{b=1}^{4} P(x|H_b,I)P(H_b|I) = \frac{1}{4} \sum_{b=1}^{4} \left(\frac{b}{5}\right)^{k_B} \left(1 - \frac{b}{5}\right)^{k_G}. \tag{2.19}$$

4. *Posterior probabilities*: Finally, we insert the results (1–3) into Equation (2.16) and find,

$$P(H_{n_B}|x,I) = \frac{\left(\frac{n_B}{5}\right)^7 \left(1 - \frac{n_B}{5}\right)^{13}}{\sum_{b=1}^{4} \left(\frac{b}{5}\right)^7 \left(1 - \frac{b}{5}\right)^{13}}. \tag{2.20}$$

The Bayesian estimates for the probabilities that n_B takes the particular value $1, 2, 3$, or 4, are given by the probabilities,

$$P(H_1|x,I) = 0.2321$$
$$P(H_2|x,I) = 0.7059$$
$$P(H_3|x,I) = 0.0620$$
$$P(H_4|x,I) = 0.0001,$$

---

[9] In a subsequent sequence of experiments, one might use different values for $P(H_b|I)$ drawing upon information obtained from previous experiments. The priors can be updated as one learns more about a system by performing more experiments.

and we can predict the most likely hypothesis H_2 with the probability of 0.7059. We also know that with probability 0.2941 this prediction will be wrong. The probabilities $P(H_b|x,I)$ are called the *posterior probabilities* of H_b, as they describe the chances of the predictor correctly guessing H_b *after* having used the information contained in the sample $x(N)$.

We can fix a significance level (for instance, $p_s = 0.1$) and then perform as many experiments as necessary to accept or reject a hypothesis. In our example, we might say at this point that a probability of 0.2941 of failing in our prediction is too high, as it is larger than the significance level $p_s = 0.1$. We might decide to perform another sequence of $N = 20$ experiments and sample the data $x'(N) = (x'_1, \cdots, x'_{20})$. In this second round of experiments, we can use the information from the first sequence $x(N)$ by identifying the prior probabilities for the new data with the posterior probabilities $P(H_b|I') = P(H_b|x,I)$ that were obtained previously. In other words, in the second experiment we use the information $I' = (x',I)$. Suppose we throw a sequence $x'(20)$ with $k'_B = 9$ and $k'_G = 11$. Following the same reasoning as before, we now get the equation,

$$P(H_{n_B}|x',I') = \frac{\left(\frac{n_B}{9}\right)^5 \left(1 - \frac{n_B}{5}\right)^{11} P(H_{n_B}|I')}{\sum_{b=1}^{4} \left(\frac{b}{5}\right)^9 \left(1 - \frac{b}{5}\right)^{11} P(H_b|I')}, \tag{2.21}$$

which updates the chances for predicting hypothesis H_b to,

$$P(H_1|x',I') = 0.0144$$
$$P(H_2|x',I') = 0.9486$$
$$P(H_3|x',I') = 0.0370$$
$$P(H_4|x',I') = 2.2 \cdot 10^{-7}.$$

We now predict H_2 with probability 0.9486. This prediction will still fail with a probability of 0.0514. If we are happy with a significance level of $p_s = 0.1$ we stop here and predict H_2 at the significance level of 0.1. If, however, we had chosen a significance level of 0.01, we may need to perform additional experiments to reach this level of confidence. If none of the possible hypotheses reaches the specified confidence level, even after many additional experiments are performed, then we have to reject all possible hypotheses and conclude that the information I was probably incorrect and that the data generation model we used to infer the probabilities was therefore inadequate.

> Bayesian reasoning makes use of the fact that the chances of making correct predictions depend on information available prior to an experiment. As experimental data accumulate, posterior probabilities from previous experiments can be used as prior probabilities for new experiments. In Bayesian reasoning, the prior information (and prior distribution functions) is updated as we gain more information about the system.

2.2.6   Bayesian and frequentist thinking

Sometimes a distinction is made between the 'frequentist' and the 'Bayesian' notion of probabilities. These are also referred to as 'objective' and 'subjective' probabilities. This can be very misleading. In the frequentist notion the probabilities of events are considered as 'real' and the data we obtain are just samples. In the Bayesian view the data are 'real' (and not probabilistic) and the probabilities of events (or of the model of the data-generating process) are uncertain. In the frequentist world, one therefore tries to infer probabilities of events from repeatable independent experiments where, with more data, one is assured of estimating the probabilities with increasing accuracy. The frequentist uses the *p-value* to reject a null hypothesis.

In the Bayesian world, one often considers reasonable data-generating models (or a parametric family of models) and then infers the probability distribution for those models that have produced the data. In this task the Bayesian tries to use any available information to infer the data-generation process and its parameters. Bayesians do not talk about the probabilities of actual events but about the probabilities of predicting them correctly. If the available information about a particular phenomenon changes, the probabilities of possible models describing the phenomenon can be updated. This allows the Bayesian to think probabilistically about events that may never have occurred before—or cannot be observed in repeatable independent trials—while the strict frequentist has no notion of probabilities of this type. The mathematical theory of probabilities, however, does not depend on whether one feels inclined towards a Bayesian or a frequentist point of view. Bayes' rule in Equation (2.15) is a necessary tool in the pocket of every twenty-first-century scientist.

2.2.6.1   *Hypothesis testing in the frequentist approach*

For completeness we recall the basic philosophy behind testing a statistical hypothesis in the frequentist view; see also Section 4.7. It follows several steps. First, the *null hypothesis* H_0 is formulated, specifying precisely what is being tested. It can be as follows. The null hypothesis H_0 is: the samples from measurement A are from the same distribution as other samples from measurement B. The basic idea is now to try to reject the null hypothesis for a given significance level α. If we can, we have shown that, for the specified significance level α, the samples from measurement A are not from the same distribution as the samples of B. To be able to reject the null hypothesis, we need the *test statistic*, which is a convenient test function $d(x)$ that measures a 'distance' between sample x_A and sample x_B. The idea is that for the test statistic the correct distribution function is known and a significance level can be specified. If the test statistic is larger than α, the null hypothesis must be rejected. Typically, null hypotheses are set up to be rejected. If they are rejected at a given significance level, this signals that there might be a statistically significant difference between A and B.

We briefly summarize this section on basic notions in probability theory.

- Probabilities can be axiomatically defined as *normalized measures* on the sample space. This is accomplished by the three Kolmogorov axioms of probability theory.

- Probabilities are abstract measures that measure our chances of observing specific events before sampling them. Histograms and relative frequencies (normalized histograms) are *empirical* distribution functions that we observe when we are sampling a process.

- Probability distribution functions can have a mean, a variance, and higher moments. Empirical distribution functions have a sample mean (average) and a sample variance.

- Probability distributions of more than one variable are called *joint probabilities*.

- A *conditional probability* is the ratio between a joint probability and one of its marginal probabilities, the condition.

- Two random variables are *statistically independent* if the distribution function of one variable does not depend on the outcome of the other.

- Frequentist hypothesis testing uses a null hypothesis and a test statistic to test if the null hypothesis for the given data can be rejected at a specified confidence level. If not, the hypothesis is not rejectable at the confidence level.

- Bayesian hypothesis testing allows one to compare the quality of one data-generating model with that of another, regardless of whether the models are correct. New data can be used to update prior and posterior probabilities iteratively. If enough data are available and the data-generating models are reasonable, the process can be iterated until a prespecified significance level is reached.

2.3   The law of large numbers—adding random numbers

How can we do algebra with random numbers? If we add random numbers, we obtain a new random number. What is the distribution function of a sum of random numbers? In this context we will encounter the mathematical notion of limits, which we summarize in the law of large numbers. This law is a statement about the convergence of sums of random numbers, given that we add infinitely many of those. We will then review the convolution product and, with its help, discuss the central limit theorem, which is at the core of classical statistics. This theorem tells us that if we add many i.i.d. random numbers originating from sources with a finite variance, the distribution function will be a Gaussian (or normal) distribution function. We will discuss the notion of α-stable processes and learn that the Gauss, Lévy, and Cauchy distributions are the only possibilities for α-stable distribution functions with a closed form.

The concept of *limits* is essential for dealing with stochastic processes. Intuitively, a limit means that elements of a sequence of 'objects' become increasingly 'similar' to a particular 'limit-object', until the elements become virtually indistinguishable from the limit object—the sequence *converges* to its *limit*. The similarity can be defined in several ways; it has to be clarified beforehand which *similarity measure* is to be used.

What is a limit of a sequence of random variables? Assume the random variables $X(n), n = 1, 2, 3 \cdots$, over the sample space $\Omega(X(n)) = \{1, \cdots, W\}$, and a random variable

Y over the same $\Omega(X(n))$. What do we mean by a limit $\lim_{n\to\infty} X(n) \to Y$ of random variables? For two real numbers one can easily visualize what convergence means.

One possibility for giving meaning to a limit for random numbers is to consider limits of distribution functions. If the probability $q_i(n) = P(X(n) = i)$ and $\bar{q}_i = P(Y = i)$, then the limit $X(n) \to Y$ can be understood in terms of the limit $\lim_{n\to\infty} q_i(n) \to \bar{q}_i$ for all $i \in \Omega$. This is the so-called *point-wise limit* of distribution functions.

> The sequence $q_i(n)$, $n = 1, 2, 3, \cdots$, converges point-wise to the limit distribution $\bar{q}_i$, if for any real number $\epsilon > 0$ there is a number n_0, such that for any $n' > n_0$, it is true that $|\bar{q}_i(n) - q_i(n')| < \epsilon$.

Note that the term point-wise convergence tells us nothing about the speed of convergence[10] or whether the speed of convergence is the same for all i. Depending on the context, different notions of convergence are sometimes used, usually for the sake of simplicity of mathematical proofs. For the most part, however, the exact notion of convergence has little or no relevance for practical questions. One of such practical questions addresses the conditions under which sampling processes converge to an average. Gerolamo Cardano noted already in the sixteenth century that sample averages improve as more measurements are taken. If we consider the random variables $X(n)$, $n = 1, 2, 3, \cdots, N$, where each of the variables $X(n)$ has an average $\langle X(n) \rangle = \mu$, does its average also converge to μ? To find out, we define the random variable,

$$Y(N) = \frac{1}{N}(X(1) + X(2) + \cdots + X(N)), \tag{2.22}$$

and $y_N = (x_1 + x_2 + \cdots + x_N)/N$, where the x_n are samples of the variables $X(n)$, and y_n is a sample of $Y(N)$. Is it then reasonable to assume that $\lim_{N\to\infty} y_N = \mu$? The *law of large numbers* provides the answer to this question. There are various versions of this law, the best known ones being the weak and the strong law.

> The *weak law of large numbers* states that if we have independent and identically distributed (i.i.d.) random variables $X(n)$ with a mean $\langle X(n) \rangle = \mu$, for any number $\epsilon > 0$ (however small) it is true that,
>
> $$\lim_{N\to\infty} P(|y_N - \mu| > \epsilon) = 0. \tag{2.23}$$
>
> A mathematician would say that $Y(t)$ converges to μ *in probability*.

---

[10] To learn more about the speed of convergence we refer to the Berry–Esseen theorem [48, 120]. For methods that can be used to derive bounds on the speed of convergence, see, e.g. Stein's method [354].

The *strong law of large numbers*[a] states that under the same conditions,

$$P(\lim_{N\to\infty} y_N = \mu) = 1. \tag{2.24}$$

A mathematician would say that $Y(t)$ converges to μ *almost surely*.

[a] The strong law indeed makes the stronger claim. It states that if one performs random sampling experiments many times with variables $Y(n), n = 1, 2, \cdots$, then the fraction of sampled sequences y_n, where $\lim_{n\to\infty} y_n \to \mu$ is not fulfilled vanishes. The weak law only implies that, no matter how small we choose the error ϵ to be, the fraction of sampled sequences that will fail to converge to μ by a margin of ϵ, will go to zero.

The law of large numbers does not tell us anything about the expected error, when estimating the average μ using finite sample averages. The theorems only tell us that if you have taken sufficiently many samples, then it is usually safe to estimate the mean μ of the i.i.d. processes from its sample average. They tell us nothing about what 'sufficiently many' means.

2.3.1   The central limit theorem

The law of large numbers does require the random variables $X(n)$ to be i.i.d., but it does not require a finite variance. If we have both finite mean and finite variance, we can state the single best-known limit theorem, the *central limit theorem*. This is at the heart of Gaussian statistics and explains the ubiquity of the so-called *Gaussian* or *normal distribution function*. Before we turn to the theorem, let us recollect some facts about Gaussian distribution functions and the notion of the *convolution product*.

2.3.1.1   *Gaussian distribution function*

The *standard normal distribution function* has the form,

$$\phi(x) = \frac{1}{\sqrt{2\pi}} \exp\left(-\frac{1}{2}x^2\right). \tag{2.25}$$

All *normal* distribution functions can be obtained by stretching and dilating the standard normal distribution,

$$p_{\text{normal}}(x|\sigma,\mu) = \frac{1}{\sigma}\phi\left(\frac{x-\mu}{\sigma}\right), \tag{2.26}$$

where μ is the mean and σ the standard deviation of the distribution.

A random variable X over the real numbers, with mean μ and standard deviation σ, is called a *Gaussian random variable*, if the probability of finding $a < X < b$ is given by,

continued

$$P(a < X < b | \sigma, \mu) = \int_a^b dx \, p_{\text{normal}}(x | \mu, \sigma).$$ (2.27)

The variable X is then said to be in $\mathcal{N}(\mu, \sigma^2)$.

We can immediately compute the moments of Gaussian random variables from Equation (2.25). As the standard normal distribution is symmetric, all odd moments with $n = 1, 3, 5, \cdots$, vanish, meaning that $\int_{-\infty}^{\infty} \phi(x) x^n = 0$. The even moments, with $n = 0, 2, 4, \cdots$, can be computed by taking derivatives repeatedly in the following way,

$$\int_{-\infty}^{\infty} dx \, \phi(x) x^{2m} = \left(-2 \frac{d}{d\alpha} \Big|_{\alpha=1} \right)^m \int_{-\infty}^{\infty} \frac{dx}{\sqrt{2\pi}} e^{-\frac{\alpha}{2}x^2} = \left(-2 \frac{d}{d\alpha} \Big|_{\alpha=1} \right)^m \int_{-\infty}^{\infty} \frac{dy}{\sqrt{2\pi\alpha}} e^{-\frac{1}{2}y^2}$$

$$= \left(-2 \frac{d}{d\alpha} \Big|_{\alpha=1} \right)^m \alpha^{-\frac{1}{2}} = 1 \cdot 3 \cdot 5 \cdots (2m-1) = \frac{(2m)!}{m!2^m}.$$ (2.28)

Here the expression $(d/d\alpha)^m$ means: apply the derivative m times.

2.3.1.2 The convolution product

The convolution product has many applications. One of them is to describe the distribution of sums of independent random variables.

The *convolution product* of two functions is defined as,

$$f * g(x) = \int_{-\infty}^{\infty} dx' f(x') g(x - x').$$ (2.29)

It has similar algebraic properties to the usual product of real numbers, $f * g = g * f$, $(f * g) * h = f * (g * h)$, and it is linear $f * (g + h) = f * g + f * h$. Here the sum of two functions is defined as $(g + h)(x) = g(x) + h(x)$.

To see how the convolution product yields the sum of two random numbers, assume that X and Y are two independent random variables over the real numbers with distribution functions $f(x)$ and $g(y)$, respectively. What is the distribution function of the random variable $Z = X + Y$? To find the probability that the sum of two random numbers is exactly z, $P(x + y = z)$, we must consider all possible realizations x and y that fulfil this condition. We do this in the following way,

$$P(x + y = z) = \int_{-\infty}^{\infty} dx \int_{-\infty}^{\infty} dy \, f(x) g(y) \delta(z - x - y),$$ (2.30)

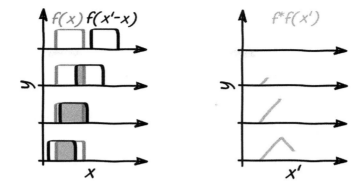

Figure 2.6 *The schematic diagram shows how the convolution product of a function can be understood as the moving overlap of two functions. Initially, we have two box functions f and start shifting the right box to the left. The area by which the boxes overlap (shaded area) is the value of the corresponding convolution product shown in the right panel. The convolution product of two box functions is a triangle. Think of the box functions as the probabilities of the outcomes of a fair dice, 1 to 6. The convolution product is the distribution function of the sum of the face values of two dice that are thrown simultaneously. It is a triangle that spans from 2 to 12. Repeated iterations of the convolution product on box functions produce functions that approach a normal distribution very quickly; see Figure 2.7.*

where $\delta(x)$ is the Dirac delta function; see Section 8.1.2. The Dirac delta ensures that $x + y = z$. We can now integrate over y, using the property of the Dirac delta that $\int_{-\infty}^{\infty} dx\, f(x)\delta(x - y) = f(y)$,

$$P(z) = \int_{-\infty}^{\infty} dx \int_{-\infty}^{\infty} dy\, f(x)g(y)\delta(z - x - y) = \int_{-\infty}^{\infty} dx f(x)g(z - x) = f * g(z), \quad (2.31)$$

where in the last step we used the definition of the convolution product Equation (2.29). We see that the distribution function of $Z = X + Y$ is given by the convolution product of the distribution functions of X and Y. For a graphical illustration of the convolution product, see Figure 2.6. As a consequence, the probability density function of the sum $X(1) + X(2) + \cdots + X(N)$, of N independent random variables $X(n)$ with the same probability density functions $f(n)$, $n = 1, \cdots, N$, is given by the convolution product $f(1) * f(2) * \cdots * f(N)$. The essence of the central limit theorem is that for large N, this iterated convolution product converges to the Gaussian distribution. This is true, no matter what f looks like, as long as f has finite mean and variance. We demonstrate this for a pathological-looking distribution function f, shown in Figure 2.7a. As it is convoluted four times with itself, it already looks 'quite' Gaussian; Figure 2.7d.

We can now state the central limit theorem more formally. As for the law of large numbers, there are various ways of stating the theorem; we use the Lévy–Lindeberg version [254].

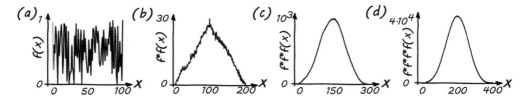

Figure 2.7 *Starting from a pathologically spiky distribution function f (a) we demonstrate how quickly the repeated convolution of f with itself converges to a Gaussian function (b and c). After four iterations, (d) the convolution product already looks Gaussian. The Gaussian distribution is the 'attractor' for an iterated convolution of distribution functions with existing mean and variance.*

> **Central limit theorem.** Assume that $X(n)$, $n = 1, 2, 3, \cdots$, are i.i.d. random variables with an existing mean $\langle X(n) \rangle = \mu$ and an existing variance $\langle X(n)^2 \rangle = \sigma^2$. Then, in the limit of large N, the variable $Y(N) = \frac{1}{\sqrt{N}} \sum_{n=1}^{N} (X(n) - \mu)$ converges to a Gaussian random variable Z, with mean 0 and variance σ^2. A mean and variance exist if they have finite values and do not diverge for large N.

The general proof is not trivial and we do not reproduce it here. However, to convince ourselves that Gaussian distributions arise in the limit of sums of i.i.d. random variables, it is sufficient to note that the sum of two Gaussian random variables is again a Gaussian random variable. This implies that Gaussian distributions are so-called *attractors* or limit distributions of the convolution product. To see this, let X and Y be two independent Gaussian random variables, both with mean zero and respective standard deviations, σ_1 and σ_2. Using Gaussians for f and g in Equation (2.31), the probability density function of $X + Y$ is found to be,

$$P(x + y = z) = \frac{1}{\sqrt{2\pi}\,\bar{\sigma}} e^{-\frac{z^2}{2\bar{\sigma}^2}}, \tag{2.32}$$

where $\bar{\sigma}$ is given by $\bar{\sigma} = \sigma_1 \sigma_2 \sqrt{1/\sigma_1^2 + 1/\sigma_2^2}$. The sum of two Gaussian random variables is again a Gaussian variable. This is explicitly shown in Exercise 2.10 at the end of the chapter. See also Equation (2.39). This proves that the Gaussian distribution is a fixed point under the convolution product, but it does not prove that the Gaussian distributions are also stable attractors for the iterated convolution product. The central limit theorem (in its various versions) proves exactly this; see for example [254]. However, we can easily convince ourselves of this fact computationally; see Figure 2.7.

2.3.1.3 *Log-normal distribution function*

The central limit theorem tells us that the expected distribution function for sums of many random numbers is the Gaussian. What about the product of random numbers, $Y(N) = \prod_{n=1}^{N} X(n)$, if the $X(n)$ are i.i.d.? If we take the logarithm of Y and define the variables $G = \log Y$ and $Z = \log X$ we get,

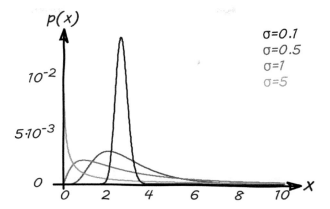

Figure 2.8 *The log-normal distribution for $\mu = 1$ and various values of σ. It is sometimes hard to decide if a given empirical distribution derived from data is a log-normal distribution or a power law.*

$$G(N) = \log Y(N) = \sum_{n=1}^{N} \log X(n) = \sum_{n=1}^{N} Z(n). \tag{2.33}$$

Recalling the central limit theorem, note that if $Z(n)$ has a finite mean $\langle Z(n)\rangle = \mu$ and variance $\sigma^2(Z) < \infty$, then $Z(n)$ fulfils the conditions for the central limit theorem, and the properly centred and scaled G is a Gaussian random number, $V = G/\sqrt{N} - \sqrt{N}\mu$, with mean 0 and variance σ^2, given, of course, that N is large enough.

If a random variable V is normally distributed with variance $\mu(V)$ and $\sigma^2(V)$, then the variable $U = \exp(V)$ is called *log-normally distributed* with $\mu(V)$ and $\sigma^2(V)$. Note that here $\mu(V)$ and $\sigma^2(V)$ denote the mean and variance of the underlying Gaussian variable V. The mean $\mu(U)$ and $\sigma^2(U)$ of the log-normal variable U can be computed,

$$\mu(U) = e^{\mu(V)+\frac{1}{2}\sigma^2(V)} \quad \text{and} \quad \sigma^2(U) = e^{2\mu(V)+\sigma^2(V)}\left(e^{\sigma^2(V)} - 1\right). \tag{2.34}$$

Now, as $V = \lim_{N\to\infty}(G(N)/\sqrt{N} - \sqrt{N}\mu)$ is a normally distributed variable with $\mu(V) = 0$ and $\sigma(V) = \sigma$, one concludes that,

$$U(N) = Y(N)^{\frac{1}{\sqrt{N}}} e^{-\mu\sqrt{N}} \tag{2.35}$$

converges towards a log-normally distributed variable U, with $\mu(U) = e^{\frac{1}{2}\sigma^2}$ and $\sigma^2(U) = e^{\sigma^2}\left(e^{\sigma^2} - 1\right)$. The log-normal variable U follows the probability density distribution,

$$\begin{aligned} p_{\text{log-normal}}(x|\mu,\sigma) &= \tfrac{1}{dx}P(x < U < x + dx) \\ &= \tfrac{1}{dx}P(\log x < \log(U) < \log(x + dx)) \\ &= \tfrac{1}{dx}P(\log x < V < \log(x + dx)) \\ &= \tfrac{1}{dx}P(\log x < V < \log x + \tfrac{dx}{x}) \\ &= \tfrac{1}{x\,d\log x}P(\log x < V < \log x + d\log x) \\ &= p_{\text{normal}}(\log x|\mu,\sigma)\tfrac{1}{x}, \end{aligned} \tag{2.36}$$

where we used $d \log x = dx/x$ and Equation (2.4). Keep in mind that μ and σ are the mean and standard deviation of $V = \log U$ and not of the variable U; see Figure 2.8.

For multiplicative random variables (and the geometric mean of random variables) the log-normal distribution plays the same role as the normal distribution plays for additive variables (and the arithmetic mean) of random variables.

2.3.2   Generalized limit theorems and α-stable processes

The central limit theorem is certainly the best-known probabilistic limit theorem. However, it is not the only one. In fact, there is an entire family of limit theorems that are tightly associated with so-called *α-stable processes*. Gaussian random processes are members of this larger family and the central limit theorem is a special case within it. The basic idea behind more general limit theorems is the concept of stable processes. In a nutshell, a process is stable if the sum of two random variables from a given distribution leads to a random variable that is of the same *type*. More precisely:

A random variable X is called *stable* if for the i.i.d. variables $X(1)$, $X(2)$, and $X(3)$ ($X(i) \stackrel{d}{=} X$ for $i = 1, 2, 3$) there are four constants a, b, c, and d such that,

$$aX(1) + bX(2) \stackrel{d}{=} cX(3) + d. \qquad (2.37)$$

Before we discuss the general case, let us focus on the Gaussian example. We have seen before in Equation (2.32) that Equation (2.37) is true for Gaussian random variables; the sum of two Gaussian random variables is again a Gaussian. To show that Gaussian random variables are stable, we can start by assuming $X(1)$, $X(2)$, and $X(3)$ to be i.i.d. Gaussian random variables with unit variance and use the definition of the cumulative distribution function,

$$P(X(n) < x) = \Phi(x) = \int_{-\infty}^{x(n)} dx' \frac{1}{\sqrt{2\pi}} e^{-\frac{1}{2}x'^2}, \qquad (2.38)$$

for $n = 1, 2, 3$, and where $x(1) = a$, $x(2) = b$ and $x(3) = c$. The computation proceeds as in Equation (2.31), except that the constraint we want to satisfy is now $aX(1) + bX(2) \stackrel{d}{=} cX(3)$, for three constants a, b, and c that satisfy $c^2 = a^2 + b^2$. Remembering that the

Dirac delta function is the derivative of the Heaviside step function, see Section 8.1.1, we compute,

$$\frac{d}{dz}P(aX(1)+bX(2)<cz)$$

$$=c\int_{-\infty}^{\infty}dx\frac{1}{\sqrt{2\pi}}e^{-\frac{1}{2}x^2}\int_{-\infty}^{\infty}dy\frac{1}{\sqrt{2\pi}}e^{-\frac{1}{2}y^2}\delta(ax+by-cz)$$

$$=c\int_{-\infty}^{\infty}dx'\frac{1}{2\pi\,ab}e^{-\frac{x'^2}{2a^2}-\frac{(cz-x')^2}{2b^2}}$$

$$=c\int_{-\infty}^{\infty}dx'\frac{1}{2\pi\,ab}e^{-\frac{x'^2}{2}\left(\frac{1}{a^2}+\frac{1}{b^2}\right)+\frac{czx'}{b^2}-\frac{z^2c^2}{2b^2}}$$

$$=\underbrace{\frac{c}{\sqrt{2\pi(a^2+b^2)}}e^{-\frac{z^2c^2}{2(a^2+b^2)}}}_{\frac{1}{\sqrt{2\pi}}e^{-\frac{1}{2}z^2}}\underbrace{\int_{-\infty}^{\infty}dx'\sqrt{\frac{a^2+b^2}{2\pi\,a^2b^2}}e^{-\frac{1}{2}\left(\frac{x'\sqrt{a^2+b^2}}{a^2b^2}-\frac{za}{b\sqrt{a^2+b^2}}\right)^2}}_{\int_{-\infty}^{\infty}dx''\frac{1}{\sqrt{2\pi}}e^{-\frac{1}{2}x''^2}=1} \tag{2.39}$$

$$=\frac{d}{dz}P(X(3)<z),$$

where in the last line we used $c^2=a^2+b^2$. This shows that if $X(1)$ and $X(2)$ are standard Gaussian distributed variables, then so is $X(3)$. The Gaussian is not the only example, however. We can now define a stable process: if a random variable X is stable, then there exist constants c_n and d_n and an i.i.d. sequence $X(n)\overset{d}{=}X$, $n=1,2,\cdots$, such that for any N,

$$\sum_{n=1}^{N}X(n)\overset{d}{=}c_NX+d_N. \tag{2.40}$$

Stable random variables can be parametrized with four parameters.

- α is the *characteristic exponent*. Parameter α characterizes how the c_n scale with n, that is, $c_n\sim n^{1/\alpha}$ for some $0<\alpha\le2$.
- β is the *skewness* and quantifies how asymmetric a distribution function is.[11]
- γ is the *scale*. If we rewrite the stability condition $aX(1)+bX(2)=\gamma X(3)+\delta$, the scale parameter γ may still be meaningful, even if the variance of a process does not exist. γ characterizes how 'stretched' $aX(1)+bX(2)$ is with respect to $X(3)$.
- δ is the *position* that quantifies how much $aX(1)+bX(2)$ is shifted with respect to $X(3)$.

---

[11] If the mean and variance of a random variable exist, the skewness of the distribution is given by the third moment of the standardized random variable $\beta=\langle((X-\mu)/\sigma)^3\rangle$. Another way of defining skewness is $\beta=\kappa_3(X)/\kappa_2(X)^{3/2}$, where κ_2 and κ_3 are the second and third cumulants of X, respectively.

We now characterize α-stable processes in terms of limits of random variables:

A random variable X is said to be in the *domain of attraction* of the random variable Y, if, for an i.i.d. sequence with variables $X(n) \stackrel{d}{=} X$, there exist constants $a_n > 0$ and b_n, such that,

$$\lim_{N \to \infty} a_N \sum_{n=1}^{N} X(n) - b_N \stackrel{d}{=} Y. \tag{2.41}$$

If there is at least one random variable X with characteristic exponent $0 < \alpha \le 2$ that is in the domain of attraction of Y, then Y is called α-*stable* [293].

2.3.2.1   *Distribution functions of α-stable processes*

To this day, only three stable random processes are known for which the distribution functions can be written in closed form, meaning that they can be written as a formula. Other stable distribution functions can only be handled numerically.

2.3.2.2   *Gaussian distribution function*

We discussed Gaussian or normal distribution functions at the beginning of this section. The normal distribution is obtained by stretching and shifting the standard normal distribution $\phi(x)$ of Equation (2.25),

$$p_{\text{normal}}(x|\mu,\sigma) = \frac{1}{\sigma}\phi\left(\frac{x-\mu}{\sigma}\right) = \frac{1}{\sqrt{2\pi}\sigma}\exp\left(-\frac{(x-\mu)^2}{2\sigma^2}\right). \tag{2.42}$$

Normal distributions are an attractor of the convolution product. They arise in countless situations, from the distribution of velocities in gases in thermal equilibrium to the distribution of errors in repeated measurements of a quantity. Measuring the weight of a tomato with your kitchen scale thousands of times will provide you with a Gaussian distribution of weights. Try it.

2.3.2.3   *Cauchy distribution function*

This continuous distribution is also called Cauchy–Lorentz or Breit–Wigner distribution. It is given by,

$$p_{\text{Cauchy}}(x|\gamma,\mu) = \frac{1}{c\pi}\frac{c^2}{(x-\mu)^2 + c^2}, \tag{2.43}$$

where c and μ are real-valued parameters. Its asymptotic power law is characterized by an exponent of two, $p_{\text{Cauchy}}(x|\gamma,\mu) \sim x^{-2}$. Note that for the Cauchy distribution, neither the mean, variance, nor any other higher moment, can be defined. In physics the distribution typically appears in resonance phenomena.

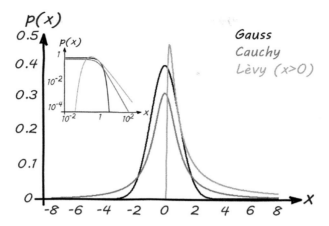

Figure 2.9 *The three families of α-stable distribution functions that can be written as a formula are the Gaussian (black), the Cauchy (dark-grey), and the Lévy (light-grey) distributions. The inset shows the distributions in a loglog plot. For large x they become increasingly linear, indicating an asymptotic power law. Power laws decay much more slowly than exponential or Gaussian distributions. They are therefore called fat-tailed.*

2.3.2.4 Lévy distribution function

This continuous probability distribution is named after Paul P. Lévy and is given by,

$$p_{\text{Levy}}(x|\gamma,\mu) = \sqrt{\frac{c}{2\pi}}\, \frac{e^{-\frac{c}{2(x-\mu)}}}{(x-\mu)^{3/2}}, \tag{2.44}$$

where c and μ are real-valued parameters and $x > \mu$ is required. Lévy distributions behave asymptotically as $p_{\text{Levy}}(x|\gamma,\delta) \sim x^{-\frac{3}{2}}$. The variance and higher moments of the Lévy distribution do not exist; they are infinite. It has applications in the area of random walks, in particular, in 'first-hit problems', where one wants to know how much time a particle in Brownian motion (random walk) needs to hit a particular target. The Lévy distribution is a special case of the inverse gamma distribution, which is important in the context of Bayesian statistics. Lévy distributions have physical applications, for instance, in the context of spectroscopy or geomagnetic reversal time statistics.

The Cauchy and Lévy distributions are shown in Figure 2.9. Both decay very slowly when compared to the normal distribution. The slow decay is referred to as a *fat-tailed* distribution, which often follows an asymptotic power law (for large x). This implies that *extreme events* (large values of the random variable) are much more likely for fat-tailed processes than they are for processes with Gaussian or exponentially decaying distribution functions. Cauchy and Lévy distributions are indeed asymptotic power laws, as seen in the loglog plot in the inset of Figure 2.9.

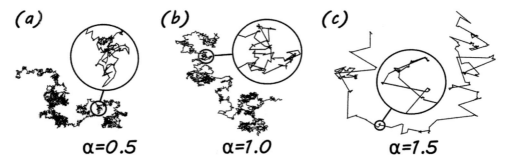

Figure 2.10 *Three examples of Lévy flights with increments are drawn from a power law distribution with three different values of* $\alpha = 0.5$, 1, 1.5. *The smaller* α *becomes, the more 'jumpy' the flights become.*

A *fat-tailed distribution* function $p(x)$ decays slowly in comparison with exponential or Gaussian distribution functions. Typically it decays as a power law for large x. The terms fat-tailed, heavy-tailed, long-tailed, scale-free, complex, or power law distribution are often used interchangeably. However, not every fat-tailed distribution is a power law.

2.3.2.5 Lévy flights

The term Lévy flight was coined by Benoît Mandelbrot to refer to random walks with fat-tailed increment distribution functions. *Lévy flights* are random walks $X_{t+1} = X_t + \Delta X_t$, where the step sizes ΔX_t are i.i.d. random variables that have fat-tailed distribution functions.[12] Lévy processes are a continuous generalization of Lévy flights and are defined as a continuous stochastic process. The corresponding distribution function of Lévy processes can be described by a generalized Fokker–Planck equation of the form

$$\frac{\partial}{\partial t}p(x,t) = -\frac{\partial}{\partial x}v(x,t)p(x,t) + \frac{\partial^{\alpha}}{\partial x^{\alpha}}D(x,t)p(x,t), \tag{2.45}$$

where v represents a flow term, D is a diffusion constant, and $\frac{\partial^{\alpha}}{\partial x^{\alpha}}$ is a so-called fractional derivative, which has a clear meaning in the Fourier transformed version of Equation (2.45). In Figure 2.10 we show several examples of two-dimensional Lévy flights with power law distributed increments $\Delta X = (\Delta X^x, \Delta X^y)$, $p(\Delta X) \sim (\Delta X)^{-\alpha}$. Various values of the exponent α are shown. Smaller values of α imply heavier tails in the increment distribution. As a consequence, extreme increments become more likely and the random walk becomes increasingly 'jumpy' for smaller α. We will say more about random walks in Sections 2.5.1.5, 6.4.3.1, and 6.5.3.

---

[12] It is a common misconception that the increments in a Lévy flight X must be from a Lévy distribution. They can be from any fat-tailed distribution.

This section has dealt with the law of large numbers and the fundamental limit theorems of probability theory.

- There are different ways of defining the equivalence of two random variables. Equivalence in distribution means that both variables have the same distribution function.
- The *law of large numbers* tells us that for large numbers of independent trials the sample mean converges to the mean of the random variable used in those trials.
- For a random variable with existing mean and variance the *central limit theorem* states that for a large number of independent trials the sum of the random variable converges to the *normal distribution*.
- Products of many random numbers under appropriate conditions are distributed according to the *log-normal* distribution.
- α-stable distributions are distributions that are 'type-invariant' under summation. This means that sums of i.i.d. random variables are random variables of the same type (with the same distribution function). Stable distributions are characterized by four parameters. Their definition as limits of sums of i.i.d. random variables leads to a generalization of the central limit theorem.
- The only stable limit distributions with known explicit formulas are the Gaussian, Cauchy, and Lévy distributions.
- *Fat-tailed* distributions concentrate much statistical weight in the tails of the distribution. Large values occur relatively often. Fat-tailed distributions appear everywhere in statistical data of complex systems.
- Lévy flights are random walks with fat-tailed increment statistics, $X_{t+1} = X_t + \Delta X_t$. This means that the increment random variable ΔX_t is drawn from a fat-tailed distribution (not necessarily the Lévy distribution).

2.4   Fat-tailed distribution functions

Gaussian distribution functions arise naturally from the central limit theorem whenever random numbers are added. This explains the ubiquity of Gaussian distributions in countless data. One understands the Cauchy and Lévy distributions, which are fat-tailed distributions, on the basis of a slightly more general version of the central limit theorem. There are, however, many more types of fat-tailed distribution functions that appear in the context of complex systems. Is there an equivalent to the central limit theorem for fat-tailed distributions other than Cauchy and Lévy? Not to our current knowledge. However, there are about five classic mechanisms that generate power laws. We discuss these mechanisms in detail in Section 3.3. They include critical and self-organized critical systems, multiplicative processes with constraints, preferential processes, and sample space reducing processes. In the following, we review the most important fat-tailed

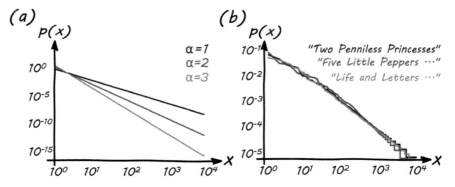

Figure 2.11 *(a) Pareto distributions are pure power laws $p \sim x^{-\alpha}$ defined for $0 < x_0 < x$. For the distribution to be normalizable $\alpha > 1$ is necessary. We show the Pareto distribution in loglog scale for $\alpha = 1, 2, 3$. (b) Empirically observed Zipf's law. The rank-frequency distribution of word occurrences from three novels are shown,* Two Penniless Princesses *by Charlotte M. Yonge (black),* Five Little Peppers and How They Grew *by Margaret Sidney (grey), and* Life and Letters of Lord Macaulay *by Sir George O. Trevelyan (light-grey). The Zipf distribution is usually associated with a scaling exponent of -2. In the rank-frequency distribution shown in (b), this becomes an exponent of -1.*

distributions that appear in the context of complex systems. We present the following distribution functions as *probability density functions* $p(x)$ where the probability of finding a realization of the random variable X in the interval $[x, x + dx]$ is $p(x)dx = P(x < X < x + dx)$. We do this because distribution density functions, when confronted with data, will frequently be useful for fitting the empirical relative frequency distributions.

2.4.1 Distribution functions that show power law tails

In this section, we present well-known distribution functions with power law tails. For historical reasons, multiple names are sometimes attached to one specific form of power law. Some of them are strongly related; some are even exactly equivalent.

2.4.1.1 *Pareto distribution*

The Pareto distribution, named after Vilfredo Pareto, is a continuous, pure power law with exponent $\alpha + 1$, given by,

$$p_{\text{Pareto}}(x|x_0, \alpha) = \begin{cases} \frac{\alpha x_0^{\alpha}}{x^{\alpha+1}} & \text{if} \quad x \geq x_0 \\ 0 & \text{if} \quad x < x_0. \end{cases} \tag{2.46}$$

The mean and variance exist if $\alpha > 1$ and $\alpha > 2$, respectively. In that case, the mean is $\langle X \rangle = \frac{\alpha x_0}{\alpha - 1}$ if $\alpha > 1$, and the variance $\sigma^2(X) = \left(\frac{\alpha x_0}{\alpha - 1}\right)^2 \frac{\alpha}{\alpha - 2}$ for $\alpha > 2$. For every higher moment m to exist, $\alpha > m$ must hold. The Pareto distribution was initially used to describe wealth distributions in economies. Pareto distributions with various values of α are shown in Figure 2.11.

2.4.1.2 *Zipf distribution*

The Zipf distribution is a discrete distribution function defined as,

$$p_{\text{Zipf}}(x_i|\alpha) = \frac{1}{Z}\frac{1}{x_i^{\alpha+1}}, \tag{2.47}$$

where Z is a normalization constant[13]. If $x_i = i$, $i = 1,2,\cdots$ then $Z = \zeta(\alpha+1)$, where $\zeta(n) = \sum_{i=1}^{\infty} i^{-n}$ is the zeta function. The Zipf distribution is a discrete version of the Pareto distribution. In the literature, the Zipf probability density distribution is often associated with a value of $\alpha = 0$, for which it cannot be normalized if the number of states i is infinite. However, in many data-related applications this problem does not arise, as all data are finite. In this case, the normalization factor is no longer the zeta function, as the sum $\sum_{i=1}^{N} i^{-n}$ is now taken up to a finite N. Often, Zipf distributions occur in *rank distributions*. Compare, for instance, Section 3.3 or [290]. If a *rank distribution* displays an exponent of -1, it is often called *Zipf's law*. This law was first noted for population sizes in cities by Felix Auerbach and later for word frequencies in written texts by George K. Zipf [419].

There are several ways of understanding the origin of this distribution. Zipf himself proposed a 'principle of least effort' [420], which did not become popular. Nowadays, more accepted ways of understanding it are *random languages*, where letters, including a white-space symbol, are randomly typed into a typewriter (a machine used in the twentieth century for writing characters similar to those produced by printers' type) [274], *preferential processes* or Yule–Simon processes [340, 416], or through *sample space reducing* processes [367]. These processes are discussed in more detail in Sections 3.3 and 6.6.2. Empirical Zipf rank distributions of word frequencies are shown in Figure 2.11 for three novels.

2.4.1.3 *Zipf–Mandelbrot distribution*

An extension of the Zipf distribution is the Zipf–Mandelbrot distribution, which is also known by the name of Zipf–Pareto distribution. It is a discrete distribution function given by,

$$p_{\text{Zipf}-\text{Mandelbrot}}(x_i|c,\alpha) = \frac{1}{Z}\frac{1}{(x_i+c)^{\alpha+1}}, \tag{2.48}$$

where c is a constant, Z is a normalization constant, and $x_i > -c$ must hold for all i. For $x_i = i$ where $i = 1,2,3,\cdots,N$, one finds that $Z = H(\alpha,s,N)$ is the Hurwitz function

---

[13] The Zipf distribution is sometimes also defined as $\sim x^{-\alpha}$. The same is true for the Zipf–Mandelbrot distribution.

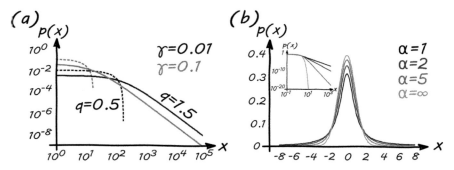

Figure 2.12 *(a) q-exponential distribution functions for two values of the scale parameter γ and two values of q in a loglog plot. For q = 0.5 < 1 the distribution has finite support and exists only in a limited region of x. For q = 1.5 > 1 the q-exponential distribution is an asymptotic power law with exponent α = 1/(1 − q). (b) Student t-distribution for various values of α. As α becomes large, the Gaussian distribution is recovered. In the inset (loglog plot) we see that the Student-t distribution is a power law for large x.*

defined as $H(\alpha, c, N) = \sum_{i=1}^{N}(i + c)^{-\alpha}$. For finite N, the distribution function is always normalizable.

2.4.1.4   *The q-exponential distribution function—Tsallis distribution*

The Tsallis distribution is the distribution function obtained by maximization of the so-called Tsallis entropy, which plays an important role in the statistics of complex systems, as we learn in Chapter 6. The Tsallis distribution is given by,

$$p_{Tsallis}(x|q, \gamma) = \frac{1}{Z}[1 - \gamma(1 - q)x]^{\frac{1}{1-q}}, \tag{2.49}$$

where Z is a normalization constant, x may be continuous or discrete $x = 1, 2, 3, \cdots$, q parametrizes the characteristic exponent of the distribution, and γ is a scale parameter. The asymptotic power law exponent is $1/(1 - q)$. The expression, $e_q(x) = [1 + (1 - q)x]^{\frac{1}{1-q}}$, is called the q-exponential. From the q-exponential one obtains the so-called q-Gaussian if one uses a squared argument, $e_q(-x^2)$. For $q=1$, the Tsallis distribution becomes the exponential or Boltzmann distribution. In this limit, the q-Gaussian converges to the normal distribution. The Tsallis distribution is equivalent to the Zipf–Mandelbrot distribution. If x has no upper bound, it is normalizable for values of $q < 2$. For values $q < 1$, the q-exponential needs to be set to zero, $e_q(x) = 0$ for $x > 1/(1 - q)$. This is a function of finite support, meaning that it extends over a limited range of x, and is zero outside. In Figure 2.12a we show q-exponential functions for different values of q and γ. Note that for $q < 1$, q-exponentials have a finite support.

2.4.1.5   *Student-t distribution*

The Student-t distribution is a continuous distribution that arises naturally in situations with small sample sizes or samples with unknown standard deviations. It is given by,

$$p_{\text{Student}}(x|\alpha) = \frac{\Gamma(\frac{\alpha+1}{2})}{\sqrt{\alpha\pi}\,\Gamma(\frac{\alpha}{2})} \left(1 + \frac{x^2}{\alpha}\right)^{-\frac{\alpha+1}{2}}, \tag{2.50}$$

where Γ is the Gamma function; see Equation (8.12). It was developed at the brewery company, Guinness, by William S. Gosset who used the pseudonym 'Student' for his publications. It was popularized by Ronald A. Fisher [129]. Its characteristic scaling exponent (it is a power law) is $\alpha + 1$. The limit $\alpha \to \infty$ recovers the Gaussian distribution function. In the original context, $\alpha + 1$ is related to the sample size. The Student-t distribution is equivalent to the Tsallis distribution. The Student-t distribution is often used in statistical hypothesis testing because it describes the distribution of the average μ of N i.i.d. random numbers around the true mean μ_0. More precisely, it describes the distribution of the standardized random variable $T = \sqrt{N}(X - \mu_0)/\sigma$. This is where the name t-distribution comes from. The Student-t distribution plays the role of the test statistic for accepting and rejecting the hypothesis that a value μ_0 represents the true mean of the variable X, $\langle X \rangle$; compare Section 2.2.6.1. In Figure 2.12b we show Student-t distributions for several values of α.

2.4.2 Other distribution functions

There are a number of distribution functions that appear frequently in the context of simple and complex systems.

2.4.2.1 Exponential- or Boltzmann distribution

Exponential distribution functions,

$$p(x|\beta) = \frac{1}{Z}e^{-\beta x}, \tag{2.51}$$

where Z is the normalization factor. Exponential distributions emerge whenever independent events i occur exactly n times in a row, given that the probability of sampling i is $q(i)$. The probability of this happening is then $q(i)^n(1 - q(i))$. Another situation where the exponential distribution occurs is the distribution of inter-event times in a Poisson process; see Section 2.5.1.4. Probably the most famous exponential distribution function to appear in physics is the Boltzmann distribution or the Boltzmann factor. It appears as a statistical factor in systems in thermal equilibrium and expresses the fact that energy states E are exponentially distributed,

$$p(E|\beta) = \frac{1}{Z}e^{-\beta E}. \tag{2.52}$$

In a physics context, β is the *inverse temperature*. Exponential distribution functions (see Figure 2.13a) have one property that distinguishes them from all other distribution functions. Changing the off-set of the distribution function (shifting it) does not change

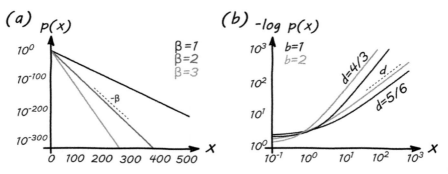

Figure 2.13 *(a) Exponential distribution functions for $\beta = 1, 2, 3$ in a semilog plot (y-axis is logarithmic). (b) Stretched exponential distributions (also called Kohlrausch–Williams–Watts functions) are plotted for different stretching exponents d and a scale parameter b. Note that $-\log(p(x)) = \log(Z) - bx^d$ is a power law. Therefore, if we plot the log of a stretched exponential distribution $-\log(p(x))$ in a loglog plot, it appears as a straight line (asymptotically).*

its overall shape, that is, $p(x|x_{\min}) = Z^{-1}\exp(-\beta(x - x_{\min}))$ remains an exponential. This is not true for any other family of distribution functions [277].

2.4.2.2 Stretched exponential distribution function

Stretched exponential distribution functions are exponential functions with a power exponent on the argument x,

$$p_{\mathrm{SE}}(x|b,d) = \frac{1}{Z}e^{-(bx)^d} \quad x > 0, \tag{2.53}$$

where b is a scale parameter and d is the exponent characterizing the 'stretching' of the exponential function. For a continuous sample space $x > 0$, the normalization constant Z is $Z = \int_0^\infty dx\, e^{-bx^d} = \frac{1}{bd}\Gamma(1/d)$, where Γ is the Gamma function. Stretched exponential distributions $p(x)$ can be identified as straight lines in plots where $-\log p(x)$ is plotted in a loglog plot, that is, in $\log x$ versus the $\log(-\log p)$. This becomes clear by noting that $\log(-\log p(x) - \log Z) = d \log x + d \log b$. A stretched exponential distribution function in this representation is shown in Figure 2.13b. The stretched exponential function (not the distribution) is used in many physical, chemical, and statistical contexts, for example, in phenomenological descriptions of relaxation processes in disordered systems.

2.4.2.3 Gumbel distribution

The *Gumbel distribution* is named after Emil J. Gumbel, who analysed biases in politically motivated murders. The Gumbel distribution appears naturally in *extreme value theory*. It describes the statistics of the extreme values recorded in finite samples of data that are sampled from exponential or normal distributions. It is also referred to as the *Fisher–Tippet*, the *log-Weibull*, or the *double exponential* distribution. The Gumbel distribution is used for predicting the likelihood of extreme events, for example, in meteorology [161]. It can be obtained by centering and scaling the so-called standard Gumbel distribution

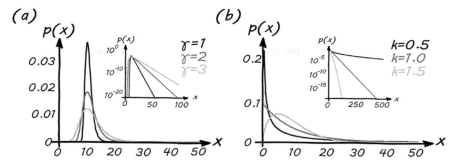

Figure 2.14 *(a) Gumbel distributions are shown for δ = 1 and the values of γ = 1,2,3. Changing the value of δ would shift the Gumbel distribution along the x-axis (not shown). The inset shows the distribution in a semilog plot. The tail of the distribution decays as an exponential for large x. (b) Weibull distributions for γ = 10, and shape k = 0.5 (black), 1 (grey), and 1.5 (light-grey). The shape parameter k is used to control lifetime distribution functions, where the probability of death increases over time: (k > 1) remains the same (k = 1), or decreases over time (k < 0).*

$p_{\text{Gumbel}}(z)$ by inserting standardized variables $z = (x - \delta)/\gamma$, where δ is a position and γ a scale parameter,

Standard Gumbel $\qquad p_{\text{Gumbel}}(z) = \frac{1}{Z}e^{-(z+\exp(-z))}, \qquad -\infty < z < \infty$

Gumbel $\qquad\qquad p_{\text{G}}(x|\delta,\gamma) = \frac{1}{\gamma}p_{\text{Gumbel}}\left(\frac{1}{\gamma}(x-\delta)\right).$

$$(2.54)$$

To fit a Gumbel distribution to data consisting of N observations, one typically plots the standardized rank $x = r/(N+1)$, where r denotes the rank of the frequency of observations of a given magnitude versus the relative frequency $p_{\text{G}}(x|\delta,\gamma)$. In Figure 2.14a the Gumbel distribution is shown for position $\delta = 1$ and scale $\gamma = 1,2,3$.

2.4.2.4 *Weibull distribution*

Weibull distributions describe the statistics of 'ageing' processes. It is named after Waloddi E.H. Weibull [403] and was developed for applications in materials testing. Lifetimes of electronic components, light bulbs, furniture, and so on, either depend on manufacturing errors or break because they wear out. Product failure at short lifetimes (often called *infant mortality*) is typically caused by manufacturing errors (the new light bulb that blows when you first turn it on), whereas for longer lifetimes, failure happens mainly due to wear and tear (the light bulb that blows after being switched on thousands of times). The Weibull distribution allows both types of cause and can tune between the importance of infant mortality versus product mortality due to wear and tear. This is done using a parameter k, where $k < 1$ means that the mortality rate decreases over time, $k = 1$ the mortality remains constant,[14] and $k > 1$ means that the mortality rate increases over time. The standard Weibull distribution is,

---

[14] For this situation the exponential distribution is a special case of the Weibull distribution.

$$p_{\text{Weibull}}(z|k) = z^{k-1}e^{-z^k}, \quad z > 0. \tag{2.55}$$

General Weibull distribution functions are obtained using the standardized variable $z = x/\gamma$, and γ is a scale parameter,

$$p_{\text{Weibull}}(x|k,\gamma) = \frac{1}{\gamma}p_{\text{Weibull}}\left(\frac{x}{\gamma}|k\right). \tag{2.56}$$

Parameter k is a stretching parameter of the exponential function. In comparison with the stretched exponential distribution, we see that k corresponds one-to-one to the stretching exponent d of the stretched exponential. In other words, Weibull probability density functions have stretched exponential tails. $k=1$ gives exponential distributions and $k=2$ yields the so-called Rayleigh distribution. A useful thing to remember is that if a random variable X follows the standard exponential distribution, then positive powers $Y \sim X^k$ are Weibull-distributed. Examples of Weibull distribution functions are shown in Figure 2.14b for several values of k and γ.

2.4.2.5   *Gamma distribution*

The Gamma distribution is a combination of a power law and an exponential. It is a two-parameter family of functions that arises in the context of waiting time distributions and Bayesian statistics as solutions of the *maximum entropy principle*; see Section 6.2.3. The exponential distribution and the chi-squared-distribution are special cases. The shape and scale parameters of the distribution are α and β, respectively. Both are positive numbers.

$$p(x|\alpha,\beta) = \frac{1}{Z}x^{\alpha-1}e^{-\beta x} \quad \text{with} \quad Z = \Gamma(\alpha)\beta^{-\alpha}. \tag{2.57}$$

In Figure 2.15a we show the Gamma distribution for various values of α. For Gamma distributed random variables X, the first moment is $\langle X \rangle = \alpha/\beta > 0$ and the variance is $\langle X^2 \rangle = \alpha/\beta^2$. The Gamma distribution belongs to the family of the so-called Wishart distributions. The negative binomial distribution is sometimes considered to be a discrete analogue of the Gamma distribution. Famous applications of the Gamma distribution are in the statistics of drought and rainfall patterns [199]. Gamma distribution functions are often used when fitting power laws with an exponential cut-off. These situations often occur as a consequence of finite size effects.

2.4.2.6   *Gompertz distribution*

The Gompertz distribution is a continuous two-parameter distribution function, named after Benjamin Gompertz. Like the Weibull distribution, the Gompertz distribution was developed in the context of lifespan statistics and demographics. More recently, the Gompertz distribution has also found applications in biology and computer science. In the theory of Erdős–Rényi networks, see Section 4.4.2, the Gompertz distribution

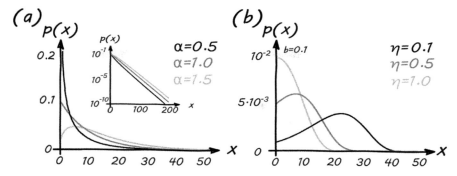

Figure 2.15 *(a) Gamma distributions for shape parameters $\alpha = 0.5, 1.0, 1.5$, and scale parameter $\beta = 1$. They are combinations of power laws and exponential functions. The asymptotic behaviour is dominated by the exponential (inset). (b) Gompertz distributions for various shape parameters $\eta = 0.1, 0.5, 1.0$, and scale parameter $b = 0.1$ are shown. Gompertz distributions appear in lifespan distributions and describe the length distribution of self-avoiding random walks on Erdős–Rényi networks.*

describes the length of *self-avoiding random walks*. The Gompertz distribution function is defined as

$$p_{\text{Gompertz}}(x|b,\eta) = \frac{1}{Z} e^{bx - \eta e^{bx}} \quad \text{with} \quad Z = \frac{1}{b\eta} e^{-\eta}, \tag{2.58}$$

where $b > 0$ is a scale and $\eta > 0$ a shape parameter. x has to be positive. In Figure 2.15b we show the Gompertz distribution for various parameter values.

2.4.2.7 *Generalized exponential distribution or Lambert-W exponential distribution*

The exponential, the power law, and the stretched exponential distribution functions can all be understood as special cases of the family of *generalized exponential distribution* functions, which we call *Lambert-W* exponential distributions.[15] These are three-parameter functions specified by a scaling parameter $0 < c \leq 1$, a shape parameter d, and a parameter r [175]. The generalized exponential distribution function is defined as

$$p(x|c,d,r) = e^{-\frac{d}{1-c}\left[\mathcal{W}_k\left(B(1-\frac{x}{r})^{\frac{1}{d}}\right) - \mathcal{W}_k(B)\right]} \quad \text{with} \quad B = \frac{1-c}{1-(1-c)r} \exp\left(\frac{1-c}{1-(1-c)r}\right), \tag{2.59}$$

where $\mathcal{W}_k$ is kth branch of the Lambert-W function; see Section 8.1.4. The factor B depends on the parameters c, d, and r.[16] Special cases of the generalized

---

[15] Note that the Weibull and the Gamma distributions are combinations of exponentials and power laws, but they do not contain the stretched exponential.

[16] The parameter r is constrained to $r > \frac{1}{1-c}$, for $d > 0$, $r = \frac{1}{1-c}$, for $d = 0$, and $r < \frac{1}{1-c}$, for $d > 0$. Sometimes, the particular choices $r = \exp(-d/2)/(1-c)$, for $d < 0$, and $r = 1/(1-c+cd)$, for $d > 0$ are useful.

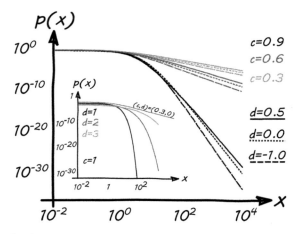

Figure 2.16 *Generalized exponential (Lambert-W exponential) distributions for several choices of scaling parameter c and form parameter d. For c = 1 and d = 1 the Lambert-W exponential distribution functions yields exponentials. For c = q and d = 0 q-exponentials are recovered; see Figure 2.12 and the stretched exponential family appears for c = 1 and d > 0. Asymptotic power law distributions are shown for c = 0.9 (black), c = 0.6 (grey), and c = 0.3 (light-grey). Different values of d are shown in dotted (d = 0), dashed (d = 0.5), and dash-dotted (d = −1) lines. Stretched exponential distributions for c = 1 are shown in the inset for d = 1 (black), d = 2 (grey), and d = 3 (light-grey).*

exponential distribution family are the exponential distributions for $c=1$ and $d=1$, the q-exponentials for $c=q$ and $d=0$, or the stretched exponential family of distributions for $c=1$ and $d>0$. This is true for any r. In Figure 2.16 we show the Lambert-W exponential distribution for several values of c and d. Generalized exponential distribution functions appear in empirical distribution functions of history-dependent processes and appear naturally in the context of statistics of strongly correlated systems; compare Sections 6.3, 6.4, and 6.5.

We briefly summarize this section on distribution functions.

- Fat-tailed distribution functions are ubiquitous in complex systems and processes. Various mechanisms are known that produce fat-tailed distribution functions. These include criticality, self-organized criticality, multiplicative processes, preferential attachment, and sample space reducing processes. These will be discussed in Chapter 3.
- We described a number of classical fat-tailed distribution functions. Many of them are related; some are even identical such as the the q-exponential, the Student-t, and the Zipf–Mandelbrot distributions. These show asymptotic power laws.
- Some important distribution functions that are not fat-tailed such as the exponential, the Gaussian, the Gamma, and the Gumbel distributions. The Weibull distribution has a stretched exponential tail.

- Gumbel distributions are important for extreme value theory and record statistics; Weibull distributions allow us to deal with rates that are increasing or decreasing over time.

- Power law, exponential, and stretched exponential distribution functions can be seen as special cases of the Lambert-W exponential or generalized exponential distribution function.

- As a practical rule of thumb, the vast majority of distribution functions are either exponentially decaying or are asymptotical fat-tailed power laws.

2.5 Stochastic processes

Processes are sequences of events that we can observe as trials. Processes evolve over time. A process can be simple, which means that the past does not influence the chances of observing future events. Or the process can be arbitrarily complex, which is typically the case if it has an underlying history-dependent, probabilistic dynamics. The notion of *memory* is essential in the context of stochastic processes. We discuss the Bernoulli process (and its multinomial statistics) and the Poisson process as examples of simple processes with absolutely no memory. We then proceed to processes where the chances of sampling the next states (events) depend only on the current state of the process. Processes that 'know' where they are, but not where they have been, are called Markov processes. In that sense they have a minimal amount of memory. We finally discuss processes where the chances of future states depend on the history of the process. These are history- or path-dependent processes. Complex systems are typically governed by path-dependent processes.

Mathematically, a random process is a map $t \to X_t$ that associates a random variable X_t with every element of a discrete (or continuous) time line t. In simple processes, all variables X_t draw their possible values from the same discrete (or continuous[17]) sample space; sample space does not evolve. Many complex systems do have co-evolving sample spaces that evolve together with the states of the process over time.

We write a process of length N as a set of random variables, $\bar{X}(N) = (X(1), X(2), \cdots, X(N))$. Each random variable $X(i)$ has its own sample space $\Omega(X(i))$. x_i is the concrete outcome of random variable $X(i)$. The probability of a process occurring is defined as the joint probability,

$$P(X(1) = x_1, X(2) = x_2, \cdots, X(N) = x_N). \tag{2.60}$$

Similarly, we denote the conditional probability to find the outcome x_{N+1} at the next time step, given all the previous observations x_i by,

$$P(X(N+1) = x_{N+1} | x_1, x_2, \cdots, x_N). \tag{2.61}$$

---

[17] The theory of processes described in continuous time and processes with continuous sample spaces requires a set of refined mathematical tools that will not be covered in this book.

2.5.1   Simple stochastic processes

The simplest stochastic processes do not have memory at all: the Bernoulli processes. Or they only have memory about the present: the Markov processes. We begin with processes without memory.

2.5.1.1   *Bernoulli processes*

Bernoulli processes are sequences of statistically independent trials. One example of a Bernoulli process $t \to X_t$ is a coin that is tossed N times. The outcome at the next time step has nothing to do with the outcomes at previous times. Processes without memory are characterized in terms of their conditional probabilities.

Memoryless processes do not depend on the past, which means that,

$$P(X(N+1) = x_{N+1}) = P(X(N+1) = x_{N+1} | x_1, x_2, \cdots, x_N). \qquad (2.62)$$

If the marginal probabilities are all identical, meaning that $P(X(t) = x_t)$ does not depend on time t, the process samples events that are independent and from identical distributions at all times—they are i.i.d. A discrete, memoryless, i.i.d. process is called a *Bernoulli process*. Sometimes, the term 'Bernoulli process' is used exclusively for a process with a binary sample space, a coin, for instance. We use the term Bernoulli process for i.i.d. processes, no matter how many discrete elements the sample space contains.

As Bernoulli processes are statistically independent, the property of Equation (2.62) holds. It can also be shown that the statistical independence of observations in a Bernoulli process follows from Equation (2.62). A *Bernoulli trial* is a single, independent observation of a random variable X_t. Observing a sequence of independent trials of identical random variables X_t constitutes a realization of the *Bernoulli process*. The X_t in the process might represent identical dice or coins.

2.5.1.2   *Binomial distribution functions*

The simplest Bernoulli processes have *binary* outcomes, {yes, no}, {0, 1}, or {success, failure}. The respective outcomes are distributed according to the *binomial distribution*. Suppose you toss a biased coin N times with probabilities q_0 of getting 'heads', and q_1 of getting 'tails'. We obtain sequences $x(N) = (x_1, \cdots, x_N)$, where x_n is the outcome of the nth trial. We denote the histogram of 'heads' and 'tails' by $k = (k_0, k_1)$, where k_0 (k_1) are the numbers of times 'heads' ('tails') appeared, $k_0 + k_1 = N$. The binomial distribution is the answer to the question: what is the probability of 'heads' occurring exactly n times? We denote this by $p_{\text{Binomial}}(n|N) = P(k_0 = n)$.

How do we obtain this function? As we know that each outcome x_n was obtained by an independent trial, the probability of sampling a sequence $x(N)$ is simply the product of the probability of sampling the outcomes x_n. It follows that the probability of sampling

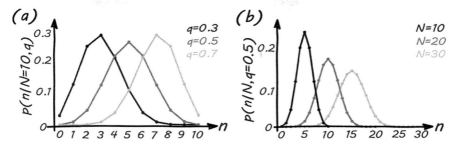

Figure 2.17 *Binomial distribution functions. (a) shows* $p_{\text{Binomial}}(n|N,q)$ *for* $N = 10$ *identical biased coins with* $q = 0.3$ *(black),* $q = 0.5$ *(grey) and* $q = 0.7$ *(light-grey). In (b) the situation for fair coins* $q = 0.5$ *and three different values of sequence lengths,* $N = 10$ *(black),* $N = 20$ *(grey), and* $N = 30$ *(light-grey) are shown. The Poisson distribution can be obtained from the binomial distribution in the limit* $N \to \infty$.

a sequence with the specific histogram $k = (k_0, k_1)$ is $q_0^{k_0} q_1^{k_1} = q_0^{k_0}(1 - q_0)^{N-k_0}$. This probability does not depend on the particular order in which the k_0 'heads' and k_1 'tails' appeared in the sequence. It means that all sequences with the same histogram are equally likely. How many sequences are there with exactly k_0 'heads' out of N observations? We know that this number is given by the binomial factor,

$$\binom{N}{k_0} = \frac{N!}{k_0!(N - k_0)!}. \tag{2.63}$$

We have derived the binomial distribution as the product of the probability of the histogram occurring times the number of possible ways of generating such a histogram in N observations.

The *binomial distribution* is defined as,

$$p_{\text{Binomial}}(n|N) = \binom{N}{n} q_0^n (1 - q_0)^{N-n}. \tag{2.64}$$

It is the probability distribution function of a random variable $K = (K_0, K_1)$ with histograms $k_0 = n$ and $k_1 = N - n$, given that there are N samples in a binary i.i.d. process $X = (X_1, \cdots, X_N)$; see Figure 2.17. If $X_n \in \{0, 1\}$; we can then write $K_1 = \sum_{n=1}^{N} X_n$ and $K_0 = N - K_1$.

The binomial distribution is normalized. By using the binomial formula $(a + b)^N = \sum_{n=0}^{N} \binom{N}{n} a^{N-n} b^n$, we see that $1 = (q_0 + q_1)^N = \sum_{n=0}^{N} \binom{N}{n} q_0^{N-n} q_1^n = \sum_{n=0}^{N} p_{\text{Binomial}}(n|N)$. Note that the histogram of N trials is a random variable $K = (K_0, K_1)$. Its possible outcomes are $k_0 = n$ and $k_1 = N - n$, with $n = 0, \cdots, N$. If $X_n \in \{0, 1\}$, then we can write $K_1 = \sum_{n=1}^{N} X_n$ and $K_0 = N - K_1$. We can also use this to compute the expectation value

$\langle K_1 \rangle_N = \sum_{n=0}^{N} p_{\text{Binomial}}(n|N)n,$

$$
\begin{aligned}
\langle K_1 \rangle &= \sum_{n=1}^{N} \binom{N}{n}(1-q_1)^{N-n}q_1^n n \\
&= \frac{d}{d\alpha}|_{\alpha=1} \sum_{n=1}^{N} \binom{N}{n}(1-q_1)^{N-n}q_1^n \alpha^n \\
&= \frac{d}{d\alpha}|_{\alpha=1}(1-q_1+\alpha q_1)^N \\
&= N(1-q_1+\alpha q_1)^{N-1}q_1|_{\alpha=1} \\
&= Nq_1.
\end{aligned}
\tag{2.65}
$$

Here, we used that $\frac{d}{d\alpha}\alpha^n = n$, if we set $\alpha = 1$. We also get $\langle K_0 \rangle = \langle N - K_1 \rangle = N - Nq_1 = (1-q_1)N = q_0 N$. The second moment is given by,

$$
\begin{aligned}
\langle K_1^2 \rangle &= \sum_{n=1}^{N} \binom{N}{n}(1-q_1)^{N-n}q_1^n n^2 \\
&= \left[\frac{d}{d\alpha}\left(1+\frac{d}{d\alpha}\right)\right]|_{\alpha=1} \sum_{n=1}^{N} \binom{N}{n}(1-q_1)^{N-n}q_1^n \alpha^n \\
&= \left[\frac{d}{d\alpha}\left(1+\frac{d}{d\alpha}\right)\right]|_{\alpha=1}(1-q_1+\alpha q_1)^N \\
&= Nq_1 + N(N-1)q_1^2.
\end{aligned}
\tag{2.66}
$$

Using that $\sigma^2(K_1) = \langle K_1^2 \rangle - \langle K_1 \rangle^2$ it follows that $\sigma^2(K_1) = Nq_1(1-q_1) = Nq_0 q_1$.

The mean and the variance of the binomial distribution $p_{\text{Binomial}}(n|N,q)$ are,

$$
\begin{aligned}
\mu &= Nq \\
\sigma^2 &= Nq(1-q).
\end{aligned}
\tag{2.67}
$$

2.5.1.3   *Multinomial processes*

Binary Bernoulli processes result in *binomial* distributions. Bernoulli processes that draw from more than two (say, W) values result in *multinomial* distributions of W different events or states. Suppose we have W independent possibilities for the outcome of a Bernoulli trial $i \in \{1,\ldots,W\}$ and the chances of the respective possibilities are $q = (q_1, q_2, \cdots, q_W)$. Assume we perform N trials in a sequence $x(N) = (x_1, \cdots, x_N)$ and record the histogram of its outcomes, $k = (k_1, k_2, \cdots, k_W)$, where obviously $N = k_1 + k_2 + \cdots + k_W$. As before, the essential question that defines the multinomial distribution is: what is the probability of observing a particular histogram k, given that we know the probabilities $q = (q_1, q_2, \cdots, q_W)$ of the underlying Bernoulli process? In complete analogy to the binomial case, one finds the following.

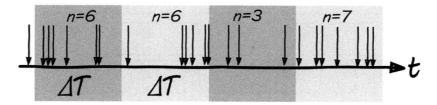

Figure 2.18 *Poisson process. Events happen randomly at a given rate λ. It is not known exactly when they appear, just that in a given time interval ΔT there are, on average, $\lambda \Delta T$ events. The probability of observing n events in a time interval ΔT is described by a Poisson distribution.*

The probability of observing a histogram $k = (k_1, k_2, \cdots, k_W)$ in a Bernoulli process of length N with W states that occur with probabilities $q = (q_1, q_2, \cdots, q_W)$ is given by the *multinomial distribution function*,

$$P(k|q,N) = \binom{N}{k} q_1^{k_1} q_2^{k_2} \cdots q_W^{k_W}, \tag{2.68}$$

where the *multinomial coefficient* is,

$$\binom{N}{k} = \frac{N!}{k_1! k_2! \cdots k_W!}. \tag{2.69}$$

Note that in this notation k is a vector. In the binomial factor k_0 was a number.

2.5.1.4 Poisson processes

Poisson processes are counting processes. They count the number of occurrences of events that happen randomly but at a given rate. One typically thinks of arrivals of events on a time line; see Figure 2.18. The Poisson distribution $p_{\text{Poisson}}(n|\lambda)$ describes the probability of observing n events in a fixed time interval, if the probability density for the events does not vary over time. For example, the number of people entering a specific building at a given rate in ten-minute intervals follows a Poisson distribution. This is no longer true if there is a nearby bus stop and the assumption of a constant arrival rate is wrong. Similarly, the number of drops of water that fall from a dripping faucet at a constant rate in three-minute intervals follows a Poisson distribution. The parameter λ controlling the average rate of events in the time window is called the *intensity* of the Poisson process.

To derive the Poisson distribution, one partitions time into non-overlapping segments of length dt; see Figure 2.19. In every segment a binary Bernoulli trial takes place. The possible outcomes are 'white' or 'grey'. We consider a time interval of length Δt. In Δt we find approximately $N \sim \Delta t/dt$ segments of length dt. We can count the number

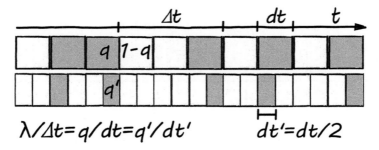

Figure 2.19 *Illustration of how the Poisson distribution can be understood as a continuous version of a binomial distribution. Each square represents one Bernoulli trial, with 'grey' outcomes having a probability q and 'white' having a probability $1 - q$. By choosing a finer segmentation of the timeline $dt' = dt/2$, one obtains twice as many Bernoulli trials within the same time interval Δt. Keeping the rate of events, that is, the expected number of 'white' events, $\lambda = q\Delta t/dt$ within a time interval Δt constant, while taking $dt \to 0$, defines the Poisson distribution.*

of times 'white' occurred in each segment dt in Δt. Given that at this level of coarse-graining the probability of observing 'white' in a segment dt is given by q, then we will on average observe $\lambda = q\Delta t/dt$ 'white' events in Δt. We then go to a finer segmentation of the time line with segments of length dt'. For instance $dt' = dt/2$. Again we will observe $\lambda = q'\Delta t/dt'$ where q' is the probability of observing 'white' in a segment dt'.

The average number of events λ observed in Δt does not depend on the coarse-graining dt, so that for a fixed intensity λ one finds that the probability of observing 'white' in a segment of length dt asymptotically behaves like $q = \lambda dt/\Delta t$, as dt goes to zero. For every scale $N = \Delta t/dt$ we can think of the Poisson distribution as a binomial distribution of observing 'white' exactly n times in N binary trials with the probability of 'white' being $q = \lambda/N$. Note that the binomial distribution at scale $dt = \Delta t/N$ is,

$$p(n|\lambda, N) = \binom{N}{n}\left(1 - \frac{\lambda}{N}\right)^{N-n}\left(\frac{\lambda}{N}\right)^n. \tag{2.70}$$

The Poisson distribution then emerges as the $N \to \infty$ limit of the binomial distribution,

$$p_{\text{Poisson}}(n|\lambda) = \lim_{N \to \infty} \binom{N}{n}\left(1 - \frac{\lambda}{N}\right)^{N-n}\left(\frac{\lambda}{N}\right)^n. \tag{2.71}$$

By using Stirling's formula $N! \sim N^N e^{-N}$, see Section 8.2.1, and the fact that $e^x = \lim_{N \to \infty}(1 + \frac{x}{N})^N$, one finds for large N that $\binom{N}{n}(\lambda/N)^n \to \lambda^n/n!$ and that $(1 - (\lambda/N))^{N-n} \to e^{-\lambda}$. Putting these two results together, one obtains the following:

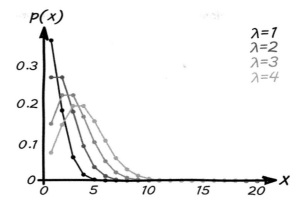

Figure 2.20 *The Poisson distribution for several values of λ, which is a single parameter that controls the distribution. The mean and variance of the Poisson distribution are both equal to $\mu = \sigma^2 = \lambda$. The distribution assigns a probability to the number of events that can be expected to happen within a unit time interval.*

The Poisson distribution is given by,

$$p_{\text{Poisson}}(n|\lambda) = \frac{\lambda^n}{n!}e^{-\lambda}. \tag{2.72}$$

It is the probability of observing exactly n independent events in a time interval Δt if the average number of events in Δt (rate) is given by the intensity of the Poisson distribution λ.

In Figure 2.20 the Poisson distribution is shown for several values of λ. It is not difficult to verify that the Poisson distribution is normalized, $\sum_{k=0}^{\infty} P(k|\lambda) = 1$. The mean and variance of the Poisson distribution also exist. To compute μ and σ^2, note that $\langle Y \rangle$ is the average number of events per unit time and we would therefore expect to find $\langle Y \rangle = \lambda$. This is indeed so,

$$\langle Y \rangle = \sum_{n=0}^{\infty} P(n|\lambda)n = \sum_{n=0}^{\infty} n\frac{\lambda^n}{n!}e^{-\lambda} = \sum_{n=1}^{\infty} \frac{\lambda^n}{(n-1)!}e^{-\lambda} = \lambda \sum_{m=0}^{\infty} \frac{\lambda^m}{m!}e^{-\lambda} = \lambda. \tag{2.73}$$

For the variance, a little algebra along the same lines yields, $\sigma^2(Y) = \langle Y^2 \rangle - \langle Y \rangle^2 = \lambda$.

For the Poisson distribution mean and variance have the same value,

$$\mu = \sigma^2 = \lambda. \tag{2.74}$$

Finally, we can compute the inter-event time distribution p_{IE} of the Poisson process. Looking at Figure 2.19 we could ask how likely it is for two 'white' events to be separated by exactly n 'grey' segments of length dt. Intuitively, it is clear that the average inter-event times $t = n\,dt$ will be inversely proportional to the intensity λ. Remember that $q = \lambda \Delta t / dt = \lambda/N$, and, as a consequence, we have $n = t/dt = Nt/\Delta t$. Again using $(1 + x/N)^N \sim \exp(x)$ for large N we find,

$$p_{\mathrm{IE}}(t|\lambda, \Delta t) = \lim_{N \to \infty} \frac{1}{Z}(1-q)^n \sim \frac{1}{Z}\left(1 - \frac{\lambda}{N}\right)^{N\frac{t}{\Delta t}} \sim \beta e^{-\beta t}, \qquad (2.75)$$

where Z is a normalization constant and $\beta = \lambda/\Delta t$.

The inter-event times of Poisson-distributed events follow an exponential distribution,

$$p_{\mathrm{IE}}(t|\lambda, \Delta t) = \frac{\lambda}{\Delta t}e^{-\frac{\lambda}{\Delta t}t}. \qquad (2.76)$$

2.5.1.5 *Markov processes*

Markov processes are processes $t \to X_t$, whose dynamics for the next time step only depend on what was realized in the last time step. In terms of conditional probabilities, Markov processes can be characterized in the following way.

Markov processes are processes for which the probability for the next time step is,

$$P(X_{N+1} = x_{N+1}|x_N) = P(X_{N+1} = x_{N+1}|x_1, x_2, \cdots, x_N). \qquad (2.77)$$

This equation states that a Markov process has no memory other than its current state or position. All older information from times t_n before N with $t_n < t_N$ are irrelevant. We discuss only discrete Markov processes.

In contrast to Bernoulli processes, which have no memory at all, Markov processes remember the last observed event. One way of constructing a Markov process is by 'summing' over Bernoulli trials. How can we see that a Markov process $t \to X_t$ can be written as a sum of a memoryless (independent) 'increment' process $t \to \Delta X_t$, where $t = 0, 1, 2, \cdots$ are elements on a discrete time line? Let us consider an example of a Markov process of this kind; see Figure 2.21. A random walker tosses a coin and walks one step to the left if he tosses 'heads' and one step to the right if he tosses 'tails'. If the walk is associated with the process $t \to X_t$, then the random variable X_t is the position of the walk at time step t, and x_n is the actual value of the position. In the next step $t + 1$, the random variable X_{t+1} can only take the values $x_t + 1$ or $x_t - 1$. We can write the random walk,

$$X_{t+1} = x_t + \Delta X_{t+1}, \qquad (2.78)$$

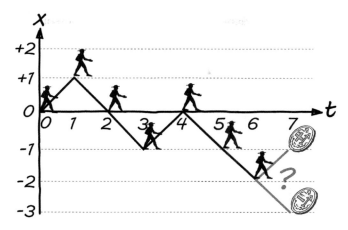

Figure 2.21 *Random walk as an example of a Markov process. A random walker is going left and right depending on the outcome of tossing a coin: one step to the left when tossing 'heads', and one to the right for 'tails'. This random walk is the cumulative sum of the outcomes of binary Bernoulli trials. The next position depends only on the current position. How the walker got there is irrelevant for the next step.*

where ΔX_n are random variables taking the values $\{+1, -1\}$ with some probabilities $q(n)$ and $1 - q(n)$. We can write $X_t = \sum_{k=0}^{t} \Delta X_k$, where $t \to \Delta X_t$ is the memoryless 'increment' process. If further $q(t) = q$ does not depend on time t, then ΔX is a Bernoulli process. To see that X_n is indeed a Markov process, all we need to show is that the probability of sampling an event at time n depends only upon events that are not older than the last recorded position x_t at time t. Clearly, the increment process $\Delta X = X_n - X_t = \sum_{k=t+1}^{n} \Delta X_k$ depends only on the outcome of memoryless trials ΔX_k with $k > t$. It follows that information about the path x_k with $k < t$ is irrelevant in terms of predicting the outcome of X_n for $n > t$.

Markov processes are widely used across all sciences. In information theory they are used to model information sources. Information sources can be thought of as random walks on a directed network; whenever a walker passes from one node to the next through a given link, the link emits a specific symbol. Discrete Markov processes with finite sample spaces are equivalent to so-called finite-state machines; see Section 6.2.2, and, in the context of linguistics, they are equivalent to *regular grammars*. Markov processes can be used to describe diffusion processes such as the Brownian motion of particles, which originally refers to the random motion of small particles suspended in liquids, that can be observed in a microscope. The motion is caused by collisions of the particles with water molecules moving at thermal speeds. Much of the area of stochastic calculus, stochastic differential equations, and discrete-time versions of stochastic processes (random walks) have been inspired by this natural phenomenon. Brownian motion was first understood by Albert Einstein on the basis of a random walk model, which led him to a new derivation for the diffusion equation [115]. It allowed him to estimate the size of molecules, which at that time were not widely believed to exist. Einstein was clever enough to formulate his random walk model of Brownian motion in discrete time. It took more than half a century to clarify the situation mathematically for continuous time—this is now possible, but mathematically rather difficult. The continuous time formulation of stochastic processes, however, does not offer deeper insights into the processes discussed in this book.

> An important criterion distinguishes Markovian from non-Markovian processes. For Markovian processes, cumulants of order two and higher vanish, while for history-dependent processes, all cumulants are non-vanishing.[a] If for a random process the third cumulant vanishes, then the process is Markovian and all cumulants of order $m > 2$ are zero.
>
> [a] Readers interested in mathematics may wish to look at the *Kramer–Moyal* expansion of stochastic processes and the *Pawula theorem*; see e.g. [143].

The random variables in random walks that describe the increments are usually statistically independent. The main body of the literature on the theory of stochastic processes is built on independent increments. However, random walks can also be built with strongly auto-correlated increments and increments with long-term memory. These will then be non-Markovian walks.

2.5.2   History- or path-dependent processes

Almost all complex systems involve history- or path-dependent processes. In general, they are poorly understood. Path-dependent processes are processes $t \to X_t$, where the probability of an outcome at the next time step is a function of the history of prior outcomes.

$$P(X_{N+1} = x_{N+1}) = f(x_1, x_2, \cdots, x_{N+1}). \tag{2.79}$$

It produces sequences $x(N)$. There are 'mild' versions of path-dependent processes that depend not on the exact order of the sequence, but only on the histogram of $x(N)$, where $P(X_{N+1} = x_{N+1}) = f(k(x(N+1)))$. Processes of this kind can be partially understood. We will discuss them in Section 6.5.2.3. If path-dependent processes also depend on the order of the sequence, then it is quite difficult to understand their statistics.

In contrast to Markov processes, path-dependent processes are frequently non-ergodic, which means in this context that time averages and ensemble averages do not coincide. What does this mean? Take N dice, throw them all at once, and generate a list of the outcomes $x(N) = (x_1, \ldots, x_N)$. In doing that, we create *ensembles* of realizations. If now we take a single dice and throw it N times in a row and record the outcomes in a sequence, $x'(N) = (x'_1, \ldots, x'_N)$, the averages of X and X' are identical. We say that the ensemble- and sequence pictures coincide. For path-dependent processes this equivalence is no longer true. Using ensemble statistics for path-dependent processes will in general lead to nonsensical results. Think, for example, of processes with symmetry breaking, where some events early on in the process determine the future of the process. Consider the silly toy process Y_t, where $Y_1 \in \{-1, 1\}$ represents a fair coin-tossing experiment, with the additional rule that after the first event, the process continues deterministically $Y_{t+1} = Y_t$

for all future time steps. The ensemble average over all possible paths[18] is $\langle Y \rangle_{\text{ensemble}} = 0$. The time average, or the sample mean, is either $\langle Y \rangle_{\text{sample}} = -1$ or $\langle Y \rangle_{\text{sample}} = 1$, each occurring with probability $1/2$. This leads us to a fundamentally important observation.

> For history- or path-dependent processes, the law of large numbers does not, in general, hold. The intuition that more samples will increase the accuracy of the average breaks down.

There are a few families of history- or path-dependent processes where scientific progress has been made. In particular, these are *reinforcement processes, processes with dynamic sample spaces*, and *driven dissipative systems*.

2.5.3   Reinforcement processes

As their name suggests, reinforcement processes reinforce their dynamics as they unfold. A typical reinforcement process is a Pólya urn, see Section 2.1.1, which we will study in detail towards the end of Chapter 6. Pólya urns are urns that contain balls of different colours. Whenever a ball of a given colour is drawn, several new balls of the same colour are added to the urn. It is a self-reinforcing Markov process that shows the rich-get-richer phenomenon or the winner-takes-all dynamics.

More general situations of probabilistic reinforcement processes are found in biology in the context of auto-catalytic cycles of chemical reactions. In auto-catalytic cycles one type of molecule can enhance (catalyse) or suppress the production or degradation of other molecules [113]. Similar cycles exist in production networks in the economy. Neural systems, where the frequent use of a particular neural pathway can modify the synaptic coupling strengths of the pathway, are examples of path-dependent processes that 'learn from experience' [186]. At a macro level, such systems are usually modelled with non-linear differential equations; at a micro level, Boolean networks are often used. It is often possible to picture these processes in terms of network models of interacting urn processes, where an event in one urn can modify the content of other urns.

A common phenomenon observed in reinforcement processes with positive and negative reinforcement mechanisms is *multistability*, which means that there are multiple attractors in the dynamics. In these systems, an event may cause a path or a trajectory to abandon its basin of attraction and to enter the basin of another attractor. These shifts are often experienced as 'regime shifts'. Such mechanisms are important for many processes that involve complex regulatory systems, such as cells or socio-economic systems. In the context of cells, regime shifts would, for example, be associated with cell differentiation. We will learn more about the statistics of simple reinforcement processes in Chapter 6.

---

[18] There are only two paths, $(1, 1, \cdots, 1)$ and $(-1, -1, \cdots, -1)$.

Complex regulatory systems can often be modelled as networks of reinforcing urn experiments, where one drawing event in one urn can increase or decrease the chances of events in dependent urns. Such processes typically show multistable stochastic dynamics. In particular, auto-catalytic systems, which we encounter in Chapter 5, can be modelled in this way.

2.5.4 Driven dissipative systems

Many systems are driven dissipative systems, which usually means that there is an energy gradient that drives the system away from equilibrium. Without that gradient, systems would relax to equilibrium. We often encounter two competing dynamics, one that drives and excites the system and the other being the process of relaxation. The resulting dynamics can be extremely rich and often shows fat-tailed statistics in many observable quantities. The reason for this is that the sample spaces of these systems are not static but evolve over time. There has been much progress recently in understanding the probabilistic nature of driven dissipative systems. Think of a pot filled with water on a stove. There are two competing thermal processes. The stove heats the water and the water cools at the interfaces with the surrounding air. Energy flows through the system from the stove through the water to the air. Flows of energy through a system make these systems do things they would never do in equilibrium, such as produce convective currents or turbulence. Systems where one process (a source) 'charges' the system and another (the sink) 'discharges' it are called *driven*. The word dissipative means that energy is flowing in and out of these systems.

With the exception of a few physical phenomena, virtually all processes that take place on this planet, including the entire biosphere and the anthroposphere, are driven non-equilibrium processes. Systems in equilibrium are the exception in nature and exist practically only in laboratories or for systems at relatively short timescales. Any machine, say an electric drill, is a driven dissipative system. An energy flow is needed to keep the machine running. If we interrupt the flow, the machine stops. The energy fed into the system will eventually be converted to work and heat, which dissipates into the environment. The energy source of the earth is the sun, and its energy sink is the approximately 3 Kelvin background radiation of the universe.

Perhaps the simplest example of a driven dissipative process is something that children might do, when home alone and bored. Take a tennis ball, take it upstairs, drop the ball, and watch it bounce downstairs; see Figure 2.22. When the ball reaches the bottom of the stairs, bring it back to the top and drop it again.[19] As the ball bounces down the stairs and comes to rest at the bottom of the stairs it dissipates energy in the form of heat produced in inelastic collisions with the stairs. What drives the process? The energy that the child

---

[19] One of the authors occasionally found this 'game' to be a fun pastime as a child. Note that if such experiments are carried out in reality, the way a ball bounces downstairs can be quite erratic. A ball does not always keep on landing on one lower stair after another. Sometimes the ball hits the edge of a stair and then bounces straight to the bottom of the stairs, and sometimes the ball gets stuck on a step and never reaches the bottom.

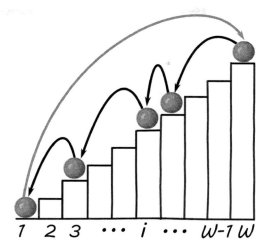

Figure 2.22 *A ball bouncing down a staircase as a simple model of a driven dissipative process. The ball bounces downwards only at random distances. Along the way it samples various steps that are below its current position; it is a sample space reducing process. Note that the process is a Markov process. The visiting distribution of the stairs is a power law; in this case, it is Zipf's law, from Section 2.4.1.*

uses to take the ball back upstairs—where does that energy go? It dissipates in the form of heat into the environment. In the following, we will consider an idealized version of this process. It will allow us to understand the origin of the peculiar statistics of driven dissipative systems; see Section 3.3.5. In particular, we will learn how the dynamics of the sample space is related to the statistics of the system. In the context of complex systems one sometimes associates driven dissipative systems with self-organized critical systems, which are usually more complicated realizations of the example of the bouncing ball.

Driven dissipative processes often consist of a dynamical sample space that increases and decreases over time. The increase comes from a driving source process that 'loads' the system into an excited state of 'high energy' with many possibilities for it to dissipate into. Sample space decrease is associated with a 'sink' process that relaxes the system towards 'low energy' states (sinks) as energy is dissipated. We will discuss the mathematics of sample space reducing processes in Sections 3.3.5 and 6.6.2.

Driven systems are characterized by a 'source' and a 'sink' process. The source process heaves a system into a 'high energy' state and is typically sample space expanding. The sink process allows the system to relax towards 'low energy' by dissipating this energy into the environment. The sink process is typically sample space reducing. Source and sink processes will create an 'energy flow' through the system, which at the microscopic level is described with a probabilistic dynamics that violates detailed balance. This means that statistical currents between two adjacent

continued

states of the system do not compensate for each other. In non-physical complex systems the role of energy in the driving process is played by other quantities.

2.5.4.1   *Processes with dynamical sample spaces*

Many stochastic processes have sample spaces that change over time. For example, they occur in evolutionary processes that involve both *innovations* and *extinctions*, such as in biological evolution or technological innovation. They have dynamical sample spaces that can shrink and expand as they evolve.

Processes with dynamical sample spaces are random processes $t \to X_t$, where the sample spaces $\Omega(X_t)$ of the random variables X_t change with time t. For the example of a ball randomly bouncing down a staircase, let us label the W individual stairs as states i. In the beginning, when the ball is at the top in state W, it can bounce to any other stair below, meaning that $\Omega_W = \{1, 2, \cdots, W - 1\}$. If the ball is later at stair $n < W$, its sample space is now smaller, as it can only reach the states $\Omega_n = \{1, 2, \cdots, n - 1\}$. At the bottom, the process needs to be restarted. If it is not restarted the process is strictly sample space reducing. Restarting, or charging, or driving usually means to expand the sample space of a stochastic system. At higher energy levels there are simply more ways of dissipating the energy, meaning that more possible states or outcomes can be reached or sampled.

It transpires that pure sample space reducing processes generically produce an exact Zipf law. If sample space reducing (relaxing) and expanding (driving) processes are balanced in appropriate ways, practically every statistical distribution function that we have discussed in this chapter can be recovered for driven dissipative systems. We will show how this is done in Section 3.3.5.

What have we learned about random processes?

- Stochastic processes are sequences of random trials.
- Bernoulli processes are time-independent and memoryless. They are sequences of i.i.d. Bernoulli trials over a finite and discrete sample space.
- Events in Bernoulli processes are multinomially distributed.
- Poisson distributions are obtained as the $N \to \infty$ limit of binomial distributions.
- Markov processes have a memory of the most recent trial only. The distribution function or the sample space for sampling the next event in a Markov process only depends on the last observation. The next step in a Markovian random walk does not depend on the details of the history of the walk, but only on the present position.
- Cumulative sums of memoryless processes yield Markov processes. In general, increments of Markov processes are random variables $\Delta X(i, t)$, which are independent but not necessarily identically distributed. They can depend on the state i they are in at time t. When at time t the Markov process is in state i, it samples the increment $\Delta X(i, t)$ to move to the next state.

- Random walks often have no memory of their increment process. Such random walks are Markov processes. However, random walks can have history-dependent or auto-correlated increment processes. This leads to path-dependent random walks.

- The challenge with history-dependent processes is that ensemble averages and sample means are no longer the same. In general, the law of large numbers does not hold for path-dependent processes; more samples will not necessarily increase the accuracy of measurements of stochastic variables.

- The Pólya urn process is a history-dependent process that is simple enough for its dynamics to be perfectly understood. Pólya urns are frequently used to model (coupled) self-reinforcing processes.

- Driven dissipative systems are at the core of many complex systems. They have a part that drives them and they have a relaxing dynamics. The driving part usually expands sample space volume; the relaxing part reduces it. The statistics of driven dissipative systems can be partially understood if the dynamics of the sample spaces is known; see Section 3.3.5.

- Evolutionary processes are processes with dynamical sample spaces, where innovation and extinction events modify the volume of sample space over time.

- In terms of understanding stochastic processes associated with the dynamics of complex systems, there is vast and unexplored terrain in front of us. Progress in the science of complex systems will, to a large extent, depend on the rate of progress in the field of path-dependent processes.

2.6 Summary

We briefly summarize the chapter and what we learned about probability and random processes.

- We learned how to handle and compare random numbers. The iterated addition of random numbers led us to the central limit theorem. The Gaussian distribution arises as an attractor of sums of random numbers. The log-normal distribution arises as an attractor for products of random numbers. We learned that two more distributions arise naturally as attractors for stable processes: the Cauchy and Lévy distributions. Both are asymptotic power laws.

- We discussed the fat-tailed distribution functions that frequently appear in many areas of complex systems. We encountered distribution functions that arise in extreme value theory and distributions that can be seen as generalizations of other distributions. In this context, we found that the Gamma distribution is a combination of exponential and power law, and the generalized exponential function (Lambert-W exponential) is a generalization of the exponential, the

continued

power law, and the stretched exponential function. The generalized exponential function will appear naturally in the context of entropies for complex systems in Chapter 6.

- We learned about stochastic processes. We divided them into classes that have no memory (Bernoulli processes), those that have a one-step memory (Markov process), and those that are history- or path-dependent. We encountered the multinomial distribution and Poisson distribution, and slightly touched upon reinforcement processes. These will be discussed in detail in Chapter 6.

- We concluded the chapter by introducing the notions of driven dissipative systems and systems with dynamical sample spaces. We will develop a theory of driven dissipative systems in detail in Chapters 3 and 6, meaning that a consistent way of computing the statistics and distribution functions for stationary driven systems will be presented. Systems with reducing sample spaces are tightly related to driven systems and allow us to understand the emergence of different distribution functions, including power laws. This will be demonstrated in Chapter 3.

- We mentioned the different philosophies with regard to hypothesis testing behind the Bayesian and the frequentist approaches.

2.7 Problems

Problem 2.1 Suppose an event A can be drawn with probability p_A. How often on average can we expect to draw A in a row? Ask yourself why the probability of drawing A exactly n times is given by $p_A^n(1 - p_A)$. Use the relation to compute the average $\langle n \rangle$. Compare the result with Equation (2.1).

Problem 2.2 Show in detail that Equation (2.17) is indeed correct.

Problem 2.3 Prove that $N!$ counts the number of ways N mutually distinct objects can be placed on N distinct positions $n = 1, \cdots, N$. Try induction on N, that is, show that $1!$ counts the number of ways in which one item can be placed on one position. Then show that if it is true for N that $N!$ counts the number of ways N objects can be placed on N positions, then it is also true for $N + 1$.

Problem 2.4 Compute in full detail the limit $N \to \infty$ of the binomial distribution to get the Poisson distribution. Use the outline for the computations given in Section 2.5.1.4.

Problem 2.5 Prove that the Poisson distribution is normalized. You may wish to remind yourself that the Taylor expansion of the exponential function is $e^x = \sum_{n=1}^{\infty} \frac{1}{n!} x^n$.

Problem 2.6 Compute the variance of the Poisson distribution with intensity λ.

Problem 2.7 Show from the definition of the expectation value, Equation (2.7), and $\sigma^2(X) = \langle (X - \langle X \rangle)^2 \rangle$ that $\sigma^2(X) = \langle X^2 \rangle - \langle X \rangle^2$.

Problem 2.8 Compute the third and fourth cumulant of a random variable X. Use the cumulant-generating function Equation (2.9) to compute the third and fourth cumulant in terms of moments.

Problem 2.9 Compute the variance $\sigma^2(k_1)$ of a binomial distribution after N trials of the underlying Bernoulli process. Use the binomial formula $(1+x)^n = \sum_{k=0}^n \binom{n}{k} x^k$, the definition of the variance, and the fact that for binomial processes $\langle k_1 \rangle = Nq$.

Problem 2.10 Completing squares can be used to solve the following integral in Equation (2.32). Note that we use the formula for the Gaussian integral, $\int_{-\infty}^{\infty} e^{-x^2}\, dx = \sqrt{\pi}$ in the sixth line.

$$\frac{d}{dy}P(X_1 + X_2 < y) = \int_{-\infty}^{\infty} \frac{dx_1\, dx_2}{2\pi\sigma_1\sigma_2} e^{-\frac{x_1^2}{2\sigma_1^2} - \frac{x_2^2}{2\sigma_2^2}} \delta(z - x_1 - x_2)$$

$$= \int_{-\infty}^{\infty} \frac{dx_1}{2\pi\sigma_1\sigma_2} e^{-\frac{x_1^2}{2\sigma_1^2} - \frac{(x_1-z)^2}{2\sigma_2^2}}$$

$$= \int_{-\infty}^{\infty} \frac{dx_1}{2\pi\sigma_1\sigma_2} e^{-\frac{1}{2}\left\{ x_1^2 \left(\frac{1}{\sigma_1^2} + \frac{1}{\sigma_2^2} \right) - \frac{2x_1 z}{\sigma_2^2} + \frac{z^2}{\sigma_2^2} \right\}}$$

$$= \int_{-\infty}^{\infty} \frac{dx_1}{2\pi\sigma_1\sigma_2} e^{-\frac{1}{2\sigma_3^2}\left\{ x_1^2 - 2x_1 z \frac{\sigma_3^2}{\sigma_2^2} + z^2 \frac{\sigma_3^2}{\sigma_2^2} \right\}}$$

$$= \int_{-\infty}^{\infty} \frac{dx_1}{2\pi\sigma_1\sigma_2} e^{-\frac{1}{2\sigma_3^2}\left\{ \left(x_1 - z\frac{\sigma_3^2}{\sigma_2^2} \right)^2 + z^2 \frac{\sigma_3^2}{\sigma_2^2}\left(1 - \frac{\sigma_3^2}{\sigma_2^2} \right) \right\}}$$

$$= \frac{\sigma_3}{\sqrt{2\pi}\sigma_1\sigma_2} e^{-\frac{z^2}{2\sigma_2^2}\left(1 - \frac{\sigma_3^2}{\sigma_2^2} \right)}$$

$$= \frac{1}{\sqrt{2\pi}\sigma_4} e^{-\frac{z^2}{2\sigma_4^2}}.$$

In the third line, we define $1/\sigma_3^2 = 1/\sigma_1^2 + 1/\sigma_2^2$. In the last line, we need to identify σ_4 with two distinct expressions, (i) $\sigma_4 = \sigma_1\sigma_2/\sigma_3$ and (ii) $\sigma_4^2 = \sigma_2^2/(1 - \sigma_3^2/\sigma_2^2)$. Show that the identifications (i) and (ii) indeed yield the same result (σ_4), that is, show that

$$\sigma_1^2\sigma_2^2/\sigma_3^2 = \frac{\sigma_2^2}{1 - \left(\frac{\sigma_3}{\sigma_2} \right)^2},$$

is true.

Problem 2.11 Suppose X is log-normally distributed, where the μ and σ^2 are the mean and variance of the normal distribution. Compute the relations between μ, σ, and $m = \langle X \rangle$ and $v = \sigma^2(X) = \langle (X - m)^2 \rangle$. Look up the integrals you need to solve in an integral table.

Problem 2.12 Show that the Zipf–Mandelbrot distribution Equation (2.48) can be written exactly as a q-exponential function as in Equation (2.49). Work out the exact relations between the various constants in the corresponding equations.

Problem 2.13 Compute the Lambert-W exponential in Equation (2.59) for (i) $(c,d) = (c,0)$ and (ii) $(c,d) = (1,d)$. Use the fact that for small $x \sim 0$ the Lambert-W function behaves like $W_0(x) \sim x$.

3

Scaling

3.1 Overview

Scaling appears practically everywhere in science. The earliest use of scaling in modern scientific thinking dates back to Galileo Galilei's observation [142] that the body mass M of animals scales as a function of their size (length), L. In particular, he found that it scales as the third power, $M \sim L^3$. Thus, if the size of an animal is doubled, $L \to 2L$, its mass is expected to be eight times larger, $M \to 8M$. To support such a mass, one would need bones that are eight times stronger. Galileo further noted that the strength of bones scales as a second power with their cross section, $S \sim R^2$, which means that if the bone radius is doubled, $R \to 2R$, its strength increases only by a factor of four, $S \to 4S$. He concluded that animals that are twice the size of other animals should have bone radii that scale with size as a power law, with an exponent 3/2. This ensures that if the size of an animal doubles, the radius scales as $R \sim L^{3/2}$, and the bone strength increases by a factor eight, $S \sim R^2 = (L^{3/2})^2 \sim L^3$, which then exactly supports the eightfold mass. That is why elephants and blue whales look more clumsy than foxes and dolphins. Galileo's graphic demonstration of his reasoning is shown in Figure 3.1. Here, scaling arguments have been used for the first time to explain physiological facts based on the single argument that bone strength should be proportional to the mass these bones have to support.

In essence, scaling works as follows. Assume that there are two functionally related quantities. If one of them is scaled by a constant (we say 'it is scaled by a scaling factor') and also causes the other to change by a constant, there is a scaling relation between the quantities. For example, take the function $f(x) = x^\alpha$. If we scale it we obtain $f(\lambda x) = \lambda^\alpha x^\alpha = Cf(x)$, where C is a constant or a factor. Changing x by a multiplicative constant changes f by (another) constant. If we have $f(x) = e^x$ and scale it, we obtain $f(\lambda x) = e^{\lambda x} \neq Cf(x)$. In the case of the exponential, the scaled function cannot be written as a factor (constant) times the original exponential, e^x. As we will see, scaling laws are always associated with power laws in the corresponding equations.

Scaling appears in many different contexts. The scaling object can be a function, a structure, a physical law, or a distribution function that describes the statistics of a system or a temporal process. Structures that exhibit scaling are often called *fractals*. They can be physical, such as a coastline or a scale-free network, or mathematical such as the

Introduction to the Theory of Complex Systems. Stefan Thurner, Rudolf Hanel, and Peter Klimek,
Oxford University Press (2018). © Stefan Thurner, Rudolf Hanel, and Peter Klimek.
DOI: 10.1093/oso/9780198821939.001.0001

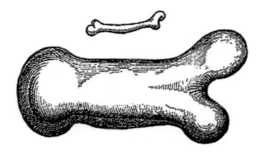

Figure 3.1 *Galileo's demonstration of the scaling law of bone radii* $R \sim L^{3/2}$ *with respect to animal size L. The large bone is about three times longer and, according to this argument, should be about* $3^{3/2} \sim 5.2$ *times thicker than the small bone. He slightly exaggerated the thickness of the scaled bone in his drawing.*

Koch curve or the Lorenz attractor. Many laws in physics are scaling laws, for example, Newton's law of gravitation, which acts between two masses m_1 and m_2, is a power law in distance r, $F(r) = Gm_1 m_2 r^{-2}$. Another example is classical diffusion, where the mean square displacement of the position of the particle scales with time $\langle (x(t) - x_0)^2 \rangle \sim t$.

In the second half of the twentieth century, the science of scaling experienced a dramatic boost, mainly through the development of a mathematical toolbox for understanding nuclear forces. This set of techniques is referred to as the renormalization group. The basic idea behind it was developed in the context of high energy physics for understanding how the elementary coupling constants change with the different energy scales at which they are observed [148]. The concept found countless applications in other areas of physics, including the understanding of critical phenomena [215, 410] and of electrical conductance [3].

Scaling laws of physics, however, are not the prime focus of this chapter. Here, we focus predominantly on scaling laws that appear in the statistical description of stochastic complex systems, where scaling appears in the distribution functions of observable quantities of dynamical systems or processes. The distribution functions exhibit power laws, approximate power laws, or fat-tailed distributions. In simple, non-complex statistical systems, such as gases, the underlying statistics is often the classical multinomial statistics that naturally leads to distribution functions that are either Gaussian or exponential.[1]

Many, if not most complex stochastic systems produce distribution functions that are fat-tailed or power laws. We discussed these in Section 2.4.1. Understanding their origin and how power law exponents can be related to the particular nature of a system, is one of the aims of the book. These questions will be discussed in some detail in Chapter 6. In this present chapter, we develop a feel for what scaling means in the context of complex stochastic systems and the forms it can take. We will define scaling

---

[1] Note, however, that in simple statistical systems that show phase transitions, non-Gaussian statistics and power laws may appear in the vicinity of 'critical points'.

and give a short account of scaling in space and time by discussing fractals and fractal time series, respectively. We will present a number of famous scaling laws in stochastic complex systems, and review the five most important mechanisms through which scaling in complex systems appears, including the following,

- criticality,
- self-organized criticality,
- preferential processes,
- multiplicative processes,
- sample space reducing processes.

We discuss several means of quantifying power laws from data in a statistically consistent way. We conclude the chapter with an example that shows—much in the spirit of Galileo—how scaling can be used to understand literally dozens of biological and physiological facts, based on a very limited number of elementary assumptions [71, 407].

3.1.1 Definition of scaling

An object is said to be scaling if it remains the same or 'invariant' (up to a factor) under scale transformations $x \to \lambda x$ (stretching or squeezing). Often, these scaling objects are given by functions $f(x)$.

A *scaling relation* is present if there are continuous and real-valued functions f and g, such that upon a scale transformation, $x \to \lambda x$, the following holds,

$$f(\lambda x) = g(\lambda)f(x), \tag{3.1}$$

where λ is a real positive number. This scaling equation has a real solution that reads,

$$f(x) = bg(x) \quad \text{with} \quad g(x) = x^c, \tag{3.2}$$

where b and c are real constants. g is called the *scaling function*.

Solutions to scaling equations are always power laws. This is obvious because the solution to equation $f(xy) = f(x)f(y)$ is always a power law, $f(x) = bx^c$. Whenever a scaling relation exists, this is reflected in the existence of a power law, or equivalently, scaling functions are power laws.[2]

---

[2] Note that one can always modulate specific periodic functions to power law scaling functions that still solve the scaling equation. In particular, for a fixed λ, the following scaling function is also a solution of Equation (3.1),

$$g(x, \lambda) = x^c \cos\left(2\pi \frac{\log x}{\log \lambda}\right). \tag{3.3}$$

3.2 Examples of scaling laws in statistical systems

Scaling laws are omnipresent in statistical systems. They appear in a large variety of systems and processes spanning physical, biological, socio-economic, and technical systems. In Table 3.1 we list a few examples. Whenever we can come up with one of the five power law generating mechanisms to generate the corresponding power law, which we discuss in Section 3.3, we indicate it.[3] Many distribution functions obtained from these complex systems are approximate power laws. Many are asymptotic, which means that only the tail (right-hand part) of the distribution is a power law, while the rest is not. We discuss asymptotic power law distribution functions in Section 2.4.1.

Table 3.1 *Examples of scaling laws in a number of stochastic complex systems and processes. The scaling appears in the corresponding (cumulative) probability density of relative frequencies or rank distributions. Whenever it is possible to indicate one of the five primary power law generating mechanisms discussed in Section 3.3 that could generate the power law, we do this. The mechanisms covered here are multiplicative (with constraints), preferential, self-organized critical (SOC), or sample space reducing (SSR). Whenever other mechanisms might be able to explain the scaling laws, this is indicated by 'other'. The references indicate works where these power laws have been demonstrated or confirmed.*

System	Mechanism	Reference
City size	multiplicative	[17, 137]
Rainfall	SOC	[308]
Landslides	SOC	[385]
Hurricane damages (secondary)		[244]
Interbank loans	multiplicative, preferential	[65]
Forest fires in various regions	SOC	[258]
Moon crater diameters	SSR (fragmentation)	[290]
Meteorite sizes	SOC, SSR (fragmentation)	[346]
Gamma rays from solar wind		[46]
Movie sales	SOC	[347]
Healthcare costs	multiplicative	[144]
Particle physics	SSR	[411]
Word frequencies in books	preferential, SSR, other	[420]

[3] This does not mean that this mechanism is the real cause of the observed power laws. We just state a plausible mechanism. It needs to be seen if these mechanisms can indeed be verified for many of the examples given.

System	Mechanism	Reference
Citations of scientific articles	preferential	[256]
Website hits	preferential, SSR	[6]
Book sales	preferential	[163]
Earthquake magnitude (Gutenberg—Richter law)	SOC	[162]
Seismic events	SOC	[326]
War intensity		[385]
Killings in wars		[212]
Size of war		[385]
Wealth distribution	multiplicative	[299]
Family names	preferential, multiplicative	[290]
Scale-free networks	preferential	[28]
Allometric scaling in biology	other	[407]
Terrorist attacks		[86]
Dynamics in cities	multiplicative, SOC	[49]
Fragmentation processes	SSR (fragmentation)	[241]
Random walks on networks		[306]
Financial markets	SOC	[153]
Crackling noise	SOC	[240]
Exponential growth with random observation times	other	[320]
Blackouts		[104]
Fossil record	SOC	[333, 409]
Bird sightings		[85]
Fluvial discharge	SOC	[322]
Anomalous diffusion		[381]
Record statistics	SSR, other	[284]
Train delays	other	[69]
Brain activity	SOC	[372]

3.2.1   A note on notation for distribution functions

When dealing with distribution functions, it is essential to know the nature of the random variable, as it determines what type of distribution function to use. Distribution functions are the result of sampling or measurement processes. There are three types of distribution functions that should be clearly distinguished.

- The *objective distribution* $q(x)$ assigns a probability to every observable state value x. Discrete sample spaces are characterized by W states $i = 1, \cdots, W$, where each state i is associated with a unique value $x = z_i$. Often, the objective distribution is unknown.
- The *relative frequencies* are given by $p_i = k_i/N$, where k_i is the number of times that state i is observed in N experiments. $k = (k_1, \cdots, k_W)$ is the *histogram* of the data. Relative frequencies can be presented in two ways:
  - if p_i is ordered with respect to its descending magnitude, this is called the *rank ordered* distribution.
  - if p_i is ordered according to the descending magnitude of the objective distribution $q(z_i)$, they are *naturally ordered* relative frequencies. For this, z_i must be numbers.
- The *frequency distribution* $\phi(n)$ counts how many states i in a sample fulfil the condition $k_i = n$.

Figure 3.2 shows the distinction between these three distribution functions. To illustrate the types of distribution functions we sampled $N = 10,000$ data points from the objective distribution $q(x) \sim x^{-0.7}$, where $x \in \{1, \cdots, 1000\}$. The objective distribution q is shown in dark-grey. The relative frequency p (in natural order) is the black line, while the rank-ordered distribution is the light-grey line. The latter exhibits an exponential decay in the tail. The inset shows the frequency distribution $\phi(n)$ for exactly the same data. Next we discuss how different sampling processes can be characterized in terms of natural order, rank order, or frequency distributions.

3.2.1.1   *Processes with natural order*

For many sampling processes, the order of the observed states is known. For example, think of x as representing the numerical values of earthquake magnitudes. Any two observations x and x' can be ordered with respect to their numerical value. We say they have a *natural order*. We can simply plot the relative frequencies $p(x)$ against the values x. If x is not discrete, it is necessary to put the data into discrete bins $x \rightarrow i$. How often a sampled value falls within a bin i is measured by the histogram k_i. The relative frequency is then obtained by $p_i = k_i/N$. If we know that distribution q, from which the samples are drawn, is a power law, this is equivalent to ranking observations according to their objective distribution q: the most likely event also has *natural rank* 1, the second most likely has rank 2, and so on.

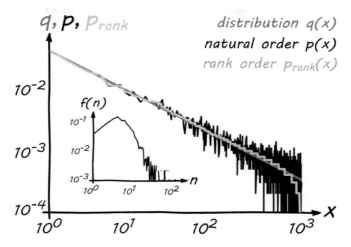

Figure 3.2 *Notation for the three different types of distribution functions. Data are sampled from a power law objective distribution $q(x) \sim x^{-\lambda}$ with an exponent $\lambda = 0.7$ (dark-grey line). The relative frequencies p_i are shown for $N = 10,000$ sampled data points according to their natural ordering (black). i here represents the numbers from 1 to 1000. The rank-ordered distribution (light-grey) is obtained by ordering the states i with respect to their observed relative frequencies p_i. The rank-ordered distribution follows a power law, except for the exponential decay that is visible for rank ~ 500 and higher. The inset shows the frequency distribution $\phi(n)$ that describes how many states i appear exactly n times. The frequency distribution has a maximum and a power law tail with exponent $\alpha = 1 + 1/\lambda \sim 2.43$. To estimate α, one should only consider the tail of the frequency distribution function.*

3.2.1.2 *Rank-ordered processes*

If the distribution q from which the samples are drawn is not known a priori because the states i have no numerical values z_i, we can only count relative frequencies p_i of states of type i, a posteriori, that is, after sampling. Imagine, for example, the case of words in a book. We cannot rank two words, but we can rank them according to the respective number of times they appear in the book. For clarity, let $k = (k_1, \cdots, k_W)$ be the histogram of N recorded states. k_i is the number of times we have observed state i, and $p_i = k_i/N$ is the relative frequency. Only after all samples have been taken, can one order states with respect to p_i, such that the rank $r1 = i$ is assigned to state i with the largest p_i, rank $r2$ to i' with the second largest $p_{i'}$, and so on. $p = (p_{r1}, \cdots, p_{rW})$ is called the *rank-ordered* distribution of the data. The natural order imposed by q and the rank order imposed by p are in general not identical for finite N. However, if data have been sampled independently, p converges towards q (for $N \to \infty$) and the rank order induced by p will asymptotically approach the natural order induced by q.

3.2.1.3 *Frequency distributions*

Frequency distributions $\phi(n)$ count how many distinct states i occur exactly n times in the data. It does not depend on the natural order of states and therefore is sometimes preferred to the rank-ordered distribution. The frequency distribution $\phi(n)$ that is associated with a power law objective distribution $q \sim x^{-\lambda}$ is not an exact power law but is a non-monotonic distribution with a maximum; see inset in Figure 3.2. Only its tail decays approximately as a power law, $\phi(n) \sim n^{-\alpha}$, and the exponents λ and α are related by,

$$\alpha = 1 + 1/\lambda. \tag{3.4}$$

If the objective distribution has an exponent λ, the tail of the associated frequency distribution has exponent α. As the frequency distribution behaves like a power law only in its tail, estimating α necessitates constraining the observed data to large values of n.

3.3 Origins of scaling

Where do these scaling laws in distribution functions come from? For non-complex stochastic systems, the central limit theorem explains why Gaussian probability distributions appear as an attractor in simple systems that are made up of non-interacting or uncorrelated components. We have seen that the asymptotic power laws of the Cauchy and Lévy distributions appear as attractors of α-stable processes; see Section 2.3.2. Note that these two power laws are associated with independent random variables. As we discussed, they appear in simple, independent, and uncorrelated systems.

Is there an equivalent theorem that can explain the omnipresence of scaling laws in stochastic complex systems? The answer seems to be that there may not be a unique theorem that does the job. It is currently believed that there are various ways of understanding the origins of power laws. In this section, we will discuss five mechanisms that lead to power laws in stochastic systems, which are very different in nature. Note that our focus is on stochastic systems. For deterministic systems, there are many explanations for the existence of power laws that are often determined by the geometry of the underlying system. Think, for example, of power laws that appear in physics, such as Newton's law, or the example of Galileo's bone strengths. In these cases the geometry explains the scaling laws. Perhaps the simplest way of thinking of the origin of power laws in deterministic processes is in the form of the dynamical equation, $\frac{d}{dt}x(t) = \lambda x^q(t)$. The solution is a q-exponential, which is an asymptotic power law; see Section 2.4.1.4.

In particular, the five mechanisms are as follows. The mechanism of *criticality* explains a physical scaling phenomenon that happens exactly at the location of phase transitions. The mechanism of *self-organized criticality* describes systems that show statistical behaviour *as if* they were at a phase transition, or critical point.[4] These systems do not need to be fine-tuned towards the critical point to show power laws in the distributions of observable quantities. Self-organized critical systems are driven systems. The mechanism of *multiplicative random processes* with constraints is a way of generating

---

[4] We use the expression *critical point* for a system that is at a phase transition. We do not, as is done in thermodynamics, use it exclusively for the end of a phase transition line in a phase diagram.

power law statistics in processes that are subject to conservation laws or other simple constraints. The oldest mechanisms have been known for almost a century; they are the so-called *preferential processes*, where events occur with probabilities that are proportional to the number of times they have occurred in the past. They are often Markov processes or self-reinforcing path-dependent processes. Finally, *sample space reducing processes* are those that reduce their sample spaces over time, as they evolve. They generically lead to power laws. If these processes are used to model driven out-of-equilibrium systems, the balance between the driving process and the relaxation process determines the distribution functions in the system [91]. This offers a self-consistent theory for the statistics of driven systems.

3.3.1 Criticality

The term criticality refers to critical phenomena, which are phenomena that happen at *critical points*. These are points at which phase transitions occur, such as the transition from liquid to vapour or from non-magnetic to magnetic in a ferromagnet. Critical points are found along the phase transition lines in phase diagrams. Think, for example, of the line that separates the liquid and vapour phase in the pressure–temperature (pT) diagram of a real gas. Critical points are associated with critical parameters, such as a critical pressure p_c or critical temperature T_c. Criticality describes the physics at critical points, which is often quite remarkable. The most important fact is that several physical quantities start to become effectively non-local at the critical point. Critical points are often associated with massive restructuring events that take place during a phase transition from an ordered to a disordered phase, or vice versa. For example, in a magnet that is in the non-magnetized phase and is approaching the critical Curie temperature T_c, spins still interact locally with their neighbours; however, clusters with aligned spins emerge. As these clusters grow, global correlations emerge; in other words, individual particles start to 'feel' the entire system—the global magnetization is felt by the individual spins. If one observes the correlation functions between the particles (e.g. their spins), one finds that the correlations *diverge* (become infinite) at the critical point, T_c. A correlation function typically quantifies the degree of similarity between some properties of particles (e.g. spin) as a function of their spatial separation. The divergence of correlation functions at the critical point often reflects a number of remarkable physical properties at the critical point. For example, in some materials the specific heat diverges, while in magnetic materials the magnetic susceptibility diverges. Often, several quantities show these divergences at the same time at the same critical parameters. The way in which the divergence of these physical quantities is realized at the critical point is a power law. The corresponding power law exponents are called the *critical exponents*. The set of several critical exponents can be used to classify different materials. If they share the same set of critical exponents, they are said to belong to the same *universality class*. The physics of different materials at the critical point with the same critical exponents is identical. This is remarkable. Note that it is not only physical materials that can be classified with sets of critical exponents, but also many of the models that describe them.

3.3.1.1   *Critical exponents*

The most famous model for studying critical phenomena is probably the Ising model. As we will see, the phenomenon of *universality* will show us that Ising models are more than toy models. The Ising model is a model of spins that can point in two directions (up or down) and that are usually arranged on a one-, two-, or three-dimensional lattice or a network. Between any pair of two spins that are nearest neighbours on the lattice, there is a potential energy, which is small if the spins are aligned and large if they point in opposite directions. The total energy is given by the Hamiltonian function,

$$H = -\mathcal{J} \sum_{(i,j)} s_i s_j, \qquad (3.5)$$

where $\mathcal{J} > 0$ is a constant (coupling constant) and s_i is the spin of particle i in the lattice. Its value can be $s_i = +1$ when it points up, and $s_i = -1$ when it points down. The sum extends over all *pairs* of neighbouring spins, which are indicated by (i,j). The Ising model in two dimensions has a critical temperature, T_c. Below this temperature all spins point in the same direction (ferromagnetic phase) and above it spins are disordered. The temperature in the model basically reflects the probability of a random spin flip. If the temperature is low, spins will tend to arrange their direction according to the majority of their neighbours' directions. When the temperature is high, they tend to point randomly in either direction, regardless of their neighbours. In other words, the phase transition is characterized by a non-zero magnetization $m = \langle s_i \rangle \neq 0$ below the critical temperature, $T < T_c$ spins align (ordered or ferromagnetic phase). Above $T > T_c$ there is no magnetization $m = 0$ (disordered or paramagnetic phase). The magnetization m is the called the *order parameter*.

At T_c several physical properties of the system diverge. The most important property is the magnetic susceptibility χ, which is the degree of magnetization that is the response when an external magnetic field is applied to the Ising system. Loosely speaking, 'susceptibility' measures how easily a material can be magnetized in a magnetic field. It is found that the susceptibility behaves as,

$$\chi(T) \sim \left(\frac{T - T_c}{T_c} \right)^{-\gamma}, \qquad (3.6)$$

which means that χ diverges at T_c and that it approaches infinity as a power law, as $T \to T_c$. γ is called the critical exponent for the susceptibility. Similarly, the specific heat C, the magnetization m, and the correlation length ξ diverge at T_c. The critical exponents are α, β, and ν, respectively. They are collected in Table 3.2, where we use the shorthand notation $\tau = (T - T_c)/T_c$. Critical exponents may depend strongly on the dimension of the system at hand. The Ising model in two dimensions has different exponents than it has in three. Many critical exponents are related, for example, $d\nu = 2 - \alpha = 2\beta + \gamma$, where d is the dimension of the system. The classic techniques for studying spin models mathematically are: the *mean field approximation* for the disordered phase or high

Table 3.2 *Critical exponents for the Ising model for two and three dimensions. Here* $\tau = (T - T_c)/T_c$.

Quantity	Symbol	Exponent	Scaling Relation	2D	3D
specific heat	C	α	$C \sim |\tau|^{-\alpha}$	0	0.11
magnetization	m	β	$m \sim |\tau|^{\beta}$	1/8	0.33
susceptibility	χ	γ	$\chi \sim |\tau|^{-\gamma}$	7/4	1.24
correlation length	ξ	ν	$\xi \sim |\tau|^{-\nu}$	1	0.63

dimensional systems (typically dimension $d > 4$); and *renormalization group methods.* The essence of the mean field approximation is the assumption that the influence on a particle from its neighbouring particles can be approximated by the *average* influence of the surrounding neighbours only. For literature on this topic, see [350]. Critical exponents are essential for defining *universality classes.*

3.3.1.2 Universality

One of the great discoveries of twentieth-century physics was the observation that critical exponents can be the same for very different physical systems and mathematical models. This allows us to characterize and classify systems and models with respect to their critical exponents. Systems with the same set of critical exponents are said to belong to a *universality class.* The word universal means that critical exponents do not depend on the particular details of the system. This is remarkable, as atoms and molecules of different materials can be very different; yet the critical exponents can be the same. This means that the physics at the critical point is *exactly* the same.

The importance of universality arises as follows. Two systems in the same universality class behave physically identically when at the critical point. If a model, like the Ising model, which is usually a crude approximation of a real magnetic material, belongs to the same universality class as a material, it is an *exact* model for the physical material. At the critical point, the model describes the properties of the material exactly. Universality allows us to connect models and real systems and at the same time gives us certainty that the model is the correct one to use. Universality is a rare phenomenon in science. It is usually not possible to know with certainty that a model describes the nature of a real system exactly.

3.3.1.3 Percolation

'To percolate' means to 'trickle through' something. This can be water through a coffee machine or viruses infecting an entire population. Percolation typically exhibits universality. In other words, it allows us to describe phenomena that are as different as coffee brewing and the spreading of an epidemic on identical mathematical grounds. For a critical system that shows percolation, think of a system that can form clusters of some kind.

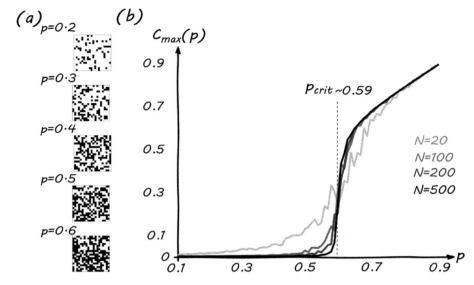

Figure 3.3 *(a) Site percolation for various levels of p. Below $p_c \sim 0.591$ clusters do not penetrate or percolate the system. (b) Size of the giant component as a function of p. The transition occurs faster, as the system gets bigger.*

Percolation is a critical phenomenon. Below a *critical parameter*, clusters are small and disconnected. Above a critical parameter, clusters massively increase until one cluster spans the entire system—it *percolates* the system. It is said that a 'giant component' emerges. Below the threshold there is no system-wide cluster and no giant component exists.

As a simple example, think of sites on a two-dimensional $N \times N$ lattice; see Figure 3.3a. The sites can be in two states, occupied (black) or non-occupied (white). Let p be the probability that a site is occupied. p plays the role of temperature T. If p is small, there will be a few occupied sites that are largely unconnected and there is no giant component. As p increases, more black sites appear and clusters start to form. At a certain critical point, p_c, the giant component emerges, at which it becomes possible to cross the system on paths of black squares only—the black squares *percolate* the system. This is the so-called percolation transition. p_c is sometimes also called the *percolation threshold*. At this transition there are essentially two critical exponents that describe the clustering behaviour around the critical point, p_c. One diverging quantity is the maximum cluster size, c_{max}. As a function of the distance from the critical point, this follows a power law,

$$c_{\text{max}} \sim |p - p_c|^{-\frac{1}{\sigma}}, \tag{3.7}$$

for large N. The other exponent comes from the correlation length, ξ. There, the average distance between two sites on the same cluster scales as,

$$\xi \sim |p - p_c|^{-\nu}. \tag{3.8}$$

The critical exponents are related through $\sigma = 1/\nu d_f$, where d_f is the fractal dimension; see Section 3.5.2. For the two-dimensional case the universality class is determined by $\sigma = 36/91$ and $\nu = 4/3$. The values of the critical exponents of a percolating system are different from those of the Ising model; however, the basic concept of criticality is otherwise exactly the same. The critical point turns out to be $p_c = 0.591$. In Figure 3.3 we show an example of a site percolation transition along different values of p for finite N.

The understanding of percolation has become relevant in network theory, particularly in the context of random networks; see Section 4.4. These networks can also show percolation transitions. To create a random network, specify a set of N nodes and link any of its possible pairs with probability p. Above the critical point $p_c = 1/N$, a connected component is forming that contains most of the N nodes. The dimensionality of the system at hand plays a role in determining the values of the critical exponents of percolation. For further literature see [58].

3.3.2 Self-organized criticality

Self-organized critical systems are driven out-of-equilibrium systems that have a critical point as an attractor. They show critical behaviour, meaning that the associated distribution functions of observable quantities are fat-tailed or power law distributions.

Self-organized critical systems are not externally fine-tuned towards a critical point in parameter space; no critical temperature, critical pressure, or critical probability is necessary. They behave as if they manage to 'find' their critical points themselves. In that sense they are said to *self-organize* towards criticality. Their macroscopic observables exhibit characteristic scaling behaviour in the spatial or temporal domain. They offer one way of intuitively and quantitatively understanding the origin of scaling laws in open, driven, and non-equilibrium systems.

The prototype model in terms of self-organized criticality is the Bak, Tang, and Wiesenfeld sand pile model [24]. Imagine that grains of sand are dropped on to a table. A sand pile starts to grow until the entire table is covered in sand. At some point, the pile reaches a *critical slope*; in other words, it is becoming too steep to absorb more grains, and avalanches are starting to occur; see Figure 1.3. These avalanches lead to a reduction of the slope. If the slope goes below the critical slope, sand can be deposited again. By accumulating sand, the slope increases again, until new avalanches are triggered, and so on. The sand pile self-organizes towards a critical slope. The process comprises two types of dynamics: the driving part, which is the external adding of grains and the relaxation, which is the toppling of the grains. Driving and relaxation might happen at different timescales. Self-organized critical systems are usually slowly driven and open. 'Slow' here means with respect to the speed of the relaxation process. The system has time to relax towards equilibrium states before the next event of the driving process occurs. In the sand pile example, slowly driven means that sand is added grain by grain, changing the system slowly, as opposed to the relaxation dynamics in which avalanches might involve thousands of grains at the same time. 'Open' means the existence of a flow through the system. It can be a an energy flow or a flow of grains on and off the table. For more literature on the topic, see [81].

Self-organized critical systems are characterized by their statistics of fluctuations, which often display approximate power laws in the corresponding probability distribution functions. This is the case for many realizations of sand pile models. Fat-tailed distributions indicate that the number of large events (avalanches) is relatively high. As some of the avalanches are large and span the entire system, the correlation lengths are large. They diverge as the system becomes infinitely big. Sand pile models only display criticality in either two or three dimensions. The insights gained from them has been carried over to other systems that self-organize towards critical states. Obvious candidates include granular media, discharge processes, cascading phenomena, and geophysics. For example, in geophysics the strain between tectonic plates is increased (driven) by their relative drift. If a critical strain level is reached, stress is released through a disruptive displacement of land masses, causing earthquakes. In this way, self-organized critical models are used to explain the Gutenberg–Richer law of size distributions of earthquakes [384]. Another example is systemic risk arising through cascading of bankruptcy events in financial systems. There, banks organize towards critical capital levels. If a bank goes bankrupt because of an external (driving) event, others might be tipped over into bankruptcy as a consequence. As the bankruptcies unfold, the system relaxes towards a new equilibrium state. Self-organized critical systems are important for understanding systemic risk [311]. Other areas of application include cosmology, evolutionary biology and ecology, physiology (e.g. heart rate, body temperature, sugar levels), forest fires, financial markets, combinatorial evolution, meteorology, economics, sociology, neurobiology, and econophysics; see Table 3.1.

3.3.3 Multiplicative processes

Multiplicative processes are processes that progress in time by successively multiplying random numbers,

$$Y_{t+1} = Y_t X_t = X_t X_{t-1} \cdots X_1 Y_0, \qquad (3.9)$$

where X_t are random variables that are drawn from a given distribution, and Y_0 is the initial value. Multiplicative processes typically contain products of many random variables. We learned in Section 2.3.1.3 that the values of the product Y_t are log-normally distributed, if t is large. This is easily seen by taking logarithms on both sides of Equation (3.9) and applying the central limit theorem; see Section 2.3.1.

If a random variable $\log X_n$ is distributed as a Gaussian, the variable X_n itself is distributed as a log-normal. The log-normal distribution function is given by,

$$p(x) = \frac{1}{x\sigma\sqrt{2\pi}} e^{-\frac{(\log x - \mu)^2}{2\sigma^2}}. \tag{3.10}$$

The log-normal distribution is not a power law; however, it decays more slowly than the Gaussian distribution and is therefore often confused with a power law. However, exact power laws can be obtained by adding simple constraints to the multiplicative processes of Equation (3.9). The simplest way of showing this has been used to explain Zipf's law in city size distributions [137]. Assume that there are N different cities, each city i with a population of C_t^i citizens at time t. Assume the growth rate for every city X_t^i to be a positive i.i.d. random number that may vary from city to city. Given that there is no migration between cities, in every city the population will change as in Equation (3.9), $C_{t+1}^i = X_{t+1}^i C_t^i$. Assuming further that the total population size in all cities together does not change means that $\sum_i C_t^i$ is a constant for all times, t. This is the crucial constraint. It implies that the overall growth rate X is effectively 1 and is therefore drawn from a distribution $p(X)$ with a mean of $\mu = \int_0^\infty X p(X) dX = 1$. The argument is now the following. We denote the cumulative distribution function of city sizes by $G_t(C) = P(C_t > C)$. The change in this distribution from one time step to the next is given by,

$$G_{t+1}(C) = P(C_{t+1} > C) = P(X_{t+1} C_t > C) = P\left(C_t > \frac{C}{X_{t+1}}\right) = E[1_{C_t > C/X_{t+1}}]$$

$$= E[E[1_{C_t > C/X_{t+1}}]|X_{t+1}] = E[G_t(C/X_{t+1})] = \int_0^\infty G_t\left(\frac{C}{X}\right) p(X) dX, \tag{3.11}$$

where $E[.]$ is the expectation value with respect to random variable X, and $1_{C_t > C/X_{t+1}}$ is an indicator function that takes the value of 1 if the condition in the subscript is fulfilled. If we assume that there is a limiting distribution $G_t \to G$, we get the equation,

$$G(C) = \int_0^\infty G\left(\frac{C}{X}\right) p(X) dX, \tag{3.12}$$

which can be solved by the Ansatz[5] $G(C) = aC^{-1}$. This shows that Zipf's law appears in the cumulative distribution function. There are several other ways of constraining multiplicative processes to obtain exact power laws. For example, exact power laws emerge if lower bounds are placed on the process in Equation (3.9), meaning that at all times Y_t is kept above a positive constant, $Y_t > c$. Figure 3.4 shows a log-normal distribution function in comparison to the power laws obtained in the population dynamics model above, which shows exact asymptotic power laws (tail).

---

[5] It can be proved that this Ansatz is the only solution to lead to a stationary cumulative distribution G.

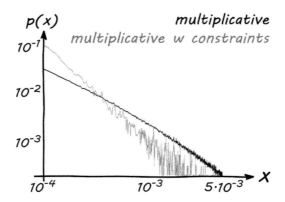

Figure 3.4 *Log-normal distribution (black) from the multiplicative process in Equation (3.9). The power law (grey) is obtained by normalizing the Y_t at every time step (multiplicative with constraints). Sometimes log-normal and power law distributions are confused when fitting data.*

3.3.4 Preferential processes

Preferential processes are self-reinforcing processes that show the *rich-get-richer phenomenon*, or the *Matthew effect*: 'For everyone who has will be given more, and he will have an abundance. Whoever does not have, even what he has will be taken from him,' [265]. Preferential processes can be seen as *the* classic way of understanding power law distribution functions. It dates back almost a century to when Udny Yule [416] first tried to understand the experimental power law distributions of species amongst genera. His basic idea is that genera with many species are more likely to experience mutations that will lead to yet more species, which triggers a self-reinforcing process.

> In preferential processes, the probability of a particular outcome occurring is proportional to the number of times this outcome has occurred in the past.

A highly intuitive way of explaining preferential processes is through a network growth model. This is called *preferential attachment*, which we also discuss in Section 4.5.4.3. Assume that at every time step you add one node to a network that initially consists of m_0 nodes, and that every new node can be linked to one of the existing nodes in the network. A node added at time $t + 1$ attaches to nodes that were present at time t. To introduce preferentiality, we introduce the probability $\Pi(k_i)$ of the new node linking to an existing node i that has a degree (number of links) k_i, as in the following,

$$\Pi(k_i) = \frac{k_i}{\sum_{j=1}^{N(t)} k_j}, \tag{3.13}$$

where $N(t)$ is the total number of nodes at time t in the network. The rate at which a particular node i acquires new links over time is $\frac{d}{dt}k_i = \Pi(k_i)$. Clearly, $\sum_{j=1}^{N(t)} k_j = 2t$, because at every time step one link is added. One link has two ends and increases the total degree by two. For large t, we have $\frac{d}{dt}k_i = \frac{k_i}{2t}$, which can be rewritten as $\frac{dk_i}{k_i} = \frac{dt}{2t}$. With the initial condition $k_i(t = t_i) = 1$, the solution is, $\log k_i(t) = \log(t)/2 + \text{const}$, or,

$$k_i(t) = \left(\frac{t}{t_i}\right)^{1/2},\tag{3.14}$$

where t_i play the role of integration constants. We now estimate the probability that node i has a degree smaller than k at time t, $P(k_i(t) < k)$. As, per construction, the model adds nodes in equal time intervals, the probability $P(t_i)$ that node i has been added at any time t_i is uniform in time, $P(t_i) = 1/(t + m_0)$. The probability that node i has a degree smaller than k therefore equals the probability that the node was added *after* a time t_i, given by $t_i = t/k^2$. We get,

$$P(k_i(t) < k) = P\left(t_i > \frac{t}{k^2}\right) = 1 - P\left(t_i \leq \frac{t}{k^2}\right) = 1 - \frac{t}{k^2(t + m_0)}.\tag{3.15}$$

Finally, we differentiate $P(k_i(t) < k)$ with respect to k to obtain the probability density function for the degree,

$$P(k) = \frac{d}{dk}P(k_i(t) < k) \sim 2k^{-3}.\tag{3.16}$$

This means that asymptotically for large times $P(k) \sim 2/k^3$ and m_0 can be safely neglected. We see that a power law with exponent three emerges from this preferential process. This particular model is called the Barabási–Albert model [28].

Preferential processes appear in many other contexts. Another example is understanding Zipf's law in word frequencies. In almost any book in almost any language, the rank distribution of words follows a Zipf distribution. This means that, counting the frequency of all words appearing in a given text, if the number of the most frequent word is $N_{\max}$, the second most frequent word appears approximately $N_{\max}/2$ times, the third most frequent $N_{\max}/3$ times, the 1000th most frequent word $N_{\max}/1000$ times, and so on. Herbert A. Simon offered an explanation in the 1950s that is based on Yule's ideas. The Simon model is slightly more general than the previously mentioned network formation model. It makes two (perhaps not very realistic) assumptions as to how a text is formed [340].

- The probability that the $k + 1$st word in a text is a word that has already appeared exactly i times is proportional to the total number of occurrences of all the words that have appeared exactly i times.
- There is a constant probability α that the $k + 1$st word will be a new word, a word that has not yet occurred in the first k words.

With these two assumptions it can be shown, with some algebra, that power laws emerge with an exponent γ that depends on α as $\gamma = (1 - \alpha)^{-1}$. Simon's basic idea rests on the observation that the ratio of Gamma functions, see Equation (8.12), yields an asymptotic power law,

$$\frac{\Gamma(x)}{\Gamma(x+k)} \sim x^{-k}. \tag{3.17}$$

Other examples of preferential processes include urn models with reinforcement, the so-called Pólya urns, mentioned in Section 2.5.3 and studied in detail in Section 6.6.1. There, the reinforcement happens through a particular replacement mechanism. Whenever a ball of a certain colour is drawn from the urn, it is replaced by δ balls of the same colour.

3.3.5 Sample space reducing processes

Sample space reducing (SSR) processes are a way of understanding the emergence of power laws in path- or history-dependent processes [93]. These are processes in which the future outcomes of the process depend on the history of past outcomes.[6] History-dependent processes are ubiquitous in natural and social systems. Many stochastic processes become more constrained as they unfold; in other words, their sample space (set of possible outcomes) reduces as they age. Whereas a newborn homo sapiens can become a composer, politician, physicist, actor, or anything else, the chances of a sixty-five-year-old physics professor becoming a concert pianist are practically zero. A characteristic feature of history-dependent systems is that their sample space is not static but changes over time. At the same time SSR processes are simple models for driven out-of-equilibrium processes, where a driving force brings the system to an 'excited' state from which it then tries to relax towards equilibrium.

SSR processes allow us to relate the details of the driving force and the details of the relaxation process to the statistics of the driven system. In this sense, they offer a theory for stationary, non-equilibrium systems.

3.3.5.1 *Zipf's law emerges*

In their simplest form, SSR processes can be depicted by the schematic process in Figure 3.5a. Imagine that there are N dice, each with a different numbers of faces. Dice number one has one face; dice number two has two faces (coin), and so on. Dice number N has N faces. We start by throwing dice number N. The possible outcomes (sample space) are the numbers between 1 and N. Assume the actual outcome is i. We then take dice number $i - 1$ and throw it. The sample space is now the interval $[1, i - 1]$, and we

---

[6] Preferential processes are a simple example of history-dependent processes.

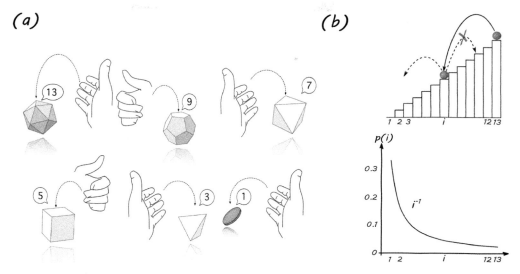

Figure 3.5 *Illustration of sample space reducing processes. (a) Imagine a set of $N = 20$ dice with different numbers of faces that range from one to twenty. We start by throwing the twenty-faced dice. Let us suppose that we get a face value of fourteen. We now have to take the thirteen-faced dice, throw it, and get a face value of, say, eight, and must thus continue with the seven-faced dice, and so on, until we reach the dice with one face. The process stops here and must be restarted, by starting with the twenty-faced dice again. If the process is repeated many times, the distribution of face values i is given by Zipf's law, $p_i = i^{-1}$. (b) The same process is depicted as a ball bouncing down a staircase in random step sizes. The ball ends up at the lowest stair (relaxation towards equilibrium) and has to be brought back up to the highest stair to restart the process (driving). After many iterations, the visiting probabilities of stairs (states) i is $p_i = i^{-1}$. If we introduce a restarting probability $1 - \lambda$ at every step, we obtain a visiting distribution of $p_i = i^{-\lambda}$. After [93].*

assume that the concrete sampled values are i'. We then take dice $i' - 1$ and continue the process until we reach dice number one. The process then halts. Alternatively, the same process can be thought of as a ball bouncing down a staircase, as shown in Figure 3.5b. Each state i of the system corresponds to one particular stair. A ball is initially placed on the topmost stair N, and can jump to any of the $N - 1$ lower stairs in the next time step with a probability $1/(N - 1)$. It never bounces upwards. Assume that the ball lands on step i. As it can only jump to steps i' that are below i, the probability of jumping to step $i' < i$ is $1/(i - 1)$. The process continues until step 1 is eventually reached. It then halts and must be restarted, by taking dice number N and repeating the procedure. The probability of visiting state i is exactly Zipf's law,

$$p(i) = i^{-1}. \tag{3.18}$$

This is seen by the following simple argument. The probability of observing a ball at state i is computed by $p_i = \sum_j^N p(i|j)p_j$, where from the previous discussion we know that the transition probability from step (state) j to i is,

$$p(i|j) = \begin{cases} \frac{1}{j-1} & \text{if } i < j \\ 0 & \text{otherwise} \end{cases}. \tag{3.19}$$

Inserting Equation (3.19) to $p_i = \sum_j^N p(i|j)p_j$, we get for the difference $p_{i+1} - p_i$,

$$p_{i+1} - p_i = -\frac{p_{i+1}}{i}, \tag{3.20}$$

If we read this equation in differential form, $dp_i/di = -p_i/i$, we see immediately that $\log p_i = -\log i$, and we get the result, $p_i = i^{-1}$.

Up to now we have assumed that every step that is lower than the present state i is visited with the same probability $1/(i-1)$. This is not necessary and we can assume a priori probabilities q_i for every state i. In this case, we get for the transition probability

$$p(i|j) = \begin{cases} \frac{q_i}{g_{j-1}} & \text{if } i < j \\ 0 & \text{otherwise} \end{cases}, \tag{3.21}$$

where $g_j = \sum_{j=1}^{j} q_j$. Remarkably, for a wide range of choices[7] of q_i, the same result for the visiting probabilities is obtained, $p_i = i^{-1}$.

> Distributions of SSR processes converge to Zipf's law as a limit distribution $p_i = i^{-1}$, regardless of the probabilities of visiting the individual states. Zipf's law acts as an attractor.

3.3.5.2   The influence of the driving rate

We can now ask what happens if we alter the driving rate of the process. In particular, if we allow the process to be restarted during its relaxation towards equilibrium. In the example shown before, we would bring the ball to its starting level at any time before it reaches the lowest state. Assume that we restart the process at any state with a probability of $1 - \lambda$, then we get for the transition rate,

$$p(i|j) = \begin{cases} \frac{\lambda}{j-1} + \frac{1-\lambda}{N} & \text{for } i < j \\ \frac{1-\lambda}{N} & \text{for } i \geq j > 1 \\ \frac{1}{N} & \text{for } i \geq j = 1. \end{cases} \tag{3.22}$$

[7] For the limits of these choices see [90].

Using the same procedure as before, we get the recursive relation $p_{i+1} - p_i = -\frac{\lambda p_{i+1}}{i}$, which yields an exact power law, where the exponent is λ,

$$p(i) = p(1) i^{-\lambda}. \tag{3.23}$$

This means that SSR processes lead to exact power law distributions in the visiting times of states, with an exponent that is one minus the driving (or reset) rate, λ. This happens if the process is driven before it reaches its equilibrium state or, in other words, if it does not have the time to fully relax before it is excited again.

The driving rate can also be interpreted as adding noise to the system. A simple way of introducing noise is to think of a combination of the SSR process and a pure random walk. For example, at every time step with probability λ, continue the SSR step (jump to one of the lower stairs) and with probability $1 - \lambda$ make a random jump to any step, higher ones, too. Again this result is robust. SSR processes lead to the exact same power laws as for a wide range of non-uniform probabilities q_i; see [90].

3.3.5.3 State-dependent driving rates and the emergence of statistics

Now assume that the driving rate depends on the current state of the system. In other words, λ becomes a function of the states, $\lambda(i)$. We assume as before that the states are ordered and that there are only transitions between states i and j if i is 'higher' than j. Following [91], let us again start with the transition probabilities, now with a state-dependent λ,

$$p(i|j) = \begin{cases} \lambda(j)\frac{q_i}{g(j-1)} + (1 - \lambda(j))q_i & \text{if } i < j \\ (1 - \lambda(j))q_i & \text{otherwise} \end{cases}. \tag{3.24}$$

The stationary distribution $p(\lambda, q)$ can be explicitly computed by observing that,

$$\frac{p_{i+1}(\lambda, q)}{q_{i+1}}\left(1 + \lambda(i+1)\frac{q_{i+1}}{g_i}\right) = \frac{p_i(\lambda, q)}{q_i}. \tag{3.25}$$

We obtain $p_i(\lambda, q) = \frac{q_i}{Z_{\lambda,q}}\prod_{1 < j \le i}\left(1 + \lambda(j)\frac{q_j}{g_{j-1}}\right)^{-1}$, where $Z_{\lambda,q}$ is the normalization constant. This equation is well approximated by,

$$p_i(\lambda, q) = \frac{q_i}{Z_{\lambda,q}}e^{-\sum_{j \le i}\lambda(j)\frac{q_j}{g_{j-1}}}. \tag{3.26}$$

If the process occurs in a continuum of states,[8] the continuum version of Equation (3.26) is given by,

---

[8] With this notation we mean that, as x becomes larger, the respective phasespace becomes smaller.

$$p_{\lambda,q}(x) = \frac{q(x)}{Z_{\lambda,q}} e^{-\int_1^x \lambda(y) \frac{q(y)}{g(y)} dy},$$ (3.27)

where $g(y) = \int_1^y q(x)\,dx$. For simplicity, let us only consider the situation for uniform $q(x) = 1/N$. To see the relation between the stationary distribution of an SSR process and the driving parameter $\lambda(x)$, we differentiate Equation (3.27) and get the relation,

$$\lambda(x) = -x \frac{d}{dx} \log p_{\lambda,q}(x).$$ (3.28)

We can now use Equation (3.28) to compute $\lambda(x)$ for any reasonable distribution function. Vice versa, we can associate any driving rate $\lambda(x)$ with a distribution function, $p_{\lambda,q}(x)$.

A state-dependent driving rate $1 - \lambda(i)$ allows us to obtain practically every distribution function that occurs in the context of complex (and simple) systems. It becomes possible to relate the details of the driving process in driven out-of-equilibrium systems *one-to-one* to the stationary distribution functions of the system (if they exist).

We now present a few concrete examples of how the state-dependent driving rate $1 - \lambda(x)$ determines the distribution function for state visits. We summarize these results in Table 3.3.

Table 3.3 *Relations of state-dependent driving rates $1 - \lambda(x)$ and distribution functions $p_{\lambda,q}(k)$ for SSR processes.*

Distribution	$\lambda(x)$	$p_{\lambda,q}(x)$
Power law	α	$x^{-\alpha}$
Exponential	βx	$e^{-\beta x}$
Power law with cut-off	$\beta x + \alpha$	$x^{-\alpha} e^{-\beta x}$
Gamma	$\beta x - \alpha + 1$	$x^{\alpha-1} e^{-\beta x}$
Normal	βx^2	$e^{-\frac{\beta}{2}x^2}$
Stretched exponential	βx^α	$e^{-\frac{\beta}{\alpha}x^\alpha}$
Gompertz	$\beta x(\alpha e^{\beta x} - 1)$	$e^{\beta x - \alpha e^{\beta x}}$
Weibull	$\alpha(\beta x)^\alpha - \alpha + 1$	$x^{\alpha-1} e^{-(\beta x)^\alpha}$
Log-normal	$\frac{\log x}{\sigma^2} - \frac{\beta}{\sigma^2} + 1$	$\frac{1}{x} e^{-\frac{(\log x - \beta)^2}{2\sigma^2}}$
Tsallis–Pareto	$\frac{\beta x}{1 - \beta x(1-q)}$	$(1 - (1-q)\beta x)^{\frac{1}{1-q}}$

Power law $p(x) \sim x^{-\alpha}$, is obtained with $\lambda(x) = -x\frac{d}{dx}(-\alpha \log x) = \alpha$. We have seen previously that a state-independent driving rate leads to exact power laws.

Exponential distribution $p(x) \sim \exp(-\beta x)$ with $\beta > 0$, is obtained with $\lambda(x) = -x\frac{d}{dx}(-\beta x) = \beta x$.

Stretched exponential and normal distribution $p(x) \sim \exp\left(-\frac{\beta}{\alpha}x^{\alpha}\right)$ with $\alpha > 0$ and $\beta > 0$, is obtained with $\lambda(x) = -x\frac{d}{dx}\left(-\frac{\beta}{\alpha}x^{\alpha}\right) = \beta x^{\alpha}$. $\alpha = 2$ corresponds to the normal distribution.

Gamma distribution $p(x) \sim x^{\alpha-1}\exp(-\beta x)$ with $\alpha > 0$ and $\beta > 0$, is obtained with $\lambda(x) = -x\frac{d}{dx}((\alpha-1)\log x - \beta x) = 1 - \alpha + \beta x$. Obviously, $\alpha - 1 \leq \beta$ is required.

Log-normal distribution $p(x) \sim \frac{1}{x}e^{-\frac{(\log x - \beta)^2}{2\sigma^2}}$, is obtained with $\lambda(x) = -x\frac{d}{dx}\left(-\log x - \frac{(\log x - \beta)^2}{2\sigma^2}\right) = 1 + \frac{\log x}{\sigma^2} - \frac{\beta}{\sigma^2}$.

Power law with exponential cut-off $p(x) \sim x^{-\alpha}\exp(-\beta x)$ with $\alpha > 0$ and $\beta > 0$, is obtained with $\lambda(x) = -x\frac{d}{dx}(-\alpha \log x - \beta x) = \alpha + \beta x$.

Tsallis–Pareto or q-exponential distribution $p(x) \sim (1 - (1-q)\beta x)^{\frac{1}{1-q}}$ with $\beta > 0$, is obtained with $\lambda(x) = -x\frac{d}{dx}\left(\frac{1}{1-q}\log(1 - (1-q)\beta x)\right) = \frac{\beta x}{1 - \beta x(1-q)}$. Note that for q-exponentials with $q < 1$ we require $\beta x < 1/(1-q)$, while for $q > 1$ no such restriction exists.

Weibull distribution $p(x) \sim x^{\alpha-1}e^{-(\beta x)^{\alpha}}$ with $\alpha > 0$ and $\beta > 0$, is obtained with $\lambda(x) = -x\frac{d}{dx}((\alpha-1)\log x - (\beta x)^{\alpha}) = 1 - \alpha + \alpha(\beta x)^{\alpha}$. Note that the standard parametrization of the Weibull distribution uses the parameter $\nu = 1/\beta$ instead of β. To ensure positivity of λ this implies that $(\alpha - 1)/\alpha \leq \beta^{\alpha}$ which is always satisfied for $\alpha < 1$. For $\alpha > 1$ we need $\beta \leq (1 - 1/\alpha)^{1/\alpha}$.

Gompertz distribution $p(x) \sim \exp\left(\beta x - \eta e^{\beta x}\right)$ with $\beta > 0$ and $\eta > 0$, is obtained with $\lambda(x) = -x\frac{d}{dx}\left(\beta x - \eta e^{\beta x}\right) = (\eta e^{\beta x} - 1)\beta x$. The restriction $e^{-\beta} \leq \eta$ applies.

In a nutshell, constant driving rates lead to exact power laws. The actual rate determines the slope. If the rate is slow enough for the process to reach a stable configuration (or a metastable state or the ground state) every time it gets excited, the exponent is -1, or Zipf's law. A rate with $\lambda(x) = x$ gives exponential distribution functions. For linear rates with an intercept $\lambda(x) = \alpha + \beta x$, power laws with exponential cut-off or Gamma functions are obtained. The Weibull distribution emerges from a driving rate that is a power law with intercept. For power law rates and for logarithms, the stretched exponential and the log-normal distributions appear, respectively. The Gaussian is recovered as a special case of the latter.

3.3.5.4 Cascading SSR processes

In the derivation of Equation (3.23) we have seen that the power exponent λ is a number between zero and one. This is perfectly in line with the interpretation of $1 - \lambda$ as the driving rate of the process. However, there is an alternative interpretation that allows us

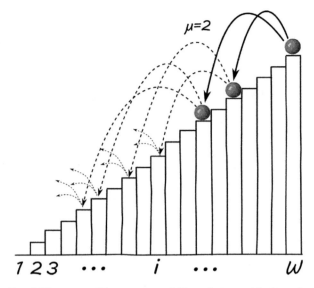

Figure 3.6 *Cascading SSR process. The process unfolds as before, with the only difference being that whenever the ball hits a state, it multiplies by a factor μ. A ball turns into μ balls, which now all continue their random jumps downwards to lower states. The process becomes a cascading process, with the visiting probability now being $p(i) \sim i^{-\mu}$. μ can be any number between zero and infinity.*

to obtain any exponent μ in the range $0 \leq \mu < \infty$. The process unfolds as before. A ball hits states randomly on its way towards a stable configuration (or a meta-stable state or the ground state). The essence of the cascading SSR process is that every time the ball hits a state it is multiplied by a factor, μ. In other words, one ball becomes μ balls whenever it hits a state; see Figure 3.6. For this case it can be shown that the visiting distribution is an exact power law; the exponent μ can be any positive number,

$$p(i) \sim i^{-\mu}. \tag{3.29}$$

It can further be shown that if one introduces a 'conservation condition', the power exponent becomes -2, regardless of the multiplication factor μ, $p(i) \sim i^{-2}$. The conservation is imposed by the following condition. Suppose a ball reaches state i. It now multiplies and the μ new balls all proceed downwards to lower states. Each new ball reaches a state $j_1, j_2, \cdots j_\mu$. The conservation law is now imposed by the requirement that the sum of all states must add up to $i = j_1 + j_2 + \cdots + j_\mu$. The intuition behind this is the concept of energy conservation. If one thinks of the cascading SSR process as a hadronization process of cosmic rays, then obviously energy is conserved at every splitting of particles. Note that for the energy distribution of cosmic rays, Enrico Fermi found a power law with exponent -2; see [126]. We could derive it here as a statistical result of an SSR process with energy conservation. For the details of the derivation, see [92]. Note that μ can also be a number smaller than 1. In that case, $1 - \mu$ is interpreted as the probability of a ball being removed from the system during the relaxation process, μ being the *survival rate*.

3.3.5.5 *Examples of SSR processes*

There are a number of applications that make clear the contexts in which SSR processes are realized in complex systems. We believe that SSR processes contain many driven non-equilibrium processes as special cases, particularly self-organized critical systems, such as sand piles. Here we mention four other examples.

Diffusion on directed networks SSR processes, like the processes discussed with the ball on the staircase, can be seen as random walks on directed networks. In these networks, the nodes are ordered so that one can reach a node with a lower order only through a directed link. If there are no cycles in the network, as a result of Equation (3.18) and due to the robustness with respect to the probabilities, q_i, we obtain that the visiting frequencies of nodes from random walkers on a directed acyclic network follow an exact Zipf law. This is indeed what is observed in many applications, such as search processes on electronic platforms and the internet [6]. In general, good search processes, where at every step in the search one is able to eliminate other states, will lead to a Zipf law.

Fragmentation processes Another example of SSR processes are fragmentation processes, which are history-dependent. Think, for example, of objects that repeatedly break at random sites. Consider a piece of dry spaghetti of length L and mark a point on it. Take the spaghetti and break it at a random position. Select the fragment that contains the mark and record its length. Then break this fragment at a random position, take the fragment containing the mark and repeat, and so on. The process is repeated until the fragment with the mark reaches a minimal length, say the diameter of an atom. The fragmentation process stops. The process is of the SSR type, as fragments are always shorter than the fragment they originate from. In particular, if the mark has been chosen on one of the endpoints of the initial spaghetti, then the consecutive fragmentation of the marked fragment is obviously a continuous version of the SSR process discussed before. Note that even though the length sequence of a *single* marked fragment is an SSR process, the size evolution of *all* fragments is more complicated, as fragment lengths are not independent of each other: every spaghetti fragment of length x splits into two fragments of respective lengths, $y < x$ and $x - y$. The evolution of the distribution of all fragment sizes was analysed in [241]. Similar examples of SSR processes in fragmentation processes also exist in higher dimensions.

Sentence formation We can treat the classic problem of understanding word frequency distributions with SSR processes and, at the same time, offer an alternative to Herbert Simon's explanation, which, though giving the correct result, rests on moderately convincing assumptions. Simon assumes that words are repeated in proportion to how often they have already appeared in the text. This is exactly what our language teachers told us *not* to do. A straightforward example of an SSR process is the formation of sentences. The first word in a sentence can be sampled from the sample space of all existing English words. Say we choose 'The cloud'. The choice of the next word is no longer as free, but is constrained by grammar and context. In our example, the second word can only be sampled from a smaller sample space, it must be a verb, and it must somehow be compatible with what clouds can do. Otherwise, we are talking out of context, we sound weird, or are being poetically experimental. As the length of a sentence

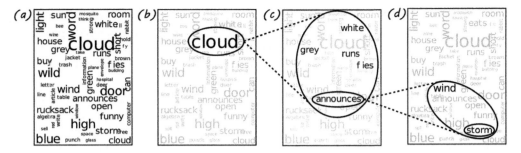

Figure 3.7 *Schematic view of the process of sentence formation. (a) Amongst all the potential N words of the English vocabulary that define the initial sample space volume, (b) we choose 'cloud'. (c) This choice restricts the sample space for the next word (circle), which has to be grammatically and semantically compatible with 'cloud'. From this set we chose 'announces', which reduces the sample space again, (d), as the next word must now be consistent both semantically and grammatically with 'The cloud announces'. The sequence of words shows a nested structure. The effect of sample space collapse is also present in the wider context of discourse formation, as both a topic and its rhetorical development impose successively nested constraints on sample space.*

increases, the size of the sample space of word use typically reduces—one has to come to the point. If we produce word sequences that do not focus in sample space, it becomes hard to convey information. At the end of every sentence, the process is repeated—it is restarted; see Figure 3.7. From time to time, sample space can also increase, which means that unexpected words or phrases are used. This will result in distribution functions of word frequencies that deviate from the exact Zipf law. Indeed, such deviations do exist and they can be understood with SSR processes [367]. This mechanism is not limited to sentence formation; it also affects processes involving a coherent progression of events, such as the succession of human activities or actions. Indications of the presence of Zipf's law in the action time series of players in massive multiplayer online games [370] and the game of Go [149] have been found.

Evolutionary processes   We will learn in Section 5.6 that combinatorial evolutionary processes can be seen as self-organized critical systems, which lead generically to power law and fat-tailed distributions. These can be understood as the result of being self-organized critical. However, they can also be understood as the result of an SSR process. The sample space of evolutionary systems at a given instance is called the *adjacent possible*, which we encounter in Section 5.1.2. Some evolutionary systems, particularly technological ones [14], which are driven in the sense that when an innovation (driving process) is made, the consequence may be a cascade of new inventions running through the system (relaxation). These evolutionary systems are driven systems, and can be described by SSR processes.

For more details of the statistics, and especially the statistical mechanics of SSR processes, see Section 6.6.2.

3.3.6   Other mechanisms

There are several other mechanisms that generate power laws. They are described in more detail in the reviews [276] and [290].

3.3.6.1   *Exponential growth with random exponentially distributed observation times*

There is a way of obtaining a power law by superimposing two exponential functions [320]. Assume you have deterministic exponential growth, $x(t) = x_0 \exp(\mu t)$, and that you sample $x(t)$ at random times t^*. The value at the sampled time is $x^* = x_0 \exp(\mu t^*)$. The cumulative distribution for the measurement times is also assumed to be an exponential, $P(t^* > t) = \exp(-\lambda t)$. The probability density function for the values x^* are then given by the power law,

$$p(x) = cx^{-\frac{\lambda}{\mu}-1}. \tag{3.30}$$

3.3.6.2   *Random typewriting*

A version of the idea of superimposing two exponential functions to obtain power laws was used by George A. Miller for a fun explanation of word rank frequency distributions [274]. This idea assumes that monkeys are typing random texts on a typewriter equipped with N letters and a space bar. The probability of hitting the space bar is q. Whenever the space bar is hit, a word is created. This can also be shown to lead to a power law in word frequency rank distributions and also in cases where all letters occur with different probabilities [89].

3.3.6.3   *Information-theoretic context*

Benoît B. Mandelbrot devised a mechanism to obtain power laws in an information-theoretic context. He assumed the evolution towards an optimally efficient language. He considered a cost function for word usage, and by assuming that natural languages have evolved to be efficient, he obtained a way of explaining Zipf's law in word frequencies by optimizing the information flow per word cost [259]. He imagines a language that consists of N words, with the cost for using the ith word in a conversation being C_i. The average cost of using these words is $C = \sum_i p_i C_i$, where p_i is the relative frequency with which i appears. The idea is to optimize the amount of *information production per word cost* in a language. One would then obtain an optimally efficient language. The average amount of information per word can be quantified with Shannon entropy, $S = -\sum_{i=1}^{N} p_i \log p_i$, and the optimization is done for S/C,

$$\frac{d}{dp_i} \frac{S}{C} = S' \frac{1}{C} - \frac{1}{C^2} SC' = -\frac{SC_i + C\log_2 p_i + 1}{C^2} = 0, \tag{3.31}$$

which yields $p_i = d2^{-SC_i/C}$, where d is a constant. If C_i is optimally coded (see Section 6.2.2.9), according to the coding theorem $\log i \leq C_i \leq \log i + 1$ holds and the power law is obtained, $p_i \sim i^{-S/C}$.

3.4   Power laws and how to measure them

Estimating power law exponents from data is not as simple as it might first appear. The simplest approach is, of course, to plot the corresponding data in loglog coordinates and then fit the portion of the data that looks straight with a straight line. The slope of that line is then the exponent. However, in reality, data represented in loglog plots rarely appear as pure power laws. Often a shoulder appears for small values; often there are only a few data points for large values, and this results in poor statistics, which is seen in the broadening of the data towards the tail of the distribution. As argued in Figure 3.8, it is hard to decide how best to fit a straight line to such data. Perhaps the best way of fitting power laws in both a statistically sound and consistent manner is to use a maximum likelihood method. How this works can be explained in a highly intuitive way in the simplest case [85], which unfortunately works only for power law exponents $\lambda < -1$. We will then show a maximum likelihood method that works for all exponents, following [167].

3.4.1   Maximum likelihood estimator for power law exponents $\lambda < -1$

Assume that a random variable X is distributed as a power law. If we then assume that the exponent λ is large enough for the distribution function to be normalized, we have,

$$p(x) = \frac{1}{Z}x^{-\lambda} = \frac{\lambda - 1}{x_{\min}}\left(\frac{x}{x_{\min}}\right)^{-\lambda}, \tag{3.32}$$

where we explicitly spelled out the normalization factor $1/Z$. Let us also assume that the concrete data values in the measured data $x(N)$, fall into W discrete bins that we label $x_i > x_{\min}$. The probability of a value falling into bin i is $p_i = p(x_i) = k_i/N$, where N is the number of data points. We can then ask what the probability is of observing the entire distribution function if the particular value of the power exponent is exactly λ. We get the answer,

$$P(x|\lambda) = \prod_{i=1}^{N} p(x_i) = \prod_{i=1}^{N} \frac{\lambda - 1}{x_{\min}}\left(\frac{x_i}{x_{\min}}\right)^{-\lambda}. \tag{3.33}$$

The probability of finding the distribution with exponent λ is called the *likelihood* of the data. To find its optimal value λ, that best fits the data, we can use Bayes' theorem; see Equation (2.15),

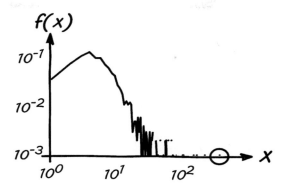

Figure 3.8 *Typical situation confronted when fitting a power law distribution from data. To the left there is a shoulder, to the right there are very few data points; the statistics for large values (for example, the values highlighted by the circle) is bad; finally, the region where a straight line occurs does not extend over many orders of magnitude. Fitting a linear line through these data points is not a good option. It is better to use a maximum likelihood method.*

$$P(x|\lambda) = P(\lambda|x)\frac{P(x)}{P(\lambda)}, \tag{3.34}$$

where $P(x)$ and $P(\lambda)$ are the prior probabilities for the data and the exponent, respectively. In the absence of any further information on λ, one has to assume that $P(\lambda)$ is uniformly distributed, meaning that it is a constant for all λ. Moreover, $P(x) = \int P(x|\lambda)P(\lambda)\,d\lambda$, and does not depend explicitly on λ anymore. Therefore it can be treated as a constant in the maximization. We therefore have the convenient situation that,

$$P(\lambda|x) = cP(x|\lambda), \tag{3.35}$$

where c is a constant. Substituting this into Equation (3.33) and taking logarithms on both sides, we have

$$\log P(\lambda|x) + \log c = \sum_{i=1}^{N} \log(\lambda - 1) - \log x_{\min} - \lambda \log \frac{x_i}{x_{\min}}. \tag{3.36}$$

To find the maximum likelihood we maximize the right-hand side of the previous expression by differentiating it with respect to λ, and then setting it to zero,

$$0 = \frac{N}{\lambda - 1} - \sum_{i=1}^{N} \log \frac{x_i}{x_{\min}}. \tag{3.37}$$

The most likely value for the power law exponent of a random variable that is assumed to be a normalizable power law ($\lambda > 1$) is,

$$\lambda = 1 + N \left(\sum_{i=1}^{N} \log \frac{x_i}{x_{\min}} \right)^{-1}. \tag{3.38}$$

This is often called the *maximum likelihood estimator*. Given the data values x_i it can computed very conveniently. The variance of the estimator is,

$$\sigma^2 = \frac{1}{N}(\lambda - 1)^2. \tag{3.39}$$

To show that the error of that estimate is indeed what we claimed before, recall that the variance of the estimate is $\langle \lambda^2 \rangle - \langle \lambda \rangle^2$. Exponentiating Equation (3.36) and using a shorthand notation $a = \sum_{i=1}^{N} \log \frac{x_i}{x_{\min}}$, we have

$$P(\lambda|x) = \frac{1}{c} e^{-a\lambda} (\lambda - 1)^N, \tag{3.40}$$

which allows us to compute the first and second moments,

$$\langle \lambda \rangle = \frac{\int_1^\infty \lambda e^{-a\lambda} (\lambda - 1)^N d\lambda}{\int_1^\infty e^{-a\lambda} (\lambda - 1)^N d\lambda} = \frac{e^{-a} a^{-2-N} (N+1+a) \Gamma(N+1)}{e^{-a} a^{-1-N} \Gamma(N+1)} = \frac{N+a+1}{a}, \tag{3.41}$$

and

$$\langle \lambda^2 \rangle = \frac{\int_1^\infty \lambda^2 e^{-a\lambda} (\lambda - 1)^N d\lambda}{\int_1^\infty e^{-a\lambda} (\lambda - 1)^N d\lambda} = \frac{e^{-a} a^{-3-N} (N^2 + 3N + a^2 + 2N + 2aN + 2) \Gamma(N+1)}{e^{-a} a^{-1-N} (N+1+a) \Gamma(N+1)}$$

$$= \frac{N^2 + 3N + a^2 + 2a + 2aN + 2}{a^2}. \tag{3.42}$$

Here the Gamma function appeared; see Equation (8.12). The variance is now,

$$\sigma^2 = \langle \lambda^2 \rangle - \langle \lambda \rangle^2 = \frac{N+1}{a^2} \sim \frac{1}{N}(\lambda - 1)^2, \tag{3.43}$$

where we expressed the shorthand notation $a = N/(\lambda - 1)$ in terms of λ from Equation (3.38). In the last step we also made the approximation $N + 1 \to N$, which is fine for large N; that is, many bins.

Note that we made two strong assumptions. First we required $\lambda < -1$, which means that we cannot use the approach presented here for $\lambda \geq -1$. This is unfortunate, as many processes show exponents larger than -1. A way out is presented in the following.

3.4.2 Maximum likelihood estimator for power laws for all exponents

It has been repeatedly stated that maximum likelihood estimates of exponents of power law distributed data can only be reliably obtained for exponents smaller than minus one, as power law distributions cannot be normalized otherwise. However, this is only true for sample spaces that are unbounded from above; in other words, infinitely large values may occur. In reality, this is rarely the case. We now show that power laws obtained from bounded sample spaces—as is the case for all data-related problems—are always free of such limitations and maximum likelihood estimates can be obtained for arbitrary powers without any restrictions. We follow the derivation in [167].

As before, assume we have N independent experiments and have obtained a data set containing measurements x_i, $x(N) = (x_1, x_2 \cdots, x_N)$. $k = (k_1, \cdots, k_W)$ is the histogram of the events recorded in $x(N)$; that is, k_i is the number of times the value x_i appears in $x(N)$. The sample frequencies are $p_i = k_i / N$. Let us assume that the process depends on a set of R parameters that are collected in the vector $\theta = (\theta_1, \cdots, \theta_R)$, and note that $\sum_{i=1}^{W} k_i = N$. As a consequence of independent sampling, the probability of sampling the entire histogram k is,

$$P(k|\theta) = \binom{N}{k} \prod_{i=1}^{W} p(x_i|\theta)^{k_i}, \tag{3.44}$$

where $\binom{N}{k} = N! / \prod_{i=1}^{W} k_i!$ is the multinomial factor. As before, Bayes' formula allows us to get an estimator for the parameters θ,

$$P(\theta|k) = P(k|\theta) \frac{P(\theta)}{P(k)}. \tag{3.45}$$

Obviously, $P(k) = \int d\theta \ P(k|\theta) P(\theta)$ does not depend on θ. Without further available information we must assume that the parameters θ are uniformly distributed between their upper and lower limits. As a consequence, $P(\theta)$ also does not depend on θ and can be treated as a constant.[9] From Equation (3.45) it follows that the value θ^* that maximizes $P(\theta|k)$ also maximizes $P(k|\theta)$. The most likely parameter values $\theta^* = (\theta_1^*, \cdots, \theta_R^*)$ are now found by maximizing the log-likelihood,

$$0 = \frac{\partial}{\partial \theta_r} \frac{1}{N} \log P(\theta|k) = \sum_{i=1}^{W} p_i \frac{\partial}{\partial \theta_r} \log p(x_i|\theta) = -\frac{\partial}{\partial \theta_r} H_{\text{cross}}(f||p(x|\theta)), \tag{3.46}$$

---

[9] Unfortunately, the same does not hold for the integration limits, $x_{\min}$ and $x_{\max}$. For those variables it turns out that $P(\theta)$ cannot be assumed to be constant between the upper and lower bounds of the respective parameter values. Bayesian estimators for $x_{\min}$ and $x_{\max}$ require the explicit consideration of a non-trivial function $P(\theta)$. Though this, in principle, is feasible, we ignore the possibility of deriving Bayesian estimates for $x_{\min}$ and $x_{\max}$ here.

for all parameters $r = 1, \cdots, R$. Here, $H_{\text{cross}}(f||p(z|\theta)) = -\sum_{i=1}^{W} f_i \log p(x_i|\lambda)$, is the so-called *cross-entropy*; see Section 6.2.3.1. In other words, maximum likelihood estimates maximize the cross-entropy with respect to the parameters θ_r. To apply Equation (3.46) to maximum likelihood estimates of power laws, the family of objective distributions is,

$$p(x|\lambda) = \frac{1}{Z_\lambda} x^{-\lambda}. \tag{3.47}$$

Note that the set of parameters θ defined here now only contains $\theta = \{\lambda\}$. The normalization constant is $1/Z_\lambda = \sum_x x^{-\lambda}$. The derivative with respect to λ of the cross-entropy, $H_{\text{cross}}(p_i||p(x|\theta)) = \lambda \sum_{i=1}^{W} p_i \log x_i - \log Z_\lambda$, has to be computed, and setting $dH_{\text{cross}}/d\lambda = 0$ yields,

$$\sum_{i=1}^{W} p_i \log x_i = \left(\sum_{i=1}^{W} x_i^{-\lambda} \right)^{-1} \sum_{i=1}^{W} x_i^{-\lambda} \log x_i. \tag{3.48}$$

The solution to this implicit equation, $\lambda = \lambda^*$, cannot be written in closed form but can easily be solved numerically.

> Maximum likelihood estimators can be used to accurately fit power laws with arbitrary exponents. Algorithms that implement the above-mentioned maximum likelihood estimates are available at the following online resource,
> `http://www.complex-systems.meduniwien.ac.at/SI2016/`
> The interested reader will find more details in [167].

3.5   Scaling in space—symmetry of non-symmetric objects, fractals

Symmetries greatly facilitate the description of objects, systems, and processes. It is often sufficient to know the symmetry of an object in order to be able to uniquely describe its basic properties. Geometry allows us to deal with symmetries of objects, for instance, translational or rotational symmetry. An object is symmetric with respect to an operation (such as a rotation) if the object looks the same before and after the operation. A cube will look exactly the same after being rotated 90° degrees along any of its three main axes. Once it is known that an object has a 60° rotational symmetry around the z-axis, it is immediately clear what that object looks like: it is a hexagonal prism. One type of symmetry that appears in functions is called periodicity. Knowing that a function or a process is periodic drastically reduces the effort of describing it. Symmetries are found not only in objects and functions, but also in physical laws. Think of time-reversal symmetry in most equations of motion. Here, one possible symmetry operation would be to replace the parameter time t by $-t$, wherever it appears in the equations. If, after that operation, the equations look exactly like they did before, they are said to be symmetric under time reversal. Symmetries are helpful mainly because they can substantially reduce the length of the description of a given system.

Many objects do not exhibit symmetries, often because it is hard to describe them with Euclidean geometry. Cows, for example, are not spheres, trees are not cubes, and mountains are not pyramids. These objects do not have apparent classical symmetries. Moreover, many physical, biological, social, and physiological processes often cannot be described using simple periodic functions or functions that show internal symmetries. In general, many aspects of complex systems cannot be described in reasonable terms by Euclidean geometry by means of lines, triangles, circles, ratios, periodic functions, and so on. At first sight, they do not seem to exhibit symmetries at all.

Fractal geometry offers a possibility of describing a wide class of seemingly non-symmetric objects and processes in an efficient and informative way. In the following, we show how the concept of symmetry can be generalized and extended to seemingly non-symmetric, erratic and disordered objects, processes, and systems. The key concepts are self similarity and fractals.

3.5.1   Self similarity and scale invariance

An object or process is called *self-similar* if a scaled version of it (enlarged or reduced) looks similar to the original object. We will see what the term 'similar' means in various contexts. Often, self-similar objects do not have a characteristic *length scale*, they are *scale-free*, or *scale-invariant*. Traditionally, most objects and phenomena involve a characteristic length or timescale. Many natural phenomena crucially depend on length scales, which must be specified before their science is discussed. The characteristic length scale in solid state physics is the distance between atoms and molecules, for lasers it is nanometres, for traffic control it is metres, and for plate tectonics it is dozens of kilometres. The fundamental forces in physics depend on characteristic scales. The strong force operates on a 10^{-15}m scale, while the weak force is relevant only at scales of 10^{-17}m. On the centimetre scale, physics is very different from the Ångstrom scale, where the quantum world comes into play.

3.5.2   Scaling in space: fractals

An atom has a characteristic length scale. A tree has no characteristic length scale. Its branches have a wide variety of different lengths. To be more precise about what we mean by 'no length scale', let us consider the Koch curve as an example. The Koch curve is obtained by iterative steps, starting from a straight line of unit length $L = 1$ (level 0). At the first step, partition the line into three equal sections and replace the middle section with two lines of length $L = 1/3$, so that a triangle-like shape appears (level 1). The length of the curve at this stage is $L = 4/3$. At the second step, partition each of the

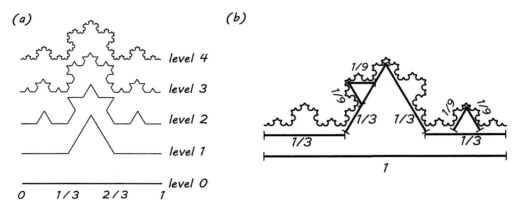

Figure 3.9 *The Koch curve is an example of a self-similar or scale-invariant object that can be created iteratively. (a) The first step is that a straight line of length 1 (level 0) is partitioned into three equal sections. The middle section is replaced with two lines of length 1/3, so that a triangle-like shape appears (level 1). The second step is to partition each of the four segments again into three equal sections and to repeat the procedure to generate higher levels. In the limit of infinitely many levels, the curve becomes the Koch curve. It is a fractal. (b) How long is the Koch curve? If we attempt to measure it empirically, its length depends on the length of the yardstick we use. For the lengths 1, 1/3, and 1/9, we find the respective lengths of the Koch curve to be 1, 4/3, and 16/9.*

four segments into three subsegments and again, replace the middle piece by two pieces of the same length as the one removed. The length at level 2 is now $L = 16/9$. At any finite number of steps n, the length of the curve is $L = \left(\frac{4}{3}\right)^n$; see Figure 3.9.

If any segment of the curve is scaled by a factor of three, it will look the same as on the previous scale. The curve is self-similar and scale-invariant; it has no characteristic length scale. If the segmentation process is continued for an infinite number of steps, the Koch curve is obtained. The Koch curve is a fractal. A fractal is often seen as an object that exists as the limit of an iterative operation. If you want to draw the Koch curve, you will find it challenging to do so. Fractals are often 'dusts', objects that cannot be drawn. They can contain points that are infinitesimally close to each other but nevertheless separated by holes. Fractals are typically objects that can no longer be characterized as one-, two-, or three-dimensional objects. The dimensionality of fractals is often somewhere 'in between' the Euclidean dimensions. The associated *fractal dimension* is a useful concept for efficiently describing a large class of objects, processes, and structures. Some famous fractals are the Cantor set, the Sierpinsky carpet, and the Mandelbrot set. The Mandelbrot set is an example of a mathematical structure that is obtained by finding all numbers c in the complex plane, for which the iterated map $f_c(z) = z^2 + c$ does *not* diverge even if applied an infinite number of times, meaning that $f_c(f_c(\dots(f_c(f_c(0))\dots)))$ remains finite.[10] Other examples of fractals are solutions (trajectories) of non-linear dynamical

---

[10] The initial value here is $z = 0$.

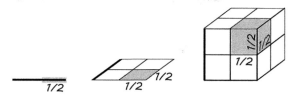

Figure 3.10 *Defining the concept of dimension as a scaling relation. By doubling the linear scale (the yardstick), the line, the area, and the volume scale by $2 = 2^1$, $4 = 2^2$, and $8 = 2^3$, respectively. The respective scaling exponent is the dimension.*

equations. Sometimes these trajectories converge to a point (fixed point), to a periodic orbit (limit cycle), or continue to explore regions of phasespace that never touch points reached previously (fractal attractor).

3.5.2.1 *Quantification of fractals, fractal dimension*

A useful concept for classifying fractals is to use their fractal dimension. Before embarking on the subject of generalized geometry and the symmetry of self-similar and fractal objects, we recall the central concept behind any geometry, namely, dimension. How are the dimensions of an object defined? Dimensions are defined as scaling relations. Imagine that we want to measure the size of objects with different dimensions. For a one-dimensional object, such as a stick, the natural measure M for describing it is its length. Assume that the stick has a length of L, and assume that we have a yardstick of length $L/2$. How does the measure change if we scale the length of the yardstick? Imagine that we scale the yardstick by a factor of two, $L/2 \to L$. Then the measure doubles, or $M(L) = 2M(L/2)$. Now we take a square with side length $L/2$. The natural measure M is now the area. How does the measure change if we scale the length of the yardstick from $L/2 \to L$? The measure (area) is obviously four times the original size, $M(L) = 4M(L/2) = 2^2 M(L/2)$. Similarly, taking a cube of side length $L/2$ and noting that its natural measure is the volume, we find that by doubling the yardstick to L, the volume behaves as $M(L) = 8M(L/2) = 2^3 M(L/2)$; see Figure 3.10. We have discovered a scaling relation similar to Equation (3.2), which defines the dimension.

> Definition of dimension. If the measure of an object M scales with its linear length L, and the scaling relation holds,
>
> $$M(\lambda L) = \lambda^d M(L), \qquad (3.49)$$
>
> then the scaling exponent d is called the *dimension* of the object. λ is the scaling factor. According to Equation (3.2) the solution to this equation is $M(L) = bL^d$, where b is a constant.

This definition is obvious for the usual Euclidean use of the word dimension. However, we can now use it to ask, for example, what the dimension of the Koch curve is. It is not really a line but neither is it a two-dimensional surface. Looking at Figure 3.9, and defining the length of the straight line as L, we see that the natural measure M of a third of that line is equal to one-quarter of that measure at the original length scale, $M\left(\frac{L}{3}\right) = \frac{1}{4}M(L)$. We now use this together with the defining equation for the dimension Equation (3.49),

$$M\left(\frac{L}{3}\right) = \frac{1}{4}M(L) = \left(\frac{1}{3}\right)^{d_{\text{Koch}}} M(L). \tag{3.50}$$

Taking logarithms on all sides, we find the 'funny' dimension of the Koch curve, $d_{\text{Koch}} = \frac{\log 4}{\log 3} = 1.262 \ldots$. This supports the insight gained earlier that the Koch curve is not a line ($d = 1$), nor does it fill a plane ($d = 2$). It has a *fractal dimension* of $1.262 \ldots$.

3.5.2.2   *Measuring fractal dimension—box counting*

Let us ask once more: how long is the Koch curve? To answer such a question practically, a measurement needs to be performed. Thus, something like a yardstick is needed. If we know the length of that yardstick, say, it is ϵ, the length of the Koch curve is best approximated by counting how often we can fit the yardstick to the curve. If it fits $N(\epsilon)$ times, the length is $L = N(\epsilon)\epsilon$. Assume that the yardstick is 1 unit long, so that it can be fitted exactly one time, and $L(\epsilon = 1) = 1\epsilon = 1$ units long; see Figure 3.9b. If the stick is of length $\epsilon = 1/3$, it can be fitted four times, and the length $L(\epsilon = 1/3) = 4\epsilon = 4/3$ units. The length obviously becomes longer as the yardstick gets shorter. Using exactly the same arguments as before, the number of times the scale fits the curve is $N(\epsilon) = \epsilon^{-d_f}$, where d_f is the fractal dimension. If we count $N(\epsilon)$ empirically with a set of yardsticks of different lengths ϵ, we can determine the fractal dimension in the limit for small ϵ,

$$d_f = -\lim_{\epsilon \to 0} \frac{\log N(\epsilon)}{\log \epsilon}. \tag{3.51}$$

Note that combining the relations $L = N(\epsilon)\epsilon$ and $N(\epsilon) = \epsilon^{-d_f}$ yields the answer to our initial question. The length of the Koch curve is,

$$L = \epsilon^{1-d_{\text{Koch}}}. \tag{3.52}$$

This idea can be generalized to any higher dimension. The Koch curve is an object that can be embedded in two dimensions. Fractal dimensions can be computed for any object that can be embedded in any integer dimension that is higher than its own. We propose using the following simple recipe.

Estimating the fractal dimension. The object of interest is a set of points embedded in a d dimensional space.

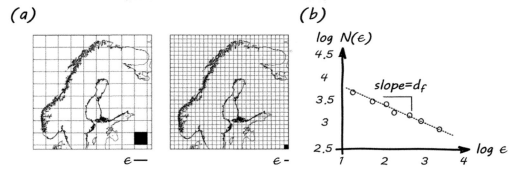

Figure 3.11 *How many boxes of size $\epsilon \times \epsilon$ are needed to cover the coast of Norway? (a) This is the number $N(\epsilon)$ of all boxes that contain a fragment of the coastline. (b) By repeating this measurement for various box sizes ϵ, we can plot $N(\epsilon)$ versus ϵ in a loglog scale to recover the fractal dimension d_f as the slope.*

- Partition the embedding region of the d-dimensional space with d-dimensional boxes of linear size ϵ.
- Count the number of boxes of linear size ϵ needed to cover the set. Assume that you count $N(\epsilon)$ boxes.
- Repeat this for smaller boxes ϵ, ideally $\epsilon \to 0$.
- The relation $N(\epsilon) = \epsilon^{-d_f}$ yields the fractal dimension, $d_f = -\lim_{\epsilon \to 0} \frac{\log N(\epsilon)}{\log \epsilon}$.

If d_f is not an integer, the set is a *fractal*.

This method is often used to empirically approximate the fractal dimension of objects. In Figure 3.11, the set of points that constitute Norway on a map is embedded in two dimensions. We therefore cover the map by two-dimensional squares of size $\epsilon \times \epsilon$ and count how many squares are needed to cover the coast. For every size ϵ we mark the number of boxes $N(\epsilon)$ in a loglog plot, as indicated in Figure 3.11b. The slope of the resulting line is the fractal dimension d_f of the coastline of Norway.

3.5.3 Scaling in time—fractal time series

Time series are obtained whenever measurements of a dynamical system are made and are recorded repeatedly along a discrete time axis. Time series are widely used in physical, biological, financial, economical, social, and technological contexts. Assume that we take measurements of a dynamical system at particular points in time t. If the corresponding data points x_t are represented in chronological order, that is, if the measurement values are presented as a function of the time at which they were taken, this is called a *time series*, $t \to x_t$; it is a discrete map or a process defined over not necessarily equally spaced time points t. Often, however, measurements are taken at equally spaced time points. The time series can be periodic or composed of superpositions of periodic functions. Many look erratic and contain considerable degrees of randomness that might reflect the nature of the underlying complex system. The time series may exhibit characteristic fluctuations over a wide range of values. Using the statistical tool sets that describe properties of stochastic time series is called time series analysis. The time series can be analysed in the time and frequency domain. The Fourier transform allows us to represent a time series in the frequency domain. If consecutive time segments of the time series are represented in the frequency domain, the term *time–frequency analysis* is used. Wavelet analysis is a special form of time–frequency analysis, that generalizes the Fourier transform and the time–frequency representation. Time series rarely show symmetries directly. Often, however, the *statistics* of time series is scale-invariant and self-similar. In this case, one talks about *fractal time series*. These are of interest here. Fractal time series analysis allows us to study how a given system is operating (statistically) at different timescales. Many complex systems do operate simultaneously on multiple time and length scales. Often, they show no characteristic frequencies and Fourier analysis is not very helpful. Scaling relations, such as the one given in Equation (3.56), are then a convenient way of describing the basic dynamical features of the underlying complex system.

The main idea behind fractal time series analysis is to describe the statistical properties of time series at different timescales. If the time axis is scaled, how does the statistics of $t \to x_t$ change? In particular, what happens to the variance and higher moments? The situation may occur that the scaling of the time axis alone does not yet lead to a 'self-similar' statistics, and so the values x_t may have to be scaled too,

$$x_t \to \lambda^H x(\lambda t), \tag{3.53}$$

where H is the so-called Hurst exponent. At this point it is necessary to clarify what the scaling of the time axis means for a discrete process. We create a new time series that now 'lives' on the time points λt. If λ is an integer and samples are spaced at equidistant time points, some locations of λt and t will coincide, and the new time series will contain only the subset of points at λt. If λ is not an integer or we have irregular sampling times, we might want to consider the last data point in every λt interval only, or to create a new time series that contains the average values over all x_t values found in the λt interval and assign it to the time point λt.

Let us now turn to the question of how to quantify self-similar, or fractal, time series. Perhaps the simplest way is to use the *mean square displacement* (MSD), which can be thought of as the 'variance' of the time series. It is defined as,

$$\mathrm{MSD}(t) = \langle (x_t - x_0)^2 \rangle, \tag{3.54}$$

where $\langle . \rangle$ is the average, taken over many different realizations, and x_0 is the first value (position) in the time series. If the time series is a random walk (Brownian motion), which progresses as $x(t+1) = x(t) + \xi_t$, where ξ_t is an i.i.d. random number, then we have the famous result that the variance of the process is growing linearly in time,

$$\langle (x_t - x_0)^2 \rangle = 2Dt. \tag{3.55}$$

Here, D is the diffusion constant. However, most complex systems do not produce time series that are simple random walks. If a particular time series does not follow the result of Equation (3.55), it is often helpful to generalize the MSD in the following way,

$$V(\tau) = \langle |x_{t+\tau} - x_t|^q \rangle_t \sim \tau^{H(q)}. \tag{3.56}$$

Here, we introduce a time shift variable τ that allows us to compute the variance as a function of different time separations between measurements, τ. $\langle . \rangle_t$ is the average over all time points t in the time series. $H(q)$ is again the Hurst exponent, which may depend on the number q. The functional form of $V(\tau)$ as a function of τ reveals much about the nature of the time series. If $H(q)$ does not depend on q, and is constant there are three important cases.

- If $H(1) = 1/2$ then we have the same situation as in Equation (3.55), which is an indication that the time series is a random walk with i.i.d. increments.
- For $H(1) > 1/2$ the time series is said to be *persistent*. Assume that at every time step a random walker can walk one discrete step to the left or to the right. Persistence in a random process means that given that the last step was to the left; the probability of taking a step to the left again is higher than taking a step to the right. Diffusion in such systems is 'faster' than normal diffusion. Note that persistent time series are not Markov processes.
- The case $H(1) < 1/2$ means *anti-persistence*. The probability of changing direction in the next time step is higher than continuing in the same direction as in the last time step.

$H(q)$ need not be a constant. If that is the case, the time series is no longer fractal but a so-called *multifractal*. For example, financial time series have been found to be multifractal [33, 125].

If $H(q)$ is a non-linear function, the corresponding time series $t \rightarrow x_t$ is a *multifractal*. The functional form of $H(q)$ sometimes reveals detailed information about the complex system behind the time series. In particular, it is possible to see that a system is simultaneously operating on various timescales, and sometimes even how it is operating.

Deviations from the scaling laws in Equations (3.54) and (3.56) sometimes signal the failure of a system. For example, the heart beat is the outcome of a complicated dynamical system that is regulated simultaneously by a number of control mechanisms that operate on different timescales. The time series of healthy heart beat intervals show scaling laws that are no longer present in the case of congestive heart failure [364].

3.6   Example—understanding allometric scaling in biology

We conclude the chapter by demonstrating an example of how a very few simple facts about physiology, and a few assumptions, allow dozens of scaling laws in biology and physiology to be derived and predicted. Sometimes, scaling laws cannot be explained as easily as in Galilei's example of bone strengths mentioned at the beginning of the chapter, and more effort is necessary to understand the origin of scaling laws. Such a case is allometric scaling in biology. We follow the arguments as they were originally presented in [71, 407]. Many scaling laws in biology are known as *allometric scaling* laws. They relate the body mass of organisms to physiological properties. One famous empirical biological scaling law relates the metabolic rate B to the body mass M of an organism,

$$B \sim M^{\frac{3}{4}}. \tag{3.57}$$

Remarkably, the power exponent is 3/4 and not 2/3, as one might believe at first sight. The natural first guess would be—very much like the reasoning of Galilei—to note that organisms burn nutrients and therefore produce heat. The heat has to be radiated into the environment, and the heat exchange happens through the surface A of the organism. One might therefore expect that if the energy production (metabolism) balances the energy loss that is proportional to the surface, $B \sim A$ holds. As the mass of any organism scales with the volume $M \sim V = A^{3/2}$, one thus one would expect, $B \sim M^{2/3}$. This is not, however, what is observed. When the metabolic rate of mammals is plotted versus body mass in loglog scale, the slope is clearly 3/4. The situation is shown in Figure 3.12. There is a whole family of other known biological allometric scaling laws, whose exponents are all multiples of 1/4. This means that for various physiological quantities Y the relation,

$$Y = Y_0 M^{b\frac{1}{4}}, \tag{3.58}$$

holds. For example, if Y is the heart rate, the experimentally observed value is $b = -1$; for the life time of animals or the length of aorta it is $b = 1$, and for the radius of the aortae of mammals it is $b = \frac{3}{2}$. We will list further examples at the end of the section.

3.6.1   Understanding the 3/4 power law

All mammals have in common a nutrient-transport system (blood circuit) that transports nutrients to all cells of the body. The three basic design principles of this nutrient-transport system are followed by all organisms.

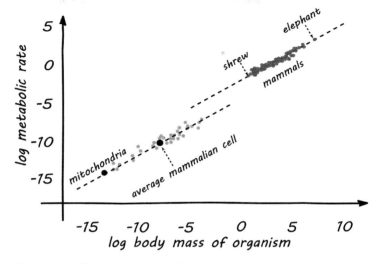

Figure 3.12 *Allometric scaling law. The metabolic rate of mammals (upper-right part) versus their body mass is a power law with scaling exponent 3/4. Practically all species are found on that line. Note that this scaling law also extends to other organisms and even cells, and stretches about twenty orders of magnitude in total. After [71].*

Design principles of the nutrient-transport network.

- *Space-filling.* As all cells in a body must be reached by the nutrients, the network must be space-filling. In mammals there is a hierarchical branching pattern in the transport network from the aorta to the capillaries.
- *Size invariance.* It is an empirical fact that across the animal kingdom the size of cells is about the same. The same is true for the lowest level of the transport network, the capillaries, which are the same size, irrespective of the organism.
- *Energetic efficiency.* Organisms evolved in such a way that the energy needed to transport materials through the circulatory system is a minimum.

We will use these principles to derive the 3/4 scaling law. For this, we employ the notation summarized in Table 3.4; see also Figure 3.13. The argument is now as follows. The more nutrients are transported at the capillary level, the more nutrients reach the cells and the higher the metabolic rate. This means that the metabolic rate B should be proportional to the rate of flow in the transport system,

$$B \sim \dot{Q}_0. \tag{3.59}$$

Table 3.4 *Notation needed in the argument to understand the 3/4 power law exponent in allometric scaling. Compare the variables with Figure 3.13.*

N	number of branchings from aorta to capillaries
k	level of branching; $k = 0$ is aorta, $k = N$ is the capillary level
n_k	number of branchings per unit at level k
l_k	length of a tube at level k
r_k	radius of a tube at level k
V_b	total volume of liquid (blood)
Δp_k	pressure difference across the tube at level k
$\bar{u}_k$	velocity of the liquid (average across cross-section) at level k
$\dot{Q}_k$	rate of flow at level k, $\dot{Q}_k = \pi r_k^2 \bar{u}_k$
N_k	total number of tubes at level k
β_k	ratio of radii at level k, $\beta_k = \frac{r_{k+1}}{r_k}$
γ_k	ratio of lengths at level k, $\gamma_k = \frac{l_{k+1}}{l_k}$

The transport liquid (blood) is not compressible and usually does not leak from the body. Total blood volume is therefore conserved in the body at all levels k. This means that,

$$\dot{Q}_0 = N_k \dot{Q}_k = N_k \pi r_k^2 \bar{u}_k = N_N \pi r_N^2 \bar{u}_N. \tag{3.60}$$

From the second design principle (size invariance), it follows that the dimensions at the capillary level are the same for all organisms, meaning that r_N, l_N, $\bar{u}_N$, and Δp_N are all independent of body size. The next step is to show that n_k, β_k, γ_k are all independent of the scale k, which means that the transportation network is a self-similar fractal. This is to a large extent experimentally verified; branching ratios in the circulatory system are similar across all levels k. For simplicity, we assume they are exactly the same. If n_k is scale-independent $n_k = n$, then it follows that $N_k = n^k$ and $N_N = n^N$. From Equation (3.60) we know that $\dot{Q}_0 \sim N_N$ and we have,

$$\dot{Q}_0 \sim N_N \sim n^N \sim \left(\frac{M}{M_0} \right)^a. \tag{3.61}$$

The last expression encodes our assumption that the metabolic rate is a power law in body mass M. Our aim is to compute the value of its exponent a. Using logarithms in the last equation we get,

(a)   (b)   (c)

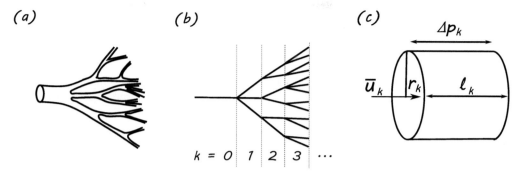

$k = 0 \quad 1 \quad 2 \quad 3 \quad \cdots$

Figure 3.13 *(a) Sketch of the nutrient-transport network in mammals from aorta to capillaries. (b) Shows the branching patterns, and (c) defines some of the variables that are needed in the example.*

$$N \sim a \frac{\log \frac{M}{M_0}}{\log n}. \tag{3.62}$$

The next step is to show the scale-independence of γ_k. From the first design principle (space filling) we have the relation, $V = N_N v_N$. If we use the fact that $V \sim N_k v_k = N_k l_k^3$, one can show that,

$$\gamma_k^3 = \left(\frac{l_{k+1}}{l_k}\right)^3 = \frac{N_k V}{N_{k+1} V} = \frac{1}{n}, \tag{3.63}$$

which demonstrates that γ is indeed scale-independent, $\gamma_k = \left(\frac{1}{n}\right)^{1/3} = \gamma$. Following exactly the same arguments, we can also show the scale-independence of β_k. As we assume the preservation of the total cross-sectional area at every scale, namely, that tubes are not elastic and that the fluid is incompressible, the total cross-section at level k is the same as at level $k+1$, $\pi r_k^2 = n\pi r_{k+1}^2$, which translates into,[11]

$$\beta_k = \frac{r_{k+1}}{r_k} = \left(\frac{1}{n}\right)^{\frac{1}{2}} = \beta. \tag{3.64}$$

Before we determine the exponent a in $B \sim M^a$, let us make the (reasonable) assumption that the total blood volume scales with body mass, $V_b \sim M$. Blood volume is the same as the tube volume at all scales,

---

[11] The full proof of this argument is more complicated and also involves design principle three, energy minimization. To perform that, one first writes the expression for the total friction (of viscous blood in the tubes) in the network, using the Poiseuille formula. The second step is to minimize this expression, subject to several constraints, like volume and mass conservation. The full argument also involves pulsatile flow and the fact that blood vessels are not rigid.

$$V_b = \sum_{k=0}^{N} N_k V_k = \pi \sum_{k=0}^{N} r_k^2 l_k n^k = \ldots = \frac{(n\gamma\beta^2)^{-(N+1)} - 1}{(n\gamma\beta^2)^{-1} - 1} n^N V_N. \tag{3.65}$$

Here, we used Equations (3.63) and (3.64) and the formula for the geometric series, where we neglect a 1, $(n\gamma\beta^2)^{-(N+1)} - 1 \sim (n\gamma\beta^2)^{-(N+1)}$. If N is large, and as $n\gamma\beta^2 < 1$, a very good approximation to Equation (3.65) will be,

$$V_b = \frac{(n\gamma\beta^2)^{-N}}{1 - n\gamma\beta^2} n^N V_N = \frac{(\gamma\beta^2)^{-N}}{1 - n\gamma\beta^2} V_N. \tag{3.66}$$

The final step is now in noting that the second design principle implies that V_N is independent of M, which means that $(\gamma\beta^2)^{-N} \sim M$, or $(\gamma\beta^2)^{-N} = M/M_0$. This gives,

Table 3.5 *Allometric scaling exponents for mammals that can be derived from the 3/4 power law. From [71, 407].*

Variable	Prediction	Observed			
Cardiovascular			**Respiratory**		
Aorta radius	3/8	0.36	Tracheal radius	3/8	0.39
Aorta pressure	0	0.032	Interpleural pressure	0	0.004
Aorta blood velocity	0	0.07	Air velocity in trachea	0	0.02
Blood volume	1	1.00	Lung volume	1	1.05
Circulation time	1/4	0.25	Volume flow to lung	3/4	0.80
Circulation distance	1/4	ND	Volume of alveolus	1/4	ND
Cardiac stroke volume	1	1.03	Tidal volume	1	1.041
Cardiac frequency	−1/4	−0.25	Respiratory frequency	−1/4	-0.26
Cardiac output	3/4	0.74	Power dissipated	3/4	0.78
Number of capillaries	3/4	ND	Number of alveoli	3/4	ND
Service volume radius	1/12	ND	Radius of alveolus	1/12	0.13
Womersley number	1/4	0.25	Area of alveolus	1/6	ND
Density of capillaries	−1/12	−0.095	Area of lung	11/12	0.95
O_2 affinity of blood	−1/12	−0.089	O_2 diffusing capacity	1	0.99
Total resistance	−3/4	−0.76	Total resistance	−3/4	-0.70
Metabolic rate	3/3	0.75	O_2 consumption rate	3/4	0.76

$$N = -\frac{\log \frac{M}{M_0}}{\log \gamma \beta^2}, \tag{3.67}$$

and to recall the previous result from Equation (3.62), by setting the two equations equal we finally arrive at,

$$a = -\frac{\log n}{\log(\gamma \beta^2)} = -\frac{\log n}{\log n^{-\frac{1}{3}-1}} = \frac{3}{4}, \tag{3.68}$$

where we used $\beta = n^{-1/2}$ and $\gamma = n^{-1/3}$. The result is the observed 3/4 power law, $B \sim M^{\frac{3}{4}}$.

From this result a number of immediate consequences can be derived. For example, the scaling laws involving the aorta are easily obtained, as follows. We know that the aorta radius is $r_0 = \beta^{-N} r_N = N_N^{1/2} r_N$. Using the result $N \sim M^a$, we arrive at $r_0 = M^{3/8}$, which is also the observed relation. In Table 3.5 we show thirty-two more allometric scaling relations that can be derived in a similar manner. The table includes results for the respiratory system that we did not cover in the above derivation. The interested reader should consult [71].

3.7 Summary

We briefly summarize what we have encountered in this chapter. Scaling arises in a wide range of contexts that range from the description of fractal objects to the statistics of complex systems. We defined scaling as the property of an object that, if scaled by a scaling factor λ, looks like the original. In terms of mathematical functions this means that $f(\lambda x) = g(\lambda)f(x)$. We learned that the solution to this scaling equation are power laws, $f(x) \sim x^c$, where c is called the scaling exponent.

In the context of stochastic complex systems, scaling appears in the probability distributions of dynamical variables. Power law distribution functions, or closely related functions, appear in practically all data that involve complex systems. We discussed the principal dynamical mechanisms that generate such scaling distributions. *Criticality* is essentially the physics of statistical systems at a phase transition point, such as the Curie temperature in magnetic systems. Power laws appear in physical quantities, such as specific heat or magnetic susceptibility, as the system parameters approach critical points. The corresponding exponents are called critical exponents. Many systems can be classified by their critical exponents. Systems that have the same set of critical exponents are said to belong to the same universality class. We mentioned the notion of percolation. *Self-organized criticality* is the phenomenon of criticality (presence of power laws) that appears without a system being fine-tuned to a critical point. The system 'finds' its critical point endogenously. The classical examples here are sand pile models that automatically self-organize towards a critical slope. Once found, the critical slope is maintained within narrow boundaries. Sand piles serve as intuitive models for a wide range of systems that show cascading and discharge phenomena, such as earthquakes, or the dynamics of crashes in financial markets. *Multiplicative processes* are stochastic processes, where products of many random numbers appear, and which are minimally constrained. This means that there are constraints that impose some conditions on the random numbers. These processes often allow us to understand the statistics of constrained growth phenomena. A classic example is the city size distribution. *Preferential processes* are self-reinforcing processes. The probability of a particular state appearing is proportional to the number of times the state has appeared in the past. These processes have been used to understand power laws for almost a century. They are used whenever 'rich-get-richer' phenomena are present. *Sample space reducing processes* are stochastic processes that reduce their sample space as they unfold. They provide a theory for driven non-equilibrium systems with stationary distributions. They allow the nature of the driving process to be related to the statistics of the process. Using the simplest form of driving, they generate power laws. However, if driving is state-dependent, it is possible to understand practically all distribution functions. Some processes have been described using more than one of these mechanisms. For example, Zipf's law of word frequencies in texts has been explained with preferential, multiplicative, self-organized critical, and SSR processes—with varying plausibility in the underlying assumptions.

We discussed ways of quantifying power law exponents from data that show scale-free statistics. We reviewed statistically sound and practical ways of estimating the exponents. Very briefly, we touched upon the subject of scaling in different contexts. In particular, we learned how scaling appears in the definition of the notion of the dimension. We learned that some objects, fractals, have non-integer dimensions. To estimate their dimension we used basic scaling arguments. Moreover, some time series can be characterized by simple scaling laws. These basically characterize the statistics of the underlying stochastic processes at multiple timescales. We concluded the chapter with an example of how one of the most famous scaling laws in biology can be understood on the basis of a few elementary observations in combination with simple scaling arguments. As a consequence, dozens of facts in physiology can be theoretically predicted: they are realized in nature with great precision.

In general, the science of complex systems is not about making exact and detailed predictions about specific events in the future. That is the business of prophets. As in statistical mechanics or quantum mechanics, understanding the *statistics* of processes correctly is considered success. Quantitative, and hopefully testable, predictions are then made on a probabilistic basis. Knowledge of the nature and origin of power law statistics is often a guideline for understanding how specific systems function. In physics there is no theory for driven non-equilibrium systems. The contributions from self-organized criticality and sample space reducing processes, which allow us to relate the observable statistics with details of the driving and relaxation processes, are a new step in the direction of control over driven out-of-equilibrium systems.

3.8   Problems

Problem 3.1 Show that both Equations (3.2) and (3.3) are solutions to the scaling relation defined in Equation (3.1).

Problem 3.2 Discuss why $f(x) = e^{-cx}$ is not a scaling function. What condition must $f(x) = \frac{\sin(cx)+b}{x}$ fulfil in order to be a scaling function? Is the q-exponential a scaling function? For the definition, see Section 2.4.1. What about Zipf's law?

Problem 3.3 Solve the differential equation $\frac{d}{dt}x(t) = \lambda x^{\mu}(t)$ and show that it is a q-exponential, as introduced in Section 2.4.1. What is the relation between q and μ? Show what happens at the limit $\mu \to 1$.

Problem 3.4 Write a short computer code to visualize the percolation transition. Initialize a two-dimensional 20×20 lattice and have every site occupied with probability p. Vary p from zero to one in steps of 0.1. For each realization count the size of the maximum cluster, $C_{\max}$, and plot it versus p.

Problem 3.5 Write a short computer code that produces distribution functions for ten products of uniformly distributed random numbers over the interval between 0 and 1. What distribution function do you expect? Next, compute distribution functions for ten products of random numbers as before, but normalize them; in other words, enforce that the sum of the ten numbers is always 1. What distribution do you expect?

Problem 3.6 Show that Equation (3.20) is indeed correct. Try to derive the result $p_i = 1/i$ without using the differential approximation. If you can, you have shown that the approximation leads to the exact result.

Problem 3.7 Show that Equation (3.30) is indeed correct. Hint: use the fact that $p(x^*)dx^* = p(t^*)dt^*$. Note that $P(t^*)$ is the cumulative distribution.

Problem 3.8 Write a short computer code that generates the time series $x_{t+1} = x_t + \xi_t$, where ξ_t is a random number from a Gaussian distribution with zero mean and unit variance. Generate 100,000 iterations. Try to plot V as a function of τ as defined in Equation (3.56). What does $H(q)$ look like? Plot it for a few values of q.

Problem 3.9 Show that Equation (3.65) is correct. Hint: use the formula for the geometric sum and the relations for γ and β.

Problem 3.10 Use the way we computed the scaling law for the aorta radius in Section 3.6.1, to show that the law for the length of the aorta is $l_0 = M^{\frac{1}{4}}$.

4

Networks

4.1 Overview

Understanding the interactions between the components of a system is key to understanding it on a quantitative and predictive basis. In complex systems, interactions are usually not uniform, not isotropic, and not homogeneous: interactions between elements can be specific. Networks are a tool for keeping track of who is interacting with whom, at what strength, when, and in what way. Networks are essential for understanding co-evolution and phase diagrams of complex systems. Networks are also convenient for describing structures of objects, flows, and data. Practically everything that can be stored in a relational database is a network. Everything that can be related to, or associated with other things is part of a network. Mathematically, networks are matrices and, as such, are just a subset of linear algebra. Their importance and value for complex systems comes from their role in dynamical adaptive systems, where networks of interactions dynamically update the states of a system, and where the dynamics of states updates the interaction networks.

Over the past decades, network theory has led to a number of practical methods that will be presented here. In this chapter, we provide a self-contained introduction to the field of network science. The aim is to expose the reader to the central concepts and techniques of the field. First, in Section 4.2, we introduce ways of representing networks mathematically by their adjacency matrices. We introduce the vocabulary and basic definitions necessary to describe the properties of networks in Section 4.3. A more comprehensive treatment of the concepts introduced in Sections 4.2 and 4.3 can be found in [204, 291]. The notion of random networks following the works of Erdős, Rényi, and Wigner will be reviewed in Section 4.4. Section 4.5 introduces the concept of complex networks. Here, we will follow the convention of defining complex networks as those types of random networks that have non-trivial topological features that do not occur in regular lattices or random networks as defined in Section 4.4. We will see that complex networks include random networks with arbitrary degree distributions, small world networks, and simple preferentially grown networks. We then turn to network concepts that are central to data analytics. Section 4.6 deals with the

Introduction to the Theory of Complex Systems. Stefan Thurner, Rudolf Hanel, and Peter Klimek,
Oxford University Press (2018). © Stefan Thurner, Rudolf Hanel, and Peter Klimek.
DOI: 10.1093/oso/9780198821939.001.0001

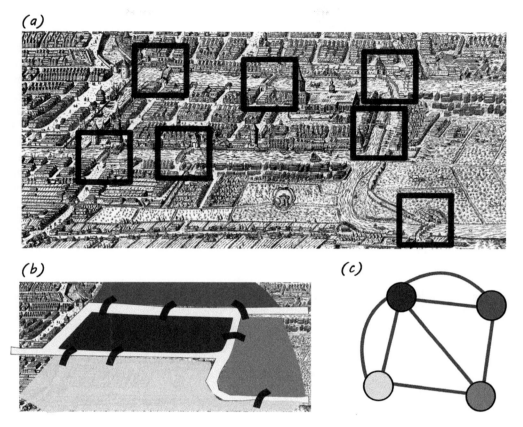

Figure 4.1 *The Seven Bridges of Königsberg problem. Is it possible to cross each bridge in Königsberg exactly once (without crossing a single bridge twice) in a single uninterrupted walk? Euler solved the problem by introducing the first network model in mathematics. The answer is no.*

problem of community detection, that is, the identification of groups of nodes that share more 'similarity' within the group than with nodes outside. Section 4.7 introduces the framework of correlation network analysis, which is a flexible procedure used for identifying the most relevant connections in large data sets. In the remaining sections of this chapter, we present network-theoretic concepts and the methods necessary to describe real world complex systems; that is, systems where dynamical states co-evolve with the system's boundary conditions that are defined by networks. Section 4.8 discusses paradigmatic dynamical processes *on* and *of* networks. Section 4.9 generalizes networks to more complicated structures; in particular to networks with multiple layers of nodes and links. We conclude the chapter with Section 4.10, which presents an example of how dynamical multilayer network concepts can be used to increase our ability to understand, describe, and manage *systemic risk* in the financial system.

4.1.1   Historical origin of network science

The study of networks dates back to the solution of the mathematical problem of the Seven Bridges of Königsberg in 1736 [337]. The River Pregel divides the city of Königsberg (today, Kaliningrad in Russia) into four different areas of land connected by seven bridges; see Figure 4.1a. The puzzle is as follows. Can every bridge in Königsberg be crossed in a single walk without any of the bridges being crossed twice? It was Leonhard Euler who answered this question [121]. The solution to what was simply a recreational problem would eventually start an entire scientific field. Euler observed that the actual route across the four individual parts of the city was irrelevant to the problem; see Figure 4.1b. He disregarded the shape of the areas, contracted them into a single point or 'node' and obtained a network; Figure 4.1c. The seven bridges are connections or 'links' between these nodes. The only information relevant to the puzzle is how the nodes are connected by these links. Next, Euler observed that to enter and leave a node in an uninterrupted walk that uses each bridge only once, the nodes need an even number of links. This holds for all nodes, except for the start and end nodes of the walk, which must have an odd number of links. The number of links attached to a node is called the degree. By observing that the degrees in the bridge network are all uneven (three nodes have a degree of three and one a degree of five), he concluded that there can be no such path.

Euler was the first to solve a real world network problem by radically removing unnecessary details from the data and by representing the simplified system as a network—a highly abstract version of reality.

4.1.2   From random matrix theory to random networks

It then took about 250 years for the field to gain momentum. In the aftermath of the Seven Bridges of Königsberg puzzle, the mathematical field of *graph theory* slowly emerged [62]. In the eighteenth and nineteenth centuries, graph theorists were mainly concerned with problems that were analytically treatable, including graph enumerations (counting networks with specific properties) [179], graph colourings (colouring nodes in a network such that no two connected nodes have the same colour) [210], or specific path-finding problems (traversing networks by only visiting each node exactly once) [406]. The next significant development was random matrix theory [271, 360]. In the 1930s, Eugene Wigner tried to understand the energy spectra of heavy atomic nuclei. His main result was that the Hamiltonians of such nuclei are so 'crowded' with interactions that the transition rates between two energy levels can be modelled with random variables [408]. He then showed that several properties of the observed spectra of such nuclei (e.g. spacings between spectral lines) can be understood with the eigenvalues of random matrices.

Wigner found that some physical systems can be understood by the statistical properties of a collection of interaction networks given by *random variables*.

The idea of random networks reached its next milestone with the work of Paul Erdős and Alfréd Rényi [117, 118]. They considered random networks where a set of links randomly connects a set of nodes. All nodes are linked with equal linking probability. Their central result was that many structural properties of such Erdős–Rényi (ER) networks (such as the ability to reach all nodes in a network in an uninterrupted walk) depend on network size in a non-trivial manner. In particular, they found abrupt transitions in many network properties as a function of the linking probability. In 1959 they remarked on the potential of their framework of random networks:

> In fact, the evolution of graphs may be considered as a rather simplified model of the evolution of certain communication nets (railway, road or electric network systems, etc.) of a country or some other unit. (Of course, if one aims at describing such a real situation, one should replace the hypothesis of equiprobablity of all connections by some more realistic hypothesis.) It seems plausible that by considering the random growth of more complicated structures (e.g. structures consisting of different sorts of 'points' and connections of different types) one could obtain fairly reasonable models of more complex real growth process (e.g. the growth of a complex communication net consisting of different types of connections, and even of organic structures of living matter, etc.) [117].

4.1.3 Small worlds and power laws

The field of network science in its modern form was triggered by two contributions in the late 1990s [28, 402]. As called for by Erdős and Rényi, these works defined classes of random networks that replaced the hypothesis of equal linking probability for all connections by more realistic assumptions. Watts and Strogatz were interested in the so-called small world puzzle, which is based on two seemingly contradictory empirical observations in many social, biological, and technological networks. On the one hand, certain networks are known to display high clustering and to consist of densely interconnected groups of nodes [339]. One would thus expect such networks to require a large number of steps to be 'traversed', as when a walker is on the network, he will more often than not get stuck in local clusters of nodes. On the other hand, however, such networks can, in reality, often be traversed in just a few steps. In social networks it has been shown with famous 'letter passing experiments' that the average 'distance' between any two individuals on the globe is about six [375]; hence the popular concept of 'six degrees of separation' [105, 401]. Watts and Strogatz reconciled these two observations by understanding 'small world networks': they showed that, starting from a network with very high clustering and very high distances, it only takes the addition of a relatively small number of random links (shortcuts) to create the small world effect. This idea had been anticipated by Mark Granovetter long before; he conjectured that social networks maintain cohesiveness on large scales due to so-called weak ties that act as bridges between well-connected groups of people [156].

Small world networks combine the properties of high clustering and a small distance between any given pair of nodes in the network [402].

The second work that started the boom of network science was contributed by Barabási and Albert [28]. Their point of departure was the observation that many real world networks, such as the world wide web, have probability distributions of node degrees that follow power laws [8]. This means that there exist nodes that are extremely well linked—hubs. They introduced a simple model for the growth of networks to account for this observation: the Barabási–Albert (BA) model [28]. Starting from a small network with only a couple of nodes, new nodes that are preferentially linked to well-connected nodes are continually added to the network.

For large networks, this preferential attachment process generates a power law in the degree distribution, meaning that the probability of finding a node with degree k is a power law, $p(k) \sim k^{-\gamma}$.

4.1.4 Networks in the big data era

If Erdős and Rényi had already anticipated these developments in 1960, why did it take more than thirty years to realize them? Naturally, the answer has to do with data availability and computer power [188]. In no other field are these developments as evident as in biology, which has essentially become a science of molecular and cellular networks [29, 200, 230]. Following the long tradition of networks in social science [398], economics [110, 190, 331], finance [5, 138, 165], and public health [75, 80] are now also starting to embrace networks. Network analysis has become one of the standard tools in data analytics [73, 291], along with regression models, time series analysis, and generative statistical models. Real networks are often much more heterogeneous than random networks such as those obtained from the ER or BA models. Links may be directed and undirected, spatially and temporally embedded, may only exist between different types of nodes that, in turn, may be related through links of more than one type; and all these properties are subject to constant dynamical changes and to the birth and death processes of nodes and links [54, 232]. Interest is currently shifting from simple and often static networks to generalized network structures that consist of multiple types of links and nodes—again, just as Erdős and Rényi predicted sixty years ago.

4.2 Network basics

In physics it is frequently unnecessary to specify who interacts with whom, as all particles in a system interact identically. For instance, all masses interact through gravity. The interaction strength between any two masses depends only on their relative distance and not on the identity of individual particles. This is in general not true for complex systems, where interactions are *specific*. Complex systems are typically characterized by multiple types of interaction that occur between specific types of elements. For instance, interactions in social networks may consist of two individuals that exchange information by meeting each other either physically or through certain communication networks, such as mobile phones or online social networks. People without access to certain communication channels consequently cannot interact through that mode of communication. Networks serve as a book-keeping system to specify who interacts with whom, in what way, how strongly, when, and for how long.

4.2.1 Networks or graphs?

Euler's investigation of the problem of the Seven Bridges of Königsberg started the field of graph theory. Much of Euler's original vocabulary for describing networks has been adopted in the graph theory of today. Graph theorists define *graphs* consisting of *vertices* that are connected by *edges*. Physicists, on the other hand, are more likely to talk about *networks* in which *nodes* are connected by *links*. For all practical purposes, the two sets of vocabulary have an identical meaning. A subtle distinction is sometimes made on the basis that by 'graphs' one is typically referring to abstract mathematical objects (ensembles), whereas networks often refer to specific realizations of networks (a social network, the metabolic network, or the world wide web) [27]. It follows that several different (physical) networks can correspond to the same graph. However, this distinction is not made by everybody and has been blurred by companies that call their networks graphs, such as Facebook's 'social graph' [387] or LinkedIn's 'economic graph' [358]. The main difference between networks and graphs is therefore mainly in the socialization of the researchers involved. Here, we will use the {network, node, link} vocabulary.

4.2.2 Nodes and links

Networks consist of a set of nodes, $\mathcal{N}$, and a set of links, $\mathcal{L} \subset \mathcal{N} \times \mathcal{N}$. A link is either directed or undirected. An undirected link is a set of two distinct nodes, say, $\{i,j\}$. A directed link is an ordered pair, (i,j) that points from node i to j. Node i is called the source of the link; j is the target.

Undirected links can be represented as two directed links pointing in opposite directions; that is, $\{i,j\} = \{(i,j),(j,i)\}$. The number of nodes, $N = |\mathcal{N}|$ is called the *order* of the network. The number of links will be referred to as $L = |\mathcal{L}|$. Note that these definitions imply that one undirected link counts as two directed links. This creates an omnipresent source of errors for factors of two.

4.2.3 Adjacency matrix of undirected networks

Simple networks consist of one type of node and one type of link. They can be conveniently represented by matrices. Each node is identified by an index, $i = 1, 2, \ldots, N$. For an undirected network, each pair of nodes is either connected by a link or not. Two connected nodes are called *adjacent*.

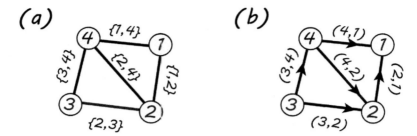

Figure 4.2 *(a) Undirected network with four nodes and five links. Links are labelled by the sets of node indices that they connect. (b) Directed network with four nodes and five links, labelled by the ordered pairs of node indices that they connect.*

Links are recorded in the $N \times N$ *adjacency matrix A*, whose elements A_{ij} indicate the presence or absence of a link. If an undirected link between nodes i and j exists, we set $A_{ij} = 1$. If no link between i and j exists, we have $A_{ij} = 0$. Undirected links impose no order on the nodes that they connect; the adjacency matrix for undirected networks is symmetric, $A_{ij} = A_{ji}$.

Figure 4.2a shows an example of an undirected network with four nodes and five links. This network can be represented by the adjacency matrix,

$$A = \begin{pmatrix} 0 & 1 & 0 & 1 \\ 1 & 0 & 1 & 1 \\ 0 & 1 & 0 & 1 \\ 1 & 1 & 1 & 0 \end{pmatrix}. \tag{4.1}$$

4.2.3.1 The adjacency matrix of directed networks

A directed link from i to j does not imply a link from j to i. The adjacency matrix of a directed network is thus, in general, not symmetric; that is, $A_{ij} \neq A_{ji}$. We use the convention that a link from i to j is represented in the adjacency matrix element A_{ji}. Figure 4.2b shows a directed network that corresponds to the non-symmetric adjacency matrix A,

$$A = \begin{pmatrix} 0 & 1 & 0 & 1 \\ 0 & 0 & 1 & 1 \\ 0 & 0 & 0 & 0 \\ 0 & 0 & 1 & 0 \end{pmatrix}. \tag{4.2}$$

Networks can also represent self-interactions of nodes. These are given by the diagonal entries in the adjacency matrix. Non-zero elements in the diagonal are called self-loops. These are links that have the same node as a start and end point.

We call directed or undirected networks without self-loops and with unweighted links, $A_{ij} \in \{0, 1\}$, *simple networks*.

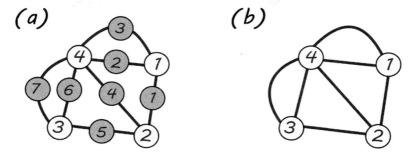

Figure 4.3 *(a) Bipartite network consisting of two types of nodes (white and grey). Links always connect nodes of different types. (b) Bipartite networks can be used to represent several other network types, including multigraphs. In these networks, nodes can be connected by more than one link.*

4.2.3.2 Bipartite networks

In many situations, networks represent relationships between different sets of nodes. For instance, these sets might correspond to patients and their illnesses [79], countries and their exports [189], or the land areas of Königsberg and their seven bridges [121]. Links in *bipartite networks* always connect elements of one type (patients, countries, land areas) with elements of a different type (illnesses, exports, or bridges). Figure 4.3a shows the Seven Bridges of Königsberg network as a bipartite network with two different sets of nodes represented by white (land areas) and grey (bridges). In general, bipartite networks can be directed or undirected.

4.2.3.3 The incidence matrix

The adjacency matrix of bipartite networks takes on a specific form. It is useful to introduce the concept of the *incidence matrix*. Incidence matrices describe the relationships between two types of element. Assume that there are N_1 nodes of type one and N_2 nodes of type two. The incidence matrix, B, is then a rectangular $N_1 \times N_2$ matrix, where the row (column) index corresponds to the index of the nodes of type one (two). Every entry in the incidence matrix is one, $B_{ij} = 1$, whenever a link between node i of type one and j of type two exists in the bipartite network, and it is zero otherwise.

The $N \times N$ adjacency matrix for a bipartite network has $N = N_1 + N_2$ nodes. By definition, all elements in the adjacency matrix that correspond to links between nodes of the same type must be zero. This results in the matrix having a block structure, with two blocks in which all elements are zero along the diagonal. Only the off-diagonal blocks contain non-zero entries. These off-diagonal blocks are given by the incidence matrix. For an undirected bipartite network with incidence matrix B, the adjacency matrix A is given by,

$$A = \begin{pmatrix} 0_{N_1 \times N_1} & B \\ B^T & 0_{N_2 \times N_2} \end{pmatrix}, \tag{4.3}$$

where $0_{N \times N}$ denotes an N-by-N matrix of zeros and the superscript T means matrix transpose. The adjacency matrix of a *directed* bipartite network also has a block structure with zero blocks along the diagonal. However, the off-diagonal blocks are no longer the respective transposed matrices. It is straightforward to generalize bipartite to *multipartite* networks. In such networks there are nodes of more than two types with links connecting nodes of different types. As before, the resulting adjacency matrix is of block form, with blocks of zeros along the diagonal and the respective incidence matrices as off-diagonal blocks.

4.2.3.4 *Multigraphs and hypergraphs*

Bipartite networks can be used to represent more general types of network. These include multigraphs, where nodes can be connected by more than one link. The Seven Bridges of Königsberg network, shown in Figure 4.1c, is a multigraph. The same multigraph is shown in Figure 4.3b with two pairs of nodes that have two links, $\{1,4\}$ and $\{3,4\}$. Each of these links corresponds to one of the grey nodes in the bipartite network shown in Figure 4.3a. Starting from a multigraph, it is always possible to identify each link with a new node of a different type. In this way, we obtain a bipartite network from a multigraph.

Another straightforward generalization of networks are *hypergraphs*. These are networks that are given by sets of N nodes and L links. Links can now connect more than two nodes. Links are not sets with two nodes as in simple networks, where $\mathcal{L} \subset \mathcal{N} \times \mathcal{N}$, but consist of an arbitrary number of nodes, $\mathcal{L} \subset \mathcal{N} \times \mathcal{N} \times \ldots \times \mathcal{N}$. Assume that a hypergraph contains a link $l = \{i_1, i_2, \ldots, i_K\}$. One way of representing this in a simple network would be to introduce a new node in the network with a link to each of the nodes contained in l. By applying this procedure to each link in the hypergraph, one again obtains a bipartite network. One set of nodes corresponds to the N original nodes of the hypergraph. The other set consists of L nodes, each representing one link of the hypergraph. Certain types of generalized network that encode multiple interactions between more than two nodes can be mapped onto simple networks and their adjacency matrix—at the cost, however, of (sometimes exponentially) inflated numbers of nodes. Hypergraphs and other generalizations of simple networks will be discussed in Section 4.9.

4.2.3.5 *Network duals and the Laplacian matrix*

Simple networks have a *dual representation*. To see what this means, consider the simple network shown in Figure 4.4a with a node set $\mathcal{N}$, a link set $\mathcal{L}$, and the adjacency matrix A. Let us transform this network into a bipartite network by introducing a node of a different type for every link (grey nodes). Replace the link between nodes 1 and 2 in Figure 4.4a with a new grey node, node 1, connected to both 1 and 2. By repeating this procedure for every link, we obtain a bipartite network with white nodes, representing the original nodes, and grey nodes, representing the original set of links; see Figure 4.4b. The bipartite network in Figure 4.4b has the incidence matrix, B,

$$B = \begin{pmatrix} 1 & 1 & 0 & 0 & 0 \\ 0 & 1 & 1 & 0 & 1 \\ 0 & 0 & 1 & 1 & 0 \\ 1 & 0 & 0 & 1 & 1 \end{pmatrix}. \tag{4.4}$$

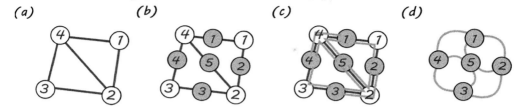

(a)   *(b)*   *(c)*   *(d)*

Figure 4.4 *Dual representation of simple networks. For each link in the simple network with white nodes shown in (a) we add a node of a different type (grey) to obtain the bipartite network shown in (b). The line graph for the original network is constructed in (c) by connecting those grey nodes in (b) that have links to the same white nodes. The network in (d) is the dual network of (a).*

Now we exchange the roles of white and grey nodes, which means that we exchange the roles of rows and columns in the incidence matrix, Equation (4.4). We now assume that the grey nodes are elements of the node set $\mathcal{N}'$ of a new simple network with adjacency matrix A'. Each white node now corresponds to an entry of the link set $\mathcal{L}'$. The nodes (links) in A' are therefore given by the links (nodes) in A. This network given by A' is called a *line graph*; see Figures 4.4c and d. A' is the dual network of A.

A simple undirected network with N nodes and L links can be represented by an $N \times L$ incidence matrix B, in which each entry states if a given link is 'incident' on a particular node. The relation between the incidence matrix B and the adjacency matrix A is,

$$A = BB^T - D, \tag{4.5}$$

where D is an $N \times N$ diagonal matrix and D_{ii} is the degree of node i. The non-diagonal entries of the matrix BB^T are identical to those of the adjacency matrix A. The diagonal entries are the node degrees. For *directed* networks, an oriented version of the incidence matrix, O, can be defined by setting $O_{ij} = -1$ whenever node i is the source of link j, or $O_{ij} = +1$, whenever node i is the target, otherwise $O_{ij} = 0$. For undirected networks, an oriented incidence matrix can be given by randomly assigning a direction to each link. The matrix $\Lambda = OO^T$ is called the *Laplacian matrix*.

In a similar way, the *line graph* of a network can be obtained as the $L \times L$ matrix $B^T B$. The line graph is a network where nodes represent the links $\mathcal{L}$ and connections indicate the shared nodes between pairs of links. Entries in the adjacency matrix of the line graph are simply given by $(B^T B)_{ij} - 2\delta_{ij}$, where δ_{ij} is the Kronecker delta; see Section 8.1.3. Note that the degree of each node in the line graph is always two, as each link connects exactly two nodes in a simple network. Starting from any bipartite network B with two sets of nodes, $\mathcal{N}_1$ and $\mathcal{N}_2$, it is therefore possible to obtain two different simple networks A_1 and A_2 that contain only nodes from $\mathcal{N}_1$ and $\mathcal{N}_2$, respectively. A link between two nodes in A_1 means that both nodes are adjacent to at least one node from $\mathcal{N}_2$ in B, and vice versa for A_2. In particular, we obtain the entries of A_1 as $(A_1)_{ij} = (BB^T)_{ij} - 2\delta_{ij}$ and for A_2 as $(A_2)_{ij} = (B^T B)_{ij} - 2\delta_{ij}$. This process of obtaining A_1 and A_2 from B is an example of a *bipartite network projection* or *one mode projection*. For example, for the bipartite network shown in Figure 4.4b, the one-mode projections A_1 and A_2 are shown in Figures 4.4a and d, respectively. Such projections can be defined in different ways, for instance, by connecting nodes in the projected network only if they share at least a minimum number of adjacent nodes of the other type.

4.2.3.6   *Weighted networks*

Links can carry additional properties such as a *weight*. Weights are real numbers that indicate how strongly two nodes are related to each other.

> *Weighted networks* are characterized by adjacency matrices that encode links as positive or negative numbers. The matrix W for a weighted network consists of entries $W_{ij} = 0$ if i and j are not connected by a link, and $W_{ij} \in \mathbb{R} \setminus \{0\}$ if there is a link between them. Weighted networks can again be undirected ($W = W^T$) or directed ($W \neq W^T$).

Real world networks are often weighted. To keep analyses simple, weighted networks are often mapped to unweighted networks by filtering links that are structurally not relevant. The filtered network is then treated as an unweighted network. Reasonable ways of implementing such mappings will be discussed in Section 4.7. If not stated otherwise, we assume that networks are unweighted.

4.3   Measures on networks

The success of network theory is partly due to the fact that many macroscopic phenomena can be understood on the basis of statistical properties of their underlying networks, in particular, when these are large ($N \gg 1$). For example, to predict whether a contagious disease will infect a significant fraction of the population of a country, it is sometimes unnecessary to have exact knowledge of the network of all physical contacts occurring over a certain time, that is, who will talk to whom, at what time, how often, and for how long. The size of an epidemic outbreak on networks with specific topologies can be statistically inferred from certain properties of the nodes and links, such as the ratio of vaccinated people and the distribution of disease-causing contacts in the population [285]. To understand the structure and function of a network, it is often enough to understand a series of statistical quantities of its nodes and links.

4.3.1   Degree of a node

The most basic property of a node is its *degree*. The degree is the number of links attached to a node. Different definitions of the degree are necessary for directed and undirected networks.

4.3.1.1   *Degree, in-degree, and out-degree*

For an undirected network with N nodes, the degree k_i of node i is given by,

$$k_i = \sum_{j=1}^{N} A_{ij} = \sum_{j=1}^{N} A_{ji}. \tag{4.6}$$

As A is symmetric, the degree can either be defined as the row sum or column sum of the adjacency matrix. Directed networks, on the other hand, have two types of degree. The in-degree k_i^{in} is the number of links that have node i as their target,[1]

$$k_i^{\text{in}} = \sum_{j} A_{ij}. \tag{4.7}$$

The out-degree k_i^{out} is the number of links with node i as their source,

$$k_i^{\text{out}} = \sum_{j} A_{ji}. \tag{4.8}$$

The degree k_i of a node in a directed network is the sum of its in- and out-degree, $k_i = k_i^{\text{in}} + k_i^{\text{out}}$.

For weighted, undirected networks, the degree is also defined as the number of links of a node (or as the number of incoming and outgoing links for weighted directed networks). However, for weighted networks, the node degrees can no longer be obtained through row and column sums of the matrices W. For weighted, undirected networks, the row or column sum gives the *strength*, s, of a node, $s_i = \sum_j W_{ij} = \sum_j W_{ji}$. For weighted, directed networks, the in-strength is defined as the sum of the weights of the incoming links, $s_i^{\text{in}} = \sum_j W_{ij}$, while the out-strength is the sum of the weights of all outgoing links, $s_i^{\text{out}} = \sum_j W_{ji}$.

4.3.1.2   *Degree distribution*

If all nodes in a network have the same degree, $k_i = c$ for all i, with some positive integer c, the network is called *regular*. If $c = N - 1$, the network is *complete* or fully connected, if $c = 0$ it is called *trivial* (the trivial network consists of isolated nodes). Real networks are almost never regular, complete, or isolated. Nodes often vary substantially in terms of their degree, sometimes even over several orders of magnitude [7]. The degrees in such networks are characterized by the probability of finding a node with exactly the degree k_i. This probability is the *degree distribution*, $P(k_i)$. The *average degree* $\langle k \rangle$ is the sample average of the degrees, $\langle k \rangle = \frac{1}{N} \sum_i k_i$. In practice, network analysis often starts by looking at the degree distribution of a network. When plotted as a function of the degree, it immediately signals if a network is likely to be an ER random network, see Section 4.4.2.2, or if it has more structure.

---

[1] In the following, we drop the summation details whenever the range of the sum is obvious.

4.3.1.3   *Nearest-neighbour degrees and assortativity*

The nearest-neighbour degree of a node i is the average degree of all of its neighbouring nodes j, $k_i^{nn} = \frac{1}{k_i} \sum_{j \text{ is neighbour of } i} k_j$. The *nearest-neighbour degree distribution* $P^{nn}(k_i)$ is the probability of finding a node with degree k_i adjacent to a randomly picked node in the network. $P^{nn}(k_i)$ can look very different, depending on whether it is evaluated over nodes with low or high degrees. This property quantifies the *assortativity*, r, of the network. Assortativity is a measure for whether nodes with a low (high) degree tend to be connected to other nodes with a low (high), $r > 0$, or high (low), $r < 0$, degree. It is typically measured by the Pearson sample correlation coefficient between the degree of nodes and their neighbours' degrees. If the vector of degrees is denoted by $k = (k_1, k_2, \cdots, k_N)$, and the vector of nearest-neighbour degrees by, $k^{nn} = (k_1^{nn}, k_2^{nn}, \cdots, k_N^{nn})$, the assortativity is,

$$r = \rho(k, k^{nn}), \tag{4.9}$$

where ρ is the Pearson correlation coefficient; see Section 8.1.10. Another way of defining assortativity is to consider two L-dimensional vectors v and w. Each entry in v and w corresponds to one link $l = \{i, j\} \in \mathcal{L}$. The values of the components are given by the degrees of the nodes connected by the link l, $v_l = k_i$ and $w_l = k_j$. Assortativity can then alternatively be defined as,

$$r = \rho(v, w). \tag{4.10}$$

Networks with $r < 0$ are called *disassortative*, networks with $r > 0$ are *assortative*. Note that in the first definition for r every node contributes equally to r, whereas in the second definition every link contributes with the same weight. The first definition is more sensitive to contributions from the (dis)assortative behaviour of low-degree nodes than the second.

4.3.1.4   *Connectivity and connectancy*

The *connectivity* κ of a network is the number of links per node, $\kappa = L/N$. Connectivity is *not* the average degree of the network. Each link contributes only once to the number of links, L, but twice to the degrees, namely, to both nodes that it connects, $\sum_i k_i = 2L$ and $\frac{2L}{N} = \langle k \rangle$. This simple fact, sometimes referred to as 'handshake lemma', introduces yet another omnipresent source of errors for factors of two.

The *connectancy* or *network density* η is the ratio of the number of actual to possible links in the network, $\eta = L/\binom{N}{2}$, where $\binom{N}{2} = N(N-1)/2$ is the binomial factor that gives the number of possibilities of choosing two out of N total nodes. Some authors use an additional factor of $1/2$ in the definition of the connectancy.

4.3.2   Walking on networks

4.3.2.1   *Walks and paths*

A *walk* on a network is an ordered set of nodes, $(a_1, a_2, \ldots, a_n)$ with the property that each ordered pair (a_i, a_{i+1}) in the walk corresponds to a link in the network. The length of a walk is the number of links that it passes. *Closed* walks have identical first and last nodes, $a_1 = a_n$, otherwise the walk is called *open*. *Simple* walks visit no node in the network more than once; they are also called *paths*.

4.3.2.2 Circuits and cycles

A walk that uses no link more than once is called a *trail*. A closed trail is called a *circuit* or a *tour*. A closed walk $(a_1, a_2, \ldots, a_{n-1}, a_1)$—where $(a_1, a_2, \ldots, a_{n-1})$ is a path—is a *cycle*. Further, a *Hamiltonian path* uses each node in the network exactly once. A network that contains a Hamiltonian path is called *traceable*. Trails that use each link of a network exactly once are called *Eulerian*. The existence of such trails makes a network *traversable*. The Seven Bridges of Königsberg problem can therefore be rephrased: is the network given by the bridges in Königsberg traversable? Does it contain an Eulerian trail?

4.3.3 Connectedness and components

A network is called *connected* if there is a path between all pairs of nodes i and j. A *subnetwork* refers to a network with a set of nodes $\mathcal{N}'$ that is a subset of the original node set, $\mathcal{N}' \subset \mathcal{N}$. An *induced* subnetwork is a subnetwork containing all the links from the original network that connect nodes that are also in $\mathcal{N}'$. The adjacency matrix of an induced subnetwork can be obtained from the original simply by deleting all rows and columns that correspond to nodes not belonging to $\mathcal{N}'$. A subnetwork is called *spanning* if $\mathcal{N}' = \mathcal{N}$. Induced subnetworks, where each node is adjacent to every other node, are called *cliques*. *Maximal subnetworks* have the property that if $i \in \mathcal{N}'$ and the network has a link between i and j, then it follows that $j \in \mathcal{N}'$.

4.3.3.1 Components of networks

Components are maximal induced connected subnetworks. There is always a path from one node of the component to each of the other nodes in the component. However, if a network has more than one component, it is not possible to walk from one component to the other. *Bridges* are links that, if removed, increase the number of components in the network. Similarly, *cut-nodes* are defined as nodes that, when removed, also increase the number of components.

4.3.3.2 Trees and forests

Trees are connected networks without cycles. Each link of a tree is a bridge. A *forest* is a network where each component is a tree. The nodes in trees are a priori not ordered. To use trees to represent hierarchies, a particular node has to be specified as a *root*. The hierarchy (or partial ordering) can then be defined as the length of the walk needed to reach a given node from the root. A *leaf* is a node in a tree with degree one.

4.3.3.3 Bow tie structure of directed networks

The definitions of paths, walks, cycles, and so on can be applied to directed networks without any modifications. However, the direction of the links has to be respected. While undirected connected networks always consist of at least one component and

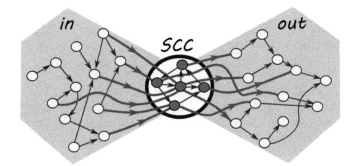

Figure 4.5 *Component structure of directed networks. All nodes in the in-component ('in') have a path to nodes in the strongly connected component (SCC). Nodes in the SCC have a path to all other nodes in the SCC and to nodes in the out-component ('out'), but not to nodes in the in-component. Starting from the out-component, it is not possible to reach nodes in the SCC.*

each node is an element of a component, for directed networks the component structure is slightly more involved. A directed network is *strongly connected* if there is a path in each direction between each pair of nodes. The maximal induced subnetworks that are strongly connected are called *strongly connected components* (SCC). Further, the set of all nodes outside SCC that have a path to at least one node from the SCC are called the *in-component*. The set of all nodes outside the SCC that can be reached with a path that starts in the SCC is the *out-component*. The resulting component structure of directed networks is sometimes called a 'bow tie'. The in-component feeds into a central 'knob' that consists of the SCC, from which the nodes in the out-component can be reached; see Figure 4.5.

4.3.4   From distances on networks to centrality

Many network properties can be understood on the basis of the length of the shortest path between a given pair of nodes. In particular, it is possible to use this length to define distances on networks, which in turn can be used to formulate a specific family of *centrality measures*. Centrality is a property of nodes that typically aims at measuring some aspect of a node's importance. One can distinguish three types of centrality measure. The simplest type is the degree itself, which sociologists sometimes refer to as 'degree centrality' [295]. This is the appropriate measure for applications where the role of a node is completely determined by the number of its connections. The second family of centrality measures is based on the idea that important nodes are those that 'reduce' the distance between other nodes in the network. This family of centrality measures includes closeness centrality [37] and betweenness centrality [134]. Finally, the third family of centrality measures can be directly related to dynamic processes on networks, in particular, to random walks on networks. These measures will be considered in detail in Section 4.8.

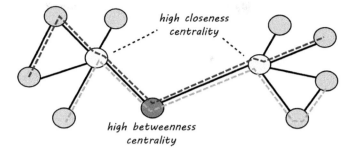

Figure 4.6 *Comparison of closeness and betweenness centrality. Closeness centrality of a node measures its average inverse distance to all other nodes, whereas betweenness centrality measures for how many node pairs their distance would increase under the hypothetical removal of the node. The white nodes have high closeness centrality, as many other nodes (grey circles) are close or adjacent to them. The dark-grey node has high betweenness centrality, as many shortest paths pass through it (several examples of shortest paths are shown as dashed lines along the links of the network).*

4.3.4.1 Geodesic paths, diameter, and characteristic distances

A *geodesic path* is the shortest path that connects two nodes i and j in a network. The *distance* $l(i,j)$ between the two nodes is given by the length of their geodesic path. If no such path exists, their distance is infinite, $l(i,j) = \infty$. The *diameter* of a network is the largest distance between any pair of nodes in the network. The *characteristic distance*, $\langle l(i,j) \rangle_{(i,j)}$, is the average distance between all pairs of nodes. It follows that the characteristic distance can be defined only for networks that consist of a single connected component. Otherwise, the distance has to be defined for each component separately.

4.3.4.2 Closeness and betweenness centrality

The closeness centrality of a node measures its 'accessibility' in the network. Different versions of this centrality exist. In its simplest formulation, the closeness of node i is the inverse average distance to each of the other nodes, $C_i^c = (N-1)/\sum_{j \neq i} l(j,i)$ [37]. Alternatively, in order to reduce the potential impact of outliers with an overly proportionally large distance, one can introduce a decay parameter $0 < \delta < 1$ and define $C_i^\delta = \sum_{j \neq i} \delta^{l(j,i)}$ [204].

Betweenness centrality measures how important a node is in terms of connecting to other nodes [134]. Let $n_i(j,k)$ be the number of geodesic paths from j to k that contain node i. The number of all shortest paths from j to k is $n(k,j)$. The betweenness centrality C_i^b of node i is then given by,

$$C_i^b = \binom{N-1}{2}^{-1} \sum_{k \neq i} \sum_{j \neq i} \frac{n_i(k,j)}{n(k,j)}. \tag{4.11}$$

In other words, C_i^b measures for how many pairs of nodes their respective distance would increase if node i were removed. Figure 4.6 illustrates the difference between these two centrality measures. A high closeness centrality is associated with nodes that are likely to distribute information 'fast' through the network. A high betweenness centrality characterizes nodes that are important for transmitting information between different regions in a network that are otherwise not well-connected. Closeness detects good 'spreaders', whereas betweenness is sensitive to 'bottlenecks' and chokepoints on the network.

4.3.5 Clustering coefficient

There are two different notions of what 'clustering' of nodes means in a network. The first states that a set of nodes clusters if they are more densely connected to each other than to nodes that are outside the set. The second definition of clustering is the probability that any two neighbours of a given node are also adjacent to each other. We refer to clusters of the first type—sets of densely interconnected nodes—as *communities* and reserve the term *clustering* exclusively for the second notion. Clustering refers to a 'microscopic' node property. It can be aggregated to a 'macroscopic' observable by averaging the probabilities over all nodes. Communities, in contrast, characterize the structure of a network on a mesoscopic level. Communities will be the focus of Section 4.6.

4.3.5.1 *From cycles and paths to clustering*

> The *clustering coefficient* of a node is the probability that any two neighbours of a node are also neighbours of each other [195].

This probability can be computed from the adjacency matrix. Note that each unique pair of neighbours (j, k) of a given node i constitutes a path of length two from j to k. If a link exists between j and k, it is a closed path of length three (i, j, k), or a closed *triad*. The problem of determining clustering coefficients can be formulated as a task of counting paths of length two and cycles of length three in the adjacency matrix. The non-zero entries in the adjacency matrix, $A_{ij} = 1$, indicate that there is a walk of length one between nodes i and j. If i and j are connected by a walk of length two, this means that there is a third node k for which $A_{ik} = 1$ and $A_{kj} = 1$, that is $A_{ik}A_{kj} = 1$. Formally, the term $A_{ik}A_{kj}$ contributes to the (i, j) entry of the second power of A; that is, to $(A^2)_{ij}$. This means that the number of walks of length two that connect nodes i and j corresponds to $(A^2)_{ij}$. This is also true in general for longer paths.

> The matrix elements $(A^n)_{ij}$ encode the number of walks of length n that connect i and j. $A^n = AA \cdots A$ means the repeated matrix product (n times) of the adjacency matrix.

The matrix element $(A^n)_{ij}$ counts the number of *walks* of length n between i and j, which is different from the number of *paths* of length n between i and j. In a path, each node is used only once. The different walks that contribute to $(A^n)_{ij}$ can use each node an arbitrary number of times. For instance, a cycle of length s counts as a closed walk of length ms for each $m = 1, 2, \ldots$. For $m > 1$ the closed walks are not cycles as they use nodes more than once.

4.3.5.2 Individual clustering coefficient

The individual clustering coefficient C_i of node i is the ratio of the number of actual versus the number of possible closed triads (cycles of length three) in i's immediate neighbourhood. The number of actual cycles is given by $(A^3)_{ii}$. If i has k_i neighbours, there are $k_i(k_i - 1)$ unique ordered pairs (j, k) of neighbours that could be connected to form the closed triads (i, j, k) and (i, k, j). The clustering coefficient C_i is then,

$$C_i = \frac{(A^3)_{ii}}{k_i(k_i - 1)}. \tag{4.12}$$

Note that this formula is applicable only for nodes with more than one neighbour, $k_i > 1$. Otherwise, C_i is undefined. Some authors use the convention that $C_i = 0$ if $k_i = 0$, and $C_i = 1$ if $k_i = 1$.

4.3.5.3 Overall and average clustering

There are two different types of clustering coefficient for the entire network, the so-called global clustering coefficients. The first is the *average clustering coefficient* of the network, $\langle C_i \rangle$, which is simply the sample average of all individual clustering coefficients,

$$\langle C_i \rangle = \frac{1}{N} \sum_i C_i. \tag{4.13}$$

The second is the *overall clustering coefficient* C of the network, which is the ratio of its three cycles to the number of paths of length two. For an undirected network with no loops $(A_{ii} = 0)$, this ratio can be written as,

$$C = \frac{\sum_i (A^3)_{ii}}{\sum_{j, k \neq j} (A^2)_{jk}}. \tag{4.14}$$

Average and overall clustering are, in general, not the same. Their values only coincide if the probability of a node i's neighbours being connected does not depend on its degree k_i. The average clustering, $\langle C_i \rangle$, puts more emphasis on contributions from low-degree nodes than the overall clustering. If there is a tendency that individual clustering C_i increases (decreases) with node degree, the average clustering coefficient $\langle C_i \rangle$ will be lower (higher) than the overall clustering [342].

4.3.5.4 Clustering in directed networks

Consider three nodes, a so-called triad, in an undirected network. There are exactly four different configurations of links, sometimes also called motifs, that can occur. These motifs for triads are: (i) all nodes are isolated, (ii) there is only one link, (iii) there are two links, or (iv) all nodes are connected in a cycle of length three. Discounting

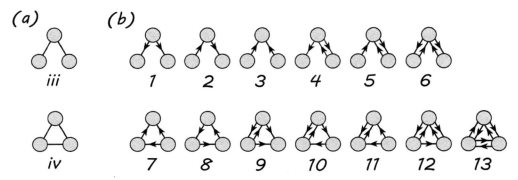

Figure 4.7 *Connected triadic motifs. (a) For an undirected network, there are two different connected triadic motifs. The clustering coefficient is defined as the ratio of the number of occurrences of these two motifs (number of closed/number or open triads). (b) For directed networks, there are six motifs where two nodes share no links and seven where all nodes are connected in at least one direction. There is no unique definition of a clustering coefficient for directed networks.*

those configurations where at least one node is isolated, (i) and (ii), only two alternative motifs remain, (iii) and (iv); see Figure 4.7a. The clustering coefficient is then simply the ratio of the number of occurrences of these two motifs. There is no straightforward generalization of this procedure for the definition of clustering in directed networks. For three nodes there are sixteen different motifs, thirteen of which have no isolated nodes; see Figure 4.7b. It is now up to the creative reader to find combinations of motifs to form ratios that are meaningful measures for clustering. The simplest way to do this is to transform the directed network to an undirected network, either by 'forgetting' about the link direction or by keeping only those links that point in both directions [123].

Note that how 'close' a directed network is to an undirected analogue can be measured by the *reciprocity*, R [145]. Reciprocity is the probability that for each link pointing from i to j, there is a link pointing in the opposite direction from j to i. As the number of bi-directional links in the network can be obtained from the number of cycles of length two, it follows that $R = \sum_i (A^2)_{ii}/L$. For $R = 1$, the adjacency matrix of a directed network is symmetric and effectively describes an undirected network.

4.4 Random networks

Random networks play a crucial role in the science of complex systems, usually in the form of null models. Consider the following example. Imagine you are interested in understanding the process by which you acquire new friends. Your hypothesis is that most of the time you meet someone new because he or she is already a friend of one of your current friends. As a trained complexity scientist, you immediately recognize that the process of becoming friends with a friend of your friend increases the clustering in the network. To confirm this hypothesis, you crawl the online social network of your choice and compute its average clustering coefficient as $\langle C_i \rangle = 0.1$. At this point, you have not understood anything about the process by which new friends are made. In principle, you could observe a value of $\langle C_i \rangle = 0.1$ for a network in which individuals choose their friends completely at random. So the research question takes on the following form: given that friendships form randomly between chosen pairs of individuals, how likely is it to observe a value of $\langle C_i \rangle = 0.1$ in such a network? To state this question in its full formal rigor, it needs to be formulated as a statistical hypothesis test. Your working hypothesis— new friendships are formed with friends of your current friends—would imply that the hypothesis that friendships form completely at random must be wrong. Only if you can reject this so-called null hypothesis does it make sense to consider alternative hypotheses. The next step is to compute an expectation value for the average clustering coefficient under the assumption that the null hypothesis is true. Assume that, as a result of this analysis, you find that your expectation value for $\langle C_i \rangle$ under the null hypothesis would be $\langle C_i^0 \rangle = 0.001 \pm 0.02$, with 0.02 being the standard deviation. Using the assumption that measurement errors are normally distributed, physicists describe the differences between expected and observed values in terms of 'sigma effects'; that is, by how many standard deviations the observed values differ from their expectations. In our case, the high clustering in the observed network would be almost a five-sigma effect, that is, there would be a 1 in 1,744,278 chance of observing $\langle C_i \rangle = 0.1$ if the null hypothesis were true. You can now conclude that, given your data, it is highly unlikely that friendships form randomly and that your model explains the observed clustering much better. What is meant exactly by 'highly unlikely' will be elaborated in Section 4.7, where we discuss statistical hypothesis tests in detail. Before doing this, we have to clarify how suitable null hypotheses can be formulated for network data, which is exactly what random networks allow us to do.

Nothing can be learned or understood from real world networks without comparing them to suitably chosen random networks that serve as null models.

4.4.1 Three sources of randomness

There are three different sources from which randomness in random networks may arise.

Probabilistic Each entry in the adjacency matrix A_{ij} is a random variable by itself.

Dynamic The adjacency matrix is subject to stochastic processes in which, over time, links and nodes are added and removed.

Static The entire adjacency matrix is drawn 'as a whole' from a probability distribution $P(A)$.

Without any further information about the nature of the network, the three types of randomness give rise to the same random networks. Given a set of realizations of a random network, it is impossible to determine if the underlying random processes are probabilistic, dynamic, or static in the above sense.

4.4.1.1   *Is a random network really a network?*

We should clarify at this point that, when we talk about random networks, we actually describe properties of a specific *ensemble* of networks. When we study properties of a real world network with a given adjacency matrix A, our null hypothesis often assumes that A has been randomly sampled from a particular ensemble of networks. This is what we mean by saying: A is a random network.[2]

4.4.2   Erdős–Rényi networks

The simplest class of random networks are Erdős–Rényi (ER) networks [117, 118] which, for reasons that will become clear immediately, are also called Poisson networks. Consider a simple undirected network A with N nodes and no loops. There exist $\binom{N}{2} = N(N-1)/2$ distinct pairs of nodes. Each pair of nodes is either connected by a link or not.

> Erdős–Rényi networks are completely specified by a single probability p, which is the probability that any possible pair of nodes are linked.

4.4.2.1   *Definition of the Erdős–Rényi ensemble*

More formally, the ER ensemble $\mathcal{E}$ is defined as $\mathcal{E} = (\mathcal{A}, \mathcal{P})$ where $\mathcal{A}$ is the set of all networks A with N nodes, $\mathcal{A} = \{A| \text{ network has } N \text{ nodes } \}$. The probability $P(A)$ of choosing a particular matrix $A \in \mathcal{A}$ is given by,

$$P(A) = p^L (1-p)^{\binom{N}{2}-L}, \tag{4.15}$$

where L is the number of links, $\sum_{ij} A_{ij} = 2L$. Note that there is an alternative way of defining ER networks. In this second definition, a fixed number of L links connect L randomly picked pairs of nodes from the $N(N-1)/2$ possible pairs. The corresponding ER ensemble, $\mathcal{E}' = (\mathcal{A}', \mathcal{P}')$ is then defined as follows. The set $\mathcal{A}'$ consists of all networks that have exactly N nodes and exactly L links. The probability measure of drawing a specific A, $P(A)$, is the same for any member of $\mathcal{A}'$,

$$P(A) = \binom{\binom{N}{2}}{L}^{-1}, \tag{4.16}$$

that is, the A are equidistributed on $\mathcal{A}'$. The difference between the two ER ensembles $\mathcal{E}$ and $\mathcal{E}'$ is as follows. The first ensemble is specified by the probability p of any pair of nodes being connected. Members of $\mathcal{A}$ might therefore vary in their total number of

---

[2] There is a deep connection to random matrix theory, which studies the properties of ensembles (sets) of matrices M distributed according to some probability distribution, $P(M)$ [271].

links L. In $\mathcal{E}'$, L is fixed and the links are randomly distributed amongst pairs of nodes. However, for $N \to \infty$ the two ensembles are identical. In the following, we define ER networks by the ensemble $\mathcal{E}$.

4.4.2.2 Degree distribution

The probability of observing a node i with degree k in an ER network is the outcome of a binomial experiment. For each of the $N - 1$ potential neighbours of a node, we decide whether to draw a link by flipping a biased coin. With probability p we draw a link, otherwise we do nothing. The degree distribution of the ER network is the probability of getting k links for a given node,

$$P(k) = \binom{N-1}{k} p^k (1-p)^{N-1-k}. \tag{4.17}$$

Obviously, the degrees in an ER network follow a binomial distribution. So why are they called Poisson networks? As shown in Section 2.5.1.4, for large N the binomial distribution approaches the Poisson distribution; see Equation (2.72). For all practical purposes, the degree distribution for sufficiently large ER networks is,

$$P(k) = \frac{(Np)^k e^{-Np}}{k!}. \tag{4.18}$$

4.4.2.3 Moments of the degree distribution

The average degree $\langle k \rangle$ is the expectation value of the node degrees, k. From the definition of ER networks as binomial experiments with $N - 1$ trials and success probability p, the expected number of successes, $\langle k \rangle$, is given by,

$$\langle k \rangle = (N-1)p. \tag{4.19}$$

To obtain the higher moments of the degree distribution, we use Equations (2.7) and (4.17),

$$\langle k^m \rangle = \sum_{k=0}^{N-1} \binom{N-1}{k} p^k (1-p)^{N-1-k} k^m \tag{4.20}$$

$$= \frac{d^m}{d\beta^m}\Big|_{\beta=0} \sum_{k=0}^{N-1} e^{\beta k} p^k \binom{N-1}{k} (1-p)^{N-1-k} \tag{4.21}$$

$$= \frac{d^m}{d\beta^m}\Big|_{\beta=0} (1-p+e^\beta p)^{N-1}, \tag{4.22}$$

where we used the properties of the polynomial expansion in the third line. For the second moment of the degree distribution, we get,

$$\langle k^2 \rangle = (N-1)(N-2)p^2 + (N-1)p \sim \langle k \rangle^2 + \langle k \rangle. \tag{4.23}$$

(a) *(b)*

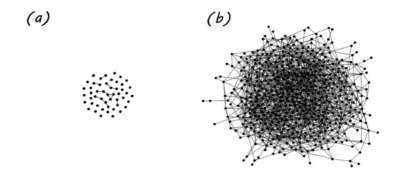

Figure 4.8 *Phase transitions in Erdős–Rényi networks. (a) ER network with N = 50 nodes and p = 0.01. There are several isolated nodes. (b) ER network with the same p but with N = 500. The network now has one component that contains all nodes.*

The variance of the degree distribution, $\sigma^2(k)$, is,

$$\sigma^2(k) = \langle k^2 \rangle - \langle k \rangle^2 = \langle k \rangle; \tag{4.24}$$

see also Section 2.5.1.4.

4.4.2.4 Clustering coefficient

The individual clustering coefficient of ER networks takes a particularly simple form. Recall that the individual clustering C_i of node i is defined as the probability that any pair of neighbours of i are also adjacent to each other. In an ER network, the probability of any pair being connected is exactly p. Therefore, all nodes in ER networks have a clustering coefficient of,

$$C_i = p. \tag{4.25}$$

4.4.3 Phase transitions in Erdős–Rényi networks

The dependence of the average degree $\langle k \rangle$ and clustering C_i on p is rather trivial. Other properties of ER networks show far more interesting behaviour. In particular, ER networks undergo a series of structural phase transitions. For example, consider in Figure 4.8 two realizations of ER networks with a different number of nodes but with the *same* value of p. The sparse network on the left contains several isolated nodes and a number of small components. The network on the right is a dense 'hairball' that consists of one single component containing all nodes of the network. Obviously, the size of the largest component depends on p, but also on the network size N.

4.4.3.1 The percolation transition of Erdős–Rényi networks

To see that ER networks have a phase transition, we need to define a suitable order parameter. On networks, phase transitions can be defined with order parameters that have a value of zero in one phase and a non-zero value in the other. The order parameter is then the conditional probability $P(X|N)$ that for a network of size N property X is true. For the example in Figure 4.8, an order parameter can be defined as the probability that the network is connected; in other words, the probability that there is a path between each pair of nodes. The formation of such a component as a function of p and N is identical to a percolation transition on networks; see Section 3.3.1.3 where percolation phenomena on regular lattices were discussed.

4.4.3.2   *Size-dependent threshold functions*

We now consider ER networks with a linking probability p that explicitly depends on the number of nodes, $p(N)$. The idea is to compare $p(N)$ with a known threshold function $t(N)$ and to make N increasingly large, $N \rightarrow \infty$; see for example [204]. The threshold function $t(N)$ has the following properties: whenever $p(N) < t(N)$, it becomes increasingly unlikely that the network has a certain property X as N increases, $\lim_{N \rightarrow \infty} P(X|N) = 0$. But whenever $p(N)$ is larger than the threshold function, $p(N) > t(N)$, the chances of the network having property X increase with N, $\lim_{N \rightarrow \infty} P(X|N) = 1$. Threshold functions can therefore be used to detect phase transitions in random networks in the limit of large N.

If for a given threshold function $t(N)$ it holds that,

$$\lim_{N \rightarrow \infty} \frac{p(N)}{t(N)} = \begin{cases} 0 \\ \infty \end{cases} \Rightarrow \lim_{N \rightarrow \infty} P(X|N) = \begin{cases} 0 \\ 1 \end{cases}, \qquad (4.26)$$

then the property X is said to have a *phase transition* at the threshold $t(N)$.

4.4.3.3   *Important thresholds for Erdős–Rényi networks*

Some threshold functions can easily be identified. For instance, the threshold for at least one link to exist is $t(N) = N^{-2}$. In their original work [118], Erdős and Rényi also showed the following thresholds,

$t(N) = N^{-2/3} \rightarrow$ at least one component containing three nodes exists,

$t(N) = N^{-1} \rightarrow$ circles exist,

$t(N) = \log(N)/N \rightarrow$ the network is connected.

4.4.3.4   *Giant components*

Giant components are components of a network that contain a non-vanishing fraction of nodes even for $N \rightarrow \infty$.[3] The threshold function for a giant component to exist can be obtained from the following considerations [55, 291]. Consider an ER network with $N - 1$ nodes and $p > 1/(N - 1)$. The network has a largest component C_M and every

---

[3] The existence of a giant component does not imply that the network is connected, which would require the giant component to span the entire network.

node belongs to this component with probability $q = C_M/(N-1)$. Add a new node, i, and draw a link to every already existing node with probability p. The new node is outside C_M if none of its neighbours are in C_M. If node i has degree k, the probability of i not being in C_M is $(1-q)^k$. As the degree k follows the ER degree distribution $P(k)$, the probability of a node belonging to C_M is $\sum_k (1-q)^k P(k)$. If this procedure converges for large N, we obtain the following fixed-point equation for the fraction of nodes outside C_M,

$$1 - q = \sum_k (1 - q)^k P(k). \tag{4.27}$$

The average degree of an ER network is $\langle k \rangle = (N-1)p$. Using the Poisson distribution for $P(k)$ we get,

$$1 - q = e^{-(N-1)p} \sum_k \frac{1}{k!} \left(p(1-q)(N-1) \right)^k = e^{-(N-1)p} e^{p(1-q)(N-1)}, \tag{4.28}$$

from which it follows that,

$$q = 1 - e^{-(N-1)pq}. \tag{4.29}$$

Equation (4.29) has a solution of $0 < q < 1$, only if $\langle k \rangle = p(N-1) > 1$ (besides the trivial solution, $q = 0$). Therefore, the threshold function for the existence of a giant component is $t(N) \sim N^{-1}$, which is also the threshold function for the existence of circles in the network. Note, that percolation transitions also exist for scale-free networks, where mean field approximations loose their applicability [87].

4.4.4   Eigenvalue spectra of random networks

The set of eigenvalues of a matrix that represents a random network is called the spectrum of the network. Although it is possible to understand several properties of these spectra analytically, such analyses can become rather complicated [271, 360]. In the following, we give an overview of some useful results for the spectral properties of certain types of random networks.

4.4.4.1   *Perron–Frobenius theorem*

A matrix A is called *positive* if all of its entries A_{ij} are positive real numbers, $A_{ij} > 0$. Similarly, we call A non-negative if all of its entries are non-negative, $A_{ij} \geq 0$.

Let A be a positive matrix, then the *Perron–Frobenius theorem* [136, 307] states that

- the largest eigenvalue λ^{PF} is a positive real number, called the Perron–Frobenius eigenvalue, which defines the *spectral radius* of A.

- all components of the eigenvector v^{PF} associated with the Perron–Frobenius eigenvalue are positive real numbers. All other eigenvectors have at least one negative or imaginary component.

Adjacency matrices of simple networks are only positive if the network is fully connected. In all other cases the adjacency matrices are non-negative. Nevertheless, the Perron–Frobenius theorem also applies to non-negative matrices, given that they are *irreducible*. Irreducible matrices are characterized by the property that there is a path between each pair of nodes in their associated network. This means that for every pair of nodes i and j there exists a natural number r, such that $(A^r)_{ij} \geq 0$. The Perron–Frobenius theorem applies to all undirected networks that are connected, and to all directed networks that are strongly connected.

4.4.4.2 Wigner's semicircle law

The adjacency matrices of ER networks are symmetric random matrices. Entries in this matrix can be seen as i.i.d. random variables that take the value of one with probability p and zero with $1 - p$, except on the diagonal where all entries are zero. The spectrum of eigenvalues of such matrices is known to follow the so-called semicircle law, that was first observed by Eugene Wigner, who tried to understand the energy spectra of heavy nuclei [408]. Wigner's semicircle law applies to real symmetric matrices S of order N with the following properties: the elements of S are i.i.d. random variables from a distribution with vanishing first moment, second moment m^2, and non-diverging higher moments. Let μ be the expected number of eigenvalues of S that are found in the interval $(\alpha N^{1/2}, \beta N^{1/2})$ for real $\alpha < \beta$. For large N, the expected number of eigenvalues in a given interval is,

$$\lim_{N \to \infty} \frac{\mu}{N} = \frac{1}{2\pi m^2} \int_\alpha^\beta \sqrt{4m^2 - x^2} dx. \tag{4.30}$$

This means that the histogram of eigenvalues of a real symmetric matrix has a semicircle as its limiting distribution.

4.4.4.3 Girko's circular law

Girko's circular law provides us with the limiting distribution of eigenvalues of non-symmetric random matrices in the limit of large N [151]. Consider random matrices with i.i.d. entries that follow a normal distribution with zero mean and a variance of m^2. The circular law states that the (complex) eigenvalues λ of a real random matrix are uniformly distributed on a disk in the complex plane with radius $m\sqrt{N}$. Note that Wigner's semicircle law applies to undirected networks, whereas the circular law applies to directed networks.

4.4.4.4 Wigner's and Girko's law for random networks

In general, ER networks do not exactly fulfil these requirements for random matrices for both the directed and undirected case. The expected value of the entries in A for ER networks is given by $\langle A_{ij} \rangle = p > 0$, whenever $i \neq j$ and $\langle A_{ii} \rangle = 0$ along the diagonal.

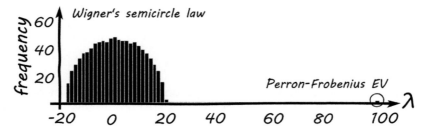

Figure 4.9 *Spectrum of eigenvalues for an undirected ER network. Here, N = 1000 nodes and p = 0.1. The Perron–Frobenius eigenvalue is found close to p(N − 1) (highlighted by the circle at λ ∼ 100). The remaining eigenvalues follow Wigner's semicircle distribution.*

Simulations nevertheless show that the spectra of ER networks do indeed have a Perron–Frobenius eigenvalue that is close to the average degree of the network and that the remaining eigenvalues form a bulk centred around −p. The frequency distribution of eigenvalues in this bulk follows Wigner's semicircle distribution for undirected networks; see Figure 4.9. For directed networks, the central bulk of the spectrum follows Girko's circular law.

4.5   Beyond Erdős–Rényi—complex networks

ER networks are the maximum-ignorance null model for random networks, for which only the numbers of nodes and links are known. In many cases, the use of ER networks as null models would be overly simplistic.

> Links in real world networks do not usually form purely by chance [204, 291]. The resulting network topology often contains non-trivial information about the dynamics, formation, and interactions of the nodes and links [7]. Such random networks are also called *complex networks*.

Complex networks vary substantially in their properties: they may show strong clustering, fat-tailed degree distributions, or be assortative or disassortative. Any single network measure alone is insufficient to capture this structural variability. However, a sufficiently long list of network measures might indeed produce an informative 'fingerprint' of the actual network structure. Some of these measures we have already encountered, such as the degree distribution, clustering, closeness centrality, and so on. Other properties of complex networks are more efficiently addressed by directly modelling the underlying network formation processes. In other words, one starts by formulating a specific hypothesis about what mechanisms drive the dynamics of nodes and links. Typically, these dynamical rules depend on specific properties of nodes and links. The aim is then to determine how these mechanisms shape the topology of the resulting network. In some cases, it is possible to understand certain topological properties analytically by using simple differential equations (e.g. network formation models such as preferential attachment [28]) or by using game-theoretic approaches (such as in strategic network formation [203]). However, analytical approaches often fail due to mathematical and combinatorial challenges. The tool of choice for analysing networks and the rules by which they evolve is the computer. We now discuss a series of complex networks that extend the concept of ER networks. We describe their properties and how they are generated.

4.5.1   The configuration model

For complex networks where the degree distribution, $P(k)$, is non-Poissonian, there are two flexible models for randomly generating networks with arbitrary degree distributions. The *configuration model* [56] assumes that the degree is a unique property of each node and that links are randomly formed between nodes to satisfy this property. The *superposition model* [2] on the other hand assumes that the observed network arises from a superposition of ER networks with different linking probabilities.

> The configuration model provides a class of random networks that are maximally random with respect to a specified degree sequence.

Suppose that you are given a network with N nodes that have a degree sequence $\{k_1, k_2, \ldots, k_N\}$, that is, for every node the degree is known. Alternatively, if only the degree distribution $P(k)$ is known but not the exact sequence, the number of nodes with a given degree can be estimated as $n(k) \sim P(k)N$, such that $N = \sum_k n(k)$. The degree sequence can then be estimated by assigning the degree k to exactly $n(k)$ nodes. The configuration model is specified as follows. Make a list of length $2L = \sum_i k_i$, allocating each index i exactly k_i times. This list will be of the form,

$$\underbrace{1, 1, \cdots, 1}_{k_1} \mid \underbrace{2, 2, \cdots 2}_{k_2} \mid \cdots\cdots \mid \underbrace{N, N \cdots N}_{k_N}..$$

Now randomly pick a pair of elements from the list and draw a link between the nodes with the selected indices. Remove those elements from the list and repeat the process until the list is empty. An example of the configuration model of a network with four nodes is shown in Figure 4.10. It is apparent that the result of the configuration model is not necessarily a simple network. In particular, there might be self-loops (when the same index is picked twice in one draw) and multilinks (when the same pair of nodes is picked twice). The procedure requires the sum of all degrees to be even. If N is large ($N \gg 1$) and if the network is sparse enough ($L \ll N(N-1)/2$), these problems can be safely neglected. Otherwise, it is necessary to post-process the results of the configuration model by using specific rewiring schemes.

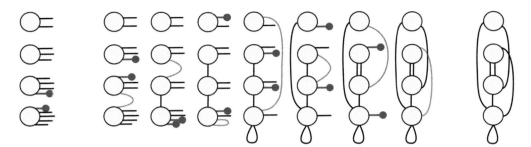

Figure 4.10 *The configuration model for a network with four nodes and a given degree sequence* {2, 3, 4, 5}. *Initially, all nodes receive a number of stubs (ends of links) that is equal to their degree. Each column corresponds to one step of the algorithm that connects two randomly chosen stubs (highlighted by dots). The resulting network, especially if it is small, will in general contain self-loops and multilinks.*

4.5.1.1 Rewiring in the configuration model

The multigraph that results from the configuration model can be rewired to become a simple network in a way that preserves the degree distribution. To do this, randomly pick a pair of multilinks or loops. Cut both links and glue the parts from different multilinks together. As a result, the number of multiple links between both pairs of nodes decreases by one. This process is repeated until there are no more pairs of multilinks.

4.5.1.2 Rewiring to erase topological information

Random rewiring of a network, while keeping the degree distribution $P(k)$ invariant, can be used to erase all topological information in a network that is not contained in the degrees. This procedure can be used to test for a structure that is not captured in $P(k)$ alone, such as clustering coefficients, centralities, or assortativity. A way of achieving this rewiring is the passing on of links. The method is particularly suitable for large networks. Note that if a node i passes on a link to node j, the degree k_i of i will decrease by one, while the degree of node j, k_j, will increase by one. If i and j are chosen such that $k_i = k_j + 1$, the passing on of a link between them will leave $P(k)$ invariant. By repeating this procedure a sufficient number of times, the network will become maximally random with respect to its given degree distribution.

4.5.1.3 The friendship paradox

The nearest-neighbour degree distribution, $P^{nn}(k_i)$, is the probability that a randomly chosen node is adjacent to a node i with degree k_i. In the configuration model, this distribution has some counterintuitive properties, known as the *friendship paradox*. It states that your friends always have more friends than you have, meaning that the expected degree of the neighbours of a node is never less than the expected degree of the node itself. This can be shown as follows. Pick a node i in the network and follow a randomly chosen link of it. What is the probability that you will end up at a node j with degree k_j? The answer is not simply $P(k_j)$ because, for instance, you have a zero chance of finding a node with degree zero in this way. To answer the question, we first observe that the probability of picking a stub from node j out of the collection of all $2L$ stubs in the configuration model is $k_j/2L$. Second, we know that there are $n(k_j) = NP(k_j)$ nodes with degree k_j in the network; hence the probability of finding one of these $n(k_j)$ nodes as a neighbour of i, which is the nearest-neighbour degree distribution $P^{nn}(k_j)$,

$$P^{nn}(k_j) = \frac{k_j}{2L} NP(k_j) = \frac{P(k_j)k_j}{\langle k \rangle}. \tag{4.31}$$

The average degree *amongst the nearest-neighbours*, or equivalently, the average nearest-neighbour degree $\langle P^{nn}(k_j) \rangle_{k_j}$ is,

$$\langle P^{nn}(k_j) \rangle_{k_j} = \sum_{k_j = 0,1,\dots} k_j \frac{p(k_j)k_j}{\langle k \rangle} = \frac{\langle k^2 \rangle}{\langle k \rangle}, \tag{4.32}$$

where we have used the definition of the expectation value. To show the friendship paradox, we compute the variance of the degree distribution, $\sigma^2(k)$,

$$\frac{\sigma^2(k)}{\langle k \rangle} = \frac{1}{\langle k \rangle} \left(\langle k^2 \rangle - \langle k \rangle^2 \right) = \langle P^{nn}(k) \rangle - \langle k \rangle > 0. \tag{4.33}$$

The average degree of the neighbours of a node is indeed larger than the expected degree of the node itself. In other words, your friends have more friends than you have.[4] All of this holds for networks with arbitrary degree distributions.

4.5.2 Network superposition model

The superposition model offers a different route to generalize ER networks to random networks with arbitrary degree distributions [2]. Recall that ER networks can be produced by a sequence of Bernoulli trials in which we flip a biased coin (probability p) for each link. In the superposition model, we have coins with different biases and randomly pick one of them for each link. The biases are random variables from a distribution $Q(p)$. The idea is similar to the concept of superstatistics in statistical mechanics [39].

4.5.2.1 Superstatistics of Erdős–Rényi networks

To generate a network in the superposition model we proceed as follows.

1. With probability $Q(p)$, pick a linking probability p.

---

[4] If the friendship paradox can teach us anything, it is that seemingly strange conclusions can arise from the uninformed use of statistics. There is no a priori reason why the arithmetic mean of the degree distribution is any more or less meaningful than its geometric mean. If we define $\langle X \rangle$ to denote the geometric mean, $(\Pi_i X_i)^{1/N}$, the paradox vanishes in this formulation, as for the geometric mean it holds that $\langle X^2 \rangle = \langle X \rangle^2$. The origin of the friendship paradox is therefore simply the Cauchy–Schwarz inequality between the arithmetic and the geometric mean, meaning that $(x+y)/2 \geq \sqrt{xy}$.

2. Randomly pick a pair of nodes and with p, link them.
3. Repeat the process until all L links have been distributed.

The degree distribution function of the resulting network is a superposition of ER networks. Let $P(k|Np)$ be the Poisson degree distribution[5] from Equation (4.18) associated with an ER network of average degree $\langle k \rangle = (N-1)p$. The degree distribution of the superposition model of ER networks is then given by,

$$P_{\text{super}}(k) = \int_0^1 dp\, P(k|Np)Q(p),\tag{4.34}$$

and the number of links is given by $L = N\sum_k P_{\text{super}}(k)k/2$. If linking probabilities are drawn from a power law distribution, so will be the resulting degree distribution $P_{\text{super}}(k)$ [2]. Irrespective of whether we consider the configuration or the superposition model, all properties except for the degree distribution will be trivial in the sense that they are maximally random. Properties like node centrality, clustering coefficient, or nearest-neighbour degree will depend only on the degree of the considered node itself. For networks that exhibit non-trivial features, such as high clustering or nodes acting as bridges, we require more sophisticated null models.

4.5.3   Small worlds

For a long time, social network analysis has been a key tool in sociology [398]. The use of social networks can be traced back for more than a century to the work of sociologists like Simmel [339] and Durkheim [109]. In its modern form, social network analysis has been strongly influenced by Mark Granovetter's publication 'The strength of weak ties' [156]. There, the main result is that the strength of an interpersonal tie (e.g. how often two people meet each other) varies directly with the overlap of the two individuals' friends, that is, with their number of common neighbours in their local network. Granovetter's result has far-reaching consequences for the structure of social networks at both the microscopic and macroscopic level. Most of the time, we interact with people through strong ties that constitute relations within a closely connected group of individuals (family, colleagues, school classes, etc.). Weak ties, however, build bridges between those groups (i.e. you serve as a bridge between your family and the group of your former school mates). How exactly do bridges shape the overall structure of societies or networks? If, most of the time, we interact only within the same group of mutually interacting people, how can information flow *between* different groups?

4.5.3.1   *Six degrees of separation*

Stanley Milgram studied the macro structure of social networks in the following experiment [375]. People from Kansas and Nebraska were given letters to be delivered to unknown target persons. They were told the name, profession, and state of residence

[5] Here Np stands for the rate λ.

of their target person. The instructions were to send the letter to the target person, if he or she were known. Otherwise, they should send the letter to an acquaintance, who might know the target person (e.g. through living in the same state, by sharing the same profession, etc.). The next person in the chain was asked to do the same until the letter eventually reached its target. Of all the letters initially distributed, between 20 and 30% actually reached their targets. The minimum length of the paths that these letters took was five, the maximum length was twelve. The phrase 'six degrees of separation' was subsequently coined by researchers who took up Milgram's idea and conducted similar experiments with ever-increasing scope. In 2001 Milgram's experiment was repeated with e-mail chains between 60,000 senders and eighteen targets [105]. In 2007, 30 billion conversations amongst 240 million people were analysed in a global instant-messaging network [251]. Both studies found that the average path length in these social networks was six. The six degrees of separation has since become a pop culture phenomenon expressing that everyone is connected to everyone else by a surprisingly small number of steps [401]. In the light of Granovetter's results on the strength of weak ties, how surprising are six degrees of separation really?

4.5.3.2   *High clustering and low characteristic distance*

Experiments of the Milgram type [251, 375, 399] estimate the characteristic distance of acquaintance networks.

Social networks show both high clustering and a comparably low characteristic distance.

As everyone has several Granovetterian weak ties, social networks exhibit a snowball effect in the number of acquaintances that can be reached in a given number of steps. If you have 100 friends and each of them has 100 friends, then there are about 10,000 friends of friends, without correcting for the number of overlapping friends.

4.5.3.3   *Watts–Strogatz model*

Watts and Strogatz formulated a simple network model that combines high clustering with low characteristic distance [402]. Their idea was to start from a regular network with high clustering and large characteristic distance. By introducing an increasing number of random links in the network, distances rapidly decrease, while a high level of clustering is maintained. The Watts–Strogatz model is formulated as follows. Start with a regular network with a high clustering coefficient, for example, by putting nodes on a ring and connecting them to their nearest m neighbour nodes.

1. Randomly pick a link of the original network.
2. Cut the link with probability ϵ. Attach the open ends to any other nodes.
3. Continue until all links of the original network have been picked once.

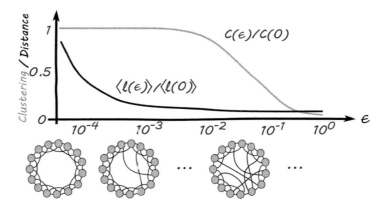

Figure 4.11 *Illustration of the Watts–Strogatz model for small world networks. Starting from a regular network with high clustering (grey line) and large characteristic distances (black line), links are randomly rewired with probability ϵ. For $\epsilon \to 1$, one obtains an ER network with low clustering and short distances. For intermediate values of ϵ, there is a range where high clustering and low distances are present at the same time—the small world regime.*

As ϵ increases from zero to one, the network changes from a regular to an ER random network at $\epsilon = 1$. To measure the small world property, we consider the characteristic distance $\langle l(\epsilon) \rangle$ and the average clustering coefficient $C(\epsilon)$ as a function of the rewiring probability ϵ. As seen in Figure 4.11, the interesting feature of this model is that the characteristic distance $\langle l(\epsilon) \rangle$ drops quickly even for small values of ϵ, whereas the clustering coefficient approaches small values (which would be expected for ER networks) only for ϵ close to one. In between there is a large range of values ϵ for which the network has both small characteristic distance and high clustering with respect to an ER network with the same number of nodes and links. Networks that combine these properties are generally referred to as *small world networks*.

The basic mechanism by which the Watts–Strogatz model generates the small world property is not exclusive to this specific model. The mechanism is as follows. Consider a network that consists of two components, each component being a clique. The distance between nodes of the same components is therefore one, whereas two nodes of different components have an infinite distance. Adding one single link between the components drastically reduces the distance between all pairs of nodes from the different components to a maximum of three. This mechanism was already anticipated by the Granovetterian weak ties; the contribution of the Watts–Strogatz model is to show the small world property already emerging for a rather small number of randomly created weak ties.

4.5.4 Hubs and scale-free networks

The structure of many real world networks is characterized by the existence of hubs [7]. Nodes vary greatly in their degree. The maximal degree in a network is often orders of magnitude larger than the average degree. The existence of a small number

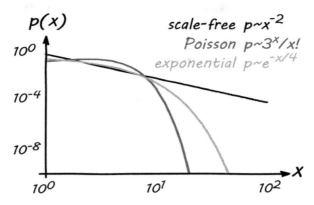

Figure 4.12 *Scale-free degree distributions appear as a straight line in loglog plot (black). The Poisson distribution (dark grey) decays much faster and shows a characteristic shoulder. For comparison, an exponential distribution is shown (light grey).*

of high-degree nodes can radically change the behaviour of dynamical processes on the network, as we will see in Section 4.8. The concept of hubs is familiar from transportation networks, where hubs refer to well-connected centres, stops, stations, or airports. For instance, the airports of Frankfurt or Dubai serve as hubs in the global air transportation network, with many smaller airports connecting to them [160]. Particularly trivial examples for networks with hubs include *star networks*, which are connected networks with N nodes and $N-1$ links that connect $N-1$ leaves to the same hub node.

4.5.4.1 Scale-free degree distributions

An important class of networks that contain hubs are *scale-free networks* [28]. These are characterized by degree distributions that follow a power law,

$$P(k) = \frac{1}{Z}k^{-\gamma},$$

(4.35)

with $Z = \sum_{k=1}^{k_{max}} k^{-\gamma}$ and $\gamma > 0$.

Hubs are often associated with scale-free degree distributions but are not exclusive to them. The term 'scale-free' originates from the fact that the quotient of the degree probabilities does not change with the scale $\lambda > 0$ of the degrees,

$$\frac{P(k_1)}{P(k_2)} = \frac{P(\lambda k_1)}{P(\lambda k_2)} = \frac{g(\lambda)P(k_1)}{g(\lambda)P(k_2)},$$

(4.36)

which is true only for power laws; see Equation (3.1). Scale-free networks are not free of any scale, but have a probability of finding a node of a given degree that is scale-invariant under a dilation (rescaling) of the degree. Scale-free degree distributions appear as straight lines in loglog coordinates. They are usually clearly distinguishable from

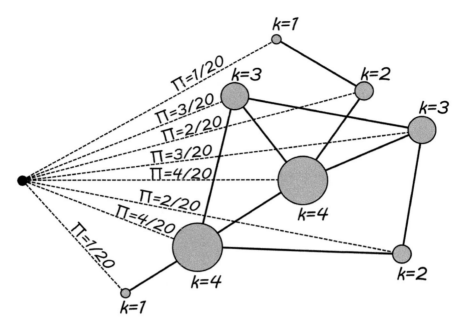

Figure 4.13 *Illustration of the preferential attachment mechanism. A new node (black) attaches to a network (grey nodes; the size represents their degree). The probability of the new node attaching to an already existing grey node i is proportional to its degree k_i.*

exponential or Poisson distribution functions, which decay much faster; see Figure 4.12. To fit them appropriately, one uses maximum likelihood estimators, as suggested in Section 3.4.

4.5.4.2   *Preferential attachment*

Preferential attachment (PA) has been suggested as a generating algorithm for scale-free networks [28]. It is a variation of the Yule–Simon preferential process that we encounter in Section 3.3.4. PA is based on the assumption that new nodes have a tendency to attach to existing nodes with a probability that is proportional to their degree. Consider the example shown in Figure 4.13, where a new node i (black) is introduced to a network (grey nodes). The size of the nodes in Figure 4.13 is proportional to their degrees k_j. PA specifies the linking probability, Π_j, of the new node i linking to an existing node j by,

$$\Pi_j = \frac{k_j}{\sum_l k_l}. \tag{4.37}$$

4.5.4.3   *Barabási–Albert model for growing scale-free networks*

Barabási and Albert formulated a model for growing networks that is based on PA and leads to scale-free networks. At each time step $t = 1, 2, \ldots$, a new node is created and the number of nodes in the network is $N(t) = t$. For the initial condition for time steps $t \leq m_0$, we assume that the first m_0 nodes form a connected network. The

Barabási–Albert (BA) model can then be specified for $t > m_0$. At every time step, $m < m_0$ links are added.

1. At time step t, add a new node with index $i = t$.
2. The probability of a new node $i = t + 1$ attaching to node j is,

$$\Pi_j(t) = \frac{k_j(t)}{\sum_{l=1}^{t} k_l(t)}. \tag{4.38}$$

3. Node i attaches to an existing nodes with this probability. One then advances to the next time step.

4.5.4.4 Degree distribution of the Barabási–Albert model

The degree distribution of the BA model was computed in Section 3.3.4. There we found,

$$P(k) \sim k^{-3}, \tag{4.39}$$

showing that it follows a power law with an exponent $\gamma = 3$. By introducing variations to the PA mechanism, the model can also generate different exponents. For example, $\Pi_i(t) \sim k_i(t) + \text{const}$, or $\Pi_i(t) \sim k_i(t)^\alpha$, with $\alpha \neq 1$, changes the exponent [64, 340].

4.5.4.5 Other properties of the Barabási–Albert model

It is possible, though beyond the scope of this book, to compute additional properties of networks generated by the BA model. For instance, for $m \geq 2$ the networks consist of a single component with a diameter that is proportional to $d \sim \log N / \log(\log N)$ almost certainly [57]. For ER networks the diameter is proportional to $d \sim \log N$. In other words, scale-free networks have a diameter that is smaller by a factor of $\log(\log N)$. For all practical purposes, BA networks therefore have the same diameter as ER networks.[6] Further, it can be shown that the overall clustering coefficient scales with network size, $C \sim N^{-3/4}$ [7]. However, the individual clustering coefficients C_i do not scale with the node degree k_i, and $C_i(k)$ as a function of the degree is constant.

4.5.4.6 Hubs in non-growing networks

Note that the emergence of hubs is not a feature that is exclusive to growing networks. Other models have been proposed [100, 371]. Davidsen, Ebel, and Bornholdt introduced a model for the emergence of hubs that was inspired by the observation of high clustering coefficients in many real world networks, particularly in social networks [100]. The network formation process in this model is shown in Figure 4.14. The main idea is that links are introduced between nodes that share common neighbours (triadic closure). The dynamics of the model is as follows.

---

[6] For example for the Avogadro number $N_A = 6 \cdot 10^{23}$, $\log(\log N_A) \sim 4$.

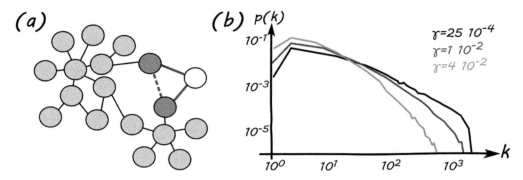

Figure 4.14 *(a) Formation of hubs in non-growing networks. A link (dashed line) is introduced between a pair of nodes (dark-grey) that have a common neighbour (white)—triadic closure. This provides an indirect mechanism for preferential attachment, as nodes with higher degrees tend to have a higher number of pairs of neighbours that are not yet connected. (b) Degree distribution in the Davidsen–Ebel–Bornholdt model for three different choices of node turnover rate* γ. *With increasing* γ, *the degree distribution approaches a power law.*

1. Initialize an ER network with N nodes and L links.
2. At each time step t, pick a randomly chosen node i and two of its neighbours. Connect these two neighbours by a link.
3. If the node i has fewer than two neighbours, add a link to a randomly chosen node.
4. With probability γ, a randomly chosen node is removed from the network along with all its links. It is replaced by a new node with a link to a randomly chosen neighbour.
5. Continue with time $t+1$ until the network reaches a steady state.

As nodes have a finite lifetime, a steady state will always be reached. Once this steady state is achieved, at each time step one link is added and, with probability γ, a node with $\langle k \rangle$ links (on average) is removed, before a new node with one link is added. With probability γ, one effectively removes $\langle k \rangle - 1$ links in each step. In the stationary state, the link addition rate equals the link removal rate and we have $1 = \gamma(\langle k \rangle - 1)$; the average degree follows as $\langle k \rangle = 1 + 1/\gamma$. The degree distribution of the model is shown in Figure 4.14b for three different choices of γ. With decreasing γ, the distribution approaches an approximate power law and hubs emerge. Note that the model uses a PA mechanism through the back door. Consider the network formation step shown in Figure 4.14a. Two nodes from the selected node i's (white) neighbourhood are chosen randomly (dark-grey nodes). The probability of any node j being in i's neighbourhood is proportional to its degree k_j. It follows that the probability of a node j acquiring a new link is proportional to k_j, which is simply PA.

Another model has been proposed for non-growing networks that shows q-exponential distribution functions. It is based on a mechanism where nodes merge. At every step in time, two linked nodes merge into a single new node that inherits the links

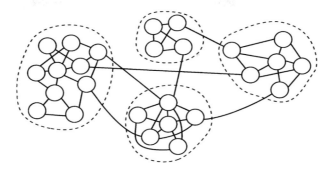

Figure 4.15 *Communities are defined as groups of nodes that are more likely to be connected amongst each other than to nodes outside of the community. The problem of community detection is to identify such groups (here highlighted with dashed lines) in a given network. This is often impossible to do in a unique way, and community detection becomes an art rather than a science.*

of the two [371]. After a merger a new node is created and randomly attached to any node in the network.

4.5.4.7 *Preferential attachment in practice*

The BA model is attractive because of its analytic tractability. However, PA requires each new node to have *global information* about each node in the network. In this formulation of PA, the new node has to know the degrees of *all* existing nodes in the network. The non-growing models show that PA can also emerge from local, microscopic interactions. There, each node needs only to have knowledge of the degrees in its direct neighbourhood [100, 390], or to have no knowledge at all [371].

4.6 Communities

> *Communities* are characterized by the property that each node is more likely to be connected to a node within the community than to a node outside; see Figure 4.15. Communities are 'noisy components'.

Along with network properties such as the small world property, fat-tailed degree distributions, and high clustering, the existence of a community structure is an essential feature of many real world networks [133]. The community structure describes the mesoscopic organization of the network. The presence of communities in a network reveals the existence of some kind of ordering principle that operates at the level of *groups* of nodes, but not necessarily at the microscopic level of individual nodes or the macroscopic level of the entire network. The problem of quantifying community structure is commonly referred to as community detection. It is rarely an exact business. Nevertheless, there are a number of approximative solutions that are useful in empirical data analysis. In the following, we give a brief overview of the most frequently used methods for community detection.

4.6.1 Graph partitioning and minimum cuts

A traditional method for the detection of communities is the minimum cut method [217]. The minimum cut of an undirected and unweighted network is a cut or *partitioning* of nodes, that is in some sense minimal, into two disjoint and non-empty subsets. The cut minimizes the number of links that connect nodes from the two different subsets. There are only non-trivial solutions for the minimum cut if the sizes of the disjoint subsets are fixed (otherwise the minimum cut would always cut out one of the nodes with the minimal degree). Often the two communities are assumed to be of equal size. An exhaustive search for the minimum cut is only feasible for small networks. A larger number of partitions can be obtained by iteratively applying the minimum cut method to the resulting partitions of the first cut. If it is not meaningful to fix the sizes of the partitions a priori, one can apply the popular method of spectral bisection [30]. The method uses properties of the network Laplacian Λ that we have encountered previously.

Network Laplacian. For an undirected simple network, the network Laplacian Λ is given by,

$$\Lambda = D - A. \tag{4.40}$$

Λ is a symmetric matrix, whose diagonal elements are the node degrees, and the off-diagonal elements are -1, if the corresponding nodes are connected, and zero otherwise; see text following Equation (4.5). A is the adjacency matrix and D is a diagonal matrix with $D_{ii} = \sum_j A_{ij}$. As all rows and columns of Λ sum to zero, it immediately follows that the vector of all-ones $\mathbf{1} = (1, 1, \ldots, 1)$ is an eigenvector of Λ with a corresponding eigenvalue of zero. Another useful spectral property of the Laplacian is that all of its eigenvalues are non-negative.

To understand this last point, remember that the Laplacian is related to the oriented incidence matrix O of the network by $\Lambda = OO^T$. For an eigenvalue λ_i of Λ with the corresponding (normalized) eigenvector v_i, we have $\Lambda v_i = \lambda_i v_i$, or $\lambda_i = v_i^T \Lambda v_i = (O^T v_i)^T (O^T v_i)$. As λ_i can be written as a scalar product of a vector with itself, we have that $\lambda_i \geq 0$ and the eigenvalues of Λ are non-negative. Assume that the network, A, has two components, C_1 and C_2. The Laplacian, Λ, is then of block diagonal form, where each block corresponds to one of the components. Each of these blocks could be described by a Laplacian of its own, again with an eigenvector of all-ones with eigenvalue zero. It follows that the eigenvector to the zero eigenvalue of Λ is two-fold degenerate, namely, $v_i^{(k)} = 1$, if $i \in C_k$, for $k = 1, 2$, and $v_i^{(k)} = 0$, otherwise. Let us assume that the separation into two components is not perfect and that there are a few links between nodes from C_1 and nodes in C_2. As Λ is non-negative, the eigenvectors $v_i^{(k)}$ will now have eigenvalues that are slightly larger than zero. The eigenvector to the smallest eigenvalue (zero) is the vector of all-ones, $\mathbf{1}$. The second smallest eigenvalue is commonly referred to as algebraic connectivity [83]. The eigenvector to the second smallest eigenvalue is orthogonal to the vector of all-ones and therefore has positive and negative entries. As the entries that belong to the same community should have almost the same value (or be identical if the communities are components) it follows that the two communities correspond to the nodes with entries below or above zero, respectively. The algebraic connectivity is a measure of how good the cut is. The closer it is to zero, the better the split is into two communities, that is, the 'closer' the two communities are to components.

4.6.2 Hierarchical clustering

Hierarchical clustering methods have the advantage over graph partitioning methods in that the number of communities does not have to be specified in advance [213]. Instead, this technique extracts a hierarchical structure of communities across several levels, even if such a hierarchy does not exist. The first step in hierarchical clustering is to define a similarity measure between two nodes i and j, s_{ij}. Popular choices for this similarity include the following measures. Let $a^{(i)}$ and $a^{(j)}$ denote the ith and jth row of the adjacency matrix, respectively.

- *Cosine similarity.* The cosine similarity is defined by the scalar product $s_{ij} = a^{(i)} a^{(j)}$.
- *Degree correlation.* The similarity s_{ij} is defined as Pearson's correlation coefficient (see Section 8.1.10) of the vectors $a^{(i)}$ and $a^{(j)}$, $s_{ij} = \rho(a^{(i)}, a^{(j)})$.
- *Hamming distance.* The similarity is defined as $s_{ij} = N - H_{ij}$, where the Hamming distance H_{ij} is the number of components in $a^{(i)}$ that differ in value from those in $a^{(j)}$.
- *Jaccard index.* Let us denote the set of all nodes that are adjacent to node i by $Nb(i)$. Similarity can be defined with the Jaccard index by $s_{ij} = \frac{|Nb(i) \cap Nb(j)|}{|Nb(i) \cup Nb(j)|}$.

This list is far from exhaustive. All these measures share the property that the similarity of two nodes, i and j, does not depend on whether they are actually connected by a link or not. In fact, all these measures can assume large values, even if $A_{ij} = 0$.

The strategies for obtaining a hierarchy of communities based on node similarity fall into two types [323]. In the agglomerative or 'bottom-up' approach one starts with each node in its own community and then extracts a hierarchy by iteratively merging two communities. In the divisive or 'top-down' approach, all nodes start in the same community, which is then recursively split until each subcommunity consists of a single node. Mergers or splits are made on the basis of a linkage criterion between two communities, derived from the similarity measures s_{ij} between the nodes from the two communities. If we consider two communities, A and B, possible choices for the linkage criterion between these communities include the definitions of the community similarity S_{AB}, mentioned here:

- *Complete linkage clustering,* where the similarity of two communities A and B is given by the lowest similarity between nodes from the two different communities, $S_{AB} = \min \{s_{ij} | i \in A, j \in B\}$.

- *Single linkage clustering,* where the similarity is given by the highest similarity between two nodes from A and B, $S_{AB} = \max\{s_{ij} | i \in A, j \in B\}$.
- *Average linkage clustering.* Here, the similarity is given by the average similarity between all nodes from the two communities, $S_{AB} = (\sum_{i \in A} \sum_{j \in B} s_{ij})/(|A||B|)$, where $|A|$ and $|B|$ are the number of nodes in A and B, respectively.

A naive bottom-up scheme for hierarchical clustering can now be given as follows [155]. Start with all nodes being assigned to a different community and apply the algorithm:

1. Find the most similar pair of communities X and Y, that is, the pair with the lowest similarity S_{XY}.
2. Merge communities X and Y into a single community. Replace X and Y by this merged community.
3. Continue with step 1, until all nodes are in the same community, then stop.

Each iteration of the above algorithm can be regarded as a hierarchical level.

4.6.3 Divisive clustering in the Girvan–Newman algorithm

Nodes that are a bridge between two communities tend to have a high betweenness centrality; see Equation (4.11). A betweenness measure can also be defined for links, the so-called *edge betweenness centrality*. It is defined as the number of geodesic paths that use a given link. Let g_{ij} be the number of geodesic paths between nodes i and j, and let $g_{ij}(l)$ be the number of geodesic paths that pass through link l. The edge betweenness centrality $C^e(l)$ of link l is defined as,

$$C^e(l) = \sum_{i,j} \frac{g_{ij}(l)}{g_{ij}}. \tag{4.41}$$

Based on $C^e(l)$, a divisive, top-down hierarchical clustering algorithm was proposed by Girvan and Newman as the following protocol [152].

1. The edge betweenness of all links in the network is computed.
2. The link with the highest edge betweenness is removed.
3. The edge betweenness is recomputed for each link in the network.
4. Steps 2 and 3 are repeated until the network has no more links.

Each iteration of this algorithm that increases the number of components in the network corresponds to a distinct hierarchical level.

4.6.4 Modularity optimization

Many community detection methods use a so-called modularity function as a central element [286, 288]. Modularity offers a step towards a more principle-based approach to clustering and community detection. It encapsulates a suitable definition of communities, allows a null model to be chosen, and quantifies the 'goodness' of the communities in the network.

The concept of *modularity* compares a given network to null models of random networks that lack community structure. Communities are revealed by comparing the actual link density of a network with the density that would be expected in a random network with no community structure whatsoever. For a simple network with L links and adjacency matrix A, assume that P_{ij} is the probability of a link between nodes i and j in the null model. Let c_i and c_j be the labels of the communities to which nodes i and j belong. The modularity Q is then given by,

$$Q = \frac{1}{2L} \sum_{ij} (A_{ij} - P_{ij}) \delta(c_i, c_j), \qquad (4.42)$$

where δ is the Kronecker delta; see Section 8.1.3.

A suitable choice for the null model P_{ij} is the configuration model of a given network A. Values of the modularity Q that are larger than zero indicate that the link density between nodes of a given community is higher than would be expected if nodes with given degrees were randomly connected. Remember that the probability of picking a stub from node i in the configuration model is $k_i/2L$. As a node j receives connections to k_j other stubs, the probability P_{ij} in the configuration model is given by $k_i k_j/2L$. This yields a modularity function,

$$Q = \frac{1}{2L} \sum_{ij} \left(A_{ij} - \frac{k_i k_j}{2L} \right) \delta(c_i, c_j). \qquad (4.43)$$

The higher the value of the modularity Q, the better the partitioning of the network into communities. Unfortunately, a brute force maximization of Q is not practical due to the combinatorial explosion of potential partitions. Modularity optimization is indeed NP hard [67]. However, a large number of computationally tractable approximations exist, some of which we mention in the following.

4.6.4.1 Girvan–Newman algorithm

One of the first attempts to maximize modularity was by means of an agglomerative hierarchical clustering approach [288]. The hierarchy of the nodes in the original network A is recorded in a different network with adjacency matrix B. Initially, all nodes are isolated in B and start in their own community of size one. In each step one link is added to B between two nodes. When nodes of two different communities become connected by a link in B, the two communities merge into a new one. This link is always chosen such that the new partition gives the maximum increase (or minimum decrease) in modularity with respect to the previous configuration. Note that modularity is always evaluated using the original network, A. After $N-1$ community mergers, the algorithm stops.

4.6.4.2 *Louvain method*

A version of agglomerative clustering that is computationally efficient, even for large networks, proceeds as follows [53]. The algorithm is initialized by putting each node in its own community. The nodes are then picked in random order. For each node i and each of its neighbours j, one computes the change of modularity that comes from placing i in the same community as j. One then assigns i to the community of the neighbour that is associated with the largest increase in modularity. After a sweep over all nodes, all nodes of a community are condensed into a supernode. The nodes of a community are replaced by this supernode, which is linked to all nodes that had a link to one of the original nodes of the community. Then the procedure starts again. Note that the Louvain method always computes the modularity of the original network (and not of the network of supernodes) when checking for improvements of new community assignments. The algorithm stops when there are no more possible increases in modularity.

4.6.4.3 *Extremal optimization*

Another way of optimizing the modularity is to treat the per-node contributions to Q as the node's 'fitness' and to proceed by extremal optimization [108]. The network is initially divided into two communities of equal size. The node with the lowest 'fitness' is then moved to the other community. The procedure stops once there are no more moves that increase the global modularity Q.

4.6.4.4 *Spectral method*

Q can be optimized using spectral methods [287]. Here we introduce a variable s_i for each node i that encodes whether the node belongs to community C_1 ($s_i = +1$) or to C_2 ($s_i = -1$). We can then rewrite the Kronecker delta, $\delta(c_i, c_j)$ in Equation (4.42) as $\delta(c_i, c_j) = (s_i s_j + 1)/2$. Let us introduce the modularity matrix, $M_{ij} = A_{ij} - \frac{k_i k_j}{2L}$, so that the modularity function Q can be written as,

$$Q = \frac{1}{4L} \sum_{ij} M_{ij}(s_i s_j + 1) = \frac{1}{4L} \sum_{ij} M_{ij} s_i s_j = \frac{1}{4L} s^T M s, \tag{4.44}$$

where we have used that the sum of all elements in a row or column of M adds up to zero. From this it follows that the vector $\mathbf{1} = (1, 1, \ldots, 1)$ is an eigenvector of M, just as in the case of the Laplacian in Equation (4.40). We can apply the same reasoning as before for the spectral bisection and replace the Laplacian Λ by the modularity matrix M. Moreover, modularity allows us to define a clear stopping criterion for the subdivision into more than two communities. Subdivisions are only acceptable if they increase the global modularity function Q.

4.6.4.5 *Which algorithm is the right one for my problem?*

There is no unique answer to that question. Each of the methods mentioned for community detection has advantages and disadvantages; we refer the interested reader to the more specialized literature [133]. As a practical guideline, the following can be stated. Each of these algorithms employs basically the same definition of what a community is. If the network has a relevant and detectable community structure, each of these methods will pick up at least some aspects of it. To validate the results, a quick sanity check is always recommended to visualize the network and check the extent to which the retrieved communities really separate. This is often enough to judge whether the network indeed shows different partitions of nodes, or whether searching for such a community structure is anyhow questionable. It is always a good idea to further validate community structure by comparing the results across several different algorithms. If one is really interested in finding an optimal solution to a community detection problem (which can be evaluated only if there is a 'ground truth' against which it can be benchmarked), it has been shown that the best solutions frequently combine several different algorithms in a quite specific way [414]. To summarize, there is no 'best algorithm' for community detection. The aim must always be to find a reasonable solution for a specific problem, and that might involve using more than one algorithm.

4.7 Functional networks—correlation network analysis

In this section, we develop the basic principles of how network analysis can be used as a data mining tool. While many data sets describe actual physical networks or social interactions, it is also possible to extract *functional networks* from data, even if interactions between nodes are not directly observable. Often such problems are of the following form. Imagine N variables that constitute the nodes in the network such as individuals, genes, stocks, and so on. For each of these N variables we have T different observations. These observations can be a time series of daily stock prices [261], repeated measurements of expression levels of genes [417], or diseases of patients [79]. In most applications, each observation will either be a real number or a categorical variable. By defining a similarity measure for any two time series of observations, it is possible to construct a *correlation network* for the N variables (nodes). In this network, two nodes i and j are connected by a weighted link that describes how strong the correlation between the observations of i and j are. A positive (negative) correlation here means that if the value of i changes, there is a tendency for j to change in the same (opposite) direction at the same time.

From a statistical perspective, this gives rise to a massive *multiple hypothesis testing* problem. For each pair of nodes we have to test the null hypothesis that measurements of the two nodes are independent of each other. Each significant link in the correlation

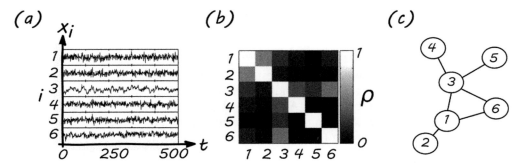

Figure 4.16 *Illustrative example of correlation network analysis. (a) We consider a data set consisting of six time series, $x_i(t)$, $i = 1, 2, \ldots, 6$. (b) The correlation matrix ρ describes how each time series is related to the others. The colour of the ith row and jth column is the correlation coefficient ρ_{ij} of $x_i(t)$ and $x_j(t)$. (c) After appropriate filtering (e.g. thresholding) of the correlation matrix ρ, we obtain a symmetric* correlation network *that can be used to extract key functional characteristics from the data.*

network indicates that the corresponding null hypothesis of independence must be rejected before we have a significant correlation or a 'discovery'. This hypothesis test has to be repeated $N(N-1)/2$ times. To extract a correlation network in which we can control the number of *false discoveries* that can be expected in the network, we need to adjust for this large number of hypothesis tests. No matter how rigorous we are in controlling the statistical significance of links in the correlation network, we always have to bear in mind that significant correlation does not imply a causal relation.

A significant link between nodes i and j in a correlation network indicates that there is some process affecting both nodes. There is no guarantee, however, that there is a direct causal interaction between i and j. They could be controlled, for example, by a third variable that is not contained within the network data. Nevertheless, the topology of the correlation network often allows us to understand which nodes are the most likely candidates for explaining specific functional properties of the network. It also allows us to see if there are tightly interrelated modules of variables such as communities.

Figure 4.16 illustrates the process of obtaining a correlation network. In Figure 4.16a we show data that consists of $N = 6$ time series, $x_i(t)$, with $T = 500$ observations each. For each pair of time series i and j we compute Pearson's correlation coefficient, ρ_{ij}; see Equation (8.19). The result is the correlation matrix, ρ, which describes how two time series i and j are related to each other, Figure 4.16b. Correlation matrices are typically fully connected weighted networks with ones along the diagonal. To extract the significant correlations from ρ, appropriate filtering techniques must be applied. This results in a correlation network, see Figure 4.16c, which can be analysed using standard network measures to identify hubs, good spreaders, communities, bottlenecks, and so on in the data.

4.7.1   Construction of correlation networks

Given a data set, our task is to construct a correlation network of N variables for which we have T observations each. These data can be represented by a rectangular $N \times T$ data matrix M. Each row corresponds to a node, each column to an observation (time). We consider cases where all entries in the data matrix, M_{it}, are either continuous (real numbers) or binary (yes or no, zero or one). The first step in constructing the correlation network is to compute the correlation coefficient between each pair of variables. The correlation coefficients that should be used depend on the variables; see Section 4.7.1.1. The following procedures are only meaningful for a suitably large number of observations. We always assume that $T \gg N$. We now have to statistically test if two variables are correlated or not. First, we formulate a null hypothesis, which assumes that the two variables are independent and not correlated. Next, we define a function of the variables that follows a known distribution (given the correctness of the null hypothesis); the so-called test statistic. This allows us to calculate how likely it is for the observed value of the test statistic to be produced by the null hypothesis. If this probability is lower than a certain prespecified threshold (significance level), then there is a significant correlation between the two variables.

4.7.1.1   *Correlation between binary variables*

Assume that all observations are dichotomous, that is, either zero or one, $M_{it} \in \{0, 1\}$. Consider the pair of nodes i and j. The ith and jth rows of M are T-dimensional vectors. Let us denote them x_t and y_t, respectively, $t = 1, \dots, T$. The null hypothesis states that the process that generates x_t is independent from the process that generates y_t. If the null hypothesis is wrong, we can reject the assumption of statistical independence between i and j, and a significant correlation exists. If this correlation is 'significant enough', we connect them by a link in the correlation network. Assume that the observations for x_t are generated by some process that does not depend on the value of y_t, and vice versa. Nevertheless, there will be cases where the actual values of x_t and y_t coincide. In general, the number of coincidences between x_t and y_t will be drawn from a particular probability distribution that is specified by the null hypothesis. Given this distribution, one can compute the probability of how likely it is that the numbers of coincidences observed in the data would occur if the null hypothesis were true. The smaller this probability—the so-called p-value—is, the more likely it is for the null hypothesis to be rejected, and i and j are indeed correlated. To compute the p-value we first need some notation. Let n_{ab} be the number of indices t for which x_t takes the value $a \in \{0, 1\}$ and y_t is $b \in \{0, 1\}$. Similarly, we define n_{00}, n_{01}, n_{10} as the numbers of joint observations of x_t and y_t, for which they take the values $\{00, 01, 10, 11\}$. Obviously, $n_{00} + n_{01} + n_{10} + n_{11} = T$. Further, let $n(x_t = 0)$ be the number of observations for which $x_t = 0$. Clearly, $n(x_t = 0) = n_{00} + n_{01}$, and $n(x_t = 1) = n_{10} + n_{11}$. Note that n_{ab} can be arranged in a matrix where a labels the rows and b the columns. This is a so-called contingency table [305]; see Table 4.1. The values of $n(x_t = a)$ and $n(y_t = b)$ give the row and column sums, respectively.

Table 4.1 *Contingency table for two binary variables, x_t and y_t. There are two columns (rows) that correspond to the possible outcomes of $x_t \in \{0, 1\}$ ($y_t \in \{0, 1\}$). Each joint observation of x_t and y_t counts towards one of the four fields, n_{ab}.*

	$x_t = 0$	$x_t = 1$	row sum
$y_t = 0$	n_{00}	n_{10}	$n(y_t = 0)$
$y_t = 1$	n_{01}	n_{11}	$n(y_t = 1)$
column sum	$n(x_t = 0)$	$n(x_t = 1)$	T

Given the null hypothesis of no correlation between x_t and y_t, as well as the row and column sums $n(x_t = a)$ and $n(y_t = b)$, we obtain the expected values for n_{ab}. The probability that $x_t = a$ for some t is $n(x_t = a)/T$ and the probability that $y_t = b$ is $n(y_t = b)/T$. Assuming the independence of x_t and y_t, the joint probability of observing $x_t = a$ and $y_t = b$ at the same time is simply the product of the corresponding marginal probabilities, $n(x_t = a)n(y_t = b)/T^2$. The expected value of n_{ab}, denoted by $\hat{n}_{ab}$, is T times this joint probability, $\hat{n}_{ab} = n(x_t = a)n(y_t = b)/T$. A chi-squared test can be used to determine whether the observed values of n_{ab} are significantly different from their expectations, given the null hypothesis, $\hat{n}_{ab}$ [304]. For this, one considers the sum of squared errors between the observed values and their expectation values. In our case, the test statistic χ^2 is defined as,

$$\chi^2 = \sum_{ab} \frac{(\hat{n}_{ab} - n_{ab})^2}{\hat{n}_{ab}}. \tag{4.45}$$

If the null hypothesis were true, the observed deviations of n_{ab} from $\hat{n}_{ab}$ would be pure chance and should be expected to follow a normal distribution. The sum of squares of independent and normally distributed random variables follows the chi-squared probability distribution, $f_{\chi^2}(x, n)$, where n denotes the degrees of freedom, which in our case is $n = 1$; see Equation (8.1.11). It follows that the probability p of observing a given chi-squared value of at least χ^2 is given by,

$$p = \int_{\chi^2}^{\infty} f_{\chi^2}(x, 1) \, dx, \tag{4.46}$$

where p is called the p-value. The function $f_{\chi^2}(x, n)$ is described in Equation (8.21). For a given value of χ^2, the value of p can be obtained from a lookup table. Given the null hypothesis of statistical independence between x and y, the p-value gives the probability of observing deviations from the expected numbers of joint observations of values for x and y that are at least as extreme as those observed in the data. For single hypothesis tests, the null hypothesis is often rejected whenever $p < 0.05$. If differences between $\hat{n}_{ab}$ and n_{ab} were indeed normally distributed, a p-value of $p = 0.05$ would correspond to a difference in observed values compared with expected values by approximately 1.96 standard deviations. For the case of binary variables, a weighted link in the correlation network between nodes i and j can be the correlation measure χ^2 or the p-value.

There is an obvious problem in the definition of the p-value if the total numbers of observations n_{ab} (which are discrete numbers) are so low that continuous distributions give poor approximations. This can be adjusted for by using so-called continuity corrections. The simplest form of such a correction is to subtract 'half an observation' from each difference between expected and observed values of n_{ab} in Equation (4.45) [415]. The Yates-corrected chi-squared value is given by $\chi_Y^2 = \sum_{ab}(|\hat{n}_{ab} - n_{ab}| - 0.5)^2/\hat{n}_{ab}$. More elaborate continuity corrections can be used to increase the accuracy of the procedure [201].

4.7.1.2 *Correlation between continuous variables*

We now consider the case where the observations are real numbers; that is, $x_t, y_t \in \mathbb{R}$. Let us think of x and y as T-dimensional vectors. The first step in measuring a potential correlation between x and y is to centre the two data vectors. We denote the sample means by $\langle x \rangle = \sum_t x_t/T$ and $\langle y \rangle = \sum_t y_t/T$. The correlation between x and y can now be measured by the cosine of the angle θ between x and y. This value is also known as Pearson's sample correlation coefficient [303],

$$\rho = \cos\theta = \frac{(x - \langle x \rangle)(y - \langle y \rangle)}{\|x - \langle x \rangle\|\|y - \langle y \rangle\|} = \frac{\sum_{t=1}^{T}(x_t - \langle x \rangle)(y_t - \langle y \rangle)}{\sqrt{\sum_{t=1}^{T}(x_t - \langle x \rangle)^2}\sqrt{\sum_{t=1}^{T}(y_t - \langle y \rangle)^2}}, \qquad (4.47)$$

which takes values between -1 and $+1$; see Equation (8.19).

The correlation coefficient states by how many standard deviations the variable x increases for every standard deviation that y increases by.

We now need a statistical hypothesis test to decide whether a non-zero correlation between x and y is significant or not. Our null hypothesis is again that x and y are independent from each other, which means that the observations (x_t, y_t) are samples of an uncorrelated bivariate normal distribution. To define a p-value upon which to reject this null hypothesis, we require a test statistic as a function of ρ. It can be shown that, if the null hypothesis is true, the variable t_S,

$$t_S = \rho\sqrt{\frac{T-2}{1-\rho^2}}, \qquad (4.48)$$

follows the Student-t distribution with $T-2$ degrees of freedom; see Equation (2.50). By considering the integral of this distribution up to the observed value of t_S, one can obtain a p-value that can be used to reject the null hypothesis at a specified significance level. As a result, we can assign a correlation coefficient ρ (or its test statistic t_S) and a p-value to each pair of nodes in the correlation network.

A shortcoming of Pearson's sample correlation coefficient ρ is that a non-zero correlation could be due to a single outlier in the data. To adjust for such cases, one can replace the values of x_t and y_t by their ranks, where the value of 1 is assigned to the largest x_t and y_t, 2 to the second largest x_t and y_t, and so on. Each pair (x_t, y_t) is then replaced by a pair $(\text{rank}(x_t), \text{rank}(y_t))$. Observations with the same values are typically assigned the same rank. The same procedure as before for computing the sample correlation coefficient can now be applied to the rank data, which leads to Spearman's rank correlation coefficient [348]. The rank correlation coefficient suppresses the influence of outliers in the data; however, it might mask small but systematic correlations in the bulk of the data. As a practical guideline, it may be beneficial to impose the restriction that both Pearson's and Spearman's correlation coefficient must be significantly different from zero to indicate a significant correlation between x and y.

4.7.1.3   *Correlation between a binary and a continuous variable*

Let us finally consider the case where one observation is a real number, $x_t \in \mathbb{R}$, and the other is binary, $y_t \in \{0, 1\}$. These situations might occur in bipartite networks. The idea here is to separate the observations x_t into two groups: group 1 where $y_t = 0$ and group 2 where $y_t = 1$. Assume that there are n_g $(g = 1, 2)$ observations in group g, with the sample mean $\langle x_g \rangle$ and the sample standard deviation σ_g^2. The null hypothesis is that observations in group 1 and group 2 are not significantly different from each other. If the values in x were normally distributed, this could be decided by the t-test. We therefore introduce the pooled standard deviation $\sigma_P = \sqrt{((n_1 - 1)\sigma_1^2 + (n_2 - 1)\sigma_2^2)/(T - 2)}$. If the data in group 1 and group 2 are sampled from normal distributions with the same mean and variance, then the test statistic,

$$t_S = \frac{\langle x_1 \rangle - \langle x_2 \rangle}{\sigma_P \sqrt{\frac{1}{n_1} + \frac{1}{n_2}}}, \tag{4.49}$$

follows the Student-t distribution with $T - 2$ degrees of freedom. If the data in x do not follow a normal distribution, Wilcoxon's rank-sum test can be applied [260]. For this, we assign to each value of x_t its rank, $\text{rank}(x_t)$. The total sum of all ranks is $T(T + 1)/2$. Let R_g be the sum of the ranks of all observations in group g, $g = 1, 2$. For each group one can define a U-value, U_g, as,

$$U_g = R_g - \frac{n_g(n_g - 1)}{2}. \tag{4.50}$$

The overall U-value of the test is defined as the smallest value over all groups, $U = \min(U_1, U2)$. Note that from $R_1 + R_2 = T(T + 1)/2$, and from $T = n_1 + n_2$, it follows that $U_1 + U_2 = n_1 n_2$. Given the null hypothesis, the expected value for U is therefore $\langle U \rangle = n_1 n_2 / 2$, with a standard deviation $\sigma_U = \sqrt{n_1 n_2 (n_1 + n_2 + 1)/12}$. For large samples, a p-value can be obtained for the null hypothesis that U follows a normal distribution with mean $\langle U \rangle$ and standard deviation σ_U.

Based on the t-test for normally distributed data or the rank-sum test for other data, an incidence matrix can be obtained with entries that are either the corresponding test statistic or the p-values. Finally, we require a criterion to decide whether the data is normally distributed or not. A suitable criterion is given by the Jarque-Bera test [207], which is based on the fact that the normal distribution has a vanishing skewness β, and a kurtosis[7] with a value of $K = 3$. The null hypothesis that the data are normally distributed implies that the sum of squared errors between the observed and expected values of the skewness and kurtosis follow a chi-squared distribution. In particular, we can define a test statistic, $\mathcal{J}B$, as

$$\mathcal{J}B = \frac{T}{6}\left(\beta^2 + \frac{1}{4}(K-3)^2\right), \tag{4.51}$$

which is chi-squared distributed under the null hypothesis for which the p-value can be computed as before.

4.7.2 Filtering the correlation network

After applying the appropriate pairwise significance tests, the correlation network is a fully connected network where every node correlates with every other node, although in most cases, these correlations will be close to zero. In the following, we assume that we have a weighted and undirected network with entries ρ_{ij} that represent the correlation coefficients. We will not consider bipartite correlation networks in the following, even though the generalization to the bipartite case is straightforward for most of the filtering procedures.

To understand the relevant relations in the data, it is necessary to identify those links that 'carry the structure' of the data. For a correlation network with a large number of nodes, we face a massive *multiple hypothesis testing problem*. If the network had only two nodes then there would be no multiple hypothesis testing problem. Given that the null hypothesis is true, the p-value corresponds to the probability that the given link has been observed by pure chance. However, for N nodes and a significance threshold of p, the expected number of false positives is then $pN(N-1)/2$. For a network with a modest size $N = 1,000$ and a reasonable significance threshold $p \sim 0.05$, one would expect $25,000$ *false positive* links to be there by pure chance. This often leads to nonsensical networks and interpretations. There are various approaches with different levels of sophistication to account for this problem.

4.7.2.1 *Naive global thresholding*

In the simplest case, one merely throws away all the links with values above or below certain thresholds. One introduces a cut-off for the correlation measure, ρ_c, and sets

---

[7] The kurtosis is based on the fourth moment of a distribution. For a random variable X, it is given by the fourth moment divided by the fourth power of the standard deviation, $K = \langle X^4 \rangle / \sigma^4$. For the Gaussian distribution $K = 3$. The skewness is defined by $\beta = \langle X^3 \rangle / \sigma^3$.

$\rho_{ij} = 0$, whenever the obtained $|\rho_{ij}| < \rho_c$. The choice of ρ_c is often arbitrary and is made on the basis of how many links should survive. For instance, for large networks it is frequently only possible to come up with meaningful visualizations if the average degree is not much larger than one, which might be used as a criterion to determine the cut-off.

4.7.2.2   *Multiple hypothesis correction*

In statistical hypothesis testing, the reduction of false positive links is addressed with multiple hypothesis testing corrections. In brief, each link (i,j) represents a distinct null hypothesis $H_0^{(ij)}$. The collection of all links forms a family of null hypotheses, $\{H_0^{(ij)}\}$. The objective is to control the number of expected false (or true) positives (or negatives) in this family of tests. Multiple hypothesis corrections can be used to determine a statistically sound global threshold. Let us define the family-wise error rate (FWER), α, which is the probability that at least one true null hypothesis $H_0^{(ij)}$ is rejected. If we perform m significance tests ($m = N(N-1)/2$ for correlation networks), then we obtain the FWER of α, if we reject a null hypothesis $H_0^{(ij)}$, whenever $p_{ij} \leq \alpha/m$. This correction is known as the *Bonferroni correction* [52]. Its advantage is that the choice of the cut-off ρ_c in global thresholding is not completely arbitrary but now has a statistical interpretation in terms of the expected number of false positives. The drawback of this correction is that the significance threshold decreases with N^2. For large networks, the Bonferroni correction might be overly restrictive. In this case, it might be more useful to control the false discovery rate (FDR) instead of the FWER [43]. A FDR of value Q means that the probability of any rejected null hypothesis being a false positive is exactly Q. If our thresholded correlation network has L links, a FWER of α means a probability of α that at least one link is a false positive, whereas a FDR of Q means that QL links are false positives. Whether it is more appropriate to control the FWER or FDR ultimately depends on the 'costs' of false positives. If a costly experiment has to be performed for each link discovered, the FWER should be preferred. For an exploratory analysis with the aim of generating a large number of hypotheses—no matter whether they are right or wrong—it might make more sense to control the FDR.

The Benjamini–Hochberg procedure can be used to control the FDR [44]. This procedure can be summarized as follows. Assign a rank $r(ij)$ to each link (i,j), such that the link with the kth lowest p-value has rank $r(ij) = k$. For a given Q, each link (or hypothesis) is assigned a value $r(ij)Q/m$. Identify the link (i,j) with the largest p-value such that $p_{ij} \leq r(ij)Q/m$. Denote the rank of this link by r^*. All links with a rank of r that is below or equal to r^* are discoveries, that is, their null hypotheses can be rejected. Note that this procedure gives satisfactory results only if all hypothesis tests are independent of each other.

4.7.2.3   *The maximum spanning tree*

A drawback of global thresholding approaches is that there is no guarantee that the filtered network will span and percolate the entire original network, or if the filtered network will be decomposed into disjoint components. This decomposition is particularly likely to occur in networks with hubs that are strongly correlated with a large number of other nodes. It is often the case that a global threshold removes either everything except for the links with the hubs, or that the filtered network still contains a large number of links so that no structure can be revealed. This issue can be addressed by considering the maximum spanning tree (MST) of the correlation network [262]. For a connected network with N nodes, the MST gives a maximally induced subnetwork with $N - 1$ links that connect all N nodes by using the links with the maximal weights. The maximum spanning tree, MST, can be constructed using Kruskal's algorithm. Consider a weighted network W that can, but need not, be a correlation network. Initially, the MST is an empty set of nodes and an empty set of links.

1. Sort all links in W by their weights in descending order.
2. Pick the link with highest weight and add it to the MST.
3. Go through the ordered list of links in W sequentially and add a link to the MST, whenever one of the following applies:
   - at least one of the end-nodes of the link is not already in the node set of the MST,
   - the link connects two unconnected components of the MST,
   - otherwise continue.
4. Stop the algorithm once W and the MST have identical node sets.

The disadvantage of this method is that there is no guarantee that links in the MST will indeed represent significant correlations. One typically needs an external evaluation criterion to decide whether the maximum spanning tree is informative about the real structure in the data. In some cases this can be done by exploiting the similarity between the construction of the MST and hierarchical clustering. If the community structure of the correlation network is known, nodes from the same community should be in close proximity to each other in the MST, meaning that they are aligned along the branches of the tree. Some authors combine the MST with a global thresholding approach [190]. In that approach, the filtered network consists of the maximum spanning tree plus all links that exceed a specific significance threshold. The former ensures that the resulting network spans the original network; the latter reveals the statistically significant correlation structure, which could have been lost if the MST alone had been used.

4.7.2.4 Network backboning with the disparity filter

The challenge of identifying the most relevant connections in large and dense networks has turned into a research field in its own right, and is sometimes referred to as network *backboning* [238]. Ideally, the 'backbone' of a network should span the entire original network and contain only links that have clear statistical meaning. One way of achieving this is to define a suitable null model for each node in the network separately, as is done in the so-called disparity filter [334]. In the following we describe the use of the disparity filter in correlation networks, where the weights are correlation coefficients, ρ_{ij}. The method applies to all weighted undirected networks and can be easily adapted for directed networks.

The disparity filter is particularly suitable for dealing with networks that contain hubs and where node strengths $s_i = \sum_i \rho_{ij}$ range over several orders of magnitude. The first step is to define a null model for each link that specifies how likely it is for a link with weight ρ_{ij} to be observed by pure chance, given the node strengths s_i and s_j. For each node i we consider the normalized weights, $w_{ij} = \rho_{ij}/\sum_j \rho_{ij}$. Let k_i be the degree of node i, that is, the number of links with non-zero weights.[8] Here the null model specifies that the k_i different normalized weights w_{ij} are randomly assigned from a uniform distribution. This means that $k_i - 1$ points are randomly drawn from the interval $[0, 1]$. Think of this interval as a line from zero to one that has been partitioned into k_i different segments of lengths w_{ij} by the $k_i - 1$ points. The null model specifies that the normalized weights of the observed network have been generated by such a segmentation process. Let us compute the probability that a link with a normalized weight of at least w_{ij} is observed in the null model. Such a link is observed whenever one of the k_i segments of the interval $[0, 1]$ has a length of at least w_{ij}. This is exactly the case when all $k_i - 1$ points fall within an interval with a length of at most $(1 - w_{ij})$. For one point the corresponding probability is $(1 - w_{ij})$. For all $k_i - 1$ points such a configuration occurs with probability $(1 - w_{ij})^{k_i-1}$. This result can be used to impose a significance threshold on the links. The p-value in the disparity filter is exactly the probability of observing a link with a given normalized weight in the null model. A link (i,j) is therefore significant at level p whenever $(1 - w_{ij})^{k_i-1} < p$.

4.7.2.5   *How should I choose my p, α, and Q?*

What is a good significance level? As already stated, for a single hypothesis test a conventional threshold for the p-value is 0.05. Assuming a normal distribution in the test statistic, physicists would call this a two-sigma effect.[9] The choice of a suitable threshold should depend on the cost of false positive links. Another strict requirement for the choice of a significance threshold is that once it has been set, it must be applied consistently throughout the entire analysis.

How to decide on good choices for the FWER, α, or the FDR, Q, in the case of multiple hypotheses is even more challenging. The vast majority of researchers use $\alpha = 0.05$ or $Q = 0.05$ as significance thresholds.[10] Ultimately, the choice of α or Q can only be based on how costly a given number of false positives or false negatives is. In situations where just one false positive link in the correlation network would be associated with substantial costs, the FWER should be controlled. The choice of α can then be made on the basis of what an acceptable risk for paying such a cost is. The FDR, in contrast, controls not the overall number of false positives but the expected proportion of false positives amongst all significant links. FDR control is particularly useful when the cost of a false negative would be significantly higher than the cost of a false positive. For instance,

---

[8] For correlation networks, k_i will often be $N - 1$.

[9] Just as a reference, in physics a five-sigma effect is required to claim discovery of a new particle. This corresponds to $p < 10^{-6}$.

[10] This choice is often motivated either by a confusion between p and α or Q by the researchers themselves, or by the desire to avoid this confusion with readers (and referees) who are not familiar with the details of multiple hypothesis testing.

one false negative in a correlation network could mean that we miss a protein–protein interaction that, after a number of more elaborate tests, could potentially serve as the basis for developing a cure for cancer. In such cases, choices of $Q > 0.05$ are often acceptable.

4.8 Dynamics on and of networks—from diffusion to co-evolution

As discussed in Section 1.5, many complex systems are characterized by a co-evolutionary dynamics of the states of their elements and the interactions between them. This co-evolutionary dynamics can be expressed in the form,

$$\frac{d}{dt}\sigma_i(t) \sim F\left(M_{ij}^{\alpha}(t), \sigma_j(t)\right)$$
$$\frac{d}{dt}M_{ij}^{\alpha}(t) \sim G\left(M_{ij}^{\vec{\alpha}}(t), \sigma_j(t)\right), \qquad (4.52)$$

with some (deterministic, stochastic, linear, or non-linear) functions F and G, that depend on the state vectors, $\sigma_i(t)$, and the interactions (of type α) in the system,[11] $M_{ij}^{\alpha}(t)$. The simplest form of this co-evolution confines states of the nodes, $\sigma_i(t)$, to real numbers or categorical variables and assumes that the interaction tensor $M_{ij}^{\alpha}(t)$ is a network A. Changes in the state vector can be understood as a dynamical process that takes place on a network. Changes in the interactions correspond to a dynamical process of the network itself. It transpires that many seemingly different real world processes take on a very similar form when expressed in terms of the co-evolutionary dynamics in Equation (4.52).

Here, we consider three different aspects of co-evolutionary dynamics with a broad spectrum of applications. First, we consider diffusion processes on networks, where substances, particles, walkers, or information packages spread on the nodes of networks by following their links [61, 83, 218, 291, 298]. Important applications of such processes include flows on networks, such as traffic, goods, or money [63]. The second aspect highlights models where nodes have internal states that characterize their activity in actively spreading information to other nodes, for example, opinions, computer viruses on the internet [204, 301] or diseases in social networks [266, 279, 285, 396]. We mention co-evolutionary versions of these models, which are referred to as adaptive network models. Some of them are remarkably well understood [158]. Finally, the third aspect considers cases where there might be more than one spreading process taking place on the network at the same time, and that these compete for individual nodes [253]. Examples of such processes include opinion formation or the adoption of novel technologies [74].

---

[11] Note that here $\frac{d}{dt}$ should be read as 'change over time'. It does not need to be a real differential.

4.8.1 Diffusion on networks

Diffusion processes that take place on networks can be conveniently formulated by using random walkers [292]. Imagine a large population of random walkers or particles that are located at the nodes in a network. At every time step these walkers jump from node to node along the links in the network. A particular diffusion process is then specified by its 'decision rule' for how the random walker chooses the next node to jump to. On directed networks, walkers also respect the directions of the links, meaning that on a directed link from i to j, they jump from i to j but never in the opposite direction. On weighted networks, the link weights typically indicate a preference on the part of a random walker to jump along the corresponding link. The number of random walkers can be conserved over time or they could be added to or removed from nodes with specific rates.

A general random walk model for diffusion on networks can be stated in terms of the following update equation. The number of walkers at node i at time t is denoted by $x_i(t)$. For the entire population we write $x(t) = (x_1(t), x_2(t), \cdots, x_N(t))$. We assume discrete time. All walkers jump from one node to another at each time step. In general, the walkers can multiply or vanish during a jump. For each walker at node j at time t, a number of $n(j \to i)$ walkers are added at the next time step, $t + 1$, at node i, and the original walker is removed from node j. The transition elements $n(j \to i)$ are often a function of the adjacency matrix of the underlying network. Jumps from node i to i are allowed. In addition, each node i is endowed with a specific birth rate of random walkers, β_i. Initially, a large number of random walkers is distributed at random across all nodes. A general random walk model for diffusion on networks is then given by the following update equation for each node i,

$$x_i(t+1) = \sum_j n(j \to i) x_j(t) + \beta_i. \tag{4.53}$$

What is the stationary distribution x^* of the random walkers across all nodes after a large number of jumps, $t \to \infty$? Stationary distributions are a specific class of centrality measure that rank the nodes according to the probability that random walkers will be found at the node; see Figure 4.17.

Depending on how the transition elements, $n(j \to i)$, and the birth rates, β_i, are specified, several centrality measures can be recovered.

- In a specific form of Laplacian diffusion, the transition elements are given by $n(j \to i) = A_{ij}/k_j$ and the birth rate is zero, $\beta_i = 0$ [83]. Laplacian diffusion and the corresponding random-walk-based centrality are discussed in Section 4.8.2.
- Eigenvector centrality is obtained from the transition elements $n(j \to i) = A_{ij}$, and a birth rate of $\beta_i = 0$ [61]; see Section 4.8.3.

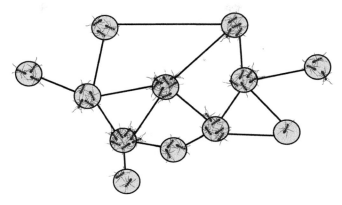

Figure 4.17 *Illustration of random walks on networks. A large number of random walkers jump at each time step from one node to another along the links of the network. Random-walk-based centrality measures estimate the probability of finding the random walkers at a given node.*

- Katz prestige defines the transition elements in the same way as eigenvector centrality, $n(j \rightarrow i) = A_{ij}$, but introduces constant birth rates $\beta_i = \beta > 0$ [218]; see Section 4.8.4.
- Google's PageRank algorithm is defined for directed networks with transition elements $n(j \rightarrow i) = A_{ij}/k_j^{out}$ and constant birth rates $\beta_i = \beta > 0$ [298]; see Section 4.8.5.

A particularly simple example of a centrality measure obtained by random walks is the node degree for unweighted networks. To see this, assume that all nodes i have birth rates of zero, $\beta_i = 0$, and that the transition elements are given by the adjacency matrix of the network, $n(j \rightarrow i) = A_{ij}$. Initially, there is exactly one random walker at each node, $x_i(0) = 1$ for all i. The node degree k_i is then the number of random walkers after exactly one time step, $x_i(t = 1) = \sum_j n(j \rightarrow i)x_j(0) = \sum_j A_{ij} = k_i$. By considering the stationary distribution x^* of random walkers for large t, one ranks nodes not only by their degree but also by the degrees of their neighbours, the degrees of their neighbours' neighbours, and so forth. This leads to a recursive definition of network centrality. In this sense, high centrality nodes are those that link to other nodes with high centrality.

4.8.2 Laplacian diffusion on networks

A simple random walk model on a network assumes that a fixed number of walkers jump randomly from node to node. To initialize the walks, we distribute a large but fixed number of walkers randomly across the nodes. At each time step, each random walker performs one jump to one of its neighbouring nodes. The probability of jumping from i to j is proportional to the weight of the corresponding link. On unweighted and undirected networks, it is A_{ji}/k_i. The process can be described with an iterative update equation for x of the form,

$$x_i(t+1) = \sum_j \frac{A_{ij}}{k_j} x_j(t). \tag{4.54}$$

The total number of random walkers is conserved in the model, $\sum_i x_i(t+1) = \sum_j (\sum_i A_{ij}) x_j(t)/k_j = \sum_j k_j x_j(t)/k_j = \sum_j x_j(t)$. In the limit of large times, we assume that there is a stationary distribution, $x^* = x(t \to \infty)$. The condition for stationarity is that the distribution does not change from one time step to the next, $x^*(t+1) = x^*(t)$. In vector notation this means that,

$$x^* = AD^{-1}x^*. \tag{4.55}$$

Reading Equation (4.55) as an eigenvalue problem, it follows that the stationary distribution is an eigenvector to the matrix AD^{-1} with the eigenvalue 1. In general, this eigenvector has entries that are not identical. As we will see next, the random-walk-based centrality measure in Equation (4.55) is actually a special case of PageRank. The random walk in Equation (4.54) can easily be generalized to directed and weighted networks with weight matrix W. In this case, the normalization factor is not given by the degree, but by the out-strength $s_i^{out} = \sum_j W_{ji}$. The iterative update equation for x is then of the form,

$$x_i(t+1) = \sum_j \frac{W_{ij}}{s_j^{out}} x_j(t). \tag{4.56}$$

The rest follows exactly as before. As a side note, we mention that there are other approaches than random walk models to introduce diffusion on networks. Physical transport processes describe diffusion as a flux of particles that is induced by a difference in concentrations between regions [128]. The flux goes from high to low concentrations; its magnitude is proportional to the concentration gradient between the regions. This is known as Fick's law. We consider such a transport process that takes place on simple networks [291]. Between adjacent nodes i and j there is a flux that is proportional to the difference between x_i and x_j,

$$\frac{dx_i}{dt} = D^{\text{diff}} \sum_j A_{ij}(x_j - x_i). \tag{4.57}$$

D^{diff} is the diffusion constant. The sum can be written as $\sum_j A_{ij}x_j - x_i \sum_j A_{ij} = \sum_j (A_{ij} - \delta_{ij}k_i)x_j$. The matrix with elements $A_{ij} - \delta_{ij}k_i$ is the Laplacian matrix $\Lambda = D - A$; see Equation (4.40). We see that the diffusion equation on networks based on Fick's law is,

$$\frac{dx}{dt} + D^{\text{diff}}\Lambda x = 0, \tag{4.58}$$

where we switched to vector notation. To solve the equation, let v_i be the eigenvectors to the eigenvalue λ_i, meaning $\Lambda v_i = \lambda_i v_i$. We know from Section 4.6 that all λ_i are non-negative and that the smallest eigenvalue is zero with a corresponding eigenvector of $v_{\min} = \mathbf{1} = (1, 1, \cdots, 1)$. From linear algebra we know that we can write x as a linear combination of its eigenvectors with the coefficients $\gamma_i(t)$, $x(t) = \sum_i \gamma_i(t) v_i$. Using this Ansatz in Equation (4.58) gives,

$$\sum_i \left(\frac{d\gamma_i}{dt} + D^{\text{diff}}\lambda_i \gamma_i \right) v_i = 0. \tag{4.59}$$

As eigenvectors are orthogonal, every term in the sum must be zero independently, and one can solve the linear differential equations, $\gamma_i(t) = \gamma_i(0)\exp(-D^{\text{diff}}\lambda_i t)$; we also have,

$$x(t) = \gamma_1(0)v_1 e^{-D^{\text{diff}}\lambda_1 t} + \cdots + \gamma_N(0)v_{\min}e^{-D^{\text{diff}}\lambda_N t}. \tag{4.60}$$

For large t, the sum in Equation (4.60) is dominated by the last term with the smallest eigenvalue (which is exactly zero, $\lambda_N = 0$), which means that the stationary distribution is obtained as,

$$x^* = \gamma_N(0)\mathbf{1}. \tag{4.61}$$

In other words, the particles are uniformly distributed across all nodes. Should the smallest eigenvalue be degenerate, which means that the network is not connected, the stationary distribution is a linear combination of the eigenvectors with eigenvalues zero.

We have considered two different approaches to diffusion: the random walker approach in Equation (4.55) and the transport process in Equation (4.58). The stationary distributions that result from these two approaches are completely different. In the first random walker approach, the stationary distribution is given by the eigenvector of the matrix $D^{-1}A$ to the eigenvalue one. In contrast, physical transport processes given by the Laplacian matrix Λ result in a uniform stationary distribution of random walkers across nodes, which is not useful as a centrality measure. Nevertheless, the two processes can be formally related to each other. For the Laplacian matrix we have that $\Lambda = D - A = D(\mathbb{I} - D^{-1}A)$, where $\mathbb{I}$ is the identity matrix with entries δ_{ij}. To see the relationship between Λ and $D^{-1}A$, introduce the random walk normalized Laplacian, $\Lambda^{rw} = D^{-1}\Lambda = \mathbb{I} - D^{-1}A$. The Laplacian matrix Λ has many useful properties. For undirected networks, Λ is symmetric, positive-semidefinite, and informative for the component and community structure of the network. The random walk normalized Laplacian allows the known properties of Λ to be connected with the properties of the random walk in Equation (4.55), which can often aid the analysis of related stochastic processes [83].

4.8.3   Eigenvector centrality

We now drop the constraint that the total number of random walkers on the network is conserved. Consider a spreading process of the following type. For each random walker at node j at time t, there will be A_{ij} incoming walkers at node i at time $t + 1$. The number of random walkers at node i evolves over time as,

$$x_i(t+1) = \sum_{j=1}^{N} A_{ij} x_j(t). \tag{4.62}$$

After t iterations, the number of random walkers is given by $x(t) = A^t x(0)$, where $x(0)$ is the initial distribution. As before, we write $x(t)$ in the basis spanned by the eigenvectors v_i of A and time-dependent coefficients $\gamma_i(t)$, $x(t) = \sum_i \gamma_i(t) v_i$. We assume that the eigenvalues are ordered by their absolute value in a descending way. It follows that λ_1 is the Perron–Frobenius eigenvalue of A with an eigenvector v_1 with non-negative entries. Using Equation (4.62) we get,

$$x(t) = A^t \sum_i \gamma_i(0) v_i = \lambda_1^t \sum_i \gamma_i(0) \left(\frac{\lambda_i}{\lambda_1}\right)^t v_i. \tag{4.63}$$

For $t \to \infty$ it follows from the properties of the Perron–Frobenius eigenvalue that $\lim_{t\to\infty} \left(\frac{\lambda_i}{\lambda_1}\right)^t = 0$ whenever $i \neq 1$. The stationary distribution, x^*, is obtained in the limit,

$$x^* = \lim_{t\to\infty} x(t) = \lim_{t\to\infty} \lambda_1^t \sum_i \gamma_i(0) \left(\frac{\lambda_i}{\lambda_1}\right)^t v_i = \lim_{t\to\infty} \lambda_1^t \gamma_1(0) v_1 \sim v_1. \tag{4.64}$$

We find that x^* is given by the eigenvector corresponding to the largest (Perron–Frobenius) eigenvalue of the adjacency matrix,

$$A x^* = \lambda_1 x^*. \tag{4.65}$$

This stationary distribution is called the eigenvector centrality of the network A [61]. The Perron–Frobenius theorem tells us that the eigenvector centrality exists and has non-negative and real entries. Eigenvector centrality basically describes a diffusion process, where walkers jump from node to node by multiplying (by a factor of A_{ij}) along each link that they follow. However, if A is directed and weighted, the eigenvector centrality has two undesirable properties. First, the overall number of random walkers diverges if the spectral radius (i.e. the Perron–Frobenius eigenvalue) is larger than one. Fortunately, this growth can always be absorbed into a global normalization constant. Second, consider a node i with an in-degree of zero. After the initial population of random walkers has left node i, there is no way for a walker to reach i. The same holds for all nodes that receive only incoming links from i after the second time step, and so forth. The eigenvector centrality of all nodes in the in-component of a directed network is exactly zero and therefore not informative. This problem is solved with the Katz prestige.

4.8.4   Katz prestige

Consider a random walk of the type given in Equation (4.62) with one important modification. To be able to assign non-zero centralities to nodes in the in-component of the network, we assume a constant in-flow or a birth rate of β random walkers at each time step and node [218]. The number of random walkers at node i then evolves as,

$$x_i(t+1) = \alpha \sum_j A_{ij} x_j(t) + \beta, \tag{4.66}$$

where we have introduced a constant α that will be chosen such that the distribution x^* is stationary. Using $\mathbf{1} = (1, 1, \cdots, 1)$, Equation (4.66) can be written in matrix notation,

$$x(t+1) = \alpha A x(t) + \beta \mathbf{1}. \tag{4.67}$$

The stationary distribution, for which the condition $x^*(t+1) = x^*(t) = x^*$ must hold, can then be obtained by rearranging terms,

$$x^* = \beta (\mathbb{I} - \alpha A)^{-1} \mathbf{1}, \tag{4.68}$$

where $\mathbb{I}$ is the identity matrix with entries δ_{ij}. This stationary distribution is called the *Katz prestige*. The parameter β is an overall constant that can be set to one without loss of generality. The parameter α should be chosen such that the number of random walkers does not diverge, which is achieved by choosing it from the range $0 < \alpha < 1/\lambda_1$, where λ_1 is the Perron–Frobenius eigenvalue of A.

In contrast to the eigenvector centrality, the Katz prestige has the nice property that it can also be used to rank nodes in the in-component of the network. Except for the constant birth of new walkers at each node, it basically describes the same diffusion process as the eigenvector centrality. A feature of this process is that the entire population of walkers at a given node is copied to each neighbouring node at each time step and is therefore multiplied. For certain types of spreading processes, this behaviour might be desirable, but there are processes for which this kind of 'copying' is a poor assumption.

4.8.5   PageRank

The world wide web is a network where nodes are pages and the links are hyperlinks. Consider the process of browsing the web. At each point in time you see a particular web page in your browser. By clicking on a link on that page you perform a walk from the current website to the page the hyperlink is linking to. Consider a population of random walkers that randomly click on hyperlinks. A model for this situation, where random walkers would 'multiply' and follow every link at once, would be a poor assumption. The eigenvector centrality and the Katz prestige are not appropriate. Instead, the population of walkers will partition itself across all outgoing links of a website. Such a diffusion process leads us to yet another centrality measure. An iterative update equation for the numbers of random walkers that follow only one of the outgoing links of their current node is given by,

$$x_i(t+1) = \alpha \sum_{j=1}^{N} A_{ij} \frac{x_j(t)}{k_j^{out}} + \beta, \tag{4.69}$$

where k_j^{out} is the out-degree of node j. This process assumes that the population of walkers at j distributes itself uniformly across all outgoing links. Node i then receives A_{ij} copies of the fraction $x_j(t)/k_j^{out}$ of random walkers. Equation (4.69) can again be written in matrix notation. For this we introduce the diagonal matrix D', with entries on the diagonal $D'_{ii} = \max(k_i^{out}, 1)$. The diffusion process is,

$$x(t+1) = \alpha A D'^{-1} x(t) + \beta \mathbf{1}. \tag{4.70}$$

Note the similarity between Equation (4.70) and the random walk in Equations (4.54) and (4.55), which are recovered in the absence of the birth rate, $\beta = 0$ and $\alpha = 1$. We obtain the stationary distribution x^* of walkers for long times, where $x(t) = x(t+1) = x^*$ holds,

$$x^* = \beta(\mathbb{I} - \alpha A D'^{-1})^{-1} \mathbf{1}. \tag{4.71}$$

The parameter β is a global constant that can be set to one. The divergence of the population at the nodes can be avoided for an appropriate choice of $0 < \alpha < 1$. This centrality measure was first proposed as a way of ranking websites according to their importance by the founders of Google [298]. The stationary distribution x^* is called the PageRank if $\alpha = 0.85$. The reasons for this particular choice of α are not clear.

4.8.6   Contagion dynamics and epidemic spreading

In the diffusion or spreading processes that we have considered to date, each node was 'susceptible' to an influx of random walkers. For many diffusion processes, this is not the case. For example, think of the spreading of infectious diseases. The spreading can be understood as a process that takes place on a social network, where links correspond to physical interactions or contacts. At each point in time, each individual can be in one of several states: she can be susceptible to the pathogen and/or she can actively spread it to other nodes. One central question in this context is whether an initial outbreak of the disease will eventually spread to a significant portion of the network and how such an outbreak might be prevented by different immunization strategies. In the following, we discuss simple diffusion models in the context of the transmission of diseases. These models also apply to other types of spreading, such as the diffusion of information or innovations, or the cascading of failures in networks.

4.8.6.1 *Epidemic spreading on networks in the SI model*

Let us assume that each node is in one of two states, *susceptible* or *infected*, the so-called SI model; see, for example, [204]. Initially, we assume that only a single node in the network is infected. If a node is infected, it spreads the disease to all adjacent susceptible nodes. The size of an epidemic outbreak can be measured as the fraction of nodes, q, that will finally be infected. For an ER network, the epidemic spreading behaviour in the SI model is closely related to the existence of a giant component that appears at the threshold $p_c \sim 1/N$; see Section 4.4.3.4. If there is at least one infected node in the giant component, the disease will spread until all other nodes in the component are infected too. For ER networks, the infected fraction q is therefore given by the size of the giant component that is given in Equation (4.29).

4.8.6.2 *SI model with immunization*

We now extend the SI model by allowing nodes to be immune to the disease (for instance, as a result of a vaccination). The probability of being immunized is given by the parameter π (immunization rate). Infected nodes can infect susceptible nodes in their neighbourhood, but not nodes that are immune. It follows that the spreading process takes place on the induced subnetwork of susceptible nodes of the original network. For an ER network of N nodes with linking probability p, this induced subnetwork is again an ER network with $(1 - \pi)N$ nodes and the same linking probability p. The question of whether a disease can spread through a large fraction of the network is then equivalent to the question of whether the induced subnetwork of susceptible nodes has a giant component. In other words, the question as to whether an epidemic outbreak will occur, depends on whether the subnetwork of susceptible nodes is above or below its percolation threshold.

The epidemic spreading threshold for ER networks in the SI-model with immunization is given by the percolation threshold for the induced subnetwork of susceptible nodes that are not immune, $p_c \sim 1/(N(1 - \pi))$. If N and p are fixed, the critical immunization rate is $\pi_c = 1 - 1/(Np)$. For immunization rates above π_c, no massive spreading will occur. The fraction q of nodes that will finally become infected can be estimated from Equation (4.29),

$$q = 1 - e^{-(N-1)p(1-\pi)q}. \tag{4.72}$$

4.8.6.3 *Epidemic spreading on networks in the SIS model*

We now consider spreading processes that allow each node to recover from an infection and to become susceptible again. Nodes undergo a life cycle from susceptible to infected to susceptible. This model is called the SIS model. Given an infected neighbour, nodes become infected at a rate η. Infected nodes recover to the susceptible state with a rate

δ. An effective spreading rate, λ, of the disease can be defined as the quotient of these two rates, $\lambda = \eta/\delta$. Without loss of generality one can set $\delta = 1$. The spreading rate λ is then the rate at which an infected node will infect one of its neighbours. As long as the density of infected nodes is low, each node will on average infect $R = \lambda \langle k \rangle$ other nodes. R is called the basic reproductive number of the disease and $\langle k \rangle$ is the average degree in the network. If $R > 1$ the number of infected nodes will multiply and the disease could, in principle, spread through the entire population. If $R < 1$ the epidemic dies out.

The reproductive number R is enough to specify the behaviour of the SIS model on networks where nodes have comparable degrees, so-called homogeneous networks. Such homogeneous networks include ER networks and the small world networks of the Watts–Strogatz model. An epidemic will persist in homogeneous networks whenever its spreading rate, λ, is larger than a critical spreading rate $\lambda_c = 1/\langle k \rangle$. Otherwise, for $\lambda < \lambda_c$, no spreading occurs. The fraction of infected nodes in the SIS model for ER networks, q_{ER}, is therefore $q_{ER} = 0$ for $\lambda < \lambda_c$. When λ is above but close to the critical spreading rate, $\lambda > \lambda_c$, q_{ER} can be estimated as $q_{ER} \sim \lambda - \lambda_c$; see [302].

For the SIS model on networks with *arbitrary* degree distributions, that is, for the configuration model, the situation is less obvious. For instance, if there is a hub with links to every other node in the network, then the disease can, in principle, reach each node within two contagion steps or less. How likely is it that such hubs are in the neighbourhood of an infected node? More formally, consider the case where one of the neighbours j of an infected node i has been infected too. What is the average number of nodes that j can infect? Clearly, the epidemic can persist only if this number, the reproductive number R', is larger than one. Remember that the average nearest-neighbour degree in the configuration model is given by $\langle k^2 \rangle/\langle k \rangle$; see Equation (4.32). The node j can, on average, infect $\langle k^2 \rangle/\langle k \rangle - 1$ other nodes (the term -1 comes from the fact that node i is already infected). Each neighbour j of an infected node can therefore infect a number of R' other nodes,

$$R' = \lambda \frac{\langle k^2 \rangle - \langle k \rangle}{\langle k \rangle}. \tag{4.73}$$

The epidemics die out if $R' < 1$ and persists for $R' > 1$. Equation (4.73) takes into account the effects of heterogeneous degree distributions on the spreading process. This result has immediate consequences for epidemic spreading on scale-free networks with a power law degree distribution $P(k) \sim k^{-\gamma}$; see [301]. For exponents $\gamma \leq 3$ the reproductive number R' in Equation (4.73) diverges as $N \to \infty$ for each non-zero spreading rate, $\lambda > 0$. The critical spreading rate for scale-free networks with $\gamma \leq 3$ is therefore zero, $\lambda_c = 0$. A single node can infect a macroscopic fraction of an infinitely large network.[12]

Scale-free networks do not have an epidemic threshold if the scaling exponent $\gamma \leq 3$.

---

[12] For $2 < \gamma \leq 3$ the second moment $\langle k^2 \rangle = \sum_k P(k)k^2 \sim \sum_k k^{2-\gamma}$ is infinite, whereas $\langle k \rangle$ is finite, so λ diverges. For $\gamma \leq 2$, the first and second moments are both infinite, but $\langle k^2 \rangle$ clearly diverges faster than $\langle k \rangle$.

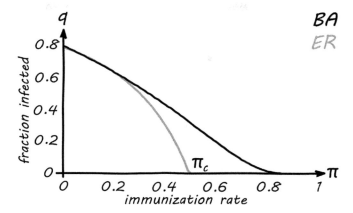

Figure 4.18 *Illustration of the epidemic threshold in the SIS model of infected and immunized nodes on ER networks (grey) and for scale-free networks from the BA model (black). For an immunization rate above π_c the fraction of infected nodes, q_{ER}, is marginal for ER networks. There is no critical π_c for the same model on a scale-free network with $\gamma = 3$. A non-zero number of infections will persist unless practically every node is immunized, $\pi = 1$.*

4.8.6.4 SIS model with immunization

It is impossible to prevent epidemic outbreaks on scale-free networks with $\gamma \leq 3$ by random immunizations. Assume that each individual has been immunized with probability π. Effectively, this rescales the spreading rate λ to $\lambda(1 - \pi)$. Unless all individuals are immunized, meaning that $\pi = 1$, the reproductive number R' in Equation (4.73) still diverges. It can be shown in the mean field limit that the fraction of infected nodes, q, in the SIS model on scale-free networks with $\gamma = 3$ depends on the immunization rate π [302],

$$q \sim 2\exp\left(\frac{-1}{\lambda(1 - \pi)}\right). \tag{4.74}$$

Equation (4.74) shows that the expected fraction of infected nodes for $\pi < 1$ for scale-free networks is always above zero, as opposed to ER networks, where a critical immunization rate π_c exists; see Figure 4.18.

The world wide web seems to exhibit a power law degree distribution over several orders of magnitude that follows an exponent of $\gamma \sim 2.1$ for the in-degree and $\gamma \sim 2.45$ for the out-degree distribution [27]. This suggests that massive outbursts of malware infections in the web are impossible to prevent unless virtually every computer is protected.

4.8.6.5 Generalized epidemic spreading, SIR models

So far we have considered epidemic spreading models where the nodes are susceptible, infected, or immune. In the spreading of real diseases, there could be more relevant states [11, 228]. Nodes could be removed after being infected for a certain time (nodes die).

Typically, models that involve these steps are named according to the possible states that nodes can assume. We have already encountered the SI model, where nodes are either susceptible (S) or infected (I) and the SIS model, where infected nodes recover and become susceptible again. Models where nodes become resistant (R) after the disease and can no longer be infected are called SIR models. In so-called SIRS models, however, nodes become susceptible again after being resistant for some time. Each of these state transitions can be specified with transition probabilities that determine the behaviour of the models. These models can be implemented as agent-based models or, if no networks are involved, by simple differential equations. For example, the fully connected SIR model with N individuals is given by the set of equations,

$$\frac{dS}{dt} = -\beta \frac{SI}{N}$$
$$\frac{dI}{dt} = \beta \frac{SI}{N} - \gamma I$$
$$\frac{dR}{dt} = \gamma I$$
$$N = S + R + I. \tag{4.75}$$

Upper-case letters denote the relative fractions of the population in a particular state. If networks are involved, things get more involved.

4.8.7   Co-evolving spreading models—adaptive networks

When networks are allowed to rewire as a response to the states of the nodes, things become mathematically interesting. Intuitively, it is reasonable that social networks might rearrange dynamically if it is known that a given person is infected and/or contagious. Social links might be temporarily removed from infected persons. Epidemic dynamics on such adaptive networks have been studied, in which nodes in susceptible states try to avoid contact with infected nodes by rewiring their links [157]. These models can be well approximated by a low-dimensional system of differential equations, which allows the phase diagrams to be understood in terms of the model parameters. These phase diagrams are very different from epidemic spreading on static networks, which has implications for the spreading of infectious diseases and related applications. An excellent overview of work in this direction is available in [158].

Without adaptive dynamics, the SI model is typically in one of two phases; either most nodes are healthy (healthy phase) or a macroscopic fraction of the nodes are infected (infected phase). Adaptive dynamics, where healthy nodes try to avoid contact with infected ones, have an additional phase [157]. The additional phase lies between the healthy and infected phases and is bistable, meaning that healthy and infected node states are stable at the same time. In some cases, the model even oscillates, with the number of infected nodes changing periodically. Oscillations arise due to two different effects introduced by the adaptive dynamics. First, the dynamics tends to isolate infected nodes. As a result, there is an effective reduction of the disease prevalence in the remaining network, and this drives the system towards a healthy state. Second, the adaptive dynamics increases the density of links between susceptible nodes. This increase drives the system towards its epidemic threshold and therefore towards the infected phase. The interplay of these two effects gives rise to oscillations between the healthy and infected phase in the following way: initially, the isolating effect dominates and the number of infected nodes decreases. The subnetwork of susceptible nodes becomes increasingly dense until it crosses its epidemic threshold. The epidemic can now spread through these susceptible nodes, and the number of infected ones starts to increase. The subnetwork of susceptible nodes dissolves as a result of this epidemic outbreak, and the isolating effect takes over again. This completes a period in the oscillation. The dynamics of the adaptive SI model has important consequences for the control of real world disease outbreaks. If the disease prevalence is low enough, adaptive dynamics makes the disease easier to control due to the ability to isolate infected individuals. However, once an outbreak occurs, the control of the disease by means of vaccinations will be complicated by the high density of links between susceptible nodes. For details on network structures close to phase transitions in adaptive network models, see [198].

4.8.8   Simple models for social dynamics

We now consider systems in which several spreading processes compete with each other. Models of this kind provide a simple mathematical representation of how collective phenomena emerge in specific classes of social dynamics. Examples include the dynamics of opinions [253, 345, 359], cultural behaviour [18, 328], language formation [353], and collective crowd behaviour in physical spaces [187]. The vast majority of these models are too simple to be credibly calibrated to real systems on a microscopic or individual-based level [74]. The value of these models is mainly to clarify whether a certain type of hypothetical microscopic dynamics is compatible with the observed macroscopic behaviour of a system. Many of these models draw their inspiration from spin models. The basic idea is that different opinions or behavioural traits can be represented as different internal states of the nodes, their 'spin'. Depending on whether two individuals seek to agree or disagree in their internal states, their interactions can be modelled as being ferromagnetic or antiferromagnetic. This analogy between social dynamics and spin models has inspired a large number of recent works, which are sometimes referred to as 'sociophysics' [78, 141]. Spin models in the context of social phenomena mainly aim to understand collective phenomena in social dynamics by borrowing methods from statistical physics.

4.8.8.1   *Voter model for opinion formation*

The voter model[13] is popular for its mathematical simplicity and its flexibility. It can easily be extended and applied to many different contexts [196]. Moreover, it can be solved exactly in any dimension [318]. The model is defined for a set of 'voters' i that

---

[13] Hardly anyone would claim that the voter model is actually a model for voters [127].

are located at the nodes of a simple network A. At any point in time, they hold one of two opinions, $\sigma_i = \pm 1$. Each node is initially assigned one of the opinions with equal probability. The dynamics of the model is specified by the update rule for nodes. At each step one randomly picks a node i and one of its nearest neighbours, say j. The state of i is updated to the state of j, $\sigma_i(t+1) = \sigma_j(t)$ and proceeds to the next step. Assuming that the network is connected, the voter model has two absorbing states[14] that correspond to $\sigma_i = +1$ and $\sigma_i = -1$, for all nodes i, respectively. Updates only occur if two neighbouring nodes differ in their states. A link that connects two such nodes is called 'active'. A suitable order parameter for describing the activity in the system is the density of active links, also called the *interface density*,

$$\rho_{\mathrm{IF}} = \frac{1}{\sum_{i=1}^{N} k_i} \left(\sum_{i=1}^{N} \sum_{j=1}^{N} A_{ij} \frac{1 - \sigma_i \sigma_j}{2} \right). \tag{4.76}$$

Initially, the system is in a maximally disordered configuration with a chance of $1/2$ of two neighbouring nodes agreeing, and $\rho_{\mathrm{IF}} \sim 1/2$. Over time, node updates tend to decrease the interface density until the system eventually reaches one of the absorbing states. Consensus is characterized by $\rho_{\mathrm{IF}} = 0$. If the network consists of more than one component, each component approaches its own consensus state. A system of finite size will always reach the consensus state. For infinite systems it has been shown that on regular lattices in more than two dimensions (and in small world networks) there is no consensus in the thermodynamic limit [357]. In these cases there is an initial decrease in ρ_{IF} (partial ordering), after which the system gets 'stuck' in metastable states. The lifetime of these metastable states diverges with increasing system size.

In the voter model there is a tendency for an individual to take the opinion that is held by the majority of his or her neighbours, as the neighbours in the majority are more likely to be picked in the update process. There are a large number of variations of the voter model that explicitly study this effect. For instance, in one variation, the majority rule specifies that a node will always choose the opinion held by the majority of its neighbours [139, 242]. Without further modifications, this rule will always lead to a consensus state. Other variants include the addition of so-called contrarians who always oppose the majority [140]. Threshold models for opinion formation assume that the view of the majority is only adopted if the fraction of neighbours with that opinion exceeds a certain threshold [234, 252, 400]. Both variants, the contrarians and the threshold voters, can prevent the system from reaching full consensus even for finite size.

4.8.8.2 *Co-evolving voter model*

As discussed in the introduction, one hallmark of complex systems is the co-evolution of internal states and interactions. For dynamical models that are defined on static networks there is no co-evolution. Co-evolution is introduced by allowing the network to become dynamic and adaptive. This step has the potential to radically alter the basic properties

---

[14] Once the system is in the absorbing state, it will remain there for an infinite amount of time.

of the system. The voter model illustrates this point exceptionally well. Consider the following modification to the update rule of the voter model that leads to the so-called *co-evolving voter model*; see [391].

1. Pick a node i at random from all nodes with a non-zero degree.
2. Pick one neighbour j of i at random.
3. With probability p remove the link between i and j and establish a new link with a node m, that is chosen at random from all nodes in the same internal state as i, meaning $m \in \{n | \sigma_n = \sigma_i\}$.
4. Otherwise, with probability $1 - p$, update as before $\sigma_i(t+1) = \sigma_j(t)$.

For $p = 0$ we recover the static voter model. For finite interaction networks A above their percolation threshold, for $p = 0$ there will always be a giant component with nodes of the same opinion. For $p > 0$ we explicitly introduce co-evolution between the node states and the link dynamics. The case of $p = 1$, where there are no updates of the internal states but only rewiring of the links, is particularly instructive. Assuming that the initial opinions are uniformly distributed, the network will segregate into two components of similar size. Each component will have a different opinion state. As the parameter p changes from zero to one, the co-evolving voter model undergoes a phase transition in which the system shifts from a giant component (all nodes have the same opinion) to a population that fragments into two opposing views. It has been shown that there is a critical transition between these two phases occurring at p_c. This is characterized by a power law for the time to reach consensus as a function of $(p - p_c)$; see [391].

The co-evolving voter model shows that collective phenomena of complex systems can only be understood on the basis of microscopic interaction rules if the interaction topology is faithfully represented in the model. More detailed models for the co-evolution of node and link states will be discussed in Chapter 5. However, most real world systems are characterized by interactions that cannot be described within the paradigm of simple networks, be they directed or undirected, weighted or unweighted, static or dynamic. To describe these systems, generalizations of the concept of networks are required that are well beyond pair-wise interactions of a single type.

4.9   Generalized networks

A comprehensive understanding of complex systems is only possible through an understanding of their full underlying interaction topology. Almost all real world systems have simultaneous interactions of many different types that might occur specifically between nodes that belong to different sets. With the growing availability of comprehensive and large-scale data sets on such systems, the interest in generalized network concepts that faithfully represent realistic interaction topologies is growing. Historically, generalized networks have been studied in mathematics in the form of power graphs and hypergraphs, or in sociology in the form of multiplex and multirelational networks; see Figures 4.19a–c. There are currently efforts to develop a fully generalized framework

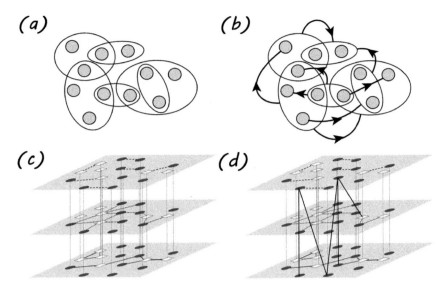

Figure 4.19 *Schematic examples of generalized networks: (a) hypergraphs, where 'links' connect more than two nodes, (b) power graphs, where links connect two sets of nodes, (c) multiplex networks, where nodes have multiple layers of links, and (d) multilayer networks, where 'links' can connect different nodes across different layers.*

for such 'networks of networks', that leads us to the concept of multilayer networks; see Figure 4.19d and [54, 232].

4.9.1   Hypergraphs

Conventional networks are given by a set of nodes $\mathcal{N}$ and a set of links $\mathcal{L}$. Links always connect exactly two nodes. It is possible to allow more general types of links that connect an arbitrary number of nodes, which leads to so-called *hypergraphs* [47]. Hypergraphs are sets of nodes and links, but the elements of the link set $\mathcal{L}$—the hyperedges—are elements of the power set of $\mathcal{N}$, $\mathcal{P}(\mathcal{N})$. The power set is defined as the set of all subsets of the original set, $\mathcal{N}$. For example, the power set of $\{1,2,3\}$ contains the elements $\mathcal{P}(\{1,2,3\}) = \{\{\},\{1\},\{2\},\{3\},\{1,2\},\{1,3\},\{2,3\},\{1,2,3\}\}$. The number of possible hyperedges is the number of elements of $\mathcal{P}(\mathcal{N})$, which gives 2^N potential links in the hypergraph with N nodes. The incidence matrix B of a hypergraph can be obtained in a similar way as for simple networks. Each hyperedge, ℓ, corresponds to one column in the incidence matrix B. The sum of the ℓth column in the incidence matrix therefore gives the number of nodes that the hyperedge ℓ connects. As for simple networks, the incidence matrix B can be related to the adjacency matrix of a bipartite network. Hypergraphs are used extensively in machine learning, where the hyperedges are used to represent communities or joint interactions [418].

4.9.2   Power graphs

Interactions that occur between *sets of nodes* need a slightly more general formulation than hypergraphs. An example of such interactions are mass action kinetics in chemistry. For instance, the reaction of H_2 and O_2 to form water, $2H_2 + O_2 \rightarrow 2H_2O$, can be represented as a link from the set $\{H_2, O_2\}$ to $\{H_2O\}$. More generally, chemical reactions map a set of elements (the reactants) to a different set of elements (the products). Topologically, this can be represented by a power graph in which links are defined between *sets of* nodes [227]. More formally, consider the power set $\mathcal{P}(\mathcal{N})$ of the node set $\mathcal{N}$. In power graphs, the nodes correspond to elements of $\mathcal{P}(\mathcal{N})$, so-called power nodes. These power nodes are connected by power links. The adjacency matrix of a power graph is therefore of dimension $2^N \times 2^N$. The mathematical representation of networks as matrices A_{ij}, is generalized to tensors with more than two indices, $A_{i_1,i_2,\dots,i_n,j_1,j_2,\dots,j_m}$, which corresponds to a link from the node set $\{i_1, i_2, \dots, i_n\}$ to the set $\{j_1, j_2, \dots, j_m\}$. Power graphs play a role, for example, in combinatorial evolution, which we will discuss in Section 5.6.4.

4.9.2.1   Bipartite representation of power graphs

To prevent the combinatorial explosion of network size, power graphs can also be represented as directed bipartite networks B, with two disjoint node sets $\mathcal{N}_1$ and $\mathcal{N}_2$. One set of nodes of the bipartite network corresponds to the original node set $\mathcal{N}$, $\mathcal{N}_1 = \mathcal{N}$. The second node set, $\mathcal{N}_2$, contains one node ρ_i for each link in the power graph. If the power link connects the node sets $\vec{i} = \{i_1, i_2, \dots, i_n\}$ and $\vec{j} = \{j_1, j_2, \dots, j_m\}$, then all nodes from $\vec{i}$ have a link pointing to ρ_i, while ρ_i points to all nodes contained in $\vec{j}$. For a chemical reaction network, each node ρ_i in $\mathcal{N}_2$ corresponds to a different chemical reaction, whereas $\mathcal{N}_1$ contains all reactants and products. Power graphs are therefore well-suited to describing chemical reaction networks in computational biology, in particular, protein–protein interactions or gene regulatory networks [103, 324].

4.9.3   Multiplex networks

Many real world systems are characterized by multiple types of interactions that are happening simultaneously. The need for networks that allow for different interaction types has long been recognized in sociology. Humans are engaged in many social interactions at the same time. They are part of social, commercial, communication, family, and other types of network. The quality of a social tie can be completely different, whether two individuals are siblings, married, colleagues, or just met once on a bus [398]. Transportation networks have a multirelational structure, where different link types correspond to different modes of transportation. Multiplex networks (sometimes also referred to as multirelational networks) allow us to represent different interaction types between nodes [54, 232]. In particular, they extend the node set $\mathcal{N}$ and link set $\mathcal{L}$ by a third set $\mathcal{A}$ that indicates the type of each link. Each link in a multiplex network is specified by the triplet $(n \in \mathcal{N}, l \in \mathcal{L}, \alpha \in \mathcal{A})$, with α being a label for the interaction type. One can think of a multiplex network as a collection of $|\mathcal{A}|$ layers of link sets $\mathcal{L}$ that share the same node set $\mathcal{N}$. Each node can be present on each of the layers. Links always connect nodes on the same layer.

4.9.3.1 Adjacency tensors

Multiplex networks can be formally described by adjacency tensors. If nodes i and j are connected in layer α, there is a non-zero entry in the tensor M_{ij}^{α}. This representation can be used as a starting point for generalizing several of the measures that we encountered in Section 4.3 on multiplex networks [34]. The consequence is often that scalar-value measures of nodes become vector-valued measures, where each entry in the vector corresponds to one specific layer. For instance, nodes i in a multiplex network have a degree vector $k_i = (k_i^1, \ldots, k_i^{|\mathcal{A}|})$, with k_i^{α} being the degree of node i in layer α.

4.9.4 Multilayer networks

The terms 'multilayer network' and 'multiplex network' are sometimes used synonymously. We will make a distinction, however, and use the term multilayer network for a yet more general network concept, from which the other generalizations can be derived [232]. Multilayer networks were proposed as a general notion of networks that includes simple networks, power graphs, hypergraphs, or multiplex networks as special cases. They are defined on a set of nodes $\mathcal{N}$ and a set of links $\mathcal{L} \subseteq \mathcal{N} \times \mathcal{N}$. As the name suggests, these networks have multiple layers, which we denote by $\mathcal{M}$. Each node can be present on any subset of layers. Each link connects a node on layer α to a node on layer β, where $\alpha, \beta \in \mathcal{M}$. Multilayer networks can change over time.

A single set of layers is not enough to describe a time dependence. To see this, consider a dynamic simple network in discrete time. The state of the network at each time step can be regarded as one layer in a multiplex network. A time-dependent simple network corresponds to a static multiplex network in which each layer describes the simple network at a given time. To describe a dynamic multilayer network, we need a sequence of sets of layers, each set in the sequence corresponding to a specific instant in time. For this, an additional index $t = 1, \ldots, T$ is introduced that labels T different sets of layers $\mathcal{M}_t$. The index t can correspond to different times at which the set of layers is observed, but it could also be used to represent other features of the data.

Multilayer networks are given by a node set $\mathcal{N}$, a link set $\mathcal{L}$, and a sequence of T sets of layers $\{\mathcal{M}_t\}_{t=1,\ldots,T}$. In component notation we can write a multilayer network as $M_{ij}^{\alpha\beta}(t)$, where the latin indices indicate nodes, and greek indices indicate layers. t indicates a further set of layers.

Finally, the following definitions in the context of multilayer networks are sometimes useful [232]. If all layers contain all nodes, the multilayer network is *node-aligned*. If each node exists in no more than one layer, the network is *layer-disjoint*. A multilayer network is *diagonal* if all links across two different layers $\alpha \neq \beta$ connect the same node i

on layers α and β. The corresponding links across different layers are also called diagonal couplings. A diagonal multilayer network is *layer-coupled* if all nodes have the same diagonal couplings between a given pair of layers. Diagonal couplings are *categorical* if each node is linked to all its counterparts in the other layers. All the generalized networks mentioned can be obtained from multilayer networks as special cases by applying specific constraints.

> In this framework, multiplex networks are node-aligned, layer-coupled, multilayer networks. A hypergraph is a multilayer network with categorical couplings in which each hyperlink represents a layer. The nodes connected by the hyperlink form a clique in the corresponding layer. By allowing directed links across layers between nodes that are connected by hyperlinks, it is possible to recover power graphs.

4.10   Example—systemic risk in financial networks

Can systemic risk, the risk that a system may collapse, be computed? Can the damages of systemic events, such as a financial crash be estimated? Can systemic risk be managed? The answer to these questions today is generally no. However, in this example, we want to show that with the use of time-resolved multiplex data of financial transactions, and with the design of appropriate centrality measures, it is possible to quantify systemic risk, which is the basis for eventually managing it.

With respect to risk in financial systems, there are three types of risk. There is *economic risk*, the risk that the investments in business ideas do not pay off. Imagine you build a prototype of a new electronic device and your grandmother supports this activity financially. Imagine that in the end no-one buys the device and all your efforts come to nothing. Your grandmother has lost her investment. In general, economic risk is not taken by granny but by the financial system. The system takes the risks and, for so doing, it expects a future payoff. In this sense the financial system is providing a service to inventors, entrepreneurs, and the economy. This service should not in itself introduce new risks—systemic risk in particular—to the system. If it does, the financial system is not well designed and is inefficiently regulated. The second type of risk is *credit-default risk*, the risk of not getting back *something* that you have lent someone. To insure oneself against the possibility of not being repaid, one usually asks for a 'risk premium' (interest rate), the amount of which will depend on the creditworthiness of the borrower. Typically, lending and borrowing take place on networks, where it can become very difficult (impossible) to assess the creditworthiness of others. At a global level, the risk in financial borrowing–lending networks is managed by demanding that every lender should maintain a liquidity (cash) cushion to enhance the probability of his/her survival in the event of the outstanding debt not being repaid. In the financial world this form of regulation is called 'Basel-type' regulation. Finally, *systemic risk* is the risk that the system stops functioning as a whole, due to local defaults and subsequent system-wide cascading of defaults. Systemic risk is the risk that a significant fraction of the financial network

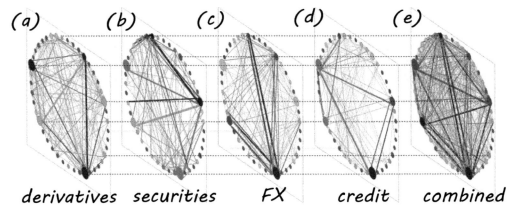

(a)   *(b)*   *(c)*   *(d)*   *(e)*

derivatives securities   FX   credit   combined

Figure 4.20 *Snapshot of the financial multiplex network of Mexico on 30 September 2013. Nodes represent the forty biggest banks; links are the netted exposures between banks (how much one of them can lose). The different layers represent different transaction types. The node size represents the total assets of banks. (a) Network layer of exposures from derivatives, (b) cross-holdings of securities, (c) foreign exchange exposures, (d) deposits and loans, and (e) the combined banking network (netted superposition of the exposures of all layers). Nodes are shaded according to their systemic impact, R_i^α, in the respective layer.*

defaults and stops functioning. Systemic risk is *not* the same as credit-default risk. The most important difference is that systemic risk spreads by *borrowing from* others, while credit-default risk spreads through *lending to* others. Imagine that there are two possible people to borrow from: one is a rich person unconnected to the rest of the world, and one is a person connected to the financial system through a dense network of financial commitments. The first person has no systemic risk—if I do not repay her, nothing will happen to the rest of the world, but if I do not repay the second person (who has a great deal of systemic risk) that could trigger a cascade of defaults. In that sense, if I borrow from the second person, systemic risk spreads from him to me.

Clearly, systemic risk arises through the networks of contracts (financial transactions) between actors. The financial system is simply a time varying multiplex network of financial transactions, $M_{ij}^\alpha(t)$. Every layer α in the multiplex network corresponds to one type of financial contract; this could be credits, derivatives, foreign exchange (FX) transactions, and so on. Every type of contract that can be traded constitutes one layer. The indices i and j label the actors (e.g. banks), and t is time. Many central banks have detailed data about the financial multiplex network of their country. In Figure 4.20 we show a historical snapshot of the Mexican multiplex banking network. Nodes represent the forty biggest banks. Their size represents the size of the banks in terms of assets. Links between banks are the mutual netted exposures (transaction-based risks) between them. Think of one layer, for example, the credit layer, as a matrix M_{ij}^{credit} that contains loans from node j to i. The exposure of node j to i on layer α is M_{ij}^α. The different layers represent the different transaction types or asset classes. The grey-levels of the banks indicate their systemic risk levels, which we will compute next.

4.10.1 Quantification of systemic risk

We would like to have a systemic risk value for every financial institution, similar to the PageRank for web pages. As we have seen, intuitively, PageRank identifies a web page as important if many important pages point to it. In the same spirit, we will consider an institution to be systemically risky if many systemically risky institutions lend to it. Why, then, do we not use PageRank as the systemic risk centrality measure? For our systemic risk measure, we would like an economically sound quantity that assigns to every bank the fraction of the total damage that it would contribute to a systemic event (crash)—should one occur. In other words, the systemic risk of a bank should be the fraction of the total *economic value* in the entire network that is affected by the bank's default. This value should be computed from the interbank credit (or exposure) multiplex network M_{ij}^{α}, and from the capitalization of the banks. The more capital a bank has, the less likely it is to collapse. Finally, it should be noted that banks only default once. By naively using the PageRank algorithm, we would allow for the possibility of multiple defaults of a single bank, and the capitalization would not be taken into consideration. The centrality measure that fulfils these requirements is the so-called DebtRank; see [35, 369].

4.10.1.1 DebtRank—how important are banks systemically?

DebtRank is a centrality measure that assigns to every node the fraction of the total economic value in the network that is potentially affected by the default of the node. DebtRank is computed by a recursive method that takes into account the capital of banks and corrects the Katz prestige for loops in the exposure networks [35]. Let us present the basic idea in the context of networks of inter-bank loans. M_{ij} is the interbank liability network at a given moment (loans from bank j to bank i). By C_i we denote the capital of bank i. If i defaults and cannot repay loans, j loses M_{ij}. If j does not have enough capital to cover that loss, j also defaults. Many central banks do have data on both, M_{ij} and C_i. The impact of bank i on bank j (in the event of a default of i) is now defined as,

$$W_{ij} = \min\left[1, \frac{M_{ij}}{C_j}\right]. \tag{4.77}$$

The value of the impact of bank i on its neighbours is $I_i = \sum_j W_{ij} v_j$. The impact is measured by the *economic value* v_i of bank i. Given the total outstanding interbank exposures of bank i, $M_i = \sum_j M_{ji}$, its economic value is defined as,

$$v_i = M_i / \sum_j M_j. \tag{4.78}$$

The total economic value is $V = \sum_i v_i$. To take into account the impact of nodes at network distance two and higher, the impact has to be computed recursively. If the network W_{ij} contains cycles, the impact can exceed one. To avoid this problem, two state variables, $h_i(t)$ and $s_i(t)$, are assigned to each node. h_i is a continuous variable between zero and one; s_i is a discrete state variable with three possible states: undistressed,

distressed, and inactive, $s_i \in \{U,D,I\}$. The initial conditions are $h_i(1) = \Psi$, for all $i \in S$; $h_i(1) = 0$, for all $i \notin S$, and $s_i(1) = D$, for all $i \in S$; $s_i(1) = U$, for all $i \notin S$ (parameter Ψ quantifies the initial level of distress: $\Psi \in [0,1]$, with $\Psi = 1$ meaning default). The dynamics of h_i is then specified by,

$$h_i(t) = \min\left[1, h_i(t-1) + \sum_{j\,|\,s_j(t-1)=D} W_{ji}h_j(t-1)\right]. \tag{4.79}$$

The sum extends over those j, for which $s_j(t-1) = D$,

$$s_i(t) = \begin{cases} D & \text{if } h_i(t) > 0; s_i(t-1) \neq I, \\ I & \text{if } s_i(t-1) = D, \\ s_i(t-1) & \text{otherwise} \end{cases} \tag{4.80}$$

The DebtRank (at time T) of the set S (set of nodes in distress at time 1), is $R_S = \sum_j h_j(T)v_j - \sum_j h_j(1)v_j$, and measures the distress in the system, excluding the initial distress. If S is a single node, the DebtRank measures its systemic impact on the network. The DebtRank of S, containing only the single node i, is,

$$R_i = \sum_j h_j(T)v_j - h_i(1)v_i. \tag{4.81}$$

The DebtRank excludes the loss generated directly by the default of the node itself and measures only the impact on the rest of the system through default contagion. Figure 4.21a shows the DebtRank R_i for the 40 largest banks in Mexico. The black line represents the total systemic risk, that is, the fraction of the total economic value in the network that would be affected in the event of i defaulting. The contributions from the individual layers, deposits and loans, foreign exchange, securities, and derivatives are shown in the shaded bars, respectively.

4.10.1.2 *Expected systemic loss*

Given the DebtRank R_S for a given set of defaulted banks S, what is the total *expected systemic loss* in a system? What is the expected size of a systemic event or a crash? To compute the expected systemic loss, first consider the simple case, where only one institution i defaults and all other $N-1$ institutions survive. In this case, the expected loss is given by $\mathrm{EL}_i^{\mathrm{syst}}(\text{one default}) = Vp_i(1-p_1)\cdots(1-p_{i-1})(1-p_{i+1})\cdots(1-p_N)R_i$, where p_i is the probability of default of institution i, and $(1-p_j)$ the survival probability of j. In the general case, we also consider joint defaults, meaning that a set of institutions S becomes distressed. Taking into account *all* possible combinations of defaulting and surviving institutions, we arrive at a combinatorial expression of the expected loss for an economy of N institutions,

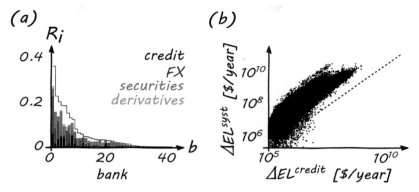

Figure 4.21 *(a) Systemic risk levels (DebtRank) R_i for the forty most relevant banks in Mexico on a given day. The systemic risk contribution from the individual layers is indicated. Note that the combined systemic risk from all layers (black line) is larger than the sum of the contributions from the individual layers. (b) Systemic risk of individual transactions. Every point represents an individual interbank liability $M_{ij}^{\alpha}(t)$ on a given layer α and given day t. Its expected systemic loss, $\Delta EL^{\mathrm{syst}}$, versus the credit risk, $\Delta EL^{\mathrm{credit}}$ is shown. Note that $\Delta EL^{\mathrm{syst}} > \Delta EL^{\mathrm{credit}}$ means that the costs for defaults of exposures not only affect the lending party alone (which suffers a maximum loss of $\Delta EL^{\mathrm{credit}}$) but also involves third parties. Typically, the public pays for the difference, $\Delta EL^{\mathrm{syst}} - \Delta EL^{\mathrm{credit}}$ in case of a systemic crash.*

$$\mathrm{EL}^{\mathrm{syst}} = V \sum_{S \in \mathcal{P}(B)} \prod_{i \in S} p_i \prod_{j \in B \setminus S} (1 - p_j) R_S, \tag{4.82}$$

where $\mathcal{P}(B)$ is the power set of the set of financial institutions B, and R_S is the DebtRank of the set S of nodes initially in distress. $R_\emptyset$, the DebtRank of the empty set, is defined as zero. As by definition of the DebtRank, $R_S \leq 1$, the value obtained in Equation (4.82) cannot exceed the total economic value, V. It is immediately clear that Equation (4.82) is computationally feasible only for a small number of financial institutions. Computing the power set and calculating the DebtRanks for all possible combinations of large financial networks is impossible. However, we can derive a practical approximation for large financial networks,

$$\mathrm{EL}^{\mathrm{syst}} \sim V \sum_{S \in \mathcal{P}(B)} \prod_{i \in S} p_i \prod_{j \in B \setminus S} (1 - p_j) \left(\sum_{i \in S} R_i \right)$$

$$= V \sum_{i=1}^{N} \underbrace{\left(\sum_{\mathcal{J} \in \mathcal{P}(B \setminus \{i\})} \prod_{j \in \mathcal{J}} p_j \prod_{k \in B \setminus (\mathcal{J} \cup \{i\})} (1 - p_k) \right)}_{=1} p_i R_i = V \sum_{i=1}^{N} p_i R_i. \tag{4.83}$$

The term in brackets sums to 1, which can be proved by induction. The approximation $R_S \sim \sum_{i \in S} R_i$ is certainly valid if the default probabilities are low ($p_i \ll 1$), or the interconnectedness is low ($R_i \sim v_i$). For details, see [310]. In Figure 4.22 we show

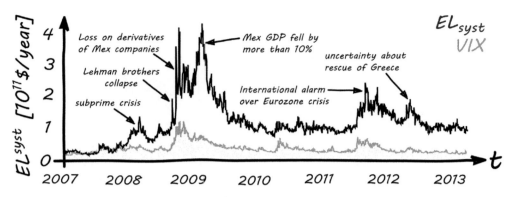

Figure 4.22 *Expected systemic losses* EL^syst *in Mexican pesos per year (black curve). Several historical events are marked. Expected systemic losses in Mexico in 2013 are about four times higher than before the crisis. For comparison, another indicator for risk in financial markets is shown; the volatility index VIX (grey) is scaled to the same value at the first time step. This market-based indicator relaxes fast to pre-crisis levels, whereas* EL^syst *does not, indicating that the expected systemic losses are indeed driven to a large extent by network topology, and are consistently underestimated by the market.*

the expected systemic loss $\mathrm{EL}^{\mathrm{syst}}$ for Mexico, computed for every day from 2007 to 2013. The $\mathrm{EL}^{\mathrm{syst}}$ in 2013 is about four times higher than before the crisis. The expected systemic loss per year quantifies the expected losses within a country if the default of a bank triggers a systemic crash, and, further, given that there are no bailouts or other interventions from the government or central banks. If this index were available for all countries, it would be possible to compare countries in terms of their systemic risk, and would allow us to see if economic policies to reduce systemic risk would work or not. Note that the *expected systemic loss* must not be confused with the *expected loss*, which is the credit default risk. It is defined for bank i,

$$\mathrm{EL}_i^{\mathrm{credit}} = \sum_j p_j \mathrm{LGD}_j M_{ij}, \tag{4.84}$$

where p_j is the default probability for bank j and LGD_j is a factor that specifies what percentage of your claims against a defaulted bank you will not able to recover. It is called the 'loss given default'.

4.10.1.3   *Systemic risk of individual transactions*

We can now compute the impact of individual transactions on credit risk and on expected systemic loss. The contribution of an individual exposure, X_{kl} (matrix with precisely one non-zero element in line k and row l, quantifying the individual transaction between banks k and l) on credit risk, is the increase in the credit risk of the bank with additional exposure (risk taken by lender),

$$\Delta \text{EL}^{\text{credit}} = \sum_{i=1}^{N} \left[\text{EL}_i^{\text{credit}}(M_{ij} + X_{kl}) - \text{EL}_i^{\text{credit}}(M_{ij}) \right]. \tag{4.85}$$

Here, $\text{EL}_i^{\text{credit}}(.)$ means that $\text{EL}_i^{\text{credit}}$ is computed from the network in the argument. The contribution of an individual transaction, X_{kl} on EL^{syst} is the difference between the total expected systemic loss with, and the total expected systemic loss without, the transaction X_{kl},

$$\Delta \text{EL}^{\text{syst}} = \sum_{i=1}^{N} p_i \left[V(M_{ij} + X_{kl}) R_i(M_{ij} + X_{kl}, C_i) - V(M_{ij}) R_i(M_{ij}, C_i) \right]. \tag{4.86}$$

Here, $R_i(M_{ij} + X_{kl}, C_i)$ is the DebtRank, and $V(M_{ij} + X_{kl})$ is the total economic value of the liability network with the specific exposure X_{kl}. A positive $\Delta \text{EL}^{\text{syst}}$ means that transaction X_{kl} increases the total systemic risk. In Figure 4.20b we plot the systemic risk contribution, $\Delta \text{EL}^{\text{syst}}$, of every transaction against its credit risk, $\Delta \text{EL}^{\text{credit}}$. Here every dot represents a single transaction of a given type. It is immediately visible that for a given credit risk, there is a huge variability in the associated systemic risk. Note the logarithmic scale. Also note that the bulk of the transactions are above the diagonal, which means that if the asset is lost in a systemic event, there is more systemic damage, $\Delta \text{EL}^{\text{syst}}$, than individual damage (the value of the asset), $\Delta \text{EL}^{\text{credit}}$. If the systemic contribution and the credit risk are equal, $\Delta \text{EL}^{\text{syst}} = \Delta \text{EL}^{\text{credit}}$, a default of the exposure would only affect the lender (who loses $\Delta \text{EL}^{\text{credit}}$), and would not involve any third party. For transactions, where $\Delta \text{EL}^{\text{syst}} > \Delta \text{EL}^{\text{credit}}$, third parties will also be affected by the default. In the case of a crash, the shortfall is usually paid from the public purse and not by the institutions that were engaged in the transaction and who caused the damage. For more details, see [310, 311].

> Systemic financial risk is an *externality*. Systemic risk is a network property to a large extent. To manage it, the underlying networks need to be rearranged.

4.10.2 Management of systemic risk

By knowing the systemic risk contribution of every transaction in a system, it is possible to manage and reduce systemic risk by effectively restructuring the network in a way that makes cascading processes unlikely. If one were to impose a tax on all transactions that is proportional to their systemic risk contribution, $\Delta \text{EL}^{\text{syst}}$, then it would be less attractive for agents to engage in the respective transaction. They would actively look for transactions that contribute little (or no) systemic risk to the system. Through this incentive scheme, the network will self-assemble in a very different way than it would otherwise do. Through this *systemic risk tax*, transaction networks rearrange themselves,

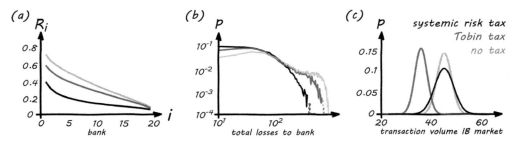

Figure 4.23 *(a) Systemic risk profile (DebtRank) R_i for twenty banks as obtained from an agent-based simulation calibrated for the Austrian banking sector at the end of the first quarter of 2006. The systemic risk profile (systemic risk contributions of individual banks) is drastically reduced when the systemic risk tax (black) is implemented. The Tobin tax (grey) reduces systemic risk slightly; however, at the cost of reducing credit volume in the market. The situation without tax (light-grey) resembles the empirical distribution (not shown). (b) Distribution of total losses to banks in a simulation for twenty banks, 10,000 independent runs, each with 500 time steps. The fat-tailed distribution of losses in the unregulated case is removed under the systemic risk tax. The remaining distribution of losses resembles the remaining economic risk from foul credits from the banks' clients, which is exogenous and cannot be removed from the system. (c) Distribution of total transaction volume in the interbank market. The systemic risk tax does not lower credit volume. The Tobin tax does, as it makes credit more expensive.*

such that cascading of defaults becomes practically impossible. It can be shown that a systemic risk tax is, in fact, able to drastically reduce systemic risk in financial markets without the efficiency of the financial network being lost and without the total volume of transactions being reduced [249, 311].

In Figure 4.23a we show the reduction of the systemic risk profile when a systemic risk tax is imposed. Results are obtained from a massive agent-based simulation of the financial economy of Austria [311]. It is clearly visible that, under the systemic risk tax (black), the profile R_i is much lower than in the unregulated case, which resembles the current situation (light-grey). In comparison, we show the effect of a Tobin tax (grey), which taxes every transaction regardless of its systemic risk contribution. The Tobin tax reduces systemic risk slightly, at the cost, however, of reducing the total transaction volume; see Figure 4.23c. In Figure 4.23b the loss-distribution for the same simulation is shown. The fat tails (large losses) are drastically reduced with the tax, while the total credit volume is not affected at all.

In summary, we have learned that it is possible to derive a centrality measure that is well-suited to quantifying the systemic risk levels of nodes in financial networks [35, 369]. This is done on the basis of knowing the exposure multiplex networks at various levels and the capitalization of the banks. The systemic risk contributions can be computed, not only for the nodes in the network (banks and institutions), but also for the links (individual transactions). We have seen that this knowledge can then be used to eliminate systemic risk allowing an incentive scheme to be designed that would make banks try to avoid systemically risky transactions. If such an incentive were installed, for example, through a systemic risk tax, financial networks would reassemble in a self-organized way towards networks that are optimal in the sense that cascading events are strongly suppressed. The system is also much less susceptible to systemic shocks, while remaining fully efficient at the same time—it has become more resilient. Further examples for using network science in applications from data analytics are found in [73].

4.11 Summary

Networks often arise in the study of complex systems. Networks are a book-keeping tool to keep track of who interacts with whom in what way, how strongly, in what direction, at what time, and for how long. A few take-home messages are the following.

- Macro properties of complex systems can often be understood on the basis of statistical properties of collections of networks. Networks can be random variables. These so-called random networks are ensembles of networks. Concrete networks are often realized from these ensembles.
- Statistical properties of networks are often informative about the underlying dynamical rules by which systems evolve.
- Networks are given by a collection of nodes and a collection of links. Links can be directed or undirected, weighted, or unweighted.
- Networks can be represented by an adjacency matrix A. A non-zero matrix element A_{ij} represents a link from node j to i.
- Another way of representing networks is through the incidence matrix, B. This is a rectangular matrix where rows correspond to nodes and columns to links. Each non-zero entry, $B_{i\ell}$ in the incidence matrix indicates that the link ℓ is 'incident' on node i.
- Weighted networks are characterized by matrices, where links are represented by positive or negative real numbers.
- The most fundamental property of a node is its number of links, the degree. The degree distribution is the probability of finding a node with a given degree in the network.
- A path on a network is an ordered sequence of nodes in which each adjacent pair of nodes is connected by a link and no node appears more than once in the sequence.
- Components are parts of the network in which there is a path between every pair of nodes.
- Distances on networks can be defined by the length of the shortest path between a pair of nodes. Nodes that have a low average distance to all other nodes in the network have a high closeness centrality. Nodes that are part of a large number of shortest paths between other pairs of nodes have a high betweenness centrality.

continued

- The clustering coefficient of a node is the probability that any two neighbours of the node are also neighbours with each other.

- ER networks are random networks that are completely specified by p, which is the probability that there is a link between any pair of nodes. ER networks are simple null models, against which properties of real world networks can be tested. ER networks undergo phase transitions, as their linking probability p changes. In particular, they have a percolation transition at $p_c \sim 1/N$. For $p > p_c$, a single macroscopic connected component emerges, the giant component.

- The Perron–Frobenius theorem states that networks with a single component have a largest eigenvalue that is a positive real number. All components of the eigenvector to this eigenvalue are also positive real numbers. All other eigenvectors have at least one negative or imaginary component.

- Complex networks are random networks where links do not form purely by chance. They display some structure, such as high clustering, the existence of hubs, assortativity, short distances, and so on.

- The configuration model specifies a class of random networks that are maximally random with respect to a given degree sequence.

- Social networks are typically small world networks. The Watts–Strogatz model is a particularly simple way of thinking of such networks.

- Scale-free networks are networks with degree distributions that are power laws. High-degree nodes are called hubs. The BA model is a simple way of obtaining power law degree distributions in growing networks through a preferential attachment mechanism.

- Communities are groups of nodes that are more likely to be connected with each other than to nodes outside the community. There is no unique and optimal way of detecting communities but there are many approximative algorithms. These often rely on the concept of modularity, which compares the link density between nodes in a given community with the density that is expected under a specific null model.

- Correlation networks can be used to infer the correlation structure of systems for which multiple observations (e.g. time series) of a number of variables is available (data). Links are given by the correlation coefficient (or related measures) between each pair of variables. Links in correlation networks indicate the existence of some process that affects both nodes. There is no guarantee that links represent actual causal interactions.

- Every link in the correlation network requires a suitable statistical hypothesis test. To obtain the entire correlation network, a massive multiple hypothesis testing problem must be solved. This problem can be tackled by controlling the FWER or the FDR of the multiple test, or by using network backboning techniques such as the disparity filter.

- Many dynamical processes are effectively diffusion processes on networks. They can be studied using models of random walkers that randomly jump along the links in specific ways. Of particular interest are the stationary distributions of random walkers after a large number of jumps. These distributions serve as centrality measures of the networks. High centrality nodes are those that link to other nodes with high centrality. Examples of dynamical random walk processes include Laplacian diffusion, eigenvector centrality, Katz prestige, and PageRank.

- Epidemic spreading is a dynamical process, where nodes change their internal states. Based on the state, nodes are either susceptible to random walkers from other nodes, or they actively spread to their neighbours. ER networks have an epidemic spreading dynamics that coincides with their percolation threshold. If the number of immunized nodes is sufficiently large, epidemic spreading stops. Scale-free networks with power law exponents of size $\gamma \leq 3$ do not have an epidemic threshold. Epidemic spreading can only be stopped if all nodes are immunized.

- Co-evolutionary network models are dynamical processes, where changes occur in the node states and in the links of the network. The dynamical rules of the nodes and links typically depend on each other.

- In many real world networks, interactions can be of multiple types and only occur between specific node types. For these cases the notion of networks has to be generalized. Multilayer networks are specified by nodes, links, and a sequence of sets of layers (one set for each point in time of a dynamic multilayer network). Links always point from node to node within a layer or between layers. Multiplex networks, power graphs, and hypergraphs are special cases of multilayer networks.

- Centrality measures can be designed for special purposes. We discussed in an extended example the design and applicability of a measure that is well suited to capturing systemic risk of nodes and links in financial networks.

- Network analysis has become a standard in data analytics and data mining.

4.12 Problems

Problem 4.1 Construct the undirected song co-writing adjacency matrix A_{ij}^Q for the band members of Queen for the album 'Innuendo', see Table 4.2. Let the indices i and j label the band members, $i, j \in \{$Mercury, May, Deacon, Taylor$\}$, and set $A_{ij}^Q = 1$, whenever i and j have co-written at least one song, otherwise $A_{ij}^Q = 0$. List the degrees k_i, $i = 1, 2, 3, 4$. What is the minimal, maximal, and average degree of the network? Draw the network.

Problem 4.2 Compute the individual clustering coefficients C_i of the network A^Q, as well as the overall and average clustering.

Table 4.2 *Track listing of the album 'Innuendo' (1991) by Queen, LP Version. The band members of Queen are (1) Freddie Mercury (vocals), (2) Brian May (guitar), (3) John Deacon (bass), and (4) Roger Taylor.*

No.	Title	Song writer(s)
1.	Innuendo	Mercury, Taylor
2.	I'm going slightly mad	Mercury
3.	Headlong	May
4.	I Can't Live with You	May
5.	Ride the Wild Wind	Taylor
6.	All God's People	Mercury
7.	These Are the Days of Our Lives	Taylor
8.	Delilah	Mercury
9.	Don't Try So Hard	Mercury
10.	The Hitman	Mercury, May, Deacon
11.	Bijou	May, Mercury
12.	The Show Must Go On	May

Problem 4.3 Compute the distance matrix L^Q for the network A^Q, where the entries L_{ij}^Q give the distance between nodes i and j. What is the diameter and the characteristic distance of A^Q?

Problem 4.4 Represent the tracks of the album given in Table 4.2 as a bipartite network. One set of nodes in this network corresponds to the band members, the other set to the tracks (you need to include only tracks with at least two different songwriters). What is the incidence matrix of this bipartite network? Compute the network dual, that is, the line graph, of A^Q.

Problem 4.5 Show that the average in- and out-degree of a directed network are equal.

Problem 4.6 Consider an undirected bipartite network with N_1 nodes of type 1 and N_2 nodes of type 2. Let the average degrees of nodes of type 1(2) be given by $\kappa_{1(2)}$. Show that $\kappa_2 = \kappa_1 N_1/N_2$.

Problem 4.7 Show that a bipartite network has no odd-numbered cycles.

Problem 4.8 Consider the configuration model of a large network with $N \gg 1$ nodes, degree sequence $\{k_1, k_2, \cdots, k_N\}$, and L links. Show that the probability p_{ij} that two nodes i and j are connected is $p_{ij} = \frac{k_i k_j}{2L}$.

Problem 4.9 Show that the expected number of common neighbours, n_{ij}, between nodes i and j in the configuration model is, $n_{ij} = p_{ij} \frac{\langle k^2 \rangle - \langle k \rangle}{\langle k \rangle}$.

Problem 4.10 Consider a directed ER network A with the linking probability p. Let p_c be the critical threshold above which a giant component forms. How does the Perron–Frobenius eigenvalue λ^{PF} change as a function of p in the regimes (a) $p \ll p_c$, (b) $p \sim p_c$, and (c) $p \gg p_c$?

5

Evolutionary Processes

5.1 Overview

Everything evolves. Physics describes the evolution of things and processes over time. However, this is not what interests us in this chapter. Here, we want to understand evolutionary *processes*. Evolutionary processes combine many of the classical features of complex systems: they are algorithmic; states co-evolve with interactions; they show power law statistics; they are self-organized critical; and they are driven non-equilibrium systems. Evolution is a dynamical process that changes the composition of large sets of interconnected elements, entities, or species over time. The essence of evolutionary processes is that, through the interaction of existing entities with each other and with their environment, they give rise to an open-ended process of creation and destruction of new entities. These entities can be biological species, molecules, ideas, languages, or products and services. Creation and destruction events result in a dynamics of diversity, which is at the core of all scientific questions related to evolution.

Systems that change their composition over time and show interesting diversity dynamics appear in many different contexts [272]. This is reflected in the number of scientific disciplines studying evolution in its various forms. Evolutionary biology studies changes in diversity and properties of species, from the fossil record to fruit fly biology [84, 154, 182, 275]. Chemical evolution tries to understand the emergence of new chemical compounds or molecules that are related to the origin of life [193, 214, 263, 315]. Evolutionary linguistics studies the origin, change, and development of text, grammar, and language in general [16, 95, 352]. Evolutionary economics deals with the appearance and disappearance of goods and services in the economy [107, 329], while technological evolution explains the emergence of new technological products from existing components [14].

Changes in diversity over time often exhibit characteristic dynamical and statistical patterns. Diversity typically changes in a strongly punctuated manner, meaning that changes in diversity happen disruptively in the form of extreme events, like 'booms' and 'busts'. These events are often separated by long time periods, where diversity changes very little and seems to have reached an equilibrium; see Figure 5.1. The diversity dynamics of evolutionary systems is said to exhibit *punctuated equilibria* [23, 154]. In statistical terms, this punctuated dynamics is tightly associated with fat-tailed or power

Introduction to the Theory of Complex Systems. Stefan Thurner, Rudolf Hanel, and Peter Klimek,
Oxford University Press (2018). © Stefan Thurner, Rudolf Hanel, and Peter Klimek.
DOI: 10.1093/oso/9780198821939.001.0001

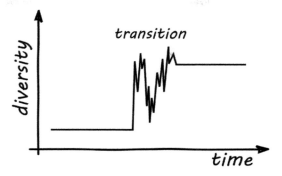

Figure 5.1 *Schematic view of a transition from one equilibrium to another. Evolutionary transition events are usually disruptive and bursty.*

law distribution functions in terms of the rates at which elements are produced and destroyed [289]. Interestingly, the same statistical patterns appear in the biological, socio-economical, technological, and linguistic contexts [85]. It is therefore natural to assume that, irrespective of their nature, evolutionary processes create and destroy elements through similar *endogenous* mechanisms, which can be understood in a general and universal framework.

Understanding evolutionary processes is central to addressing several challenges in science and society. These range from the resilience of ecosystems to the stability of socio-economic systems, including financial markets. However, we are still far from a satisfactory quantitative understanding of evolutionary processes. In this chapter, we review a selection of the classic approaches and sketch a way towards a more comprehensive theory of evolutionary processes. We present a few directions that might lead us there. Progress in understanding evolutionary processes is tightly linked with progress in the theory of complex systems.

5.1.1 Science of evolution

The science of evolution is an attempt to understand evolutionary processes in quantitative terms. It starts in its modern form with the Darwinian narrative [99], which is not yet quantitative but sets the stage.

Darwinian narrative. Individual elements in a given population produce offspring that are not identical, but show slight variations. Given an environment with limited resources, some of these new individuals will 'do better', which is reflected in their reproductive rates—their *fitness*. The variations that do well reproduce more than others. Over time, the variations that did best will dominate the population. The variations that did badly will have vanished. Which elements 'do well' depends on the environment. Different environments can favour different variations. Such environments are called *ecological niches*.

The Darwinian narrative further explains that new species emerge as variations and that some species are driven from existence because of limited resources. It does not tell us anything about the collective dynamics of evolutionary processes, in particular, under what conditions and to what extent, diversity changes appear, and how they influence diversity changes in the future system. The Darwinian narrative does not help us to understand a number of facts that characterize most evolutionary processes. These include:

- The existence of punctuated equilibria: the fact that creation and destruction of entities happens in bursts, which are separated by periods of quasi-equilibrium.
- The abundance of power laws and fat-tailed distributions in the statistics of measurable quantities, such as the size of extreme events, the duration of equilibrium periods between extreme events, the duration of extreme events, and so on.
- Self-organization.
- The endogenous mechanisms that explain bursts of diversification, such as the Cambrian explosion or industrial revolutions.
- The origin of the high proneness to collapse and of the systemic risk of evolutionary systems.

Perhaps the main shortcoming of the Darwinian narrative is that it fails to provide a satisfactory explanation of what fitness is. According to the narrative, species with a high fitness survive, reproduce, and dominate the scene. However, as fitness is defined as the reproductive success (reproduction rate), this leads to a somewhat tautological situation of *what* fitness *is*. This fitness is not a quantity that can be used to make future predictions; it is a quantity that tells us something about reproductive rates now and in the past. Creation and destruction events can massively change the reproductive success of other species. As long as it is impossible to describe the mechanisms by which fitness co-evolves with the interactions between species and their interactions with the environment, the narrative remains a verbal narrative and does not serve as a quantitative and predictive scientific theory. The very essence of a quantitative and predictive approach to evolutionary processes is to understand the dynamics of the co-evolution of fitness of species through their interactions amongst each other and with the environment.

Traditional evolutionary models have taken a first step in this direction. We will review some of them in Section 5.3.1. There, fitness is interpreted as a function that depends on the abundances of all other species and their interactions. However, most of these models assume static fitness functions and cannot therefore explain co-evolutionary dynamics. Models of this kind work well for timescales that are short enough for no species to be introduced and for interactions between the existing species not to change significantly. They are able to model the equilibrium phases between punctuations. However, they are largely incapable of understanding the open-ended, endogenous, and bursty dynamics of collective creation and destruction. One aim of this chapter is to derive a fully co-evolutionary and open-ended framework that allows us to understand the previously mentioned characteristic features of evolutionary processes. We present the basic framework in Section 5.3.3 and discuss co-evolving models in Sections 5.5 and 5.6. The framework is based on the notion of evolution as an algorithmic three-step process [368].

5.1.2 Evolution as an algorithmic three-step process

Evolution is a three-step process.

- Step 1: a new entity (species, product, idea, word) is created and placed in a given environment.
- Step 2: the new entity interacts with the environment. The environment consists of all already existing entities and (perhaps) a physical environment, together with all their interactions. As a result of these interactions, the new entity is either destroyed or it survives and proliferates.
- Step 3: if the new entity survives, it becomes *part of the environment*. The environment changes and becomes a new boundary for all existing and newly arriving entities. Within this new boundary, existing entities will try to relax towards a new equilibrium.

This process is clearly a *driven, dissipative non-equilibrium process*; it is discussed in Section 2.5.4. The driving force is realized in step 1, where a new entity is placed into the system. In this step, the system is driven away from a (potentially) existing equilibrium. Because of the changes in the environment some—potentially even all—existing entities start to relax towards new equilibrium states. This takes place in a relaxation process that then starts to unfold.

Step 2 means that entities i interact with other entities or with the physical environment, j. The properties of entities are represented in their state $\sigma_i(t)$. For example, the state can be the abundance x_i of species i, $\sigma_i(t) = x_i(t)$. The possible states of the physical environment j, such as energy sources, are encoded as states in the same state vector. They are also labelled as $\sigma_j(t)$. For example, they could quantify the availability of energy at source j. Interactions are described by time-dependent interaction matrices $M_{ij}^{\alpha}(t)$. If more than two entities interact, we need interaction tensors $M_{ijk...}^{\alpha}(t)$. Index α denotes the specific *type* of interaction. The dynamical state changes of the species depend largely on these interactions. Formally, the evolution equation of the states is,

$$\frac{d}{dt}\sigma_i(t) \sim F\left(M_{ijk...}^{\alpha}(t), \sigma_j(t)\right),\tag{5.1}$$

where F is some function that depends on the state vectors and the interactions.[1] In this notation, creation means that if at time t an entity did not exist $\sigma_i(t) = 0$, at time $t + dt$ it does, $\sigma_i(t + dt) > 0$. Destruction means the opposite, $\sigma_i(t) > 0$ and $\sigma_i(t + dt) = 0$.

---

[1] Here $\frac{d}{dt}\sigma_i(t)$ means 'change of states'. It does not need to be a real differential.

Step 3 means that as a result of new entities arriving or of entities being eliminated, interactions will change over time. To describe this, we need an evolution equation for the interactions,

$$\frac{d}{dt}M_{ijk\ldots}^{\alpha}(t) \sim G\left(M_{ijk\ldots}^{\alpha}(t),\sigma_j(t)\right), \qquad (5.2)$$

where G is some function that depends on states and interactions. The combined dynamics of Equations (5.1) and (5.2), we call *co-evolutionary dynamics*; compare with Equation (4.52). States and interactions update each other. An analytical treatment of this co-evolutionary dynamics is not usually possible because the boundary conditions (here in the form of M^{α}) co-evolve with the states. However, these systems can be well-described as *algorithms*. This is exactly why we introduce a general evolution algorithm, which we will use throughout the chapter. In Section 5.3.3 we develop a practical way of handling this co-evolutionary framework in an algorithmic and fully open-ended way.

How do new 'things' come into being? Often, new entities enter from 'outside'. A new species arrives on an island, a word from a foreign language enters the vocabulary, or a new idea in physics is taken over from biology. In many cases, however, new entities arrive as a consequence of the combination of things that already exist. This is particularly obvious for the case of technological evolution, where technical components are combined to create new objects or new components. It is also true in biology, where new species arise through genetic variations; this can be seen, for example, in the combination of a piece of DNA with a gamma photon to cause a mutation. Evolution driven by new entities that come into existence through the process of combination of existing things is called *combinatorial evolution*. Evolution is almost always driven by *combinatorial* processes [14, 170]. Formally, combinatorial evolution is possible if two existing entities give rise to a third entity,

entity 1 + entity 2 → entity 3

In our framework of Equation (5.1) this combinatorial process to create entity 3 would read, $\frac{d}{dt}\sigma_3(t) \sim M_{312}\sigma_1(t)\sigma_2(t)$, where $M_{312} > 0$. We will call M_{312} the 'rule' that entity 3 can be created from a combination of entities 1 and 2. The interaction tensor $M_{ijk\ldots}$ is the collection of all such rules. We call it the *rule table*.

Combinatorial evolution has a long tradition and dates back to over a century ago. It was pioneered verbally by Robert Thurston [373] in the context of the history of the steam engine, and by Joseph A. Schumpeter[2] [329, 330], who first described economic development as a bursty evolutionary three-step process. Schumpeter postulated that economic systems evolve in *gales of creative destruction*, where entrepreneurs (today, we

---

[2] It is important to note that Schumpeter aspired to be three things, namely, the greatest horseman, lover, and economist of all time [392]. Of these three aspirations, he claimed to have fulfilled two, but he refused to specify which two. However, he did note that there was an abundance of really good horsemen [327].

use the term 'start-ups') seek to become monopolists through innovations that disrupt currently existing industrial structures. In particular, he argued that macro-economic equilibrium represents a 'normal mode of economic affairs' that is constantly perturbed by entrepreneurs launching innovations (Step 1). The successful market introduction of a new product or a new production process disturbs (drives) the normal flow of economic life (Step 2). If successful, they are a threat to already existing technologies, which stand to lose their position within the current economy (Step 3).[3] For this three-step process Schumpeter coined the term *creative destruction*. Examples of creative destruction include the four industrial revolutions. The first was the transformation from manual production to machines, the second to mass production, and the third to automation. Manufacturing is currently undergoing a transformation that is driven by digitalization, and is sometimes referred to as *Industry 4.0* [247].

Schumpeter also noted that economic development occurs in a strongly punctuated way, where disruptive innovations trigger massive changes that transform the economy from one relatively stable set of states to another. Dynamics, where initially small modifications in the microscopic constituents of a system, such as the addition of a single new product, may trigger massive macroscopic changes, is the essence of critical transitions in complex systems. These are often directly associated with power law statistics. Evolution is a *critical* process of this kind.

It is important to note that even though mainstream economics [135] involves evolutionary game theory [283], it is not co-evolutionary, combinatorial, or critical. Mainstream economics and finance today are theories of equilibrium processes. Economics deals with static fitness functions called utility functions or production functions.

We have mentioned previously that evolutionary dynamics is radically different from traditional physics in at least two aspects. The first is that boundaries of the system co-evolve with the dynamic variables, which forces us to switch from an analytic to an algorithmic approach. Algorithmic approaches allow us to perform quantitative and predictive science in exactly the same way as analytic approaches do. The second aspect is that evolution is open-ended, meaning that the phasespace is not well defined. In physics or statistics this situation typically means: end of story! If it is not clear what the possible outcomes are in the next time step, how can one make predictions? The good news is that, typically, phasespace does not evolve erratically and its evolution is restricted by specific constraints. It evolves from one time step to the next by creating new entities as a combination of existing ones, and by destroying existing entities. If the number of interactions is not too large, there is hope that the dynamics of the phasespace itself can be understood, at least in statistical terms. In this context, the notion of the *adjacent possible* becomes useful [219].

---

[3] In Schumpeter's own words [329]: (Step 1) 'The fundamental impulse that sets and keeps the capitalist engine in motion comes from the new consumer goods, the new methods of production or transportation, the new markets, the new forms of industrial organization that capitalist enterprise creates (Step 2) that incessantly revolutionize the economic structure from *within*, and (Step 3) incessantly destroying the old one, incessantly creating a new one. Creative destruction is an essential fact of capitalism. It is what capitalism consists of and what every capitalist concern must live in. Stable capitalism is a contradiction in terms.'

Assuming that time is discrete, the *adjacent possible* of a system at time t is the set of all potential states (or configurations) of that system that can be reached in the next time step $t+1$.

For example, for a traveller who finds herself at a fork in the road, the adjacent possible consists of two possibilities: turn left *and* turn right. In classical physics, the adjacent possible is a very small set, often just one point. The possible next steps for the positions of a falling stone are fully determined by the equations of motion, and the adjacent possible at $t+dt$ is $x(t+dt)$, which is a single point. In quantum mechanics, the adjacent possible of a particle is already an extended object before the collapse of the wave function. In biology, the adjacent possible is generally a huge set, including, for example, all the phenotypes that *might* be created by mutations in the next time step. Algorithmic co-evolutionary processes often describe *how* a particular system explores its adjacent possible. Exploration mainly occurs through combination of already existing things: two molecules rearrange their bonds to form a new molecule; smartphones were invented as a combination of telephones and touch screen displays; mice and men have in common that they procreate by combining genes in sexual reproduction. Systems that produce new entities through combination can be described in a statistical framework that allows us to understand how fast and under what conditions systems explore their adjacent possible [169]. We discuss the adjacent possible in more depth in Sections 5.4 and 5.6.2. The main focus of this chapter is to derive a practicable version of a co-evolving, combinatorial, and critical evolution model that evolves into a well-defined adjacent possible at every time step. We call it the CCC model. It is based on the prevously mentioned three-step process, which is summarized by the combined dynamics of Equations (5.1) and (5.2). The essence of Section 5.6.4 is to derive a practical non-linear model that is both computationally efficient and can be calibrated with actual data.

5.1.3   What can be expected from a science of evolution?

If it cannot be prestated what things will be present in the future, and much less, how these things will interact with one another, how can we make predictive and quantifiable statements about diversity dynamics? There are famous examples, where the creation of a single new entity dramatically changed the boundary conditions for a vast number of other entities. Photosynthesis came on to the stage in the form of filamentous photosynthetic organisms about 3.4 billion years ago and turned the planet into an oxygen-rich hothouse that led to the development of complex life forms [245]. In 1904 John Ambrose Fleming accidentally manufactured the first vacuum tube when trying to produce a light bulb [116]. The consequences are still playing out. As of 2015, global sales in the semiconductor industry have grown to USD335.2 billion and semiconductors serve as the enabling technology for industries representing roughly 10% of world GDP [332]. In 1973 Vint Cerf and Bob Kahn developed a protocol that allows error-controlled transmissions between different types of university networks to speed up computations [76]. Today, the world wide web has penetrated virtually all aspects of life for practically all human beings. A science of evolution would have failed miserably to predict the appearance and consequences of these singular events.

A science of evolution will never be able to make detailed predictions about future states of the world and the implications they have. What is expected from a science of evolution is that it gets the statistics right—much like in statistical mechanics, where no-one expects predictions about which gas particle bumps into another. In statistical mechanics, correct statistical predictions about pressure, temperature, mixtures, surface tensions, and phase transitions are expected. These statistical predictions can be of amazing precision. In a similar way, from a science of evolution we expect answers to questions like: how many of the innovations will have systemic consequences? Under what circumstances can we trigger or control the systemic consequences of innovations? How big will the changes in diversity typically be? How long do transitions between equilibria last? How long do equilibrium periods last? How stable and resilient are ecosystems with respect to removal of entities? How often will there be cascading failure events in economies or financial systems? What do phase diagrams of evolutionary systems look like, how many phases do they have, and what are the relevant parameters?

We will conclude the chapter with two concrete examples of a combinatorial, co-evolutionary, critical evolutionary approach in the context of economics. In the first example, we discuss how interaction rules for manufacturing goods in combination with the adjacent possible of a country's product space determine economic growth. The second example shows how the critical combinatorial co-evolution model from Section 5.6.4 allows us to predict the evolution of the economic diversity of individual countries. These examples can help us understand the reasons why some poor countries remain poor over long periods, while other countries are able to improve economically. The example in Section 4.10, at the end of Chapter 4 deals with the systemic risk of financial networks. It is another example of a co-evolutionary, critical—however, not combinatorial—evolutionary system. We recommend having another look at it after reading Section 5.7.

In summary, any coherent theory of evolutionary systems must be able to provide quantitative and predictive answers to the phenomena of punctuated equilibria, the abundance of power law statistics, driven out-of-equilibrium criticality, and open-ended, combinatorial co-evolution. To ensure open-endedness, any reasonable framework should allow us to work with infinite numbers of entities and infinite interaction tensors. In Section 5.6.4 we introduce a critical combinatorial co-evolution model, the CCC model, that fulfils all these criteria simultaneously [169, 233, 235, 368]. The model is based on a statistical framework that continually produces innovations and explores its adjacent possible through combinatorial interactions. The generic dynamics of the model is self-organized critical and characterized by periods of equilibria that are disrupted in 'gales of creative destruction'. The model can be calibrated to empirical time series data from evolutionary systems.

5.2   Evidence for complex dynamics in evolutionary processes

There is an impressive amount of empirical evidence that evolutionary processes are indeed critical, combinatorial, and co-evolutionary. Criticality is closely related to the occurrence of punctuated equilibria. Creation and extinction events often occur in bursts that are well separated in time. Most of the time, the system is in a temporary equilibrium, where diversity changes are relatively small. However, whenever transitions to other equilibria occur, this causes large diversity changes, such as mass extinction or mass speciation events that radically restructure the entire system. Punctuated equilibria are therefore associated with fat-tailed (sometimes exact power law) distributions in many observables, such as the size of diversity changes, creation or extinction rates, or lifetimes of entities. In most cases, evolution creates new entities as a combination of existing ones. In biology, most organisms procreate through the combination of genetic material in sexual reproduction. Technologies are introduced as new combinations of existing modular components and processes. Entities can be combined with other entities to catalyse the creation of new entities. This is the process that drives evolutionary change. As new entities themselves become building blocks for new combinations, the set of all possible entities—the adjacent possible—*co-evolves* with the set of all existing entities.

5.2.1   Criticality, punctuated equilibria, and the abundance of fat-tailed statistics

Evolutionary systems typically have more than one equilibrium state. The latter are loosely defined as attractor states to which the system returns after small perturbations. Large perturbations might drive the system far away from its initial equilibrium, so that it settles into another equilibrium. Due to interactions between entities, the transition from one equilibrium to another often involves many entities. Massive change occurs in bursts with large fluctuations. In the following, we refer to a selection of evolutionary observables that are known to exhibit fat-tailed statistics.

5.2.1.1   *Evidence from the fossil record*

Life on Earth has witnessed several mass extinction events [317]. The so-called 'big five' occurred during the Cretaceous, 65 million years ago (mya), Triassic (208 mya), Permian (245 mya), Devonian (367 mya), and the Ordovician (439 mya); see Figure 5.2. It has been estimated that a series of extinction events spread over a few million years at the Permian–Triassic boundary wiped out a combined total of around 96% of marine species and 70% of terrestrial vertebrate species [166, 397]. These results have been derived from data sets containing all known times of origination and extinction of all biological species, the so-called *fossil record*. The distribution of the percentage changes in biodiversity has a fat tail; see Figure 5.3a. Similar fat tails also appear in the distribution functions of species lifetimes, Figure 5.3b, and the number of species per genus (not shown). The five mass extinction events were preceded by a significant

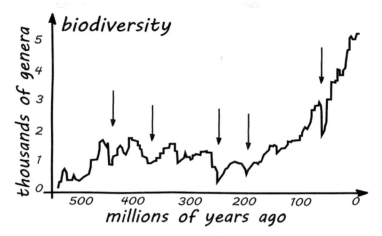

Figure 5.2 *Biodiversity over the past 550 million years. The 'big-five' mass extinctions are marked with arrows. Extinction events are often followed by massive speciation events, where diversity drastically increases—it 'explodes'.*

mass speciation event 541 million years ago, the so-called Cambrian explosion [119]. During the Cambrian explosion, almost all blueprints for multicellular eukaryotic species appeared during the relatively short time interval of 20–25 million years. The existence of fat-tailed distributions empirically rules out the possibility of evolution being driven by simple stochastic processes. The simplest of these would be that the number of biological species that go extinct or come into existence follow a Poisson distribution. However, the statistics of extinction and endogenous speciation events cannot be understood as the result of uncorrelated random processes.

5.2.1.2 *Evidence from economics*

Fat-tailed distributions appear in many socio-economic time series [368]. The number of patents is a proxy for the number of new inventions that are made in a country. The annual percentage changes in patent numbers awarded in the USA follow an approximate power law [388]; see Figure 5.3d. The number of corporate bankruptcies is an indicator for economic extinction. The year-to-year percentage changes in the number of bankruptcies in the USA since 1800 shows a fat-tailed distribution [388]; see Figure 5.3e. Similar observations hold for a number of closely related economic time series, including firm sizes [19] and the rates of innovation [338]. On an aggregated level of all economic activities, the percentage changes in the GDP in most economies exhibit a fat-tailed distribution; see Figure 5.3f.

5.2.1.3 *Open-endedness*

One consequence of punctuated equilibria is that its evolution is *open-ended*. Equilibria do not last for ever; they are metastable. Evolution progresses from one metastable

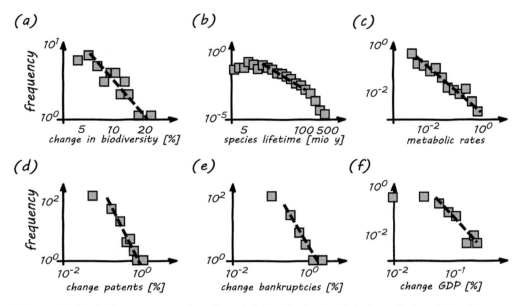

Figure 5.3 *Evolutionary systems show fat-tailed distributions in their dynamical variables. Here are some examples: (a) The distributions of the percentage change in biodiversity from the fossil record. (b) The lifetimes of individual species in the fossil record. (c) The frequencies of metabolic reaction rates in cellular networks of living organisms. Many economic time series show fat-tailed distributions. For example, the annual change (d) in the number of registered patents in the USA, (e) in business failures in the USA since 1800, and (f) in GDP in the UK since 1950. Clearly, some of these distributions are not exact power laws; however, all show fat tails. The statistics cannot be explained by Gaussian or Poisson statistics.*

equilibrium to the next. To our knowledge, there has been no scientific evidence compiled to date, that shows evolutionary processes to be running towards prespecified ends [101].

5.2.2   Evidence for combinatorial co-evolution

Evolutionary processes are driven by a stream of appearances of new entities and species. How these new entities appear—how inventions are made—has been studied in many different contexts. It turns out that inventions, novelties, and innovations are made on the basis of combinatorial interactions.

5.2.2.1   *Evidence from evolutionary biology*

At first sight, the fact that sexual reproduction is so widespread in biology seems paradoxical from an evolutionary perspective [32, 296]. After all, sex is costly. Sexually reproducing species have to produce twice as many offspring on an individual basis (the females) than asexual species, as only one sex is able to produce the offspring. Individuals also have to invest considerable time and energy into the mating process. The ubiquity of sexual reproduction has posed a key problem to evolutionary biology

[180]. An important piece in this puzzle is the so-called Hill–Robertson effect [181, 191], which explains how combinatorial interactions can improve natural selection through an increased variability: consider an asexual population with a genome of length two, the genes being x and y. Assume that both genes have advantageous mutations, X and Y, respectively. If we introduce individuals with genotypes xY and Xy to the population, these individuals will be in direct competition with each other and one advantageous mutation will be selected out. If the species were to reproduce sexually, there is a chance that individuals with a genotype of XY would appear, which would then outcompete all other variations, xy, xY, and Xy.

Recent experiments have further clarified the interplay between combinatorial inter-actions and an increased ability to co-evolve with changes in the environment. In an experiment, two strains of *E. coli* that differed substantially in terms of their rates of com-bining genetic material from different bacterial cells were exposed to a *changing* environ-ment [82]. During the experiment, the concentration of nutrients was gradually shifted towards a solution in which none of the ancestral strains would have been viable. Under certain conditions, the highly combining strain significantly outperformed the strain with lower combination rates. A full resolution of the puzzle of the ubiquity of combination in biology requires the effects of changing boundary conditions to be taken into account.

5.2.2.2   *Evidence from technological development*

Technologies are not able to reproduce sexually—yet. How is a new technology derived from the set of existing technologies? Take any piece of technology, say, a smartphone and open it; you will find an assembly of components (or modules) such as a battery, a display, a memory card, and so on. If you open one of these components, you will find different assemblies of components that are again made of other components. This also works in the other direction: connect your smartphone to a router in your home to control heating and ventilation, and your smartphone becomes a component of a larger technology.

Technologies are combinations of other technologies.

The essence of technologies is that they are modular [14]. Through this modularity, technical components can be combined to form technologies with novel functions. The more technologies that are available, the more possible combinations of these technologies exist, and the more inventions are possible. In other words, the set of possible inventions *co-evolves* with the current set of existing technologies. This idea was introduced by William Fielding Ogburn [294], who tried to understand why cultures differ in their level of technological progress. Technologically less-developed cultures cannot simply transition to modern technologies without mastering the necessary ingredients and capabilities needed for those technologies. There is recent empirical evidence from world trade data that technological diversity is indeed essential for further technological progress and economic success [189]. W. Brian Arthur argues that one arrives at a plausible mechanism for the evolution of technology by putting together the following 'principles':

- New technologies are combinations of existing technologies.
- The arrival of new technologies accelerates the arrival of further new technologies.

Arthur calls this process combinatorial evolution [14]. In this view, early primitive components can be combined into simple technologies, like wheels, axes, hides, and so on. These become components for new technologies, which again become building blocks for further new technologies. Over time, more and more technologies form from a few initial primitive components. Arthur describes this process as a collection of technologies that 'bootstraps itself upwards from the few to the many and from the simple to the complex. We can say that technology creates itself out of itself' [14]. Again, world trade data provides empirical evidence for a general drift towards more complex products over time [183, 233].

5.2.2.3   *Evidence from economic development*

A successful invention is more than a new combination of things. A new technology should be 'useful' for some purpose; it should have some *function*. What determines the usefulness of an invention? Schumpeter observed that economies change purely from within [330]. Entrepreneurs try to introduce new products, new services, or new ways of production through new combinations of already existing technologies. *Useful* goods and services are those that can be used as components for further combinations. Useful things can therefore lead to cascades of inventions. Useful innovations become a substitute for, or replace, currently used technologies, and, in so doing, can seriously disrupt the current state of the economy. Schumpeter called these cascading processes 'gales of creative destruction'. The example he gives is the invention of the car, which drove the horse-based transportation industry out of business. Economic change can be understood as 'the process of industrial mutation that incessantly revolutionizes the economic structure from within, incessantly destroying the old one, incessantly creating a new one' [330]. Creative destruction has been confirmed empirically in world trade data [233].

> In summary, there is ample evidence that evolutionary processes are critical, combinatorial, co-evolving, and open-ended. Evolution creates new entities through combinations of existing ones, which leads to highly modular structures. These new entities can then be used as building blocks for yet newer entities. The set of all *possible* novelties at a given point in time—the adjacent possible—constantly co-evolves with the set of *available* entities.

5.3   From simple evolution models to a general evolution algorithm

Traditionally, evolutionary processes were extensively studied with the so-called replicator equation. It describes the dynamics of populations in terms of changes in species abundances that arise through different forms of species–species interactions and, sometimes, stochastic factors. The replicator equation is well-suited to systems with fixed and small numbers of species. As a consequence, it is of limited use for describing the open-ended, critical, and punctuated nature of evolutionary processes. To account for these aspects, co-evolution between species, their interactions, and their boundary conditions must be explicitly expressed in a more general and more complete theory of evolutionary processes. In this section, we derive such a framework in the form of a general evolution algorithm. It provides a suitable mathematical framework to describe all the characteristics of evolutionary processes that we have mentioned here. This section will set the stage for a number of important classic evolution models that are introduced in subsequent sections of the chapter. Some of them will appear as special cases of the general framework presented here.

5.3.1   Traditional approaches to evolution—the replicator equation

Darwinian evolution is based on the 'principles' of reproduction, mutation, and selection [99]. Populations as a whole reproduce in a way that introduces variations through mutation. Selection acts upon these variants and changes the species' composition of the system over time. The evolutionary dynamics of such systems can be described with simple first-order differential equations. Several authors have noted that the seemingly different aspects of reproduction, mutation, and selection can be combined into a single framework [132, 297]. In the following, we present this framework in the context of biological or molecular species. However, all findings also apply to evolutionary processes in different contexts. Consider a system of N interacting species $i = 1, 2, \ldots, N$ described by their relative abundances $x_i \in [0, 1]$. $x = (x_1, x_2, \cdots, x_N)$ is the vector that contains these relative abundances for all species. Relative here means that $\sum_i x_i = 1$.

The *replicator equation* describes a general continuous deterministic dynamics of an evolutionary system in terms of species abundances,

$$\frac{d}{dt}x_i = x_i \left[f_i(x) - \phi(x) \right], \qquad \text{where} \qquad \phi(x) = \sum_{j=1}^{N} x_j f_j(x). \qquad (5.3)$$

Here $f_i(x)$ is the *fitness function* of species i.

The fitness function of one species may depend on the abundance of every other species. This is called frequency-dependent replication. The term $\phi(x)$ is the weighted population average of the species' fitnesses. It ensures normalization of the abundance vector x at all times. For each species i, Equation (5.3) compares the fitness of i with the population average. Species with fitness larger (smaller) than the population average proliferate (die out) exponentially. In physics, one would call Equation (5.3) a master equation. In this view, replicator equations are a set of first-order differential equations that describe the time evolution of the probability of occupying a specific set of states. The fitness functions $f_i(x)$ describe the observed transition rates to state i. The functional form of $f_i(x)$ depends on the specific types of evolutionary interactions. Many aspects of evolutionary population dynamics and evolutionary game theory can be understood by choosing specific forms for $f_i(x)$. These may include stochastic, spatial, or individual-based components. We now discuss the evolutionary properties that can be described with simple choices for f_i in the replicator equation.

5.3.1.1   *Replication*

In the trivial case, the fitness function does not depend on the frequencies of the other species, and $f_i(x)$ is a constant $f_i(x) = f_i$ for each species. Let us denote the presence of one individual from species i by y_i. f_i is then the rate at which this individual replicates and is replaced by two individuals of the same kind. In a 'stylized' reaction equation this reads,

$$y_i \xrightarrow{f_i} y_i + y_i. \tag{5.4}$$

The master equation, Equation (5.3), for the species abundances is the frequency-independent replicator equation [97],

$$\frac{d}{dt}x_i = x_i\,(f_i - \phi). \tag{5.5}$$

Species with a fitness lower than the average fitness ϕ have a negative replication rate and hence die out. As a result, the average fitness in the system increases over time and approaches the maximum fitness f_i in the system. In the long run, only the species with the maximum fitness survives in Equation (5.5).

5.3.1.2   *Competition*

The simplest non-trivial case for frequency-dependent replication is linear dependence, where two species i and j are in competition with each other. When an individual from species i meets an individual from species j, y_i replicates with probability p_{ij},

$$y_i + y_j \xrightarrow{p_{ij}} y_i. \tag{5.6}$$

The master equation for linear interactions is the linear replicator equation [97],

$$\frac{d}{dt}x_i = x_i \left(\sum_j p_{ij}x_j - \phi \right), \tag{5.7}$$

where ϕ, again, ensures normalization. Equation (5.7) is also called the 'game-dynamical equation' [192].[4] The interest in these equations arises mainly from their relation to game-theoretic settings. Of particular interest are the stable solutions of Equation (5.7). So-called evolutionary stable states and evolutionary stable strategies have been analysed extensively and will not be covered here. The interested reader is referred to [192]. Conceptually, the remarkable feature of the competitive interactions in Equation (5.7) is 'cooperation'. This means that several species might now co-exist in mutual interdependence.

5.3.1.3 *Mutation*

Mutation can be understood as a type of interaction in which a novel species j arises from species i purely by chance. A constant rate, q_{ij}, determines how individuals from species j transition (mutate) to species i,

$$y_j \xrightarrow{q_{ij}} y_i. \tag{5.8}$$

The master equation for a system of species (replicators) that interact through competition and mutation is given by,

$$\frac{d}{dt}x_i = \sum_j q_{ij}\left(x_i \sum_k p_{jk}x_k\right) - x_i\phi. \tag{5.9}$$

This is a special case of a replicator-mutator equation with linear frequency-dependent replication [164]. A special case of Equation (5.9) is the so-called quasi-species equation, which describes asexual organisms or self-replicating macro-molecules [114].

5.3.1.4 *Combination*

We now consider the case where the introduction of new species occurs not merely by chance, but where the presence of a species k *catalyses* the transition from j to i. In stylized terms, this corresponds to a non-linear interaction where j and k can be combined to create i,

$$y_j + y_k \xrightarrow{\alpha_{ijk}} y_i. \tag{5.10}$$

For example, chemical reactions and sexual reproduction can be described by these combinatorial interactions. The associated master equation is given by the catalytic network equation [21, 349],

---

[4] The matrix p_{ij} is sometimes called a payoff matrix. The entries p_{ij} indicate the 'payoff' that individuals i receive from encounters with individuals j in terms of increased or decreased reproductive success.

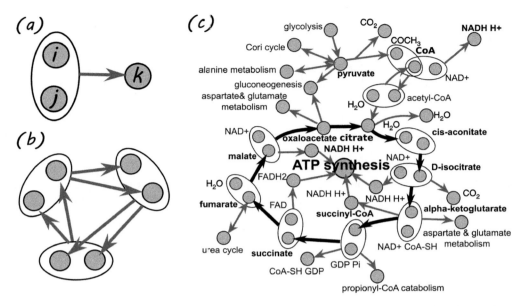

Figure 5.4 *Evolutionary systems can be modelled as entities and rules of how to combine them to form new entities. (a) Entities i and j can be combined to create k. The 'rule' that both entities i and j are necessary to produce k is shown as an oval symbol around them. The situation is similar to logical AND gates, where both inputs need to be present to produce a non-zero output. (b) By linking rules into cyclic structures, one obtains catalytic sets that can sustain themselves: auto-catalytic sets. (c) An important example of a real catalytic set is the Krebs (or citric acid) cycle.*

$$\frac{d}{dt}x_i = \sum_{j,k} \alpha_{ijk}x_j x_k - x_i \phi. \tag{5.11}$$

It is immediately clear that the linear replicator equation, Equation (5.7) is a special case of the catalytic network equation, Equation (5.11). So is the replicator-mutator case of Equation (5.9). It follows that replication, competition, and mutation are special cases of combinatorial interactions. In this sense, combinatorial interactions are the fundamental evolutionary processes.

The catalytic network equation opens the possibility of self-sustaining groups and 'organizations' of species—so-called *auto-catalytic sets*. Self-sustaining auto-catalytic sets are defined as a set of combinatorial interactions, where all elements on the left-hand side of Equation (5.10) (reactants) are also produced by another reaction in the system, that is, they also appear on the right-hand side of at least one other reaction (products).

We refer to systems with species that have combinatorial (or catalytic) interactions as *catalytic sets*. The basic combinatorial interaction is shown in Figure 5.4a. The system consists of three entities i, j, and k, with the combination rule that i and j can be combined to create k. The fact that both i and j need to be present to form k is encoded in $\alpha_{ijk} > 0$ in Equation (5.11), which is depicted by the oval symbol. The dynamics of this system is simple. If i and j are present at the same time, they will combine to form k with rate α_{ijk}. Eventually, the concentrations x_i and x_j will be used up and the dynamics stops. If one of the reactants is not present, nothing happens. Note that these rules are much like logical AND gates. If both inputs are one, the output is one. If one or both inputs are missing, the output is zero.

Things become more interesting in systems with more than one rule. Figure 5.4b shows an example of a system with three rules and six entities. Here, the rules are linked together to form a cyclic structure of interactions. Note that each entity in this 'cycle' is produced as a combination of other entities in the cycle. In principle, such systems are able to self-sustain over an extended period of time. They are called *auto-catalytic sets*.

> An *auto-catalytic set* is defined as a catalytic set in which each species is produced from a combination of other species from the catalytic set.

In general, the topology of such auto-catalytic sets can become quite complicated. Figure 5.4c shows catalytic sets that are linked in a cyclical way without forming an *auto-*catalytic set (as not all entities are produced from within the catalytic set). It shows the most famous example of a catalytic set, the Krebs cycle. The Krebs cycle is a series of chemical reactions used by all aerobic organisms. It is one of the cornerstones of life on this planet.[5] Self-sustaining auto-catalytic sets cannot be observed in linear- or frequency-independent replicator equations. The natural network representation of chemical reactions are power graphs; see Section 4.9.2.

5.3.2 Limits to the traditional approach

The description of evolutionary processes by frequency-dependent replicator equations has limitations of both a technical and conceptual nature.

5.3.2.1 *Size and initial conditions*

Equation (5.11) is a non-linear differential equation. To solve it one has to specify initial conditions. No matter how precisely we determine these initial conditions, because of the non-linearity we must expect chaotic dynamics, meaning that after a short while, the trajectories will become meaningless in terms of their predictive power. For open-ended evolution, it is necessary to operate with large numbers of species. For millions of species, equations like Equation (5.11) are hard to compute, let alone interpret.

---

[5] The Krebs cycle, or citric acid cycle, generates chemical energy in the form of ATP molecules through the oxidation of acetyl coenzyme A, which is derived from carbohydrates, fats, and proteins. To understand the origin of life it is necessary to understand how evolutionary systems assemble and bootstrap these complex interaction structures from simple chemical reactions.

5.3.2.2   *Well-stirred reactor*

In Equation (5.3) it is assumed that the probability of any two species meeting and interacting depends only on their abundances. In chemistry this assumption is equivalent to a 'well-stirred reactor', where a homogenous spatial distribution of all individuals is assumed. It is further assumed that substrates (reactants), raw materials, and other resources necessary for replication and production are always abundant. These assumptions can be relaxed to a certain degree by introducing substrates and raw materials explicitly as species in the model or by introducing spatial or individual-specific constraints.

5.3.2.3   *Open-endedness*

The framework of replicator equations can only be used in a meaningful way if the set of species and interactions is fixed and can be prestated. Assume for a moment that it would be possible to measure the initial conditions and fitness functions in Equation (5.3) with perfect accuracy for all existing species and then to simulate with infinite precision the resulting dynamics for as long as we wish. Even under those highly ideal circumstances, there is no way of measuring the fitness functions of species that do not yet exist. Without an algorithm for specifying the fitness functions of as-yet unobserved species, Equation (5.3) loses its meaning with the arrival of the first new species. However, the very essence of evolution *is* the continual creation of new species. Replicator equations can lead to remarkably complicated dynamics. Nevertheless, they are always confined to a closed system with a fixed number of species. There is no way of capturing open-ended evolution in Equation (5.3).

5.3.2.4   *Co-evolution and separation of timescales*

Equation (5.3) is applicable to dynamics on timescales that are short enough to safely neglect diversification and extinction events. Replicator equations essentially describe equilibrium situations with fixed sets of species and interaction rules. Real evolutionary processes often take place on different timescales. Population dynamics happens at short timescales; the arrival of new species and the associated change in interactions between them unfolds on longer timescales. The dynamics on short timescales corresponds to the situation in metastable equilibria. The transitions between these equilibria happen on comparably long timescales. For instance, it took evolution several million years to produce the first homo sapiens from the earlier apes (long timescale). This arrival triggered a mass extinction of other species that played out over a couple of thousand years (short timescale). Models aiming to include and explain punctuated equilibria have to include mechanisms for innovation and extinction dynamics that operate on different timescales from those of population dynamics. In other words, the interaction rules have to be dynamically updated on a long timescale to allow for co-evolutionary dynamics.

5.3.3   Towards a general evolution algorithm

We have learned that combinatorial interactions in the catalytic network equation are good at describing the situation in equilibrium and close to equilibrium. We will therefore keep the basic elements of Equation (5.11) such as the entities and rules. Entities can be molecules, ideas, goods, biological species, services, and so on. We keep the notion of abundances x_i; however, we generalize them to become more general state vectors σ_i that allow us to describe more states than abundance alone. We will keep the concept of interaction rules, α_{ijk}. Rules specify which entities can be combined to produce new entities. For instance, there is a rule that hydrogen combined with oxygen forms water. There is a rule that a male and a female can produce a baby; however, there is no rule that a fish and a dog can produce a chicken. There is also no rule in physics that two blocks of uranium 235 can be welded into one big block of uranium. We generalize rules and introduce interaction tensors $M^\alpha_{ijk...}$, such that interactions between more than two entities can be described (power graph). More than one interaction type α will be possible. To ensure open-endedness we work with infinite dimensional species vectors σ and interaction tensors M.

We keep combinatorial interactions for three reasons: first, it allows us to explain replication, competition, and mutation as fundamental components of evolutionary dynamics. Secondly, it provides us with the possibility of generating auto-catalytic sets as fundamental dynamical components, and thirdly, combinatorial interactions are a way of creating new entities and of producing open-endedness. An important new step in a general evolution algorithm is the introduction of temporal update rules for interaction tensors to describe co-evolutionary dynamics. Interaction rules will change as a function of the states of the entities. The essential question now is how to specify the interaction tensors.

Biologists integrate the replicator equations to understand several aspects of the dynamics of evolutionary systems. We will not present the general evolution algorithm as a set of differential equations, but as an algorithmic update equation. We will also not integrate the general equation but let it evolve and study the *statistics* of its output.

5.3.3.1   Co-evolution between species and their interactions

How can interaction tensors M be specified? It is impossible to prestate an infinitely large number of interactions between an infinite number of species. However, a general framework needs somehow to specify how the set of species and the set of interactions co-construct each other in an algorithmic way. There are two ways of achieving this. First, one can specify the interaction tensor as the realization of an ensemble of random interaction tensors. For example, one can characterize M as an infinite-dimensional random tensor with a given linking rate, similar to an Erdős–Rényi network. In this case, the linking rate would capture the probability of there being a rule for combining a given pair of entities to create a new element. By creating and destroying species over time, the system also effectively explores specific regions of the infinite and static interaction tensor. Which interactions are 'active' depends on the current species configuration. We refer to this type of co-evolution of species configurations and active interactions as *endogenous*. Secondly, one can describe the dynamics of the interaction tensor by means of external ad hoc assumptions about how to specify the interactions of newly arriving species. For example, one can assume that the interaction terms for a new species are drawn from a specific distribution. The way in which these samples are drawn may or may not depend on the current state of the system. The co-evolutionary dynamics is then given by a dynamic interaction tensor. We will consider several examples of both these types of co-evolution model. Those in which co-evolution is modelled based on ad hoc assumptions about the dynamics of the interaction tensors will be discussed in Section 5.5. The endogenous co-evolution of the interaction tensors will be studied in Section 5.6.4. The framework that we are about to develop will accommodate both options for modelling the co-evolution between species and their interactions. We can now formulate the general evolution algorithm.

5.3.4   General evolution algorithm

As we discuss in Section 5.1.2, any evolutionary system follows a three-step process.

- Step 1: a new entity is created and placed in a given environment. The environment consists of other entities and a physical environment.
- Step 2: the new entity interacts with its environment. As a result of these interactions, the new entity is either destroyed or it survives.
- Step 3: if it survives, it becomes part of the environment. The environment changes and becomes the new 'boundary' for all existing and newly arriving entities.

We now have to specify what entities and interactions are.

Entities The properties of entities i are completely specified by their state vector, $\sigma_i(t)$. For example, one property can be the abundance x_i, $\sigma_i(t) = x_i(t)$, or any function of abundance.[6] If the entity is not present at time t, $\sigma_i(t) = 0$. The dimension of the state vector $\sigma(t)$, N, is typically large. To describe open-endedness in evolutionary processes, we allow infinitely large systems, where $N \to \infty$. It is convenient to include all possible states of the physical environment, such as energy sources, substrates, or raw materials, in the state vector. For example, the corresponding entries $\sigma_j(t)$ describe the availability of a given resource. Entities can interact with these resources in a combinatorial way, exactly as they interact with other entities.

Interactions Interactions are specified by dynamic interaction tensors $M^{\alpha}_{ijk...}(t)$, which describe the creation, production, or destruction of an entity i from the combination of entities $j, k, ...$. The discrete index α denotes the specific type of interaction.

We have now collected all the necessary building blocks to state a general algorithm for evolutionary processes that follows these three steps. By construction, it is co-evolutionary and combinatorial. What then remains to be shown is that the algorithm leads to evolutionary dynamics that is critical and punctuated and that the statistics produces fat-tailed distributions of the size of diversity change, restructuring times, lifetimes of species, occurrence times of transitions between equilibria, and so on.

---

[6] State vectors can also, of course, incorporate other properties than abundance. However, for simplicity we think of abundance here.

General evolution algorithm.

Step 1: new entities come into being at time t through one of two ways. The first option is through a spontaneous creation (new species enters an island). This is reflected by a spontaneous (or exogeneous) state change from $\sigma_i(t) = 0$ at time $t + dt$ to $\sigma_i(t + dt) > 0$. For spontaneous destruction or extinction events, the opposite holds, $\sigma_i(t) > 0$ and $\sigma_i(t + dt) = 0$. The second option is through the action of one or several combinatorial processes, that is, through the catalytic update equation, $\sigma_i(t + dt) = \sigma_i(t) + M_{ijk...}^{\alpha}(t)\sigma_j(t)\sigma_k(t) \cdots$.

Step 2: the new entity i interacts with other existing entities or the physical environment j through combinatorial interactions. These interactions lead to dynamical state changes that are given by the update equation,

$$\sigma_i(t + dt) \sim \sigma_i(t) + F\left(M_{ijk...}^{\alpha}(t), \sigma_j(t), \sigma_k(t), \cdots\right), \qquad (5.12)$$

where F is some function that depends on the state vectors and the combinatorial interactions M. F can be a deterministic or stochastic, a linear or non-linear function. However, F does not have to be a function in a mathematical sense. In general, it will be a set of descriptions and conditions of how entities are combined to form new ones or how existing ones are destroyed.

Step 3: if the new entity survives, it becomes part of the environment. The environment is simply the collection of all possible interactions that are recorded in M. The arrival of a new entity will change the space of all interactions taking place with the existing entities and the physical environment. Formally, this is captured in the evolution update equation of the interactions,

$$M_{ijk...}^{\alpha}(t + dt) \sim M_{ijk...}^{\alpha}(t) + G\left(M_{ijk...}^{\alpha}(t), \sigma_j(t), \sigma_k(t), \cdots\right), \qquad (5.13)$$

where G is some other function that depends on states and interactions. Co-evolutionary dynamics of the entities and their dynamical interaction rules are described by simultaneous updates of Equations (5.12) and (5.13). Each new entity, innovation, novelty, or mutation has the potential to introduce new rules for each other entity, which in turn might lead to the creation or extinction of entities. The spaces of entities and rules co-construct each other.

We will show in Section 5.6.4 that by using minimal assumptions for the functions F and G, critical phenomena, punctuated equilibria, and correct fat-tailed statistics emerge endogenously. The existence of critical behaviour can be understood analytically in a mean field approximation. The algorithmic description also works at the limit of infinitely large systems, which makes it possible to describe open-ended evolution. We will show in Section 5.7 how the model can be calibrated to real world data.

We formulated the combinatorial rules in terms of an interaction tensor M^{α}. What are the different layers α? For example, one layer can be thought of as containing the rules for constructing entities, and another as encoding destructive interactions. For example, if, in the words of Schumpeter, you combine the entity 'combustion engine' with the entity 'petrol', you destroy the entity 'horse and carriage industry'.

> We summarize that combinatorial interactions play a fundamental role in evolutionary processes. They allow us to describe replication, competition, mutation, and catalytic dynamics. Traditionally, evolutionary dynamics describes species abundances with frequency-dependent replicator equations. This approach, however, cannot capture the co-evolutionary and open-ended aspects of evolution. To overcome the limitations, we need to describe (algorithmically) how entities and interactions dynamically co-construct each other in an open-ended way, namely, in a way that is feasible even for infinite numbers of species $N \to \infty$. A general evolution algorithm for combinatorial co-evolutionary dynamics is presented in Equations (5.12) and (5.13).

Large parts of the remaining sections in this chapter are devoted to the description of specific models of evolutionary processes and their essential characteristics. All these models can be understood as a special case of the general evolution algorithm. Obviously, the replicator dynamics is a special case of the general evolution algorithm in differential form.

5.4   What is fitness?

Evolutionary systems create new entities in an ongoing sequence of combinatorial events. What is it, exactly, that determines whether a new entity survives, proliferates, and becomes part of the environment? Traditionally, the rate at which entities reproduce is associated with their *fitness*. In this section we try to understand what fitness really is. We briefly review traditional, widely used definitions of fitness. Typically, each entity is embedded in a high-dimensional landscape that assumes some sort of metric for all possible states of the entity. In evolutionary biology, these states might correspond to all possible genotypes of the species. The distance between two genotypes is their number of different genes. The 'height' at each point in this genotype landscape is given by the entity's fitness or reproductive success. This space is called the *fitness landscape* of the entity.

The essence of evolutionary processes is that the fitness landscape of each entity co-evolves with changes in the composition of all entities and interactions in the system. Several models attempt to make this co-evolution of fitness landscapes and entity space explicit. In this context we will encounter the NK model. Models of this type turn out to be spin glasses—models that describe frustration phenomena in spin systems. The NK model reveals a deep connection between co-evolution and critical phenomena. However, we will also see that the NK model will not help us understand what actually determines the fitness of individual entities, as fitness landscapes in NK models are externally given. We will conclude this section by discussing a concept that will enable us to go beyond this limitation: the concept of the *adjacent possible*. In brief, the adjacent possible is the set of all possible states that the system can reach within the next time step. For a given entity, there will be some states in the adjacent possible where the entity will do well and some where it will not. The larger the number of states in the adjacent possible where the entity does well, the higher its chances for reproductive success in the future. These properties of the adjacent possible will lead us to more of a principle-based understanding of what the concept of fitness really is.

5.4.1 Fitness landscapes?

In popular language, evolution is sometimes associated with the 'survival of the fittest'. What determines who is fittest? We have not yet provided a satisfactory answer to this question. Frequency-dependent replicator equations as in Equation (5.3) assign a fitness function $f_i(x)$ to every entity i. Fitness functions are phenomenological descriptions of the amount by which the presence of other entities or environmental factors impacts the entity's proliferation rate. Fitness is defined as the observed growth rate in the abundance of a species in a system with a specific set of other species and interactions. This implies that 'survival of the fittest' essentially means the 'survival of those that are most likely to survive in the current environment'.

> Fitness, when interpreted as phenomenological proliferation rates, is a tautological concept that does not increase our understanding of the mechanisms that drive evolutionary processes.

Despite this conceptual shortcoming, fitness functions of this kind are nevertheless important tools for understanding how (usually small) systems with a fixed set of species evolve over time.

5.4.1.1 Geometrical interpretation of fitness functions—fitness landscapes

Evolutionary dynamics described by the replicator equation, Equation (5.3), can be interpreted in a geometric way. Consider the space of all possible configurations of species abundances x in a replicator equation. For N species, this species space is an $N - 1$ dimensional simplex. The fitness function for species i, $f_i(x)$, assigns a numerical value (the reproductive success of i) to every point in this space x. This map from an N-dimensional space to a scalar value is called the *fitness landscape*. As the fitness function of one entity may depend on the abundance of all other entities in the system, fitness landscapes are typically high-dimensional objects. Given a fitness landscape, evolution can be pictured as a dynamic process where entities (or sets of entities) move in their fitness landscape towards peaks of their optimal reproductive success. This view was first proposed by Sewall Wright in 1932 when trying to describe a mechanism for natural selection [412]; see Figure 5.5.

5.4.2 Simple fitness landscape models

Frequency-dependent replicator equations describe a trivial form of co-evolution between species space and fitness landscape. As the fitness functions in Equation (5.3)

Figure 5.5 *The first description of a fitness landscape by Sewall Wright [412]. He projected the high-dimensional genotype space into two dimensions and identified fitness peaks (+) and valleys (-) through 'equiadaptivity' lines in this space (dotted lines), which represent contours with respect to adaptiveness.*

and its variations do not explicitly depend on time, the fitness landscape is static. Co-evolution is then reduced to adaptive changes in entity space in the direction of the gradients of the fitness functions. The following example shows how a fitness landscape can be constructed and how it determines the dynamics of a frequency-dependent replicator equation.

5.4.2.1   Static fitness functions: Lotka–Volterra dynamics

Lotka–Volterra dynamics describes a system with two species, prey and predator, where $x_1(t)$ is the number of prey and $x_2(t)$ is the number of predators at time t. The (un-normalized) frequency-dependent replicator equations for this system have linear fitness functions,[7]

$$\dot{x}_1 = x_1\,(p_{12}x_2 + f_1),$$
$$\dot{x}_2 = x_2\,(p_{21}x_1 + f_2). \tag{5.14}$$

---

[7] Note that one can always normalize such systems by adding an additional third species ($N = 3$). For this let $z = \sum_{i=1}^{N-1} x_i$, then a normalized replicator equation for the Lotka–Volterra system is obtained by the transformation of variables, $y_i = \frac{x_i}{1+z}$, and $y_N = \frac{1}{1+z}$.

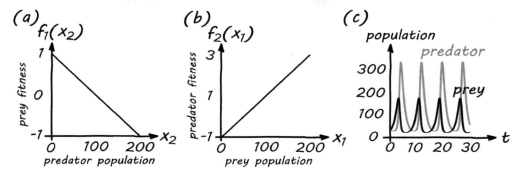

Figure 5.6 *The Lotka–Volterra system for predator and prey dynamics. (a) The fitness function of the prey population, $f_1 = f_1(x_2)$, decreases linearly with the predator population. (b) The fitness function of the predator population, $f_2 = f_2(x_1)$, grows linearly with the size of the prey population. (c) The dynamics of the abundances of both populations show periodic oscillations.*

The fitness parameters f_1 and f_2 are real numbers and represent the replication rates. p_{12} and p_{21} are the competition rates. To model predator–prey relations, we assume that $f_1 > 0$ and $f_2 < 0$, and $p_{12} < 0$ and $p_{12} > 0$. If species 1 exists alone, it can survive (e.g. because there are enough resources to eat, like grass), whereas species 2 will vanish if it cannot prey on species 1. When species 1 and 2 co-exist, the predator population grows at the expense of the prey population. The fitness landscapes for both species are shown in Figure 5.6a and b for $f_1 = 1, f_2 = -1, p_{12} = -0.01$, and $p_{21} = 0.02$. The fitness function f_1 for the prey population x_1, decreases linearly with the predator population. For the predator population, x_2, fitness increases linearly with the size of the prey population. At each point in time, the abundance of species i changes according to the value of the fitness landscape $f_i(x)$, taken at the position of the current configuration in species space, (x_1, x_2). An increase in the number of predators leads to a decrease in the number of prey, which might at some point lead to a decrease in the number of predators. The Lotka–Volterra system has periodic solutions for the species abundances, as shown in Figure 5.6c.

The Lotka–Volterra dynamics illustrates that even small systems can show non-trivial dynamical patterns in entity space. However, the dynamics of the fitness landscapes is trivial—they are static. Replicator dynamics on static fitness landscapes is not open-ended (the system has a simple limit cycle as an attractor), nor is it punctuated, critical, or co-evolving.

5.4.3   Evolutionary dynamics on fitness landscapes

Co-evolutionary dynamics of fitness landscapes was pioneered by Sewall Wright and Stuart Kauffman and collaborators [220, 225]. Their basic idea is that species are represented by their genotypes, and each genotype is assigned a fitness value. Over the course of the dynamics, species adapt their genotypes so as to maximize their fitness. Wright presented the basic picture first. He assumed a high-dimensional space of genotype frequencies that, for the sake of simplicity, he reduced to two dimensions [412]; see Figure 5.5. To each point in this space he assigned a fitness value that represents the reproduction rate of a particular genotype. A species can then be described as a collection (cloud) of similar genotypes, localized in some area in the two-dimensional genotype space. The regions of this area with the highest fitness values reproduce the fastest, and thus—over generations—the area (cloud) that represents the species moves towards the peaks in the fitness landscape.

One can think of a situation where the fitness of a species changes drastically under small variations of its genotype. In that case, fitness landscapes are not smooth surfaces, but *rugged*. They show many local maxima of comparable heights in which the evolution towards a global maximum can become stuck. This situation is reminiscent of so-called frustrated systems in condensed matter physics. Frustration phenomena, which are characterized by highly degenerate ground states, are at the heart of many problems in statistical physics, including metastable configurations, glassy positional disorder, and irregular atomic bond structures [273, 389, 395]. We will mention in this section that a mathematical analogy can be drawn between frustration phenomena in physics and important aspects of evolutionary processes.

5.4.3.1 NK model

Stuart Kauffman proposed a model of a fitness landscape for a single species with tunable ruggedness: the NK model. The name is derived from its two parameters, N and K. For simplicity, let us assume that each gene has two different states, zero or one. A species is completely represented by its genotype. There is no distinction between phenotype and genotype in the NK model. The genotype is a sequence of N binary genes. The set of all possible genotypes is the set of all possible binary strings of length N. The ith letter in sequence s, s_i, is called a gene.

The first assumption of the NK model is that the fitness of an organism with sequence s, $F(s)$, is equal to the average value of the fitness functions $f(s_i)$ of its genes s_i,

$$F(s) = \frac{1}{N} \sum_{i=1}^{N} f(s_i). \qquad (5.15)$$

The second assumption is that the fitness of each gene, $f(s_i)$, is determined by 'epistasis', which is the biological phenomenon that the effect of one gene might depend on the entire genetic background of the organism. In the NK model the genetic background of each gene s_i is given by exactly K other genes. Whether two genes interact is recorded in a network A. If gene s_j has an epistatic influence on s_i then $A_{ij} = 1$, otherwise $A_{ij} = 0$. The fitness function $f(s_i)$ for gene s_i is given as,

$$f(s_i) = R(s_i, \{s_j | A_{ij} = 1\}), \qquad (5.16)$$

where R is some function that maps the state of genes s_i and $\{s_j | A_{ij} = 1\}$ to a real number between zero and one, $R : \{0, 1\}^{K+1} \to [0, 1]$.

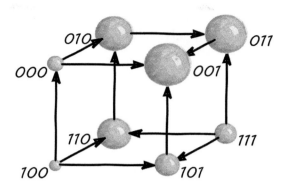

Figure 5.7 *Fitness landscape for a species in the NK model with N = 3, K = 1, for an epistatic interaction network given in Equation (5.17), and fitness values for the genes as in Table 5.1. Each node of the cube represents a genotype. The size of each node is proportional to the fitness of its genotype, F(s). Adjacent nodes are linked by arrows. The direction indicates a step from lower to higher fitness. In this example, the species will always reach the most fit genotype 001.*

A metric for genotype space The most common biological mutations are single point mutations. In the context of the NK model, a single point mutation corresponds to a 'bit flip' of a single gene, $s_i \to 1 - s_i$. Single point mutations can be used to define a biologically meaningful distance between two genotypes, s and s'. The distance between two genotypes is the minimal number of bit flips that transform s into s', or vice versa. A metric for the space of all genotypes can be conveniently defined as the Hamming distance, which is the number of differing bits in two bit strings. The genotype space of the NK model can be represented as an N-dimensional hypercube. Every node in the hypercube represents one genotype (string). Each genotype is connected to N other genotypes that differ by a single gene; see Figure 5.7.

Fitness landscapes in the NK model Each genetic sequence has an associated fitness $F(s)$. The fitness landscape in the NK model is defined over an N-dimensional hypercube. Each node has a fitness value. A local maximum (minimum) in the fitness landscape at the genotype s means that all neighbouring genotypes have a lower (higher) fitness than $F(s)$. The question remains as to how to specify $R(s_i, \{s_j | A_{ij} = 1\})$ in Equation (5.16). The fitness of gene $s_i, f(s_i)$, depends on the state of $K + 1$ genes. There are 2^{K+1} possible states of these $K + 1$ other genes. Here we assume that the values of $f(s_i)$ are drawn randomly from a uniform distribution between zero and one. Note that we are not drawing the values for $f(s_i)$ independently; they must be consistent with the influence of the K genes (see next paragraph). The fitness landscape is static; it does not change during the run of the model.

Epistatic interaction networks and fitness For the simple case of $N = 3$ and $K = 1$, the fitness landscape of a species can be visualized. The species is described by an epistatic interaction network A of order $N = 3$. In this network each node represents a gene. Epistasis is represented by the directed links in the network. Each gene has $K = 1$

incoming link(s) from other genes. For example, assume that the epistatic network interaction network is given by the adjacency matrix,

$$A = \begin{pmatrix} 0 & 0 & 1 \\ 0 & 0 & 1 \\ 0 & 1 & 0 \end{pmatrix}. \tag{5.17}$$

A non-zero entry A_{ij} means that the fitness of gene s_i (row index) depends on the state of gene s_j (column index). In Table 5.1 we show a possible outcome for the random choices of fitness functions $f(s_i)$ for each possible genotype with the epistatic interaction network given in Equation (5.17). Note how the contribution of gene s_i to the overall fitness $F(s)$ remains constant across genotypes that differ only in terms of genes that have no epistatic interactions with s_i. For example, there is no epistatic interaction between gene 1 and 2. Consequently, the fitness of gene 1, $f(s_1)$ does not change under a bit flip of gene 2, and vice versa. Gene 1 interacts epistatically with gene 3. It follows that the contribution to the overall fitness of gene 1 changes whenever the state of either gene 3 or gene 1 itself changes. The fitness landscape for the species is shown in Figure 5.7.

5.4.3.2 *Dynamics of the NK model*

We assume initially that the species consists of a large number of individuals with an identical genotype. The dynamics of the NK model includes a separation of timescales. On the long timescale, the dynamics of the model is driven entirely by single-point mutations. Each gene mutates (bit flips) at the same rate. After a mutation has occurred, population dynamics on a short timescale takes over. Assume that the new mutant genotype s' has a higher fitness than the original genotype s, $F(s') > F(s)$. Individuals with genotype s' will reproduce exponentially faster than individuals of type s. The relative abundance of the species gradually shifts from node s on the fitness hypercube

Table 5.1 *A specific random choice for the fitness values $F(s)$ and the individual gene contributions $f(s_i)$ for the epistatic interaction network given in Equation (5.17).*

genotype	$f(s_1)$	$f(s_2)$	$f(s_3)$	$F(s)$
000	0.814	0.278	0.157	0.416
001	0.913	0.546	0.970	0.810
010	0.814	0.957	0.421	0.730
011	0.913	0.964	0.485	0.787
100	0.632	0.278	0.157	0.356
101	0.097	0.546	0.970	0.538
110	0.632	0.957	0.421	0.670
111	0.097	0.964	0.485	0.515

to node s'. After a sufficiently long time, practically all individuals will be of genotype s'. In contrast, if the new genotype s' has a lower fitness than s, $F(s') < F(s)$, mutated individuals will immediately die out.

The NK model assumes that the time between two mutations (long timescale) is much longer than the short timescale on which the population dynamics takes place. What happens then is that after an advantageous mutation, $F(s') > F(s)$, all individuals of a species move 'as a whole' from node s on the fitness hypercube to node s'. On the long timescale the species in the NK model can always be represented by a single genotype. All individuals of the species are always located at the same node in the fitness hypercube. Initially, the species is assigned a random genotype s. In the model this relates to choosing one node at random. The NK model is specified by the following protocol.

1. At each time step, a single point mutation occurs. The new genotype is s'.
2. If the new genotype s' has higher fitness than the previous genotype s, $F(s') > F(s)$, the new genotype replaces the old one. This means that the species moves to the new position on the hypercube. Otherwise, nothing happens, and the species stays at the previous position in the cube.

In the example shown in Figure 5.7, the species will always end up in the most fit genotype 001. The generic behaviour of the NK model is that the species will get stuck in a local fitness peak. For a given N, the time to reach a local maximum is inversely related to K. To see this, consider the case of $K = 1$. The resulting fitness landscape will be smooth with a low number of maxima and minima, as two adjacent genotypes will differ only in terms of contributions from one gene. By increasing K the landscape becomes increasingly rugged. Starting from a random initial position, individual species have shorter distances to find their fitness maxima. The maximum value for K is $K = N - 1$. At this point, the fitness values of two adjacent genotypes are completely independent from each other. The fitness landscape is maximally rugged; the number of maxima and minima reaches a maximum.

As a dynamical model the NK model is not very spectacular. It takes a population and walks it up—mutation by mutation—to the next (local or global) fitness maximum. Once there, it stops. If the fitness landscape is rugged, the dynamics will, most likely, get stuck in a local maximum; if it is smooth it will reach the global optimum. In its basic formulation, the NK model is a way of conceptualizing fitness landscapes in terms of genomes and the influences of other genes on the fitness, rather than an interesting dynamical model.

5.4.3.3 NKCS model

Things become much more interesting when more than one species is considered in the NK model. Consider now several species, each having its own NK fitness landscape. For simplicity, assume the same N and K for each species, but different epistatic interaction networks and resulting fitness landscapes. Each species s is now coupled to S_s other species. The coupling is modelled by epistatic interactions. Each gene in species s is coupled to C randomly chosen genes from S_s randomly chosen species. For simplicity, we assume that each species depends on the same number of other species, $S_s = S$ for all s. The resulting model is known as the NKCS model [224]. For example, for $C = 1$ and $S = 2$ each gene from s is coupled to one randomly chosen gene from two different randomly chosen species. In other words, the network of species–species interactions is a directed regular random network in which each species has an in-degree of S.

Dynamics of the NKCS model Initially, each species is assigned a random genotype. The model dynamics is then given as follows.

1. At each time step, sequentially pick each species once.
2. Once picked, the species undergoes a single point mutation.
3. If the resulting genotype has higher fitness than before, the new genotype replaces the old one. Otherwise, nothing happens.

In the NK model, the species always reaches a local fitness maximum. In contrast, for the NKCS model to reach a steady state, each species needs to reach a local fitness maximum *at the same time*. This is only possible under certain conditions.

Co-evolution in the NKCS model The NKCS model captures co-evolution between species. Each species evolves so as to increase its own fitness. In doing so, it changes one of its genes, which can also change the fitness contribution of a gene in another species. Therefore, the fitness of other species may change with every single point mutation. The further a species has to evolve to find a maximum in its landscape, the more such co-evolutionary adaptations will occur before the species reaches its maximum. As the distance to the next fitness maximum depends on K, so too does the expected number of co-evolutionary adaptations. Consequently, the relevant parameter of interest in the NKCS model is again K.

Co-evolutionary avalanches After a single point mutation, a certain (perhaps infinite) number of adaptations might be triggered—a co-evolutionary avalanche. The size of this avalanche can be measured as the number of update steps it takes for the system to reach a configuration where each species has reached a (local or global) fitness maximum. Computer simulations show that if K is large enough, co-evolutionary avalanches are of finite size. As K decreases and the fitness landscape becomes increasingly smooth, the shortest distance to the next maximum increases for every species and the avalanche sizes increase. There is a critical value, K_c, below which avalanche sizes diverge. The NKCS model is therefore always in one of two regimes. For $K > K_c$, the system is in a frozen regime, where all species always find a fitness peak. For $K < K_c$, the system enters a chaotic regime, characterized by infinitely large co-evolutionary avalanches. Interestingly, this critical value turns out to be given by the product of the parameters, $K_c \sim CS$.

Criticality in the NKCS model At $K = K_c$ the NKCS model undergoes a continuous phase transition. Precisely and only at K_c, the avalanche sizes follow a power law distribution, which is a classical critical phenomenon that we encounter in Section 3.3.1. Computer simulations confirm that the average species fitness is maximized exactly at K_c. At this maximum, the number of update steps is also maximized along which each species can improve its fitness before it reaches a steady state. This can be understood using the following argument. For $K < K_c$, the trajectories of species are chaotic, which means that they spend little time in the vicinity of their local fitness maxima. For $K > K_c$, from Equations (5.15) and (5.16) we see that random fluctuations in species fitness will be 'averaged out' by the law of large numbers. Above (below) K_c, the average species fitness monotonously decreases with increasing (decreasing) K, which means that K_c is a maximum. Note that many of these properties hold for different choices of species fitness functions. The model dynamics depends critically on the ruggedness of the fitness landscape. Note also that there is no a priori reason why $F(s)$ must be the arithmetic mean of the individual gene contributions and not, for example, the geometric mean or simply the sum.

The existence of a critical threshold K_c for a wide variety of NK type models has fuelled speculation about the role of critical phenomena and scaling relations in evolutionary processes. Suppose that species had the possibility of adapting their CS values (meaning their inter-species dependencies) so as to maximize their average fitness. The ecosystem would then tune itself to the point of maximum species turnover, in other words, to a point with the highest probability of co-evolutionary avalanches. If the events in which a species adapts its relations to other species are interpreted as 'extinction events', this could, indeed, explain power laws observed in the fossil data. However, the NKCS model itself provides no explicit endogenous mechanism for extinction. Another reason for the interest in NK-type models today is that they provide a simple example of an optimization problem that is NP complete for $K \geq 2$; see [404]. Yet another reason for sustained interest is their proximity to spin glasses.

5.4.3.4 NK models and spin models

NK models are closely related to spin models used in statistical physics to model, for example, magnetic materials. A particularly instructive example of a spin model is the Ising model. Despite its conceptual simplicity, its phasespace has a rich structure and shows phase transitions. The Ising model describes a set of N entities with internal states σ_i and a set of rules given by the interaction matrix M. The entities represent magnetic dipole moments of atomic spins. In the Ising model each spin i can take one of two internal states, up and down, $\sigma_i \in \{-1, +1\}$. The entries in M describe interactions between pairs of spins. If two spins interact, it is either energetically favourable for them to align with each other (ferromagnetic case) or to point in opposite directions (antiferromagnetic case). Ferromagnetic interactions between spins σ_i and σ_j are modelled with positive entries, $M_{ij} > 0$. Antiferromagnetic interactions correspond to negative entries, $M_{ij} < 0$. If two spins do not interact, $M_{ij} = 0$. Interactions are always symmetric, $M_{ij} = M_{ji}$. M is a weighted undirected adjacency matrix that is often modelled as a d-dimensional lattice. In the presence of external magnets, each spin i interacts with an external magnetic field h_i that is usually the same for all spins. The energy $H(\sigma)$ of a configuration of states σ is,

$$H(\sigma) = -\sum_{i=1}^{N}\sum_{j>i} M_{ij}\sigma_i\sigma_j - \sum_i h_i\sigma_i. \tag{5.18}$$

The probability of observing a specific configuration σ at inverse temperature $\beta = 1/(kT)$ depends on the energy $H(\sigma)$,

$$p(\sigma | \beta) = \frac{\exp(-\beta H(\sigma))}{\sum_\sigma \exp(-\beta H(\sigma))}, \qquad (5.19)$$

where $\sum_\sigma$ denotes the sum over all possible configurations σ. An important order parameter of the Ising model is the magnetization, m, which is the average value of the expectation value of each spin, $m = \langle \sigma \rangle$.

Continuous order–disorder transition in the ferromagnetic Ising model   Consider the case of a ferromagnetic Ising model, $M_{ij} \geq 0$ for all i and j. Interactions take place on a d-dimensional regular lattice. At finite temperature and in the absence of an external magnetic field ($h_i = 0$) there are two competing effects. First, thermal fluctuations drive the system towards a disordered state where the internal states of each spin are completely random. If the temperature T is high enough (β low enough) each spin is equally likely to be in state $+1$ and -1. The expectation value of each spin, and therefore the magnetization, is zero, $m = \langle \sigma \rangle = 0$. Ferromagnetic interactions compete with thermal fluctuations. Ferromagnetism tends to align spins in the same direction. If the temperature is low enough (β high enough) the spins arrange themselves such that they point in the same direction. The magnetization is then non-zero and the system is in the *ordered phase*, $m \neq 0$. As the temperature approaches zero, the system becomes fully ordered, meaning that $m = \pm 1$.

In the high temperature regime the ferromagnetic Ising model is in a disordered state, whereas in the low temperature regime the system becomes ordered. Between these two regimes there is a continuous phase transition from order to disorder as temperature increases. An exception is the Ising model on a one-dimensional lattice that has no phase transition. On d-dimensional lattices with $d > 1$, the existence of phase transitions can either be shown exactly, by using mean field approaches [159], or by numerical simulations.

Spin glasses   Generalizations of Ising models can describe spin glasses. These are partially disordered or frustrated systems that can get 'stuck' in stable configurations that are different from the overall energetic minimum. A toy model for spin glasses is the Edwards–Anderson model [111], which is an Ising model with ferromagnetic and antiferromagnetic interactions. Interactions take place on a d-dimensional lattice described by the interaction matrix M. The non-zero entries in M are sampled from a normal distribution with a given mean and variance. For sufficiently high temperatures, the Edwards–Anderson model is in a disordered state due to thermal fluctuations. As interactions can now be ferromagnetic and antiferromagnetic, one might expect to find a stable phase with zero magnetization for sufficiently low temperatures. This is not what happens. For low temperatures, a new phase appears that is different from both the ordered and the disordered phase—the so-called *glassy phase*. This phase is characterized by the following property. Assume that an external field is switched on, which increases the magnetization, *m*. Then the external field is switched off. If the system were in a disordered or stable phase, its magnetization would decay exponentially fast. In the glassy phase, however, this decay can take orders of magnitude longer than in the disordered or stable phase [159]. The reason for this strange and slow decay behaviour is that the energy landscape of the Edwards–Anderson model becomes increasingly rugged for low temperatures. During relaxation, spin glasses get stuck in local energy valleys of non-zero magnetization that are higher than the global energy minimum. Thermal fluctuations drive the 'glass' from one local energy valley to the next. This form of relaxation dynamics in spin glasses is very similar to the dynamics of NK models on rugged fitness landscapes.

The similarity between spin glasses and NK models can be expressed in more formal terms. Recall that NK models describe the fitness function of a genotype *s* of length *N* by the fitness function $F(s)$, which is the arithmetic mean of the individual gene fitness contributions $f(s_i)$; see Equation (5.15). The fitness functions for each gene s_i are given by a random function R of Equation (5.16). We can now specify the function R such that we recover spin-model-like interactions. Genes s_i are described as binary states, $s_i \in \{0,1\}$. Using the transformation of variables $\sigma_i = 1 - 2s_i$, the gene states can be mapped to spin-like states $\sigma_i \in \{+1,-1\}$. Assume that the epistatic interaction network of the NK model is a *d*-dimensional lattice. Interactions between genes *i* and *j* are assumed to be of the form $M_{ij}\sigma_i\sigma_j$, with $M_{ij} = M_{ji}$ being sampled from a normal distribution with a given mean and variance. The fitness function for each gene $i, f(\sigma_i)$, is then,

$$f(\sigma_i) = -\frac{1}{2}\sum_j M_{ij}\sigma_i\sigma_j - h_i\sigma_i, \tag{5.20}$$

where the term $-h_i\sigma_i$ describes how the fitness of gene *i* depends on its own internal state, σ_i. The fitness of the entire genotype, $F(\sigma)$ is,

$$F(\sigma) = \frac{1}{N}\sum_{i=1}^N f(\sigma_i) = -\frac{1}{N}\sum_i\sum_{j>i} M_{ij}\sigma_i\sigma_j - \frac{1}{N}\sum_i h_i\sigma_i, \tag{5.21}$$

where we have used the symmetry of *M* in the last step. Up to a global constant *N*, Equation (5.21) is identical to the energy of the Edwards–Anderson spin glass. In this sense, NK models are concrete realizations of spin glasses. There is a difference, however, in how the dynamics of spin models and NK models are specified. The NK model is deterministic and has no temperature. A state change is *always* accepted if the fitness of the mutant genotype increases and *never* if the overall fitness were to decrease. In contrast, spin models can be nicely formulated at non-zero temperatures. The probability of accepting a state change increases exponentially with the associated decrease in energy (scaled by the inverse temperature).

Phase transitions in spin models are usually studied at finite temperature, whereas NK models do not involve a temperature. However, the relaxation dynamics of spin glasses at low temperatures is remarkably similar to the dynamics of the NK model. Both dynamics lead to frustrated configurations due to the ruggedness of their underlying fitness and energy landscapes.

5.4.3.5 NK models for technological change: Wright's law of technological progress

There is a natural correspondence between evolutionary optimization in the NK models and models of technological development. A number of phenomenological 'laws' that characterize technological development have been proposed in the literature [281]. Perhaps the most famous one dates back to an observation by Gordon Moore [278] that the density of transistors in integrated circuits doubles each year. Moore's observation was also found in other areas of technological development, for example, in memory capacity, quality-adjusted microprocessor prices [70], and the number of pixels in digital cameras [280]. As a simple consequence, Moore's law states that the unit cost of a technology, $y(t)$, decreases exponentially over time t,

$$y(t) \sim \exp{(-\alpha t)}, \tag{5.22}$$

with $\alpha > 0$ a constant. Let us denote the cumulative number of all units of technology produced until t by the cumulative production, $n(t)$. It has been observed empirically that $n(t)$ often grows exponentially in the early phases of development [325],

$$n(t) \sim \exp{(\lambda t)}, \tag{5.23}$$

with a constant growth rate $\lambda > 0$. By eliminating time t from Equations (5.22) and (5.23), we obtain that the unit cost $y(t)$ depends on the cumulative production $n(t)$ in the form of a power law,

$$y(t) \sim n(t)^{-\gamma}, \tag{5.24}$$

with $\gamma = \alpha/\lambda > 0$ [325, 281]. Equation (5.24) is another famous phenomenological 'law' in technological development. It was Theodore Wright who observed that the unit cost of a technology $y(t)$ decreases as a power law of the cumulative number of items that have been produced [413]. Equation (5.24) is called the learning curve of a technology. It describes that the more we produce of a certain thing, the more we learn about how to improve its production, namely, learning by doing [362].

5.4.3.6 *Trivial case of single-component technologies*

There is an intricate relation between Moore's law, Wright's law for learning curves, the complexity of technological components, and evolutionary processes as described by NK models [270]. To understand it, consider a simple technology that consists only of a single component or one single building block. At each time step, one unit of the technology is produced, and an attempt at innovation is made in order to reduce its unit cost, y. The cumulative production n grows by one unit at each time step. Assume that the unit cost $y(n)$ of a technology is rescaled such that it takes values between zero and one. For clarity, we suppress time dependence of y in the following. Each innovation attempt changes the unit cost to a different random value y', drawn from the uniform distribution $P(y)$. The probability of drawing a new cost y' that is smaller than the current cost y is denoted by $P(y' < y)$. It is given by $P(y' < y) = \int_0^y P(y')dy' = \langle y \rangle$, where $\langle y \rangle$ is the expectation value of y at a given time n. If a cost reduction occurs, each value of y' between zero and $\langle y \rangle$ is equally likely to be the new cost. On average, the cost will be reduced by an amount of $\langle y \rangle / 2$. The rate of change of the unit cost, dy/dn, can be approximated in the following way. A cost reduction by an amount of $\langle y \rangle / 2$ occurs with probability $P(y' < y) = \langle y \rangle$. The rate of change is proportional to,

$$\frac{dy}{dn} = -\frac{\langle y \rangle}{2} P(y' < y) = -\frac{\langle y \rangle^2}{2}, \tag{5.25}$$

which is solved by a power law,

$$y(n) = \frac{y(0)}{1 + \frac{y(0)n}{2}} \sim n^{-1}. \tag{5.26}$$

We recovered Wright's law with a power exponent of -1. Real technologies typically have different exponents [270]. Real technologies, however, also consist of more than one component.

5.4.3.7 *Technological complexity—design structure matrix*

We now consider technologies that consist of more than one component. The dependencies between the individual components of a technology can be represented in a design structure matrix, D [72]. Figure 5.8 shows D for automobile engines [267]. D represents a technology as an unweighted, directed network. Nodes correspond to components. A directed link from component j to i (black square in row i and column j in Figure 5.8) means that changes in the design of component j make changes in the design of i necessary, $D_{ij} = 1$. If j is redesigned to reduce its cost, it is also necessary to redesign components i for which $D_{ij} = 1$. Consequently, changes in the cost of j are likely to change the costs of the connected components i. Assume that each component has a cost, y_i. The cost of the entire technology, y, is the sum of the cost of its components,

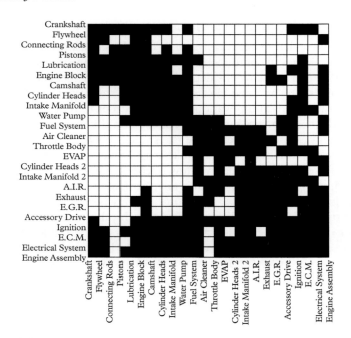

Figure 5.8 *The design structure matrix D for automobile engines. Each row and column corresponds to a component of the engine. A white square in row i and column j indicates that a change in the design of component j does not require any changes in the design of component i. A black square in row i and column j, $D_{ij} = 1$, means that a change of the design of component j requires a change in the design of i. In this case it follows that a change in the cost of j goes together with a change in the cost of i. For instance, changes in the design of the engine block also require changes in the design of the pistons, but do not, however, require changes in the air cleaner. Data from [267].*

$y = \sum_i y_i$. Given that an innovation occurs, the new cost of each component y_i' is a function of the costs of all components on which i depends,

$$y_i' = f(y_i, \{y_k | D_{ik} = 1\}). \tag{5.27}$$

At each time step a single randomly chosen component, say, i, is updated (redesigned). It is assigned a new cost from a fixed probability distribution, and so is every component that depends on i. The update is only accepted if it leads to a cost decrease of the entire technology. Solving for $y(n)$ is more involved than for a single component, but it is still possible [270]. The result is that technological costs decline with cumulative production n as,

$$y(n) \sim n^{-\frac{1}{d^{\min}}}, \tag{5.28}$$

where $d^{\min} = \min(\{\sum_j D_{ij}\})$ is the minimal out-degree of matrix D. The result is again Wright's law, but now with an exponent that is, in general, different from one. Empirical tests of Equation (5.28) are complicated by the fact that it is hard to measure the rate at which attempts to improve a given technology are made. However, this model provides a directly falsifiable prediction. In particular, this exercise suggests that technological improvements can be accelerated by reducing the connectivity in the design structure matrices.

NK models can also be formulated to describe technological development. Equation (5.27) closely resembles the fitness functions of the NK model in Equations (5.15) and (5.16). The design structure matrix plays the role of the epistatic interaction network. In both cases we find that the rate of change and successful improvements can be understood on the basis of the connectivity of these networks. Both models produce power law statistics and show punctuated equilibria. The model for technological improvements allows us to understand the power laws in the learning curves of technical progress, which is a specific form of Moore's law.

5.4.4 Co-evolving fitness landscapes—the Bak–Sneppen model

The Bak–Sneppen model is a stylized toy model for evolutionary processes. It serves as a 'model of models' for co-evolutionary dynamics on fitness landscapes. The aim of the model is to identify the fundamental mechanism by which more complex fitness landscape models, such as the NK model, generate critical behaviour. The Bak–Sneppen model was amongst the first to provide a mechanism for the emergence of punctuated equilibria. The model shows self-organized criticality. Its observables exhibit power law behaviour independent of the concrete parameter choices in the model, which is in contrast to NK models that have to be fine-tuned to show critical behaviour. The Bak–Sneppen model has versions that are exactly solvable [23].

The narrative is as follows. Consider a fixed number of $N \gg 1$ species and their fitness functions. Species spend most of their time at their fitness maxima. Initially, each species i resides at a local fitness maximum of height F_i. To reach another maximum with $G_i > F_i$ in the fitness landscape, the species has to pass through a barrier of depth B_i. For a given mutation rate, λ, which is the same for all species, the time-to-mutation T_i of species i goes as,

$$T_i \sim \exp \frac{B_i}{\lambda}. \tag{5.29}$$

This means that time-to-mutations are exponentially separated by their barriers-to-mutation, B_i. The species with the lowest barrier is exponentially more likely to mutate than any other species in the system. We are, therefore, interested in the species with the lowest barrier to formulate the model dynamics.

5.4.4.1   Dynamics of the Bak–Sneppen model

Every species i is initiated with a random fitness barrier $B_i \in [0,1]$, which is, in the simplest case, chosen from a uniform distribution. A mutation of i is modelled as the assignment of a new random fitness barrier B_i chosen from the same distribution. Each species is assumed to be coupled to S other species. The model is given by the following protocol.

1. At every time step, t, the species i with the lowest barrier to mutation is chosen to mutate.
2. A new, randomly chosen barrier B_i is assigned to i.
3. Each species j that is coupled to i will also mutate and is assigned a new random barrier, B_j. Species that are coupled to j do not mutate.
4. The model proceeds to the next time step, $t + 1$.

Initially, any fitness value between zero and one has the same probability of appearing for all species. In the first update, the lowest barrier-to-mutation B_i is therefore close to zero. The new barrier for species i is very likely to be higher than its old value. The S species that mutate due to their interactions with i are also unlikely to receive significantly lower fitness values on average in the update. It follows that in the first time steps of the dynamics, the average fitness in the system increases as the lowest barrier to mutation increases over time. As time progresses, an increasing number of mutating species will be assigned a new barrier that is smaller than the lowest barrier to mutation in the step before. Those species are very likely to be chosen within the next mutation steps. Average species fitness in the system will no longer increase, but will fluctuate around a steady-state level. Computer simulations confirm that for $S = 2$ this is the case when the lowest barrier approaches a critical value of approximately $B_c \sim 2/3$.

The Bak–Sneppen model is closely related to the fitness landscape models that we discussed in Section 5.4.3. One can identify the barrier to mutation B_i with the unit costs of technologies y in the model for technological change. The Bak–Sneppen model basically assumes that the design structure matrix or the epistatic interaction network is an ER network with average degree S. Besides these similarities, there are differences in the model dynamics. The Bak–Sneppen model accepts the mutation of a species even if its fitness does not improve. However, such newly generated low-fitness species are also very likely to be removed within the next model updates, and thus play a negligible role.

5.4.4.2   Self-organized criticality and punctuated equilibria

The Bak–Sneppen model leads to the formation of co-evolutionary avalanches. Their formation is not obvious, as at every time step t only one species mutates. This timescale t must not be confused with the scale T_i in Equation (5.29), which gives the approximate time between two mutations of species i. On timescale t there is one mutation at each time step. On timescale T_i, however, the time between two mutations depends exponentially on B_i. If in the steady state a mutation happens to assign low new fitness values to several species, the subsequent mutations occur over relatively short time periods T; see Equation (5.29). After a certain number of mutations, the lowest barrier approaches its steady-state level B_c again. On timescale T this means that there are no mutations over long stretches of time. The model shows long periods of practically no activity (high values of the lowest B_i), followed by co-evolutionary avalanches that involve multiple species updates over a short period of time. Punctuated equilibria emerge. It can be shown that the numbers of mutations per time interval in T indeed follow a power law [130]. The Bak–Sneppen model is an example of a self-organized critical system. There is a direct correspondence between the parameters of the Bak–Sneppen and sand pile models [24], see also Section 3.3.2. The slope in the sand pile model plays the role of the lowest barrier B_i in the Bak–Sneppen model. Both parameters initially increase over time until they approach a steady state where the system becomes critical.

In the Bak–Sneppen model the species with the lowest barrier B_i is chosen to mutate. The result of this selection rule is that the system is always driven towards a critical state. The NK model, in contrast, updates species sequentially. Despite its elegant explanation for the emergence of punctuated equilibria, the Bak–Sneppen model is missing several ingredients to explain evolutionary processes in a comprehensive way. For example, it does not involve speciation or external stress. These issues have been addressed in several extensions in order to understand empirical observations from the fossil record, such as the species lifetime distributions or the observed number of species per genus; for a review see [289]. However, in these model variants there are also no explicit, endogenous extinction events.

5.4.5 The adjacent possible in fitness landscape models

In combinatory evolutionary systems at every time step, there are a large number of entities that could be produced or destroyed, given the entities that currently exist in the system. The adjacent possible consists of the entirety of all these possible states [219]. The fitness landscape models that we have discussed in this section are specified by algorithmic rules for how the fitness of a newly created (or updated) entity is determined. These algorithms provide explicit construction rules for the adjacent possible.

For example, the adjacent possible of a species in the NK model consists of the set of all genotypes that could, in principle, exist in the next time step.[8] At any given time t the species occupies exactly one node in its fitness landscape, which is an N-dimensional hypercube. At time $t + 1$, either a mutation has occurred that moves the species to one of its N neighbours, or no advantageous mutation took place and the species remains at the same node. For a species with genome length N, the adjacent possible consists of the $N + 1$ genotypes that have a Hamming distance of one or zero to the current genotype. Of these $N + 1$ possibilities, one will be realized. The actual one is determined by the value of the predetermined random fitness function.

---

[8] Remember that a species in the NK model has exactly one genotype at each point in time.

The adjacent possible in the Bak–Sneppen model is different. At each time step t one species and its S neighbours are chosen to mutate. The internal state of each species i is specified by its lowest fitness barrier, $B_i \in [0, 1]$. The $S + 1$ species that mutate have a sample space $[0, 1]^{S+1}$. In principle, each combination of $S + 1$ different species from the total N species could be chosen to mutate. There are $b = \binom{N}{S+1}$ such combinations, where b is the binomial coefficient. The adjacent possible in the Bak–Sneppen model is therefore of size $[0, 1]^{b(S+1)}$. Which of these possibilities is realized is determined by a random number. Therefore, the NK model and the Bak–Sneppen model both provide little insight into *how* evolutionary systems explore their adjacent possible. Both models construct their fitness landscapes essentially at random.

In this section, we described evolutionary processes as dynamical processes on fitness landscapes. The classic model in this context is the NK model, which assigns a random but fixed fitness value to each genotype. Its dynamics consists of species undergoing single point mutations that might increase their fitness. From the perspective of the general evolution algorithm, NK models describe situations with static fitness landscapes, as the interactions in Equation (5.13) do not change, $M(t + dt) = M(t)$. The state dynamics, Equation (5.12), in NK models is essentially that of spin glasses. NK models explain how the formation of large update cascades depends on the ruggedness of the underlying fitness landscape. The principle of rugged fitness landscapes can be used to understand phenomenological laws of technological development and to derive the learning curves for technological progress that are power laws, which can be directly related to Moore's law. The Bak–Sneppen model illustrates a mechanism by which rugged fitness landscapes generate self-organized critical behaviour.

5.5 Linear evolution models

In this section, we will learn about evolution models that assume linear interactions in the general evolution algorithm. These models will be linear in the sense that the interaction tensor M in Equations (5.12) and (5.13) is replaced by an interaction matrix. Linear models cannot describe combinatorial interactions, but serve as a useful starting point for exploring the co-evolution between entity space and interactions. These models are particularly suited to understanding how initially random configurations of species with low diversity co-evolve into structured organizations of interconnected auto-catalytic sets, able to maintain a high species diversity over long time periods. Linear models are also helpful in understanding the conditions under which highly diverse systems collapse into states of low diversity. We will see how the emergence of auto-catalytic sets and their systemic risk of collapse are related to the topology of the underlying interaction networks. In this sense the risk for collapse becomes measurable.

5.5.1 Emergence of auto-catalytic sets—the Jain–Krishna model

A fantastic feature of combinatorial interactions in catalytic network models is that they explain how entities that are non self-replicating, can form replicating, self-maintaining organizations: so-called auto-catalytic sets. The question of whether the origin of life was a historical accident and how probable its occurrence was is closely related to the likelihood of the emergence of auto-catalytic sets under prebiotic conditions. The dynamics of combinatorial co-evolution often involves a separation of timescales into 'ecological' and 'evolutionary' dynamics. Ecological population dynamics happens on the fast timescale. The system approaches the attractor configuration of a given fixed composition of species relatively quickly. The evolutionary dynamics unfolds on a much slower timescale. It changes the composition of species through creation and extinction events. The combination of the dynamics on both timescales allows us to understand a plethora of evolutionary phenomena. We will see that with the combination of the two dynamics, the formation of simple auto-catalytic sets can happen even in models with linear interactions.

5.5.1.1 *Catalytic reactions of polymers and the origin of life*

J. Doyne Farmer and colleagues tried to understand how complex metabolic reaction networks that sustain living organisms have evolved from simpler molecules. In particular, they studied a model of catalytic reactions in polymer chemistry [22]. They found that, under equilibrium conditions, systems favour homogeneous distributions of short polymers. When the system is driven out-of-equilibrium by an inflow of chemical substrates, a self-sustaining auto-catalytic set emerges to 'digest' the inflow—a 'metabolism' emerges spontaneously. In a different numerical experiment the researchers introduced mutations that take place on longer timescales [21]. These mutations spontaneously created species by reactions that involved only the species itself or already existing elements of the metabolism. The chemical kinetics of the catalytic reaction network provides a selection mechanism for the newly created species. The system is driven towards a state of punctuated equilibria with long periods during which no mutation is selected. Eventually, a new auto-catalytic set is implemented that radically rearranges the internal pathways of reactions in the model. Similar phenomena are observed in other models of artificial chemistries and their computational frameworks. Examples include algorithmic chemistries such as the Turing gas [131], the theory of chemical organizations [264], and chemical graph replacement and transformation models [45]. In the following, we focus on a linearized model of catalytic networks that is extremely transparent and illustrates some of the essential dynamics of evolutionary dynamics in general [205, 206].

5.5.1.2 *Linear reaction dynamics—the fast timescale*

The idea is to model an idealized primordial soup that contains N chemical species, y_i. The species may catalyse each other with probability, p. The 'population dynamics' of the chemical species is given by a linearized version of the catalytic network equation in Equation (5.11). To see how the linearization is obtained, assume two reactants A and B with a respective number of molecules, n_A and n_B. Think of these reactants as simple molecules, such as methane, ammonia, or hydrogen. Without a catalyst the reactants do not react. In the presence of a catalyst j, A and B react to produce a molecule i,

$$A + B \xrightarrow{j} i. \tag{5.30}$$

The rate equation for this reaction is given by a catalytic network equation of the form,

$$\frac{d}{dt} y_i = k \left(1 + v y_j \right) n_A n_B - \phi y_i, \tag{5.31}$$

where y_i denotes the concentration of i, k is the rate, v is the catalytic efficiency, and ϕ is the dilution flux. Under the assumptions that the reactants A and B are buffered, which means that n_A and n_B are very large compared to the concentration of the catalyst, and assuming that the spontaneous reaction is substantially slower than the catalysed reaction, the rate equation, Equation (5.31), depends only on the catalyst population, $\frac{d}{dt} y_i = M y_i - \phi y_i$.

Note that i might become a catalyst itself for another reaction. If we assume that there are many catalysed reactions in our primordial pond, and if we denote the relative abundances of the catalysts by $x_i = y_i / \sum_i y_i$, we obtain the linear equation,

$$\frac{d}{dt} x_i = \sum_{j=1}^{N} M_{ij} x_j - \phi x_i, \tag{5.32}$$

which describes the chemical network of catalysts that drive the chemical reactions in the system. The directed interaction network M_{ij} encodes the rate at which i is produced by a reaction that is catalysed by j. Note that, here, non-linear combinatorial interactions take place only between buffered reactants, A and B in Equation (5.30). Interactions between the catalysts y_i in Equation (5.31) are always linear. It follows that Equation (5.32) describes the emergence of auto-catalytic sets, which are not exclusively based on combinatorial interactions.

To proceed, we have to specify M_{ij} in Equation (5.32). For a simple example, assume that initially at $t = 0$, we define M as a directed Erdős-Rényi network (ER). With probability p we set $M_{ij} = 1$. There are no self-loops, meaning that $M_{ii} = 0$. If two species (catalysts) interact, they interact constructively and always with the same intensity. Note that the model is meaningful only if M is initialized such that the network is below the ER percolation threshold p_c; see Section 4.4. As soon as the interaction matrix M is specified, we can solve the model in Equation (5.32). As it is linear, the formal solution[9] in matrix notation is,

---

[9] Note that the solution does not depend on the dilution flux ϕ, which simply rescales all abundances.

$$x(t) = e^{Mt}x(0), \tag{5.33}$$

where e^{Mt} only makes sense in the series expansion of the exponential, $e^{Mt} = 1 + Mt + (1/2)M^2 t^2 + \ldots$. Note that x is a vector, $x = (x_1, x_2, \cdots, x_N)$. Let λ_i be the eigenvalue associated to the eigenvector v_i of M. We can write $x(0)$ as a linear combination of the basis vectors v_i with the coefficients α_i, $x(0) = \sum_i \alpha_i v_i$ and use this in Equation (5.33) to obtain,

$$x(t) = \sum_i \alpha_i e^{Mt} v_i = \sum_i \alpha_i e^{\lambda_i t} v_i. \tag{5.34}$$

M is a positive semi-definite matrix. From the Perron–Frobenius theorem we know that M has a real largest eigenvector, λ_1, and that all entries in the eigenvector, v_1, that is associated with λ_1 are real numbers that are positive or zero. For large t, $x(t)$ will be dominated by the term $e^{\lambda_1 t}v_1$. This exponentially dominates all other terms. The attractor configuration x^* is therefore found to be,

$$x^* = \lim_{t \to \infty} x(t) = v_1. \tag{5.35}$$

This means that the solution of the catalytic equation, the concentrations of the catalysts, is simply the eigenvector to the largest eigenvalue of M—the eigenvector centrality from Section 4.8.3. The Perron–Frobenius theorem provides us with a relation between species concentrations in linearized catalytic network systems and the topology of their interaction networks.

5.5.1.3 Extinction dynamics—the long timescale

Extinctions take place on a timescale (T) that is much longer than the timescale on which the population dynamics takes place (t). In many cases, the time between two extinction events can be assumed to be long enough for the catalyst concentrations to reach their attractor configuration x^*, given by Equation (5.35). The catalyst with the smallest entry in x^* is identified as the least-fit species. Say that this is species k. It will be eliminated. We remove it together with all its in- and outgoing links, meaning that we set $M_{ki} = M_{ik} = 0$ for all i. A new species is then introduced and takes the place of k. The new species interacts with the already existing species with probability p, meaning that we set $M_{ki} = 1$ and $M_{jk} = 1$ with probability p, for all values of $i, j \neq k$. Now the interaction matrix M has changed, and x^* is in general no longer the attractor configuration for the new M.

5.5.1.4 Combined dynamics—the Jain–Krishna model

The idea of the Jain–Krishna model is to repeatedly solve the linear catalytic equation, Equation (5.32), obtain the attractor solution of the concentrations, eliminate the least-abundant species, and then solve the catalytic equation again. First, initialize the interaction matrix M as an ER network with no self-loops. The protocol of the Jain–Krishna evolutionary algorithm then reads:

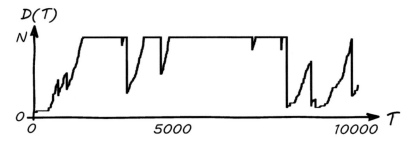

Figure 5.9 *Diversity over time in the Jain–Krishna model. Initially, the model is in a low diversity regime. With the emergence of the first auto-catalytic cycle a diversification boom sets in. The model enters a highly diversified regime. After several thousand iterations, diversity crashes and recovers shortly afterwards. Booms and busts can be directly related to the co-evolving cyclical structures in the interaction network.*

1. At time T, solve the linear catalytic equation, Equation (5.32), to obtain x^*.
2. Identify the least-fit species in x^*. Say this is k.
3. Remove the least-fit species k by setting $M_{ik} = M_{ki} = 0$ for all i.
4. Introduce a new species with random interactions: $M_{ik} = 1$ and $M_{kj} = 1$ with probability p for all $i, j \neq k$ (no self-loops).
5. Solve the linear catalytic equation for the new M, and repeat the process at $T + 1$.

The dynamics of the model is clearly co-evolutionary in the sense of the general evolution algorithm from Section 5.3.4.

The diversity dynamics of the Jain–Krishna model for $N = 100$ species is shown in Figure 5.9. The diversity is the number of non-zero entries in the population, $D(T) = \sum_{i=1}^{N} \Theta(x_i(T))$, where Θ is the Heaviside function; see Equation (8.1). Initially, the model is in a low diversity regime. With the emergence of the first auto-catalytic cycle that sustains a growing auto-catalytic set, a massive diversification event takes place and the model enters a highly diversified regime. After some time the diversity crashes due to the removal of a keystone species, which breaks the auto-catalytic cycle. Booms and busts can be directly related to cyclical structures in the interaction network, M.

5.5.1.5 *Diversity, eigenvalues, and auto-catalytic cycles*

The existence of cycles in the interaction network has immediate consequences for the diversity of the system. An auto-catalytic cycle is a set of reactions that form a closed directed loop in the reaction network M. The simplest auto-catalytic cycle is a closed path of length two. Consider two species i and j that form a cycle, $M_{ij} = 1$ and $M_{ji} = 1$. Note that this auto-catalytic cycle is an auto-catalytic set. Any species k with an incoming link from i or j is also a member of the auto-catalytic set but is not necessarily a member of the auto-catalytic cycle. A species l with a link to i or j is a member neither of the cycle nor of the auto-catalytic set. In general, for each component C of the directed interaction network M, the following holds. If C contains only one species (isolated node), the species is a member neither of an auto-catalytic cycle nor of an auto-catalytic set. If C contains more than one species, then each species in C is typically a member of an in-component, out-component, or strongly connected component; see Section 4.3. Species in the strongly connected component of C form an auto-catalytic cycle (or interconnected cycles). Together with species in the out-component of C, they form an auto-catalytic set. If there is no strongly connected component in C, then there is no cycle and no auto-catalytic set. It follows that each auto-catalytic set contains at least one auto-catalytic cycle. Species in the in-component are neither part of an auto-catalytic cycle, nor of an auto-catalytic set.

> A set of catalysts $\{i_1, i_2, \ldots, i_K\}$ forms an auto-catalytic set if each of its members is the product of another catalyst of the set. In terms of the interaction network M, every directed cycle is an auto-catalytic cycle and every auto-catalytic set contains at least one cycle.

There is a relation between the Perron–Frobenius eigenvalue, λ_1, and the existence of auto-catalytic cycles. It can be shown with elementary linear algebra that if the interaction network M has no cycle, then $\lambda_1 = 0$, and if M has at least one cycle, then $\lambda_1 \geq 1$. If no cycles exist, the interaction network is a tree or a forest. In this case, all species concentrations 'flow downstream' towards the nodes at the ends of the trees (leaves). If the longest path in M is r steps, each species sitting at a leaf can receive at most r times support from the other species. The diversity in an interaction network with no cycles is equal to the number of leaf species. For $\lambda_1 \geq 1$, where cycles exist, the abundance of each species i that is part of an auto-catalytic set will grow exponentially as $x_i(t) \sim \exp(\lambda_1 t)$. Note that the normalization of the species concentrations ensures that in the attractor configuration, x^*, only species that are part of an auto-catalytic set have non-zero support. All other species will be exponentially suppressed. The diversity in the model is then the number of species that reside in one of the strongly connected or out-components of M.

5.5.1.6 *Understanding the emergence of diversity*

The model is initialized with an average node degree of $m = p(N - 1) \ll 1$, far below the ER percolation threshold. The initial interaction network is sparse and almost certainly has no cycles. Diversity is therefore given by the number of leaf species until the first auto-catalytic set emerges. Let us assume that this happens at time $T = T_1$. The first set will most likely be a cycle of length 2, which occurs approximately with probability $p^2 T_1$ after T_1 iterations. At T_1 we can therefore expect a jump from $\lambda_1 = 0$ to $\lambda_1 \geq 1$. Each species in the auto-catalytic set at time T_1 will not be in the set of least-fit species, as their $x_i^* > 0$. As soon as the first auto-catalytic set comes into existence, the species that has been replaced will almost certainly not be from that set. From time to time newly introduced species will join the auto-catalytic set. As a consequence, these new species will also no longer belong to the set of least-fit species. The growth of the auto-catalytic set continues until *all* species are part of it. This happens at T_2. As all species are in the auto-catalytic set, one of them will be the least-fit and will be replaced. This might break the cycle and reduce the largest eigenvalue.

5.5.1.7 Collapse—understanding mass extinctions

The dynamics of the Jain–Krishna model is non-trivial in the fully diversified regime. For $T > T_2$, the node that is replaced (mutating node) has to be chosen from the auto-catalytic set. The auto-catalytic cycles in the network will remain intact if the node is chosen from one of their out-components. However, removals of species in the strongly connected component might break one or several auto-catalytic cycles. This can have dramatic system-wide consequences. To quantify how systemically important individual species are, that is, how important they are for high levels of diversity, we can define the *keystone index*. Consider the hypothetical removal of species i along with all its links, which would lead to the adjacency matrix M'. In matrix M', all entries in the ith row and the ith column are equal to zero, but M' is otherwise identical to M. Let λ_1' be the Perron–Frobenius eigenvalue of M'. The keystone index K_i for species i is defined as,

$$K_i = \frac{\lambda_1 - \lambda_1'}{\lambda_1}. \tag{5.36}$$

Large values of K_i indicate that the species belongs to an auto-catalytic cycle. If that species is removed, the largest eigenvalue of the system is dramatically reduced and high levels of diversity may no longer be sustainable. Species i is then called a *keystone species*. Low values of K_i indicate that species i contributes relatively little to the diversity dynamics dominating the current state of the system. Most of the time the least-fit species have a low keystone index K_i. This means that the current auto-catalytic cycle remains intact and that the system remains diversified. The simulation shown in Figure 5.9 shows that full diversity can be maintained over extended periods. However, it may occur that the least-fit species in a fully diversified population is a keystone species. If M contains a single cycle that is destroyed by the removal of the least-fit species, which happens to be a keystone species, the network becomes acyclic. Diversity collapses to levels before the emergence of the first auto-catalytic set. The removal of the keystone species has triggered a mass extinction. This is seen in Figure 5.9.

5.5.1.8 Introducing competition—emergence of cooperation

In the absence of destructive interactions in the Jain–Krishna model, auto-catalytic cycles form relatively easily. We now introduce competition to the model [206]. For every species i with probability p, an interaction term M_{ij} will be non-zero. In that case, the actual entry M_{ij} is drawn uniformly from the interval $[-1,+1]$. Now self-loops are allowed. Left by themselves, species i degrade with a rate $D_i = M_{ii}$, which is a random number chosen from $[-1,0]$. Both these changes in the Jain–Krishna model make it harder for self-maintaining organizations to appear and to persist. With the modified definition of M, the competitive version of the model is specified by the following population dynamics,

$$\frac{d}{dt}x_i = \begin{cases} f_i & \text{if } x_i > 0 \quad \text{or} \quad f_i \geq 0 \\ 0 & \text{if } x_i = 0 \quad \text{and} \quad f_i < 0 \end{cases}, \tag{5.37}$$

where $f_i = \sum_j M_{ij} x_j - x_i \phi$. The rest of the model follows exactly the same steps as the original Jain–Krishna model, the sole difference being that the updates of M_{ij} are now performed with uniform random numbers from $[-1, +1]$. The matrix M is no longer positive semi-definite and the Perron–Frobenius theorem can no longer be applied. The dynamics of the competitive model variant exhibits the same phenomenology in diversity terms. Numerical simulations confirm that similar mechanisms are indeed at work. However, interesting new features appear.

As the first auto-catalytic cycle appears in the model with competition, the species in this set will proliferate exponentially and survive the extinction dynamics. Keep in mind that mutant species now show a variation in the number of positive and negative links. In the diversity growth phase, mutant species with more positive links are more likely to attach themselves to the auto-catalytic set and therefore are more likely to survive. There is a bias in the extinction dynamics in favour of selecting species (with positive links) that tend to cooperate with other species. Mutualism emerges. Simulations confirm that the sudden explosion of diversity in the growth phase is characterized by a sharp increase in the number of positive links in the network. Over longer timescales, the competitive model again shows punctuated equilibria with an ongoing alternation of phases with high and low diversity. Crashes result from the removal of keystone species and the subsequent loss of cycles in the system. In the diversified phase, the system has the same structural vulnerabilities as before, due to its dependence on keystone species to achieve self-sustainability.

The Jain–Krishna model provides a simple explanation for the mechanism by which the environment determines whether a mutant species will be successful or not. Species fitness is related to network properties, the eigenvector centrality, and the Perron–Frobenius eigenvalue. The more positive feedback loops a species is involved in, the higher its chance of survival and success. Note that fitness here is *not* given just by the sum of all positive influences that a species receives from the other species, as occurs in the frequency-dependent replicator equation, Equation (5.3). Fitness in the Jain–Krishna model is a dynamical and emergent phenomenon of the entire network.

5.5.2 Sequentially linear models and the edge of chaos

A surprising feature of evolutionary systems and, more generally, of complex systems, is that they combine a remarkable degree of stability with the ability to adapt to changing environments. Self-organized criticality is a popular attempt to explain the seemingly contradictory combination of stability and adaptability. The narrative is that organisms live at the *edge of chaos*, where they have access to two different dynamical regimes [222, 246]. The stable (or regular) regime allows the organism to maintain stable dynamical configurations, which are used in times of normal environmental conditions. Access to the disordered regime allows the organism to switch to a mode where large parts of the possible phasespace can be sampled in a short amount of time, which boosts the chances of the organism to adapt to changing environments. This narrative is intriguing, but it provides no explanation for the mechanisms by which evolutionary systems self-organize towards this magical edge of chaos and how they switch between stable and disordered regimes. In the following, we show how such a switching behaviour generically emerges in an extremely simple model that is again based on linear interactions between species. The only non-linear ingredient in the otherwise linear dynamics is the constraint that species abundances must be non-negative. This minimal non-linear assumption of positive concentrations is enough to generate surprisingly rich dynamical behaviour, including self-organized criticality [171, 172, 356]. The model also shows that the edge of chaos is an extended region in the parameter space of evolutionary processes, and that it is, by no means, an exception.

5.5.2.1   *A minimally non-linear evolution model*

We consider a system of N species described by their abundances x_i. Species–species interactions are given by the interaction matrix M. If a species j promotes the proliferation of i, we have $M_{ij} > 0$. If $M_{ij} < 0$ then species j suppresses the production of species i. No interaction means $M_{ij} = 0$. For the off-diagonal matrix elements we assume random entries at locations ij with probability p. Their values are drawn from a uniform or a Gaussian distribution. Species in isolation, meaning that they are not promoted by others, decay with a specific rate $D_i = -M_{ii} > 0$. For simplicity, we assume that all decay rates are the same $D_i = D$. Each species is produced with an externally given production (or driving) rate, Φ_i. Think of this rate as species flowing in or out of the system. The resulting evolutionary dynamics takes the following form,

$$\frac{d}{dt}x_i = \sum_{j=1}^{N} M_{ij}x_j + \Phi_i. \tag{5.38}$$

The stationary state of the system, for which $\frac{d}{dt}x_i = 0$, is denoted by x^*. It is implicitly given by $\Phi_i = -\sum_j M_{ij}x_j^*$, and Equation (5.38) can be rewritten as,

$$\frac{d}{dt}x_i = \sum_{j=1}^{N} M_{ij}\left(x_j - x_j^*\right). \tag{5.39}$$

Equation (5.39) describes the abundances of species x_i that must be non-negative numbers. As M is not positive semi-definite, the Perron–Frobenius theorem does not apply and there is no guarantee that the solution x^* to Equation (5.38) will have only positive or zero entries. On the other hand, we require concentrations to be strictly positive (or zero)—never negative. We therefore add the explicit constraint that all abundances must be positive or zero at all times,

$$x_i(t) \geq 0 \quad \text{for all species } i \text{ and for all times } t. \tag{5.40}$$

The positivity constraint in Equation (5.40) has severe consequences for the linear dynamics of Equation (5.38). Assume that after specifying initial concentrations, we let the model evolve. Obviously, some concentrations will reduce over time, $\frac{d}{dt}x_i(t) < 0$. At some time t_0, one of these concentrations will hit zero, $x_i(t_0) = 0$. Equation (5.40) ensures that $x_i(t)$ remains exactly zero as long as the derivative of species i is negative, $\frac{d}{dt}x_i(t) < 0$. Assume that at some later time t_1 the derivative turns positive again, $\frac{d}{dt}x_i(t_1) \geq 0$. The species i will be produced again, and the positivity constraint will no longer affect $x_i(t)$. The dynamics is then again described solely by Equation (5.38).

In the following, we say that species i is *active* at time t, whenever $x_i(t) > 0$, and *inactive* if $x_i(t) = 0$. The *active set* of species contains all species that are currently active. Inactive species have zero concentrations and do not influence the production or suppression of other species. Inactive species can be produced by active species. As soon as this happens, they will no longer be inactive. Over time, species will typically change their status from active to inactive and back. Let us index by T_i ($i = 1, 2, \cdots$) those instances in time where the composition of the active set changes. If S_0 is the initial active set, one observes a sequence of active set changes of the form,

$$S_0 \xrightarrow{T_1} S_1 \xrightarrow{T_2} S_2 \xrightarrow{T_3} \ldots . \tag{5.41}$$

For each active set S_i there is a different active interaction matrix M^{S_i}. The active interaction matrix has the property that $M^{S_i}_{mn} = M_{mn}$, whenever species m and n are both active, and $m, n \in S_i$, and $M^{S_i}_{mn} = 0$, otherwise. The interaction matrix effectively reduces its dimension whenever a species becomes inactive. The sequence of active sets is therefore equivalent to a sequence of active interaction networks given by,

$$M^{S_0} \xrightarrow{T_1} M^{S_1} \xrightarrow{T_2} M^{S_2} \xrightarrow{T_3} \ldots . \tag{5.42}$$

The dynamics of the concentrations in the active set in each time interval $[T_m, T_{m+1}]$ is described by a linear dynamical equation,

$$\frac{d}{dt}x_i^{S_m}(t) = \sum_{j \in S_m} M_{ij}^{S_m} \left(x_j^{S_m}(t) - x_j^{*S_m} \right), \tag{5.43}$$

where $x_i^{S_m}$ is the abundance vector of the active species. The dynamics of Equation (5.38), together with the positivity constraint of Equation (5.40), leads to a sequence of *linear* models, a so-called *sequentially linear model*. The sequential linearity simplifies the analysis substantially. Equation (5.43) allows us to understand the attractor configuration for each active set in terms of its eigenvalues and (strictly positive) eigenvectors[10] of the active submatrices M^{S_m}. These attractor configurations can be stable or unstable.

---

[10] A strictly positive eigenvector means that all its entries are positive real numbers.

To understand this we have to recall the concept of the Lyapunov exponent, which measures how fast two initially infinitesimally close trajectories diverge over time.

> The Lyapunov exponent $\bar{\lambda}$ is defined as the asymptotic ($t \to \infty$) exponential rate of divergence between two trajectories $x(t)$ and $x(t) + \delta x(t)$,
>
> $$||\delta x(t)|| \sim e^{\bar{\lambda} t}||\delta x(0)||, \tag{5.44}$$
>
> where $||\delta x(t)||$ is the Euclidean distance between the trajectories at time t and $||\delta x(0)||$ is their infinitesimal separation at the initial configuration. For every dimension in the system there is one Lyapunov exponent. $\bar{\lambda}$ is the maximum of all exponents. If $\bar{\lambda}$ is positive, the system is called *chaotic*, or strongly mixing. If $\bar{\lambda}$ is negative, trajectories in the system converge to an attractor, which can be a point (fixed point), a line (limit cycle), or a fractal object. If the system approaches a limit cycle, the system is periodic. An interesting case arises when the exponent $\bar{\lambda}$ is exactly zero. This is the edge of chaos.

Consider an N-dimensional system given by the set of linear, ordinary differential equations of the form,

$$\frac{d}{dt}x = Mx, \tag{5.45}$$

with x being an N-dimensional vector and M an $N \times N$ matrix with constant coefficients. The N Lyapunov exponents $\bar{\lambda}_i$ of this system are closely related with the eigenvalues of M. If M has N different complex eigenvalues, denoted by $\lambda_1, \lambda_2, \ldots, \lambda_N$ (in descending order of their real parts), the Lyapunov exponents are simply the real parts of these eigenvalues, $\bar{\lambda}_i = \Re(\lambda_i)$. These N Lyapunov exponents form the so-called Lyapunov spectrum. In general, the Lyapunov spectrum depends on the initial conditions of the trajectories. One is usually interested in the values of the spectrum in the vicinity of an attractor x^*, meaning that $x(0) = x^*$ in Equation (5.44). For $t \to \infty$, the divergence of two trajectories will be exponentially dominated by contributions from λ_1. The maximal Lyapunov exponent $\bar{\lambda}$ is therefore given by the real part of λ_1. The attractor x^* is called unstable (stable), whenever its maximal Lyapunov exponent $\bar{\lambda}$ is positive (negative).

For the sequentially linear model in Equation (5.43) each active set has its own attractor configuration that is either stable or unstable. The active set S_m is associated with a stable (unstable) attractor, if the Lyapunov exponent $\bar{\lambda}^{S_m}$ of the active interaction matrix M^{S_m} is smaller (larger) than zero.

5.5.2.2 *A simple example*

In the following, we consider a simple example with $N = 4$ species to illustrate the dynamics of the sequentially linear model. Species have fixed point concentrations $x_i^* = 100$ and decay rates $D_i = M_{ii} = -0.23$ for all species $i = 1, \ldots, 4$. We specify the linear interaction network to be,

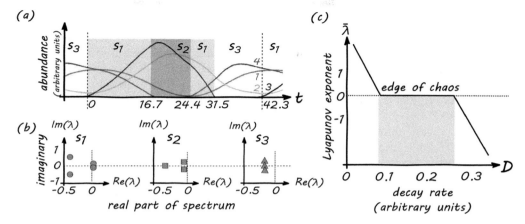

Figure 5.10 *Example of the dynamics of a sequentially linear system. (a) The dynamics of species abundances follows a periodic sequence of four time domains with three different active sets (S_1, S_2, and S_3). (b) Each domain is associated with the spectrum of eigenvalues of its active set, indicating whether the dynamics is characterized by a stable or unstable fixed point. (c) The system self-organizes towards the edge of chaos. The Lyapunov exponent $\bar{\lambda}$ of the minimally non-linear model is zero for a wide range of decay rates D. Each point where $\bar{\lambda} = 0$ can correspond to a different global attractor of the dynamical system.*

$$M = \begin{pmatrix} -0.23 & -0.1 & 0 & 0.1 \\ 0 & -0.23 & 0.2 & 0 \\ 1 & 0 & -0.23 & -1 \\ -0.8 & -0.8 & 0.1 & -0.23 \end{pmatrix}. \tag{5.46}$$

In this example, the model undergoes a series of switches between different active sets occurring in a periodic sequence; see Figure 5.10a. We find four domains of active networks that are associated with three different active sets. For each of the four time domains, the complex eigenvalue spectrum is shown in Figure 5.10b. The first domain with active set S_1 has a positive maximum real part of the eigenvalue, which means that the fixed point of the dynamics is unstable. In the first domain there is a species (species 4, dark-grey) that hits zero at time 16.7. There it is deactivated by the positivity condition. At this point, the system switches to the second active set S_2 with a stable fixed point (negative maximum real eigenvalue). As the system approaches the new fixed point of the second domain, species 4 is activated again at 24.4 and the system switches to the third time domain. Domain 3 has the same active set as the first domain (S_1), but different initial conditions, so that another species (species 3, black) is deactivated at time 31.5 where the system switches to the fourth domain, which has a stable fixed point. In approaching the fixed point of the fourth domain, species 3 is produced again at time 42.3. Once this species is produced, the system is back to its initial active set with the same initial conditions at $t = 0$ and the next period starts. We call the set of periodic trajectories a 'global attractor'.

5.5.2.3   *Sequentially linear models and the edge of chaos*

In general, the dynamics of sequentially linear systems can be described as follows. The trajectories converge to different *attractors*, including fixed points and limit cycles. Which attractor is initially 'found' by the system is determined by the initial conditions. Sequentially linear systems can have many different limit cycles and fixed points. At every point in time, the system tries to converge to an attractor. The attractor may not be *accessible*, however. This means that the trajectory, on its way towards the attractor, hits the positivity boundary. At this point, the positivity condition kicks in and effectively switches off some interactions. Now, with the interaction matrix changed, the system might be attracted towards a different attractor. If the system does not converge to an accessible fixed point, it is either unstable, meaning that some concentrations x_i diverge or that the system keeps passing through the same sequence of active sets, over and over again. This periodic sequence effectively defines the 'global attractor' of the system. Small perturbations of the trajectories $x(t)$ on the global attractor will die out, which shows that any dynamics that does not converge to a fixed point has to be periodic. However, if perturbations are large, the system can switch to a different global attractor. The existence of multiple different global attractors was shown for a wide range of system sizes N, connectivities of the interaction matrix, and decay rates [356, 172].

Self-organization towards the edge of chaos   A particularly interesting feature of sequentially linear models is that they self-organize towards the edge of chaos. Assume that the system is initially prepared in an unstable state, meaning that its attractor has a positive maximal Lyapunov exponent. Eventually, one species will hit the zero abundance threshold and be deactivated. The dynamics switches either towards an accessible fixed point or towards a limit cycle, as described before. The system has 'found' an active set, where the dynamics is stable again and the maximal Lyapunov exponent has been reduced to zero, or slightly below. This way, the switching dynamics leads to a self-organization towards the edge of chaos.

Robustness of the edge of chaos in parameter space   We can also understand that the self-organization towards the edge of chaos happens for a large range in parameter space. Here, the free parameters are, for example, the decay rates D. According to Girko's circular law, for large systems with random interactions in M, one expects the eigenvalues to be uniformly distributed in a circle in the complex plane with radius ρ. Girko's law further specifies that the radius of this circle is proportional to the variance of the off-diagonal entries in M; see Section 4.4. Increasing the decay rates D shifts the circle along the negative real axis. Assume for a moment that we drop the positivity constraint in Equation (5.40) and consider only the linear dynamics in Equation (5.39). The system will be stable (unstable) whenever the decay rates D are larger (smaller) than ρ. Without the positivity constraint, the edge of chaos (the regime where the maximal Lyapunov exponent is exactly zero) is given by a single point, where D matches the spectral radius ρ of the interaction matrix. However, enforcing the positivity constraint changes the robustness of the model in a spectacular way. If a species is deactivated, one effectively reduces the dimensionality of the interaction matrix M. This means that the link density in M gets smaller, which effectively reduces the radius ρ of the circular distribution of eigenvalues. Further, those species with the highest reaction rates—the largest contributions to the variance of M—will be deactivated first. The model selects species with lower variance in the interaction terms. The system self-organizes into a critical state at $\lambda^{S_m} = 0$, even when decay rates are smaller than the radius ρ. As a consequence, the maximal Lyapunov exponent of sequentially linear models has zero-values for an entire *range* of decay rates D; see Figure 5.10c. The edge of chaos has been inflated to a window of high adaptability, $D^{min} < D < D^{max}$. Below D^{min} the dynamics diverges for some species and the system is unstable. Above D^{max} the system reaches a periodic limit cycle.

The minimally non-linear model shows amazing features of self-organization towards the edge of chaos for a large range of parameter space, and not just a specific choice of D. This self-tuning, as well as the inflation of the edge of chaos, could have consequences for understanding the multistability of interaction networks in biology. For example, one can imagine organisms that have evolved the ability to control the degradation rates of molecular species. Using a set of fixed interactions (chemical rules), the organism could then switch between different expression modes (global attractors) of the network by adjusting the decay rates. This is shown in detail in [171]. Control of intracellular degradation rates could then become a mechanism that drives cell differentiation.

5.5.3 Systemic risk in evolutionary systems—modelling collapse

Evolutionary models with linear interactions cannot describe how new entities, innovations, and novelties emerge through the combination of existing entities. Linear models can be quite accurate, however, for understanding extinction dynamics. In this section, we try to better understand systemic risk—the risk of an evolutionary system collapsing. Linear interactions are a good starting point to model the risk of collapse. We have seen in the Jain–Krishna model how the removal of a single species can lead to the collapse of an entire ecosystem. The concept of systemic risk tries to quantify the risk of collapse of complex systems. In particular, we want to understand the likelihood that an initially small and localized event (removal of a single species) might lead to macroscopic consequences. For this reason, we discuss a linear evolutionary model that addresses the systemic risk of collapse.

5.5.3.1 *Extinction in the Solé–Manrubia model*

Solé and Manrubia analysed a model of extinction that essentially maps the replicator-mutator dynamics of Equation (5.9) to an update equation for species diversity [343]. Species i are represented by binary variables $\sigma_i(t)$, that indicate whether a species i exists or not. If the abundance is positive, $x_i(t) > 0$, we have $\sigma_i(t) = 1$, otherwise $\sigma_i(t) = 0$. Species–species interactions are encoded in the interaction matrix M_{ij}.

A positive (negative) value of M_{ij} indicates that species i increases (suppresses) the reproductive success of species j. For simplicity, all entries in M_{ij} are chosen from a uniform distribution $[-1, 1]$. The Solé–Manrubia model assumes that a species becomes extinct if its net support from all other species drops below a certain threshold h. For N species the dynamics is given by a discrete update equation for σ_i,

$$\sigma_i(t+1) = \Theta\left(\sum_j M_{ji}\sigma_j(t) - h\right), \tag{5.47}$$

where $\Theta(x)$ is the Heaviside step function, see Equation (8.1). Note the obvious similarity to Ising-type spin models. The system is now driven by two processes that mimic mutation and speciation.

Mutation and speciation   Mutation is modelled by random changes in the interaction matrix M. Species that fall below the support threshold die out and are eliminated. Extinct species are immediately replaced by copies of other, surviving species. The idea is that an extinct species leaves an empty ecological niche that is immediately filled by an invading species, which needs to adapt to the new niche. Adaptation is modelled by adding noise terms to the species–species interactions. In particular, if species j replaces species i, then the kth row and column in the interaction matrix M change as,

$$\begin{aligned} M_{ik} &= M_{jk} + \xi_{ik} \\ M_{ki} &= M_{kj} + \xi_{ki}, \end{aligned} \tag{5.48}$$

where ξ is a random variable drawn from a uniform distribution $[-\epsilon, \epsilon]$, for a fixed value of ϵ.

Dynamics in the Solé–Manrubia model   The dynamics of the Solé–Manrubia model is completely specified by the following protocol.

1. *Extinction:* update the species variables $\sigma_i(t)$ according to Equation (5.47) for all species in parallel.
2. *Mutation:* for each species i, choose an incoming connection in M at random and assign it a new value chosen from the interval $[-1, 1]$ with uniform probability.
3. *Speciation:* replace each extinct species with a copy of a randomly chosen surviving species by using Equation (5.48).

Extinction cascades   For a threshold of $h = 0$, the dynamics of the Solé–Manrubia model leads to extinction cascades and punctuated equilibria. After an initially transient behaviour, mass extinctions are observed in the model that wipe out up to 90% of the species. Such events are often followed by comparably long phases of little activity. The distribution of the number of extinctions within a given time interval follows an approximate power law distribution with an exponent close to -2. For other choices of *h*, the power law in the extinction cascades disappears. The model has a critical point, but it is not self-organized critical.

A note on universality As is the case for the NK, NKCS, and the Bak–Sneppen model, the power law in the distribution functions of extinction sizes in the Solé–Manrubia model is quite robust with respect to the details of the model. For instance, one can choose the entries in M to be ± 1, or entries can be zero with a given probability. One can exclude mutation ($\epsilon = 0$) and still observe similar extinction cascades. This kind of robustness is a common feature of models that show critical phenomena; we called it *universality* in Section 3.3.1. In the vicinity of critical points, the properties of a system can be entirely characterized by macroscopic long-range correlations. The dynamics of the system becomes independent of the microscopic details of the interactions [351]. If universality did not exist (or if one did not believe it existed), the interest in stylized spin models for evolution would be purely academic.

The presence of fat-tailed and power law distribution functions in real evolutionary systems makes it plausible, to a certain extent, that critical phenomena may indeed be involved. If this is the case, we do not need to get the details of the interactions right. All we need to do is determine the *universality class* of real evolutionary systems. This would help to narrow down the classes of realistic models.

Ecology or evolution? In the Solé–Manrubia model, speciation occurs on the same timescale as population dynamics; see Equation (5.47). Typically, whether or not an invading species can survive in a given environment is determined within a few generations. Speciation, however, tends to occur on much longer timescales. One could argue that the model thus describes *ecological* processes rather than evolutionary ones. Other ecological models that describe species interactions through foodwebs are discussed, for example, in [4, 10].

5.5.3.2 *Financial networks and cascading failure*

The financial crisis of 2007 briefly increased the level of public awareness that financial systems are subject to systemic risk. In particular, it seems to be understood by now that financial crises are far more frequent than once-in-a-lifetime events. The risk of collapse is created at the system level and cannot be understood as being a property of its components, such as banks. Systemic risk arises through a particular form of spreading dynamics in which the failure of one entity might induce the failure of another.

One can model financial collapse by the following assumptions. There are N banks, connected to each other through financial contracts M_{ij}, which are generally nothing other than promises of future payments (credits, derivatives, etc.). Every entity is characterized by a wealth level $x_i(t)$. If the wealth of entity i drops below a given threshold h, the entity is bankrupt and is eliminated by setting $M_{ij} = 0$ for all entities j, meaning that they will not get the promised payments in the future.[11] These losses are subtracted from the wealth $x_j(t + 1) = x_j(t) - M_{ij}$. If that value is below the threshold, j is also bankrupt at time $t + 1$. Depending on the topology of M, this might lead to cascading dynamics of defaults that are often characterized by approximate power laws in terms of losses. The situation is similar to the risk of collapsing in ecological systems, where, for example, future payments could be replaced by trophic dependencies in food webs. An important difference is, however, that while biological or ecological systems have naturally evolved the ability to remain relatively stable over long time periods, the same cannot be said of financial networks. In fact, the policy of 'too big to fail' implies that financial systems follow the rule of survival of the fattest, rather than survival of the fittest [165]. This lack of evolutionary selection should of course be compensated for by intelligent regulation, which does not yet exist [312], but could exist [311]. For intelligent regulation it would be necessary to develop a network centrality measure that performs the same function as the keystone index in the Jain–Krishna model: to quantify the systemic risk contribution of individual entities. How this can be done in the financial context is shown in the example in Section 4.10.

To summarize, in this section we studied evolution models with linear interactions between entities. Linear models often assume a separation of timescales, where the dynamics of the linear species–species interactions occurs much faster than the dynamics of innovations and extinctions.

- The *Jain–Krishna model* considers a linear population dynamics, where co-evolution of entities and interactions is implemented through the replacement of the least-fit entity by one with randomly chosen new interactions. The model shows how system-spanning auto-catalytic sets drive the dynamics of massive speciation and collapse. Fitness emerges as a non-trivial topological property of the linear interaction network. The more cycles an entity is a part of, the greater its chances of survival.

- The *sequentially linear model* is a model of co-evolution between entities and linear interactions amongst them. It is a linear population dynamics model with the non-linear constraint that all abundances of the entities must be zero or positive at all times. The interactions of entities that hit zero abundance are switched off. The dynamics proceeds with a reduced interaction matrix of all entities with positive abundances. The system might approach a fixed point, where inactive entities are switched on again along the path. The sequentially linear model has a rich behaviour in terms of stability. It can switch between many periodic attractors that represent different modes of the dynamics of the same system. For a large region in parameter space the system is able to self-organize towards the 'edge of chaos',

---

[11] For example, think that M_{ij} is the amount that i has borrowed from j.

a regime that allows for the co-existence of stable periodic trajectories and the inherent ability to adapt to changing environments.

- The *Solé–Manrubia model* explains extinction events. The least-fit entity is invaded by mutated variants of surviving entities. The combination of mutation and speciation events with the invasion of empty ecological niches can amplify mutual dependences amongst entities, and this increases the likelihood of macroscopic extinction events.

- *Systemic risk* of financial contagion was shortly mentioned. Cascading events can unfold on interaction networks, where entities are linked through financial contracts. Whenever an entity becomes bankrupt this could affect the liquidity of connected entities, which could also drive them into default. Parallels can be drawn with the risk of ecosystem collapse. Financial systems are co-evolutionary critical systems, where the wealth of the entities influences future contracts and contracts influence future wealth.

5.6 Non-linear evolution models—combinatorial evolution

In this section, we discuss evolutionary models that are special cases of the general evolution algorithm, Equations (5.12) and (5.13), with non-linear or combinatorial interactions. Non-linear interactions allow us to describe the arrival of new entities through the combination of existing ones. We start with the simplest case, where systems evolve exclusively through combinations. We show that such systems exhibit a 'creative' phase transition in their diversity dynamics. In particular, we clarify the conditions for a sustained diversity phase in terms of initial diversity and the density of combination rules (connectancy of the interaction tensor, M). The model describes open-ended combinatorial evolution. We will discuss the Arthur–Polak model, a model for the evolution of technologies, to learn that the existence of specific selection rules, in combination with the creative phase transition, is enough to produce technologies with a surprising level of sophistication in relatively short times. The essential feature of these evolutionary processes is that newly created entities can be combined with other entities, which leads to the entities having a highly modular structure. Joseph A. Schumpeter observed that successful innovations have the potential to disrupt and destroy entire industries. Destructive interactions play a central role in evolutionary processes. Following the Schumpeterian narrative, we will discuss a combinatorial co-evolution model that combines constructive and destructive combinatorial events. We show that the resulting model, the critical combinatorial co-evolution model, or CCC model, essentially contains all the defining components of evolutionary systems that we have encountered so far. The CCC model generically leads to critical phenomena, punctuated equilibria, and power law statistics of evolutionary observables. The critical properties of the system can be understood analytically in a mean field approximation that also applies to the open-ended case of infinite numbers of entities and interactions. The CCC model shows how co-evolution between species and their interactions endogenously emerges from the combination of elementary production and destruction rules. Most of the evolutionary models described so far can be recovered as special cases or simple modifications of the CCC model.

5.6.1   Schumpeter got it right

The central mantra in Schumpeter's work is that radical innovation is the key mechanism driving economic development [329, 330]. In the absence of innovations, the economy would relax into a stationary state characterized by an equilibrium of supply and demand. In reality, entrepreneurs continually introduce new goods, services, or new means of production. A successful innovation might replace existing goods or services and thereby substantially impact other industries. Eventually, innovations can trigger cascading processes that transform entire industries in a disruptive process that Schumpeter called *creative destruction*. In other words, a new good can be combined with already existing goods to produce yet other new goods that, in turn, replace (and destroy) existing goods. Examples of this process abound. A recent example is that the printed media are currently losing their once market-dominant position due to digitalization.

Creative destruction can be understood as an evolutionary three-step process in the context of technological innovations. Economic development is an evolutionary, out-of-equilibrium process that is driven by the appearance and disappearance of goods and services. Appearance and disappearance of goods occur as a result of combinatorial events.

Schumpeter phrased his ideas on economic development in non-mathematical terms. He did so for a good reason. To talk about mathematical formulations for driven, out-of-equilibrium evolutionary processes that are based on non-linear, combinatorial interactions would have been considered crazy at the time. Only very recently has it become possible to approach such systems—usually numerically. This section studies models of combinatorial co-evolution that are derived from the general evolution algorithm, Equations (5.12) and (5.13).

5.6.2   Generic creative phase transition

To formalize innovation dynamics one has to pinpoint how combinatorial interactions can be described in a meaningful, statistical, and open-ended way. The general evolution algorithm contains only *entities* and *rules*. For simplicity, we assume that the abundance of the entities is one, whenever they exist, and zero otherwise. For combinatorial interactions, the interaction tensor M^+ in the general evolution algorithm, Equations (5.12) and (5.13), has three indices. Again, in the following, we assume binary entries; if j and k can be combined to form i, we set $M^+_{ijk} = 1$. Otherwise, we set $M^+_{ijk} = 0$. The *adjacent possible* is the set of all entities that might be produced in an evolutionary system within the next time step, based on the set of entities that exist now. Combinatorial rules constrain the adjacent possible in specific ways. The adjacent possible consists of all entities i that can be produced by combining entities that currently exist. Whenever $M_{ijk}^+ = 1$, there is a production rule for entity i. The size and structure of the adjacent possible can be quantified under certain assumptions. For simplicity, we assume that the location of non-zero elements in M is purely random. M^+ defines a random tensor network topology (two elements combined point to a third). Once it has been sampled, it is fixed. To quantify the adjacent possible, the only thing that matters is which entities are available for combinatorial interactions.

5.6.2.1 *Quantifying the adjacent possible*

The generic dynamics of the adjacent possible for combinatorial systems can be defined in terms of two parameters. The first is the *density of productive pairs*, denoted by r. This is the probability that if you pick two entities at random, there is a production rule that allows you to combine them to produce a third entity. The number of productive pairs in the entire system, which is equal to the number of non-zero entries in M^+, is rN. We denote the number of entities that exist at time t (diversity) by n_t. The second parameter is the initial size of the system in terms of existing entities at $t = 0$, n_0.

We start with n_0 elements. How many *new* entities can we produce in the next time step? In other words, what is the size of the current adjacent possible? A question that is easier to answer is: how many entities can we produce, even if they already exist? This number is given by the number of possible pairs in n_0, multiplied by the probability that a given pair is a *productive* pair. The number of productive pairs is rN; the number of all pairs in the system is $N(N-1)/2$. The number of entities that can be produced at $t = 0$ is therefore,

$$\frac{n_0(n_0-1)}{2}\frac{rN2}{N(N-1)} = \frac{rn_0(n_0-1)}{N-1} \sim \frac{rn_0^2}{N}, \tag{5.49}$$

where we used that $n_0 \gg 1$ and $N \gg 1$. How many of these new entities are already contained in n_0? The answer is n_0/N. These need to be subtracted from all entities that could have been produced, and we obtain for the size of the adjacent possible at $t = 0$,

$$\Delta n_0 = \frac{rn_0^2}{N}\left(1 - \frac{n_0}{N}\right). \tag{5.50}$$

The new number of elements after the first time step is therefore $n_1 = n_0 + \Delta n_0$. Let us iterate Equation (5.50) for $t > 1$. The diversity dynamics of systems with productive combinatorial interactions can then be mapped to the recurrence equation,

$$n_{t+1} = n_t + \Delta n_t,$$
$$\Delta n_{t+1} = \frac{r}{N}\left(1 - \frac{n_{t+1}}{N}\right)\left(n_{t+1}^2 - n_t^2\right), \tag{5.51}$$

with the initial condition n_0. Note that a transformation of variables $n \to n' = n/N$ scales out the dimensionality N in Equation (5.51), so that we get,

$$n'_{t+1} = n'_t + \Delta n'_t,$$

$$\Delta n'_{t+1} = \frac{r}{N^2}\left(1 - n'_{t+1}\right)\left(n^2_{t+1} - n^2_t\right) = r(1 - n'_{t+1})\left(n'^2_{t+1} - n'^2_t\right), \tag{5.52}$$

where we introduced $\Delta n'_t = \Delta n_t / N$. This means that only one parameter remains in the model, the rule density r. The model is independent of size.

> The model is fully valid even for infinitely many entities and interactions (production rules). In this sense it describes an open-ended situation.

The asymptotic diversity n_∞ is a function of the parameter r and the initial condition n_0. To compute it, let us define $c_t = \Delta n_{t+1}/\Delta n_t$ and denote its asymptotic value by $c = \lim_{t\to\infty} c_t$. On the one hand, we can use the recurrence relation Equation (5.51) in c_t and take the limit $t \to \infty$ to obtain,

$$c = 2r\left(1 - \frac{n_\infty}{N}\right)\frac{n_\infty}{N}, \tag{5.53}$$

where we used that $n^2_{t+1} - n^2_t = \Delta n^2_t + 2n_t\Delta n_t$. On the other hand, we can estimate n_∞ by,

$$n_\infty = \sum_{t=0}^{\infty} c^t n_0 = \frac{n_0}{1 - c}, \tag{5.54}$$

where we used the formula for the geometric series. Merging Equations (5.53) and (5.54) leads to the cubic equation,

$$c^3 - 2c^2 + c\left(1 + 2r\frac{n_0}{N}\right) + 2r\frac{n_0}{N}\left(\frac{n_0}{N} - 1\right) = 0, \tag{5.55}$$

which can be solved for c by using, for example, Cardano's method. Discarding the complex solution, we find the two real solutions,

$$n^{\pm}_{\infty} = \frac{N}{2}\left(1 \pm \sqrt{1 - \frac{2c}{r}}\right). \tag{5.56}$$

5.6.2.2   *The generic phase transition in combinatorial evolutionary systems*

We obtain two solutions, n^+_∞ and n^-_∞. The actual solution has to switch from n^-_∞ to n^+_∞ in order to be monotonic in n_0. Below a critical value $r < r_c$ this happens smoothly. For $r > r_c$ the solution is forced to jump. We plot the result for the final diversity, the self-consistent solution to Equation (5.56), in Figure 5.11. The discontinuous transition is clearly visible. There exists a critical line that separates the two phases, the unpopulated low-diversity phase and the almost fully diversified phase. The final diversity n_∞ is shown as a function of r in Figure 5.12 for various choices of the initial diversity n_0. The lines in Figure 5.12

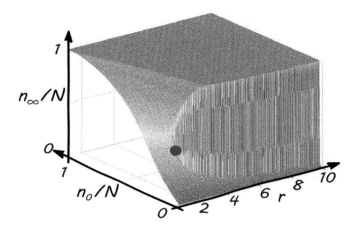

Figure 5.11 *Phase diagram of a combinatorial evolution system with positive interactions only. Evolutionary systems generically exhibit a creative phase transition. If the rule density r or the initial diversity n_0 is too low, the system does not substantially increase its diversity over time. For a rule density r below a critical point $r < r_c$, the asymptotic diversity n_∞ changes smoothly as a function of the initial diversity, a_0. The black dot marks the onset of the creative phase transition (critical point). Above this critical rule density, the system exhibits a discontinuous phase transition.*

are identical to the Maxwell lines that describe the thermodynamic phase transition of a van der Waals gas. The nature of the creative and the thermodynamic transition is equivalent, both being described by cubic equations. For more details, see [169, 170].

The creative phase transition can be seen as a generalization of the percolation transition in Erdős–Rényi networks to random interaction tensors, as described by M^+. To see this, assume for a moment that we would allow only linear random interactions in which an entity j is produced by a single other entity i. In this case, M^+ would be a directed Erdős–Rényi network. Starting from a single initial entity, the maximal asymptotic diversity is the size of the strongly connected component plus the size of its out-component. For linear interactions the critical point of the percolation transition is identical to the critical point of the creative phase transition.[12] For non-linear combinatorial interactions, however, the creative phase transition has an additional quality with no analogue in the linear case: for a fixed rule density above the critical point $r > r_c$, there is a discontinuous jump in the asymptotic diversity n_∞ as a function of the initial diversity, n_0; see Figure 5.11. This phenomenon is absent in the linear case.

---

[12] For linear interaction matrices M^+ we recover the SI model for epidemic spreading on networks; see Section 4.8. Susceptible nodes can be identified with entities that have not yet been produced. Infected nodes are available entities. The asymptotic diversity is the infected fraction of nodes, q. There is no critical dependence of q on the number of initially infected nodes in the SI model. Given that there is at least one infected node in the strongly connected or the in-component, all nodes in the strongly connected and out-component get infected too.

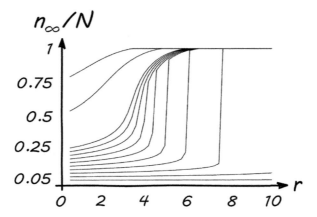

Figure 5.12 *Final diversity as a function of pair density r. A discontinuous transition between the final diversity and the rule density is seen in the cuts through the creative phase diagram of Figure 5.11, for several values of n_0. The resulting lines are identical to the Maxwell lines that describe the phase transition of a van der Waals gas in a pV diagram.*

The creative phase transition is universal for all models that employ combinatorial interactions. It is a topological property of systems that are driven by combinatorial interactions. It can be seen as a generalization of the percolation transition in Erdős-Rényi networks to random interaction tensors M^+.

5.6.3 Arthur–Polak model of technological evolution

Arthur and Polak formulated an interesting idea to understand the combinatorial evolution of technologies. The model describes a particular realization of combinatorial interactions that leads to a creative phase transition [15]. The entities in the Arthur–Polak model are elementary logical gates and logical primitives (constant logical functions). Rules specify that these logical gates are randomly combined (wired) to create new logical gates. In this way, a technology of more and more complicated random gates form over time—a technology 'bootstraps itself' from simple elements [14].

Initially, the Arthur–Polak model contains two types of entities in a pool: first, there are binary constants, zero and one. These constants can be seen as a logical function $f : \{0,1\} \rightarrow \{0,1\}$ that maps the value of one bit to itself, $f(0) = 0$ and $f(1) = 1$. The function f is computed by a primitive logical gate. This gate has one pin for the input and one for the output of f. The value at the input pin of the constant function f is always equal to the value at the output pin. The second type of entity in the Arthur–Polak model are NAND gates that have two input pins and one output pin. NAND gates produce an output of 1 if the inputs are either (00), (01) or (10) and produce an output of 0, if the inputs are (11). Here (AB) means input on pin A and input on pin B. The set of in- and outputs that results from all the possible inputs are called the 'truth table'. Technologies in the model are constructed from these two entities by randomly wiring the pins of the entities. Each resulting technology is again a logical circuit that has a certain number of input and output pins. The utility (or fitness) of technologies is determined by an externally given list of desirable logical functions. These include negation and implication, k-bit adders, k-bit-equal, k-bit less, as well as complex gates such as n-way-xor or m-bitwise-and for several values of k, n, and m. The function of every technology is represented by its truth table.

The dynamics in the Arthur–Polak model is as follows. Imagine a pool of logical NAND gates and logical constants. These are the simplest technologies. At each time step, between 2 and 12 technologies are randomly selected from the pool and randomly wired together at their pins. The truth table for the new technology is computed and checked to see if it appears in the list of desired logic functions. If it does, the new gate is able to perform one of the desired functions and is adopted, which means that it is added to the pool as a new entity that can be used now for further combinations. If it does not perform any useful function, the technology is discarded. Over time, more and more useful technologies appear. Under the minimal guidance of the external list of useful functions, the model 'develops' random technologies by chance that are able to perform relatively complicated logical operations. The amazing feature of the model is that the development occurs extremely fast. The dynamics unfolds in a punctuated way. At certain points in time, keystone technologies are 'invented' that can be immediately combined with existing functions such that the new combinations satisfy a large number of the desired logical operations. Many useful inventions happen within short time periods. Creations and extinctions occur in bursts. Consider the arrival of a new technology that computes one of the desired functions better (say with fewer components) than the current technologies. The new technology is then adopted and replaces one of the existing ones that performs the same function. The replaced technology may now be obsolete. It may be that certain technologies in the pool were only needed as building blocks to create the technology that was replaced. These technologies are now also obsolete and go extinct. The removal of a single entity may trigger a cascade of removals. The new technology will now be used in further combination steps. For example, a better two-bit-adder might facilitate the fast invention of a better four-bit-adder. In this case, it will only take a short time for the current four-bit-adder to be replaced. We see creative destruction at work.

Within 250,000 iterations, the simple model usually generates logical gates, such as eight-way-xor, eight-way-and, three-bitwise-xor, and eight-bit adders. Note that such configurations are extremely unlikely to be sampled randomly. For a circuit with n input and m outputs there are $(2^m)^{2^n}$ possible realizations. An eight-bit-adder has sixteen inputs and nine outputs. The Arthur–Polak algorithm effectively samples only a tiny region of phasespace that contains $10^{177,554}$ possibilities. After only 250,000 iterations it finds useful technologies with sixteen inputs and nine outputs. However, complicated desired functions, such as the eight-bit-adder, can only be produced if specific intermediate steps (two- and four-bit-adders) are actively selected for. The key message in the Arthur–Polak model is that technological development is extremely efficient if it works with modular building blocks and if simple steppingstones are selected for. Similar observations have been made with respect to biological evolution. The evolution of complicated features requires the existence of simpler steppingstones [250].

The Arthur–Polak model has been criticized for needing an exogenous list of desired functions to generate complicated useful logical operations, and that, as a result, the model is not self-organized. However, for a model of technological evolution, where homo sapiens is constantly selecting for useful technological building blocks, the model is perfectly valid. The environment in the model with which the new technologies 'interact', is essentially determined by the list of desired logical functions. The Arthur–Polak model for technological evolution does not address the question of how the environment co-evolves with successful new entities. Without co-evolution the desired functions will eventually be satisfied and evolution stops.

5.6.4 The open-ended co-evolving combinatorial critical model—CCC model

The hallmark of the general evolution algorithm given in Equations (5.12) and (5.13) is that it describes how changes in the composition of entities co-evolve with changes in the interaction tensors. The general evolution algorithm is extremely general. Here, we try to derive its simplest form, that is open-ended, co-evolving, combinatorial, and critical. We will refer to it as the CCC model. We use quadratic interactions, where combinatorial interactions are defined for pairs of entities. Quadratic interactions will naturally lead to co-evolution between entities and their environment (interactions) in a self-organized critical way [368]. We have mentioned that any comprehensive model for evolutionary processes should be:

- *Co-evolutionary.* The space of entities and the space of interactions co-evolve and co-construct each other. As a result, non-trivial dynamics of species and their fitness landscapes should appear.

- *Combinatorial.* Entities emerge spontaneously or through the combination of existing things.

- *Critical.* Critical phenomena appear as empirical facts in evolutionary systems. Fat-tailed distribution functions appear as a consequence. Meaningful models must be able to understand the origin of these phenomena.

These properties have been addressed separately in some of the models that we have discussed so far in this chapter. However, none of them combines *all* three properties simultaneously in an open-ended setting. For instance, the NK model describes co-evolution between species and fitness space, but it needs to be fine-tuned for critical behaviour; it has no explicit combinatorial interactions. The Jain–Krishna model does provide a mechanism for co-evolution and displays punctuated equilibria, but it does not include the emergence of new species or innovations as combinatory events. The simple combinatory model that we used to illustrate the creative phase transition provides no explicit extinction mechanism, and the Arthur–Polak model for technological development explores dynamics with externally given fitness functions that do not co-evolve. In the following, we try to formulate the simplest possible 'unified' model that puts together the three strands [368]. We focus on the question as to how macroscopic properties of evolutionary processes emerge from elementary microscopic combinatorial events. On the basis of a microscopic model, we wish to understand the statistics of diversity change that is characterized by punctuated equilibria. Evolutionary processes are well understood in the metastable quasi-equilibrium regimes, where replicator equations might be applicable. The nature and the statistics of the transitions between equilibria is a challenge that has to be tackled. In particular, we want to understand the distribution functions of the following dynamical variables that are associated with these transitions:

- size of evolutionary events,
- duration of restructuring events,
- duration of equilibrium phases,
- size of restructuring events, and
- burstiness of the transitions between equilibria.

5.6.4.1 Combinatorial interactions in the CCC model

We start with quadratic combinatorial interactions given in Equation (5.12), that describe how two entities, j and k, impact the state σ_i of entity i,

$$\sigma_i(t+1) \sim \sigma_i(t) + F\left(M_{ijk}^{\pm}(t), \sigma_j(t)\right). \tag{5.57}$$

This is a discrete catalytic network equation where $\sigma_i(t)$ is an N-dimensional state vector that, for simplicity's sake, represents the abundance of species, $x_i(t)$. Non-zero entries in the interaction tensor $M^{\pm}$ are 'production rules'.

Productive combinatorial interactions For the simplest case, we assume that the state of entity i at time t is described by a binary variable $\sigma_i(t)$. If i has non-zero abundance at time t, $x_i(t) > 0$, we set $\sigma_i(t) = 1$. If i does not exist at t we have $\sigma_i(t) = 0$. In the simplest case, the interaction tensor (or production table) M^+ is also binary. If two entities j and k can, in principle, be combined to create i, we set $M_{ijk}^+ = 1$, otherwise $M_{ijk}^+ = 0$. One can think of the production table in various ways: (i) it could represent a stylized chemical reaction between chemicals j and k that react to produce i. (ii) It could describe a situation where species k exerts selective pressure on species j to mutate into species i, or (iii) it could depict technological components j and k that, when combined, produce component i. We call j and k the *productive set* of i. Formally, such a production process can be expressed as,

$$\sigma_i(t+1) = M_{ijk}^+ \sigma_j(t)\sigma_k(t). \tag{5.58}$$

A production process is called *active* if, in addition to an existing production rule, entities j and k are indeed available, meaning that $\sigma_j(t) = \sigma_k(t) = M_{ijk}^+ = 1$. In general, a specific entity might be produced by several different productive sets. The number of active productive sets for species i is denoted by $N_i^{\text{prod}}(t) = \sum_{jk} M_{ijk}^+ \sigma_j(t)\sigma_k(t)$.

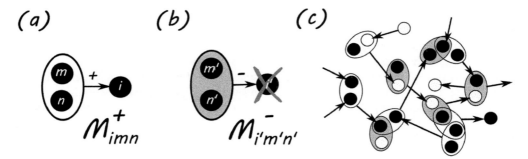

Figure 5.13 *(a) A constructive interaction M_{imn}^{+} is shown where two entities are combined to form a new one. (b) A destructive interaction $M_{i'm'n'}^{-}$ means that if entities m' and n' are combined, entity i' is destroyed as a consequence. (c) A section of the infinite tensor network $M^{\pm}$. Existing entities are marked as black circles; non-existent ones as white circles.*

Destructive combinatorial interactions   Schumpeter described situations where the combination of two entities produces an entity that will effectively destroy another existing entity. For example, the combination of the internal combustion engine and petroleum effectively destroyed the horse-drawn carriage industry over time. Destructive interactions are encoded in the *destruction table*, M^{-}, which is structurally the same as the production table. If the combination of two entities j and k has a destructive influence on i, we set $M_{ijk}^{-} = 1$, otherwise $M_{ijk}^{-} = 0$. Here the pair j and k is the *destructive set* of i. Formally, we have that, given $\sigma_i(t) = 1$,

$$\sigma_i(t+1) = 1 - M_{ijk}^{-}\sigma_j(t)\sigma_k(t). \tag{5.59}$$

The number of active destructions of i at time t is $N_i^{\text{destr}}(t) = \sum_{jk} M_{ijk}^{-}\sigma_j(t)\sigma_k(t)$. In Figure 5.13 we show a constructive (a) and a destructive (b) event. Figure 5.13c shows a part of the interaction tensor network $M^{\pm}$, and the state vector σ_i. Existing entities are marked as black circles; non-existent ones as white circles.

Dynamics of the CCC model   In general, every entity can be associated with several productive and destructive sets that produce it or have a negative influence on it. The dynamics of the CCC model is specified by the assumption that an entity i can survive at any point in time if it has more active productive than destructive sets. On the other hand, if the destructive interactions outweigh the productive ones, the entity will go extinct in the next time step. This leads us to the following update rules,[13]

$$N_i^{\text{prod}}(t) > N_i^{\text{destr}}(t) \rightarrow \sigma_i(t+1) = 1$$
$$N_i^{\text{prod}}(t) < N_i^{\text{destr}}(t) \rightarrow \sigma_i(t+1) = 0$$
$$N_i^{\text{prod}}(t) = N_i^{\text{destr}}(t) \rightarrow \sigma_i(t+1) = \sigma_i(t). \tag{5.60}$$

---

[13] Many other rules are possible. Many lead to very similar results; see what follows.

In addition to these fully deterministic rules, we assume that from time to time spontaneous events occur that create or destroy a particular entity. At every time step any entity is either annihilated or spontaneously produced with probability $p \ll 1/N$. For higher values of p the stochastic updates (driving) may dominate the dynamics. Despite its small value, p plays an important role as a driving force.

The algorithm   The update rules of the CCC model can now be summarized in the following algorithm.

1. Pick entity i at random (random sequential update).
2. Sum all productive and destructive influences on i, and compute a 'fitness',

$$f_i(t) = \sum_{j,k=1}^{N} (M_{ijk}^{+} - M_{ijk}^{-})\sigma_j(t)\sigma_k(t). \tag{5.61}$$

If $f_i(t) > 0$, set $\sigma_i(t+1) = 1$, for $f_i(t) < 0$, set $\sigma_i(t+1) = 0$. For $f_i(t) = 0$, do not change the state, $\sigma_i(t+1) = \sigma_i(t)$.
3. With probability p, switch the state, $\sigma_i(t+1) = 1 - \sigma_i(t)$.
4. Continue until all entities have been updated once, then go to the next time step.

5.6.4.2   *Critical behaviour in the CCC model*

Random interaction tensors   We still have to specify the interaction tensors. The simplest choice is to use random tensors, which can be parameterized by two numbers, r^{+} for the constructive tensor and r^{-} for the destructive tensor. $r^{\pm}$ is the average number of productive or destructive sets per entity, respectively. The probability of a particular entry in $M^{\pm}$ being one is therefore $r^{\pm}\binom{N}{2}^{-1}$. Once chosen, the interaction tensors remain fixed throughout the simulation.

Figure 5.14 shows the result of a generic run of the CCC model. Every column represents a time step; every row corresponds to an entity. A pixel at column t in row i represents the state of entity i, $\sigma_i(t)$. If entity i exists, the pixel is white; if it does not exist, it is black. One immediately observes the presence of punctuated dynamics. There are relatively long periods during which the composition of entities practically does not change (quasi-equilibrium). Phases of stability appear as horizontally striped patterns. There appears to be no unique stable attractor. Each equilibrium phase is characterized by its unique configuration of active entities. These phases may last for a few or for hundreds of iterations. In the simulations (settings as in Figure 5.14) we find that the distribution of lifetimes of equilibria is an approximate power law with an exponent -2.5. The episodes of metastable equilibria are interrupted by periods of disruptive change, where massive restructuring takes place in the system. This is seen in the chaotic patterns. During these periods, many entities change their state from one time step to the next, the composition of the set of entities is massively restructured, and diversity fluctuates strongly. The size distribution of the percentage changes in diversity is an approximate power law with exponent -1. The chaotic phases last for anything between

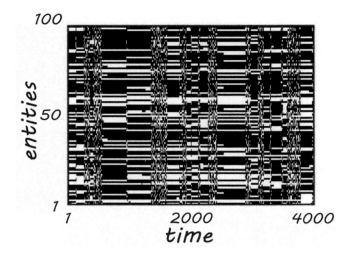

Figure 5.14 *Generic trajectory of the evolution in the CCC model. Each row corresponds to an entity that is either present (white) or absent (black) at a particular point in time (column). The model is self-organized critical, and shows punctuated equilibria and fat-tailed cascade sizes. Periods of metastable equilibria appear as horizontal stripes. They are punctuated by phases of disruptive change, where the composition of the population of entities is massively restructured (chaotic regions). Approximate power laws appear in the distribution functions for the size of diversity changes, lifetimes of equilibria, and lifetimes of the restructuring phases. The generic patterns of this dynamics persist for a large number of parameter settings and particular choices of $M^{\pm}$. The simulation shown was done with $r^+ = 10, r^- = 15$, $p = 0.0002$, and $N = 100$. For details see [368].*

a few iterations to several hundred iterations; the distribution of their lifetimes is again an approximate power law with exponent -2.5.

Punctuated equilibria in the CCC model appear generically for a wide range of choices of the model parameters; see Figure 5.15. For each parameter setting we show the number of transitions between equilibrium phases as a function of r^+ and r^- for a fixed value of p. Black indicates that either no stable phases exist and that chaotic phases dominate the dynamics, or that the system remains in an equilibrium state throughout the simulation. The latter is particularly likely for $r^+ \gg r^-$, where the system is fully diversified, or for $r^- \gg r^+$, where diversity is limited because of the strong presence of destructive events. In between there is a wide range of values of $r^{\pm}$, for which the dynamics produces patterns of the type shown in Figure 5.14.

Specific interaction tensors   The generic appearance of punctuated equilibria in the CCC model persists for many different choices and implementations of the interaction tensors. This has been demonstrated for the following model variants of Equation (5.61); see [368].

- *Variable set sizes.* Productive and destructive sets may contain an arbitrary number of entities. In general, combinatorial interactions can be of the form $i_1 + i_2 + \ldots + i_K \rightarrow j$ with some $K > 1$, without changing the behaviour of the CCC model

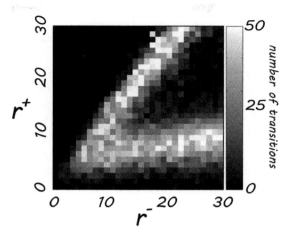

Figure 5.15 *The number of transitions between equilibria observed in the CCC model is shown as a function of the model parameters r^+ and r^- and for $p = 0.0001$. Punctuated equilibria in the model appear generically for a wide range of parameters. No fine-tuning is necessary.*

qualitatively [235]. The set sizes can be allowed to vary within the same system. If all sets are of size $K = 1$ the system does not display punctuated equilibria—it is linear.

- *Finite lifetimes of entities.* Entities can be introduced that, without active productions and destructions, decay with rate λ. This does not change the behaviour of the model. In fact, λ effectively serves as a stochastic driving force, and the model does not freeze to a static configuration, even if one sets $p = 0$.

- *Scale-free production and destruction tensors.* The number of productive or destructive sets per entity are chosen from a power law distribution. This leads to punctuated equilibria with a prolonged duration of the equilibrium phases.

- *Asymmetry in productions and destructions.* The generic model behaviour does not change if destructive influences have a different impact than productive ones. One can modify the algorithm such that $f_i(t)$ is set to a negative value whenever there is a single active destructive influence. Or one can introduce a threshold $m > 0$ such that an entity can only be produced if $f_i(t) > m$.

- *Limited resources.* Model variants with limited resources can be formulated. The number of entities that can be sustained is then limited by a finite amount of substrate. Such modifications effectively introduce an upper bound of the diversity that the model can sustain. In the accessible region of phasespace, the system basically shows the same properties as in the case of unlimited resources.

- *Update procedure.* The occurrence of punctuated equilibria is independent of the update procedure. Parallel or sequential updates in random or deterministic order all lead to the same generic patterns that are observed in Figure 5.14.

- *Modular interaction tensors.* One can imagine a structure like stochastic block models for the interaction tensors $M^\pm$. The system then consists of several 'modules' of entities that are characterized by different values of $r^{\pm}$. In these model variants there is a tendency for the overall system behaviour to be driven by the most densely connected module.

- *Competitive replacement.* Destructive interactions are not assigned randomly, but depend on productive interactions. For instance, a given entity i may compete with any of its 'precursor' or 'ancestral' entities that are (with a certain probability) either directly in one of i's productive sets or in a productive set of one of the entities in i's productive sets. This choice leads to the same dynamics as in Figure 5.14.
- *Bounded rationality.* Similarly, one can introduce bounded rational actors into the model by accepting an update only with a given probability q, whereas with probability $1 - q$ one sets $f_i(t) \rightarrow -f_i(t)$. This again introduces a stochastic driving force that prevents the system from freezing, even without external driving $p = 0$.

> The CCC model is robust under changes of its model parameters, the topology of the involved interaction tensors, and details of the specific update rules. This is an indication that the appearance of punctuated equilibria and the associated approximate power laws could be *universal* for models that combine productive and destructive combinatorial interactions.

The CCC model as a sand pile   The robustness of punctuated equilibria and their associated criticality raises the question as to the nature of the underlying mechanisms behind the massive restructuring events in species space. The baseline model is a driven non-equilibrium model. When the driving force is turned off, $p = 0$, the model relaxes into a static rest state. Assume that the model is prepared in such a rest state and we flip the state of one randomly chosen entity. There is a chance that this perturbation activates a productive or destructive set. This will potentially then activate other productive or destructive sets, and a cascade of secondary events may be triggered. Eventually, such a cascade will die out again, and the system relaxes into a new rest state. In sand pile models analogous rest states are reached when no more sand grains can topple. The size of these cascades can be quantified in terms of their size and duration. Cascade size is the number of entities that change their state during the cascade. The duration is the number of time steps needed for the system to reach its next rest state after the perturbation. The distribution functions for the size and duration of cascades are shown in Figure 5.16. Both follow an approximate power law. This is an indication that the CCC model is self-organized critical and closely related to sand pile models, which are also characterized by approximate power laws in the distributions of cascade size and duration. Sand pile models and the CCC model show similar exponents.

The cascade dynamics of the CCC model is also similar to the dynamics of the Bak–Sneppen model in Section 5.4.4. After a transient, the system settles into a stable configuration characterized by a balance of active productive and destructive interactions, as expressed in the 'fitness' $f_i(t)$ of Equation (5.61). A single update may destroy the balance and trigger a cascade of adaptations. In contrast to the Bak–Sneppen

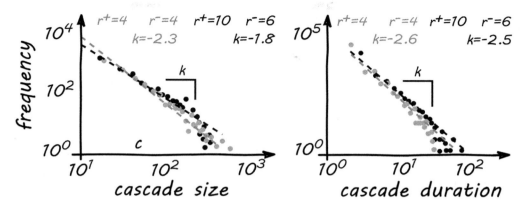

Figure 5.16 *The distribution functions of cascade sizes and durations follow approximate power laws. Cascades are triggered by a single perturbation. The slopes can be varied slightly with the model parameters. This demonstrates the close relatedness of the CCC model to self-organized critical systems like sand pile models, which are very similar in their avalanche size and duration distributions.*

model, the CCC model provides an explicit mechanism for the mutual influence between entities and fitness functions $f_i(t)$, as well as a mechanism for extinction events.

The CCC model as a spin model The CCC model can be formulated as a spin model, and a mean field solution for its asymptotic diversity can be obtained analytically. It can be shown that the mean field approximation of the CCC model is able to reproduce actual data from the fossil record, the metabolic activity in bacteria, or the GDP time series of the UK. It is also possible to demonstrate the existence of a discontinuous order–disorder phase transition within the mean field approach. The demonstration is technical, but the skilled physicist is invited to check details in [235].

5.6.4.3 Co-evolution of species and fitness landscapes in the CCC model

The production and destruction tables $M^\pm$ in the CCC model are collections of static rules that remain fixed over time. One can think of these rules as something like the 'laws of chemistry', where the entities are atoms or molecules of different types that, when combined, can react to produce other molecules. What can be produced and what cannot is given in the production tables. The complete production tables are in general not known; in chemistry, we have not yet invented *all* possible compounds (combinations of molecules). However, the rules of what reacts with what to form what are there and are given. They are unchangeable, whether we know them or not. Three hundred years ago, no-one knew that hydrogen and oxygen could produce water; nevertheless, the chemical possibility (the rule) existed. What is observable and known, however, is a subset of the rule tables that are part of the currently active productions.

This situation is shown in Figure 5.17. The production and destruction tables $M^\pm$ are not generally known. However, they do exist and are fixed; Figure 5.17a. A production is denoted by an oval symbol and an arrow that points to the output. The entities that exist populate the production tensor network. In Figure 5.17b we see the combined information of the static production table and the dynamical state vector $\sigma(t)$.

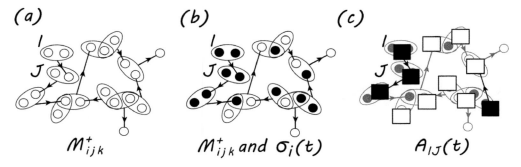

(a)   *(b)*   *(c)*

M^+_{ijk} M^+_{ijk} and $\sigma_i(t)$ $A_{IJ}(t)$

Figure 5.17 *Knowns and unknowns in the CCC model. Entities are shown as circles. Production rules are shown as oval symbols with an arrow that points to the output entity. (a) The production and destruction tables $M^\pm$ are usually not observed, but they exist and are fixed. (b) Combination of the information in $M^\pm$ and $\sigma(t)$. At each point in time, a different set of entities can be active (grey circles), $\sigma_i(t) = 1$. A production rule is active if all input entities are active. (c) What can be observed are the active productions. If the output of an active production is the input of another active production, the two productions can be linked in a directed network—the active production network A_{IJ}.*

The existing entities are black. Active productions are, of course, observable. Productions that are inactive might be known or not. The information of the active productions can be mapped into a production *network*.

Active production networks The time-dependent state of the system, $\sigma(t)$, in combination with its static rules, $M^\pm$, can be mapped onto a dynamic set of *active productions*. Let us denote the states of the productive sets by π_K, where K labels the productions. A production is either active or inactive; if it is active $\pi_K = 1$, otherwise it is 0. In Figure 5.17c we show the active productions as black squares, the inactive ones as white. The state of a production is given by $\pi_K(t) = \sigma_i(t)\sigma_j(t)\sigma_k(t)$. The set of all productions that are currently active can be drawn as a network. Here, the nodes are the productions. A link A_{IJ} points from active production I ($\pi_I = 1$) to another active production J ($\pi_J = 1$), if I currently produces an output that serves as an input for J (regardless of whether J produces anything with it). In this case we set $A_{IJ} = 1$ (black arrows in Figure 5.17c), otherwise $A_{IJ} = 0$. In this way, we have mapped the interaction tensor network in combination with the state vector $\sigma(t)$ to a temporal network $A(t)$ that we can study in terms of auto-catalytic cycles and keystone species.

Auto-catalytic cycles in the CCC model In the Jain–Krishna model discussed in Section 5.5.1, we have seen that the different diversity regimes are characterized by the absence or presence of auto-catalytic cycles in the underlying interaction networks [205, 206]. Periods with high species diversity are characterized by the existence of cycles. They can simply be identified by non-zero Perron–Frobenius eigenvalues. A similar situation holds for the production networks, A, of the CCC model. If the system 'finds' a set of productions that mutually sustain each other, the diversity increases strongly. The Perron–Frobenius eigenvalue of the production network correlates directly with the diversity of the system. The likelihood of destroying productions that belong to an

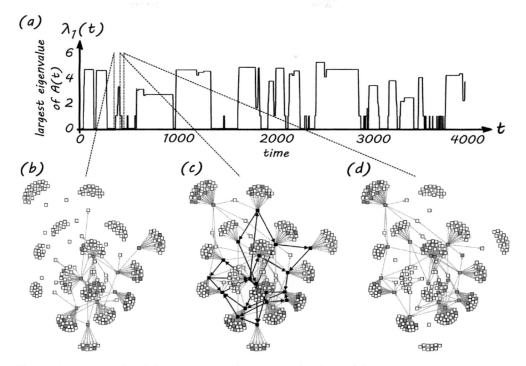

Figure 5.18 *Dynamics of the active production network $A(t)$ in the CCC model. (a) Time evolution of the maximum eigenvalue of $A(t)$ for the simulation shown in Figure 5.14. The eigenvalue is zero when the system is in the chaotic phase. Vertical dotted lines indicate times for which the active production networks are shown in (b)–(d). (b) A snapshot of $A(t)$ in the chaotic phase. Active (inactive) production processes are shown as grey (white) squares. The network is acyclic, that is, its largest eigenvalue is zero. (c) Snapshot of $A(t)$ in an equilibrium phase. Keystone production processes are shown as black squares. They are members of a cycle (black arrows) that dominates the model dynamics. (d) Eventually, a perturbation will hit one of the keystone productions and destroy the auto-catalytic cycle. The system enters a chaotic phase again.*

auto-catalytic cycle is substantially decreased, as destructive influences need to outweigh the productive interactions, which are over-represented in equilibrium phases. Regime shifts from a chaotic phase to an equilibrium phase are indeed accompanied by a jump in the largest eigenvalue of the active production network from zero to one; see Figure 5.18a.

The role of keystone productions The system is in a chaotic phase if there are no auto-catalytic cycles in the production network $A(t)$; see Figure 5.18b. Once an auto-catalytic cycle in the production network is established, the system typically enters a metastable equilibrium; see Figure 5.18c. However, due to the stochastic driving process, $p > 0$, the system is constantly exposed to small perturbations. While most perturbations trigger cascades of negligible size, they occasionally percolate through the entire system. This happens whenever a keystone production is deactivated, and is visible in a sharp decrease in the largest eigenvalue; see Figure 5.18d. In the equilibrium phase, keystone productions are typically members of cycles. If a perturbation hits one of the keystone productions, there is a chance that a cycle will be broken. The system then either undergoes a shift towards dynamics governed by a second cycle that already exists in the network, or it will enter the chaotic phase. The chaotic phase lasts until the next cyclic production network is established. At the end of the restructuring period, the system enters a new equilibrium dominated by a different auto-catalytic cycle and keystone productions. The shifting between phases can be interpreted as business 'cycles' in the sense of Schumpeter. The process of creative destruction is realized in the CCC model by the arrival of spontaneous innovations triggering a cascade of destruction processes that finally causes the shift to a new equilibrium.

5.6.5   CCC model in relation to other evolutionary models

In the following, we briefly mention how some of the models discussed in this chapter can be interpreted as special cases of the CCC model.

5.6.5.1   *Creative phase transition of random catalytic sets*

Recovering the model that we used to illustrate the creative phase transition in Section 5.6.2 is straightforward: there is no destructive dynamics, $r^- = 0$, and the random perturbations, $p = 0$, in the CCC model are turned off.

5.6.5.2   *Linear interaction models*

With a little bit of bending, it is possible to interpret evolution models with linear interactions that we studied in Section 5.5 in the light of the CCC model. This is particularly so for the Jain–Krishna model, the Solé–Manrubia model, and the sequentially linear models. As these models evolve, they effectively change their interaction networks over time. In the Jain–Krishna model, the least-fit species is replaced by a randomly linked species, the Solé–Manrubia model replaces the least-fit species by a copy of an existing one, and the sequentially linear model uses the positivity constraint to change the effective interaction matrix. The CCC model, in contrast, does not explicitly update its interaction rules dynamically. Co-evolutionary dynamics arises because existing entities constantly sample different regions of the interaction tensor network. The existing entities 'populate' the infinite interaction tensor network like a cloud that drifts through the tensor network. As we have seen, this effectively leads to very similar dynamics to that in the Jain–Krishna model. Before this becomes apparent, however, we have to project the tensor network to a network of productions. Auto-catalytic cycles and keystone species then dominate the diversity dynamics in exactly the same way as in the Jain–Krishna model.

5.6.5.3   *NK models*

Both the CCC model and the NK models in Section 5.4.3, can be interpreted as spin glasses. In both cases, it is possible to solve them analytically within a mean field approach that allows us to understand the phase diagrams of the two models.

In summary, in this section we encountered specific realizations of the general evolution algorithm given in Equations (5.12) and (5.13). We studied how evolutionary systems explore their adjacent possible and discovered that non-linear combinatorial evolution has a generic creative phase transition. Systems with rule densities or initial sizes below a critical value are unable to sustain a substantial level of diversity. Above the critical values, auto-catalytic cycles emerge that lead to a high diversity of entities. Endogenous evolutionary processes without exogenous selection rules cannot be understood without destructive interactions. Models with constructive and destructive rule tables generically show the phenomenon of punctuated equilibria. They are inherently self-organized critical—no fine-tuning of model parameters is needed. Criticality appears over a wide range of parameter space. Besides criticality, these models show non-trivial co-evolution between species and their fitness landscapes. We discussed the (perhaps) simplest possible realization of such a model, the *combinatory, co-evolutionary, critical,* CCC model. This model includes all the characteristics of evolutionary processes discussed in this chapter. The existence of punctuated equilibria can be understood by the endogenous emergence of auto-catalytic cycles and their destruction through the removal of keystone species. The model can be mapped to a spin model, where the existence of an order–disorder phase transition can be analytically understood in a mean field approach. The CCC model is open-ended in the sense that it works for infinite sets of entities and interaction rules.

5.7 Examples—evolutionary models for economic predictions

We conclude this chapter with two examples. Both illustrate how the concepts we have elaborated can be used in empirical data analysis. Both examples are related to economic development. The reason why we focus on economics is data availability. It is much simpler to obtain time-resolved high-quality data on microscopic interactions for national economies than for trophic and symbiotic interactions between species that lived a hundred million years ago. The first example shows how the 'fitness' of countries can be empirically measured as a structural network property of their economy [94, 189, 190]. The set of products that a country exports to a certain degree reflects its economic structure. While some countries are able to export a highly diverse set of specialized products, other countries export only a few ubiquitous products. Fitness of a national economy can be measured in terms of the 'complexity' of the products that it is able to export.[14] Based on this definition of fitness, under certain conditions it becomes possible to predict the long-term economic growth of countries to a certain extent. Fitness determines a certain level of economic output that a country can sustain structurally. Countries with a current output below that level are likely to grow faster than countries with an output that cannot be sustained by their economic fitness.

---

[14] A 'complex' product is a product that is hard to produce. Typically, these are highly specialized products that are made only by a few countries.

In the second example, we tackle a long-standing puzzle in economics: why is economic growth so unevenly distributed across countries? Economic development is usually understood as a process of moving from labour-intensive products (garments, textiles, agricultural products) to more capital-intensive products (pharmaceuticals, microelectronics, etc.). In principle, labour and capital can flow freely between countries. Why then is the shift from labour- to capital-intensive products, and the associated economic growth, unevenly distributed? While some countries, for example, in Southeast Asia, have been able to catch up to (if not overtake) the Western world in terms of economic growth, this trend is not observed in many African countries. We show that this different growth dynamics can be explained in terms of a critical transition in a version of a CCC model that is appropriately calibrated to world trade data [233]. In the model, all countries, in principle, have access to the same production and destruction rules, but they differ in terms of their initial diversity of products. Each country is either below, at, or above the critical values that determine the creative phase transition. Whether a country is able to reach a regime of high growth depends on whether its initial product diversity is large enough to start the engine of creative destruction and thereby continually restructure its economy towards a high diversity of goods.

5.7.1 Estimation of fitness of countries from economic data

To produce a product one needs components and a set of skills to enable the components to be combined. While components can usually be bought on global markets, skills are harder to obtain. Not all inputs necessary to produce a product can be brought in from abroad [189]. Property rights, regulations, infrastructure, labour skills, and so on, represent non-tradable *capabilities* that reside within a country. In the long run, the productivity of a country is determined largely by its specific combination of capabilities and skills, and it is these that enable the production of goods or services. The presence of a certain capability within a country can often not be assessed in a systematic way. Hausmann and Hidalgo suggested that these capabilities can be indirectly observed through the composition of all a country's exported goods. By examining what products a country exports, one can infer what capabilities reside within that country. How economic growth is related to the presence of those capabilities can then be tested. The fitness of a national economy is a function of its available capabilities. By using exports as a proxy for those capabilities, it is possible to estimate the fitness of a country [189].

Export data of countries is available from international trade data. Trade data from the import and export of products is structured into product categories, such as beverages, raw materials, pharmaceuticals, or microcircuits. Figure 5.19 shows the exports of 123 countries (rows) in 732 product categories (columns) in 2009. It is immediately apparent that some countries have substantial exports in almost all product categories, whereas other (usually poor) countries only export a small number of ubiquitous products. Poor countries often export similar types of products. In Figure 5.19 the countries are sorted by diversity, with the most diverse countries in terms of exported products being shown on the top. The products exported by poorer countries appear to be subsets of the products that are also exported by the richer countries.

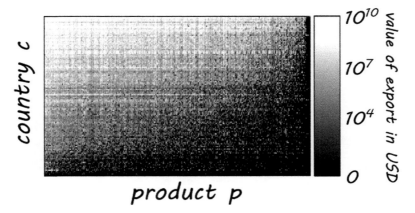

Figure 5.19 *The world export matrix of 2009 shows which countries (rows) export which products (columns). The shades of grey represent the value of the exports in USD. Countries are sorted according to their product diversity from top to bottom. Some countries have substantial exports in almost all product categories (top rows). Countries at the bottom have a much lower diversity of exports. They typically export products that are also exported by many other countries.*

Let us denote the exports (measured in USD) of product category p from country c in a given year by $x(c,p)$. To see if a country has a proportion of exports of a product p that is higher than the proportion of world exports of product category p, one can define the so-called *revealed comparative advantage* [25],

$$RCA(c,p) = \frac{\frac{x(c,p)}{\sum_p x(c,p)}}{\frac{\sum_c x(c,p)}{\sum_{c,p} x(c,p)}}. \tag{5.62}$$

A value of $RCA(c,p) > 1$ (< 1) indicates that country c exports more (less) of product p than the average country.

5.7.1.1 *Product diversity and economic growth*

Whether a country c exports an above-average quantity of product p can be encoded in an unweighted, undirected bipartite network, $E_{cp} = 1$, if $RCA(c,p) \geq 1$, otherwise $E_{cp} = 0$. The effective product diversity of a country is given by the degree of this network,

$$k_{c,0} = \sum_p E_{cp}. \tag{5.63}$$

The degree of the product nodes reflects the number of countries that export a given product,

$$k_{p,0} = \sum_c E_{cp}. \tag{5.64}$$

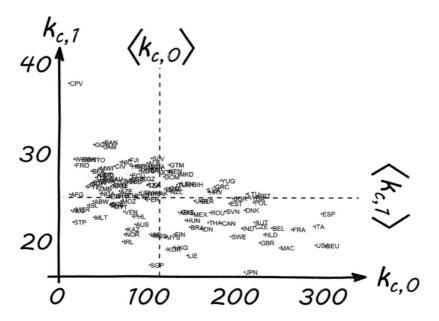

Figure 5.20 *The product diversity of countries, $k_{c,0}$, is negatively correlated with the ubiquity of their products, $k_{c,1}$. Rich countries are characterized by a high diversity and exclusive products (lower-right region). Poor countries have a low diversity of highly ubiquitous products (upper-left region).*

Hidalgo and Hausmann propose a measure for the ubiquity of a country's products. Ubiquity is high if many other countries make and export the same products. Low ubiquity is generally a sign of highly specialized products that are hard to manufacture. A measure of ubiquity is based on the notion of generalized degrees that are defined in recursive equations,

$$k_{c,N} = \frac{1}{k_{c,0}} \sum_p E_{cp} k_{p,N-1},$$

$$k_{p,N} = \frac{1}{k_{p,0}} \sum_c E_{cp} k_{c,N-1}. \tag{5.65}$$

For even N, $k_{c,N}$ are generalized measures of product diversity. For odd N they describe generalized measures for the ubiquity of a country's exports. The degree of ubiquity of a country's exports can be estimated by $k_{c,1}$. Product diversity, $k_{c,0}$, and ubiquity, $k_{c,1}$, for the 123 countries are shown in Figure 5.20. They appear to be negatively correlated. Countries in the upper-left region are non-diversified and produce standard products that can be found in a large number of other countries. The lower-right region shows fully diversified countries with exclusive products. Hidalgo and Hausmann argue that $k_{c,N}$ can be used to rank countries according to their economic fitness in terms of their economic capabilities. Fit means to have access to a set of capabilities that enables them to produce a large variety of exclusive products. Intuitively, a country with high economic fitness should have a high potential for economic growth. This hypothesis can be tested empirically. To this end, one can ask whether $k_{c,N}$ at time t correlates with future growth of the country, measured at time $t + \Delta t$. Economic growth is measured as the growth rate of the per capita GDP from time step t to $t + \Delta t$, $\mathrm{GDP}(c, t + \Delta t)/\mathrm{GDP}(c, t)$.[15] To test the hypothesis that economic fitness correlates with future growth, Hidalgo and Hausmann considered the regression,

$$\log\left(\frac{\mathrm{GDP}(c, t + \Delta t)}{\mathrm{GDP}(c, t)}\right) = a + b_1 \mathrm{GDP}(c, t) + b_2 k_{c,N}(t) + b_3 k_{c,N+1}(t). \qquad (5.66)$$

The economic fitness values $k_{c,N}(t)$, measured in $t = 1985$, are indeed strongly correlated with changes in GDP within the next $\Delta t = 5$ to 20 years [189].[16]

5.7.1.2 Two regimes in economic development

Based on the idea of using bipartite centrality measures to predict economic growth, such as those suggested in Equations (5.65), Pietronero and collaborators suggested a slightly different version for country fitness [94]. The idea is the following. It may happen that products appearing to have low ubiquity (high exclusivity) with respect to $k_{p,N}$ are also produced by countries with low diversity. In such cases the products are unlikely to be exclusive because they require expensive capabilities. A fitness measure for countries should take these cases into account by penalizing exclusive products that are also exported by low-diversity countries. Pietronero and collaborators, therefore, replace the generalized degrees $k_{c,N}$ and $k_{p,N}$ by recursive update equations for what they call *country fitness* $f_{c,N}$, and *product complexity* $q_{p,N}$,

$$f_{c,N} = \sum_p E_{cp} q_{p,N-1},$$

$$q_{p,N} = \frac{1}{\sum_c E_{cp} \frac{1}{f_{c,N-1}}}, \qquad (5.67)$$

with initial conditions $f_{c,0} = 1$, for all c and $q_{p,0} = 1$, for all p. Over time, each country describes a trajectory in the space spanned by its fitness $f_{c,N}$ and its GDP. These trajectories are shown in Figure 5.21 in an aggregated way. It is clear that countries approach a region along the diagonal in the figure. Phenomenologically, this region represents the GDP that is sustainable with a given country fitness. If a country is below that diagonal, GDP is likely to grow. Some countries can be seen to have a relatively low fitness but growth in GDP to levels that are far above the diagonal line. These are countries that exploit raw materials and do not invest in increasing their fitness,

---

[15] The GDP is adjusted for differences in purchasing power across different countries.

[16] For instance, for $\Delta t = 20$ years and $N = 18$, they find regression coefficients $b_2 = 38.8$ (significantly different from zero at a p-value of $p < 0.001$), and $b_3 = 1127$ ($p < 0.001$), while b_1 is not significantly different from zero ($p > 0.1$); the intercept is $a = -23801$ ($p < 0.001$). Similar results are obtained for $\Delta t = 5$, 10 and choices of $N \geq 8$.

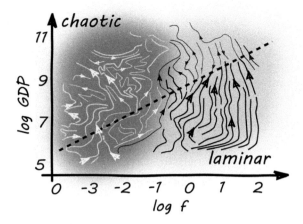

Figure 5.21 *Schematic view of the trajectories of countries in the fitness-GDP plane from 1995 to 2010. Over time, high-fitness countries (black) tend to approach a region (broken line), which can be interpreted as the levels of GDP that are sustainable with a given fitness. Some countries are seen to first increase f and then rise to high GDP levels. Others just exploit raw materials and do not invest in fitness (upward lines). These flows are 'laminar'; they have a relatively high level of predictability. Below a critical threshold of fitness, the trajectories become erratic. After [94].*

for example, Bahrain, Saudi Arabia, and the United Arab Emirates. There is a group of countries that have massively invested in increasing fitness, initially without much gain in GDP. However, once high levels of fitness have been reached they can grow much bigger and faster towards the diagonal. Countries with this type of dynamics in Figure 5.21 include China, India, and Thailand. These trajectories have a high level of predictability. Increasing fitness usually is achieved by increasing capabilities. There seems to be a critical value of fitness, below which the dynamics towards the diagonal is not realized. Instead, an erratic behaviour of the trajectories in the fitness-GDP plane is observed, and predictability is lost. The next example tries to answer the question as to whether the increase or decrease in product diversity of countries can be predicted. This will enable us to relate the diversity dynamics of countries to the creative phase transition.

5.7.2   Predicting product diversity from data

National economies are evolutionary systems characterized by a continuous stream of creation and destruction of goods and services. Entrepreneurs introduce new goods and services. Some of these have the potential to disrupt other more traditional economic activities [329, 330]. Some innovations may replace existing goods and, as a result, cause devastating effects on seemingly well-established industries. The addition of a single innovation can trigger a cascading process of replacements called *creative destruction*. For instance, the invention of the combustion engine rendered obsolete not only horse-drawn carriage manufacturers, but also all those industries that depended on the demand for carriages, such as farriers, horse groomers, and blacksmiths. Millions of jobs were destroyed; millions created. The economic landscape underwent a massive restructuring event from the equilibrium of a horse-based industry to the equilibrium of an automobile-based industry. Innovations occur whenever new capabilities (new labour skills, new institutions, infrastructures, or new technologies) become accessible—usually through the combination of existing capabilities. For instance, the combination of knowledge stocks like cloud computing, smart sensor networks, and decentralized decision-making with current production facilities is now giving rise to upgraded industrial infrastructure—a trend referred to as industry 4.0. New capabilities can lead to a large number of new products, services, and new means of production that result in existing capabilities being superseded by new ones and then abandoned. All products and services that relied on the abandoned capabilities disappear from the market.

The process via which countries acquire and abandon capabilities is the three-step process of combinatorial co-evolution, where new capabilities (i) are introduced, (ii) compete for survival in the existing world, and, if successful, (iii) become part of the environment and thus stimulate the creation and destruction of other capabilities. The product diversity of countries is driven by the process of combinatorial co-evolution. Changes in product diversity occur in cascades of creation and destruction events in the wake of the presence of new capabilities [233]. In this example, we show how the CCC model from Section 5.6.4 can be calibrated to world trade data in order to predict changes in product diversity within countries.

5.7.2.1 *The multicountry CCC model*

Within the framework of the CCC model, the process of how countries acquire novel capabilities and products can be described; see [233]. Equations (5.65) and (5.67) show how the country–product matrix E can be used to determine the fitness of a country. Countries can increase their fitness by increasing their diversity of non-ubiquitous products. Making such products requires the development of new capabilities. This evolution of capabilities is the focus of the multicountry CCC model. Relations between capabilities and products are encoded in a rectangular matrix Π. If a product p requires capability i to be produced, we set $\Pi_{pi} = 1$, and zero otherwise. In the model Π is fixed, meaning that each product always requires the same capabilities in order to be produced.

Capabilities are either present or absent in a country. Different countries have access to different sets of capabilities. In the multicountry CCC model, capabilities are the evolutionary entities (production technologies, knowledge stocks, etc.). The capabilities i of country c are represented in a time-dependent, binary state vector $\sigma_{i,c}(t)$. $\sigma_{i,c}(t) = 1$ means that country c has access to capability i at time t, otherwise $\sigma_{i,c}(t) = 0$. We assume that a country always exports a product if it has all the necessary capabilities for producing it. Countries upgrade their capabilities, that is, change their state vector $\sigma_{i,c}(t)$. As capabilities are time-dependent, so too is the country–product matrix $E = E(t)$,

$$E_{cp}(t) = \begin{cases} 1 & \text{if } \sum_i \Pi_{pi} = \sum_i \Pi_{pi}\sigma_{i,c}(t) \\ 0 & \text{otherwise} \end{cases}. \tag{5.68}$$

New capabilities can be acquired through the combination of existing capabilities. This is modelled with a random rule table M^+ with rule density, r^+. Note that here the evolutionary rules specify the development of new capabilities (not products). Capabilities can also be abandoned. We assume that it is likely for an upgraded capability to replace its precursor. For example, smart manufacturing replaces non-smart factories. The loss of capabilities is modelled with a destructive rule table, M^-, now a two-dimensional matrix that records the suppression rules. If j and k are combined to produce i, $M^+_{ijk} = 1$, then with probability r^- we have that either $M^-_{ki} = 1$ (i replaces k) or $M^-_{ji} = 1$ (i replaces j). If a capability k is abandoned, all products p that required k as an input (meaning that $\Pi_{pk} = 1$) will disappear from the market. A capability is present whenever its productive influences outweigh the destructive ones. Productive and destructive interactions are summed, then,

$$f_{k,c}(t) = \sum_{i,j} M^+_{ijk}\sigma_{i,c}(t)\sigma_{j,c}(t) - \sum_i M^-_{ik}\sigma_{i,c}(t). \tag{5.69}$$

The dynamics of the CCC model within each country is specified by Equation (5.61). We assume a small but non-zero rate p of spontaneous acquisitions or abandonments of capabilities (for instance, a company moves from one country to another). Note that each country is modelled by the same $M^\pm$, which means that all countries can, in principle, develop the same capabilities. Differences between countries can therefore arise only from differences in their initial conditions in terms of the present availability of capabilities. The CCC model is highly history-dependent.

5.7.2.2   *Calibration of the model*

World trade data allows us to obtain real world export diversities for the 99 countries that existed during the entire 1984–2000 timespan, across 732 different product categories [233]. The diversity profiles across countries in 1984 and 2000 are shown in Figure 5.22a and b, respectively. The CCC model is initialized with the actual number of products in every country in 1984. Each national economy is modelled by the same fixed production and destruction rule tables $M^\pm$ with rule densities $r^\pm$ that can be chosen as a free parameter. Another free parameter is the probability $\bar{r}$ of non-zero entries in the product-capability matrix Π. $M^\pm$ and Π are initialized as random tensors and networks and remain fixed throughout the simulation. To match model time with real time, the number of products appearing and disappearing in the data per time interval can be considered. The timescale in the model can then be scaled to match the rate of product (dis)appearances in the data.

The model result is shown as the black lines in Figure 5.22a and b. The change in product diversity over the 1984–2000 timespan is shown in Figure 5.22c for the data (grey) and the model (black). Over the sixteen years an overall trend towards increased diversity is visible. The increases in diversity are not distributed evenly across the countries. There are two sets of countries that show almost no change: those with a low initial diversity and those that are fully diversified. The countries in between are 'transitioning' countries that are in the process of catching up with the highly diversified countries.

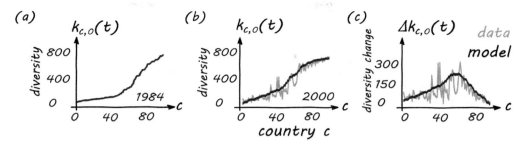

Figure 5.22 *(a) Diversity profile of countries in 1984. Countries are shown on the x-axis, ranked by their initial diversity in 1984. Poorly diversified countries are to the left, highly diversified on the right. Product diversities in 1984 for the data (grey) and the model (black) are identical, as the model was calibrated to this empirical initial condition. (b) Diversity profile in 2000. There is an overall tendency to increase in diversity. The black line shows the diversity profile as obtained from the multicountry version of the CCC model. (c) Diversity increases are spread unevenly across countries, as can be seen in the difference in diversity between 2000 and 1984, $\Delta k_{c,0}(t) = k_{c,0}(2000) - k_{c,0}(1984)$. Three types of country can be identified: (i) low-diversity countries that remain at low levels of product diversity, (ii) countries that are fully diversified and remain that way, and (iii) countries in between that are in a critical regime. The model is able to predict these changes quite well.*

It can be seen that countries in the third quartile of initial diversity have the largest diversity increase. The change in the diversity profile is predicted well with the CCC model.

The CCC model allows us to provide an explanation for why some poor countries are able to catch up with rich ones, while others are unable to do so. Faster growth is more likely to occur if the initial diversity is high enough to trigger a substantial number of creative destruction processes. For a fixed rule density $r^{\pm}$, which we ensure for all countries, the initial diversity ($n_0 = k_{c,0}(1984)$) has to be high enough for the economy to cross the creative phase transition; see Section 5.6.2. Countries have an initial diversity that is either below, at, or above this critical value of the initial diversity. For the small system sizes considered here, just a couple of hundred products and capabilities, this critical value is, of course, not a sharp point as in Figure 5.11, but smeared out over an interval. Figure 5.22c suggests that countries with an initial product diversity of around $k_{c,0}^{\mathrm{crit}} \sim 100 - 200$ are close to this critical value. Accordingly, countries belong to one of the three types:

1. Supra-critical countries are highly diverse countries. Most of them belong to the Western world. They are fully diversified and able to maintain their high income through their economic complexity.

2. Critical, low-diversity countries. These include transitioning economies around a critical diversity. Some are able to achieve convergence with high-income countries in terms of economic diversity; some are not.

3. Sub-critical, low-diversity countries are caught in a poverty trap. The trap is a literal trap: they are stuck below the critical initial density and cannot reach the diversified phase. They simply do not have enough capabilities to start the engine of creative destruction.

We discussed two examples of evolutionary processes in the context of economic development. In the first we showed how a 'fitness' for a country can be conceptualized. The concept of fitness rests on the assumption that each product requires a specific set of capabilities (knowledge, production infrastructure, regulations, laboratory skills, etc.) to be manufactured. The larger the set of capabilities in a country, the more diverse its products. High fitness means that a country is able to manufacture many products that are hard to produce because many (expensive) capabilities are required. Fitness is empirically measured based on actual export data available from world trade databases. Technically, fitness is a network centrality measure that generalizes random-walk-based centrality to bipartite networks. Countries tend to converge to a level of productivity (measured by GDP) that is determined by their fitness. Fitness allows for a surprising predictability of economic growth. This is no longer true for countries below a critical value of fitness. There, growth trajectories remain unpredictable.

In the second example, we showed how the CCC model of Section 5.6.4 can be used to predict changes in the product diversity of individual countries. Economic development can be understood as an evolutionary process driven by the appearance of new goods, services, and capabilities. Innovations usually occur as a result of combinations of already existing goods, services, and capabilities. The arrival of innovations can have devastating effects on already established industries, but it can also facilitate cascading events for further new innovations. The evolution of product diversity can be modelled by entrepreneurial activities in which available capabilities are combined to create new products as substitutes for existing ones. Using the fact that the CCC model has a creative phase transition, it is possible to understand that countries with a low initial diversity cannot increase their diversity. In low-diversity countries below the critical initial diversity, cascading innovations are very unlikely to occur. However, low-diversity countries close to the critical diversity are substantially more likely to experience bursts of innovations. The initial product diversities, as well as the product creation or destruction rates in the model, can be calibrated to world trade data. We showed that using the calibrated model, it is indeed possible to predict changes in product diversity to a high degree.

Both examples illustrate how a general framework for evolution dynamics can be transformed into a data analytics tool for making quantitative and falsifiable statements about real world evolutionary systems.

5.8 Summary

In this chapter we discussed evolution as a dynamical process that changes the composition of large sets of interacting entities over time. The essence of evolutionary processes is that they engender a continuous stream of entity creation and destruction events. Once a new entity is introduced, it changes the environment for potentially all the other entities. The existing entities will adapt to the changing environment. These adaptations can facilitate the emergence of new entities or can cause the extinction of others. There is an impressive amount of empirical evidence of the dynamics of the diversity of evolutionary systems being characterized by punctuated equilibria. This means that there are relatively long time periods when the system seems to be in equilibrium. From time to time, the system changes from one equilibrium to another in disruptive events, during which the composition of entities is massively rearranged. Equilibria are punctuated by disruptive events. Consequently, the statistics of diversity changes is often dominated by fat-tailed and power law distribution functions. It seems impossible to predict future events in evolutionary systems. This, however, is not the goal of the science of evolutionary processes. The challenge there is to understand the underlying statistics. This has been the main focus of the chapter. Our aim was to develop a general mathematical framework describing the combinatorial co-evolution of species and their environments, able to understand the statistics of evolutionary processes, including the sizes and durations of the disruptive restucturing events, the durations of the equilibrium phases, the origin and details of cascades in diversity changes, systemic risk, and innovation rates. The general framework is based on a three-step process that any non-trivial evolutionary system follows.

1. A new entity is created and placed in a specific environment.
2. The new entity interacts with its environment. These interactions determine whether the new entity survives or dies out.
3. A surviving entity becomes part of the environment and introduces new interactions with existing entities and may also change existing ones. These interactions can eventually lead to cascading creation or extinction events.

A mathematical framework to describe this three-step process is presented in the *general evolution algorithm*. There, the properties of all entities at time t are collected in a state vector, $\sigma(t)$. The ith entry in the vector $\sigma_i(t)$ describes the state of entity i, for instance, its abundance. The environment of entity i is defined as the collection of all interactions that i has with other entities or the physical environment that is also encoded in the state vector. Interactions are recorded in a time-dependent interaction tensor $M_{ijk\ldots}^{\alpha}(t)$, where α denotes the specific type of interaction, and the set of indices $ijk\ldots$ indicates the interactions between species. The general evolution algorithm can be expressed as a co-evolving dynamical system, that updates itself according to,

$$\sigma_i(t+dt) \sim \sigma_i(t) + F\left(M_{ijk\ldots}^{\alpha}(t), \sigma_j(t)\right)$$
$$M_{ijk\ldots}^{\alpha}(t+dt) \sim M_{ijk\ldots}^{\alpha}(t) + G\left(M_{ijk\ldots}^{\alpha}(t), \sigma_j(t)\right), \tag{5.70}$$

where F and G are some functions (deterministic or stochastic, linear or non-linear) that depend on both the state vectors and the interactions. An analytical treatment of the algorithm is in general not possible, except for trivial cases of F and G.

To discuss the various aspects of the dynamics of evolutionary processes, we reviewed several classic models in the light of the algorithm in Equation (5.70). We learned that in fitness landscape models, evolutionary processes take the form of optimization problems on landscapes. These models allow us to understand the possible origins of (self-organized) critical dynamics and punctuated equilibria; however, they do not provide an adequate way of thinking about co-evolution of entities with their environments. We encountered models with linear interactions between entities, where the interaction tensor is reduced to an interaction matrix. Co-evolutionary dynamics and criticality can be understood within the linear framework, which also shows how 'fitness' emerges dynamically as a topological property of the species–species interaction network. We learned about the role of auto-catalytic cycles. We mentioned that linear co-evolution models are useful in terms of considering the risk of evolutionary system-wide collapse. The topological properties of the underlying interaction networks play a central role. Finally, we studied co-evolutionary diversity dynamics in a framework that focused on non-linear, combinatorial interactions—the combinatorial, co-evolutionary, critical, or CCC model. The interaction tensor M^α is given by two layers of a multiplex tensor network. The layers describe which species can be combined to act productively or destructively on other species. We showed that the dynamics of the CCC model generically leads to punctuated equilibria and critical dynamics. The former can be understood through the formation of auto-catalytic cycles. The removal of keystone species can lead to collapse and mass extinction events. The dynamics of the CCC model can be shown to be identical to a spin model with a discontinuous order–disorder transition that can be analytically understood in a mean field approximation. All the models discussed, including the NK model, Bak–Sneppen model, Solé–Manrubia model, Jain–Krishna model, Arthur–Polak model, and the CCC model, can be seen as special cases of the general evolution algorithm.

We concluded the chapter with two examples illustrating how the general evolution algorithm, Equation (5.70), and the CCC model, in particular, can be used to arrive at quantitative and falsifiable predictions for the growth and product diversity dynamics of countries.

We collect a few take-home messages from our exploration of evolutionary processes.

- Punctuated equilibria are ubiquitous in evolutionary systems. Phases of metastable equilibria are interrupted by massive restructuring events that lead to massive changes in the composition of systems. As a result, critical phenomena are omnipresent in evolutionary processes.
- Combinatorial interactions are fundamental for evolutionary processes. They can be formalized in catalytic network models.
- Co-evolution of species with their fitness landscapes is the key component for understanding the dynamics of evolutionary systems.
- A comprehensive model for evolutionary processes must include a possibility for combinatorial interactions. These naturally describe critical phenomena as a consequence of co-evolutionary dynamics of species and fitness landscapes.
- Evolutionary processes are fully characterized by the general evolution algorithm.

- To understand endogenous selection, it is necessary to study models of adaptive and co-evolving fitness landscapes. These often show generic (self-organized) critical dynamics.

- Combinatorial and co-evolving systems with positive and negative interactions strongly resemble spin glasses, which describe frustrated systems. Transitions between metastable configurations in spin models correspond to regime shifts between equilibria in evolutionary systems. These cause mass extinctions, booms of diversification, revolutions, and so on.

- Fitness can be understood as an emerging dynamical and topological property of the tensor networks of species–species interactions. Fitness is tightly related to 'membership' in auto-catalytic sets, which are themselves related to the existence of positive feedback cycles.

- Evolution explores the *adjacent possible* of a system. Generically, evolutionary systems have a creative phase transition (generalized percolation transition) that separates a high- from a low-diversity phase. In the low-diversity phase the existence of closed and interconnected feedback loops is extremely unlikely. Only systems in the high-diversity phase are able to sustain their diversity over extended periods of time.

- Economic development and technological change are prime examples of man-made evolutionary processes. Schumpeter's *creative destruction* is a consequence of combinatorial dynamics with positive and negative interactions.

- All the statistical features of evolutionary processes discussed can be understood in a simple realization of the general evolution algorithm, the critical combinatorial co-evolution model, or the CCC model.

5.9 Problems

Problem 5.1 Consider the two-dimensional Lotka–Volterra dynamics, Equation (5.14). What are the two fixed points ($\frac{d}{dt} x_i = 0$ for $i = 1, 2$) of Equation (5.14)?

Problem 5.2 Compute the Lyapunov exponents of each fixed point of the Lotka–Volterra Equation (5.14).

Problem 5.3 The generalized Lotka–Volterra equation for N species with abundances $x_i(t)$ is given by

$$\dot{x}_i = x_i \left(f_i + \sum_{j=1}^{N} p_{ij} x_j \right), \tag{5.71}$$

where f_i are frequency-independent replication rates and p_{ij} is the payoff matrix. Show that the generalized Lotka–Volterra equation with $N - 1$ species is identical to a replicator equation with N species.

Problem 5.4 Consider the NK model for a fixed $N = 10$. How does the average number of mutations that a species experiences until it reaches a fitness peak depend on K? Hint: implement a computer simulation of the NK model and plot the average number of mutations until the dynamics stops as a function of $K = 1, 2, \ldots, N - 1$.

Problem 5.5 For a given N, C, and S, how does the average species fitness in the NKCS model depend on K? Consider twenty-five species on a regular network, where the fitness of each species depends on $S = 4$ other randomly chosen species, and where $C = 1$. Hint: implement a computer simulation of the NKCS model with $N = 10$ and plot the average species fitness, $F(s)$ in Equation (5.15) (averaged over all species), as a function of $K = 1, 2, \ldots, N - 1$.

Problem 5.6 How many states does the adjacent possible of the NKCS model contain?

Problem 5.7 Implement the Bak–Sneppen model in a computer simulation with $S = 2$ and $N = 1000$ species. Give the average value of the lowest barrier to mutation, B_c, to which the model self-organizes. Hint: after the initial transient phase of the simulation, monitor the lowest barrier to mutation over time and plot its histogram.

Problem 5.8 Consider the Jain–Krishna model with positive species–species interactions. Show that switches to a fully diversified regime and crashes to a state of low diversity are always accompanied by jumps in the Perron–Frobenius eigenvalue. Hint: implement the Jain–Krishna model in a computer simulation with $N = 100$ and $p = 0.0025$. Plot the diversity in the model and the Perron–Frobenius eigenvalue over time T.

Problem 5.9 Consider a system with species–species interactions given in a directed network with adjacency matrix A with elements $A_{ij} \in \{0, 1\}$. Show that the system contains no auto-catalytic cycles if all eigenvalues of A are zero.

Problem 5.10 Assume that A from the previous exercise has at least one auto-catalytic cycle. Show that the Perron–Frobenius eigenvalue of A is at least 1.

Problem 5.11 Consider the Jain–Krishna model with positive *and negative* species–species interactions ($N = 100$ and $p = 0.0025$). Let $L^+(T)$ $(L^-(T))$ be the number of positive (negative) links in the interaction network at time T. How do $L^+(T)$ and $L^-(T)$ change during booms and crashes of the diversity?

Problem 5.12 In Section 5.6.2 we studied the generic creative phase transition of a system in which pairs of entities can be combined to create new entities. How does the creative phase transition in Figure 5.11 change if entities are created as combinations of three entities? Replace the tensor M_{ijk}^+ by a tensor M_{ijkl}^+ in which non-zero entries indicate that entities j, k, and l must be combined to create entity i. Perform computer simulations to compute the asymptotic diversity n_∞ as a function of the initial density n_0 and the rule density r.

6

Statistical Mechanics and Information Theory for Complex Systems

6.1 Overview

Many complex systems are stochastic in nature. This can mean two things. The first possibility is that states of the elements of a system, or the interactions between them, are not following deterministic laws, but evolve randomly. Randomness can have endogenous causes, such as the quantum mechanical nature of particles or the free will of humans, or it can be exogenously brought into the system. The second possibility is that the system itself is evolving deterministically but that we do not have complete information about it. This situation can arise, for example, due to limited data resolution or computational constraints. In that case, the trajectories of states and interactions cannot be inferred with certainty, and a degree of stochasticity is created by the *description* of the system or the process. Our state of ignorance forces us to a *probabilistic* description of a coarse-grained dynamics. This means that we have to assign probabilities to the occurrence of states, interactions, and the transitions between states. Depending on the nature of the underlying system, there are three main directions for understanding a system based on probabilistic knowledge. All three involve a notion of entropy, which is a functional that depends on the probabilities of states and transitions between them. If we deal with physical systems we use *statistical mechanics* to understand its systemic properties; if we are interested in quantifying the information that is produced by a system we use *information theory*; and if we want to infer distribution functions from statistical systems, we use the concept of the *maximum entropy principle*. These three conceptually very different approaches use entropy in different ways. In this chapter, we will carefully dissect these three aspects or 'faces' of entropy and show how they can be used meaningfully to better understand and predict the statistics of strongly interacting and path-dependent systems. In particular, we explore a conceptually coherent mathematical framework for an entropy of complex systems.

Introduction to the Theory of Complex Systems. Stefan Thurner, Rudolf Hanel, and Peter Klimek,
Oxford University Press (2018). © Stefan Thurner, Rudolf Hanel, and Peter Klimek.
DOI: 10.1093/oso/9780198821939.001.0001

6.1.1 The three faces of entropy

6.1.1.1 Physics and entropy

In many systems, the number of states is countable, which means that there is a clear notion of what a state is and how many different states exist. Such a state is called a *microstate*. Particular collections of microstates are often associated with a so-called *macrostate*, which typically refers to a systemic or global property of the system. For example, a particular microstate describes the positions and momenta of a number of gas molecules. The macrostate associated with that microstate includes global variables, such as the temperature or pressure of the gas in a given container. Many microstates are often associated with a single macrostate. The macrostate can be seen as a very coarse-grained description of a system or a process. If all microstates occur with the same probability, the likelihood of encountering the system in a given macrostate is proportional to the number of its associated microstates. This provides a link between the microscopic (e.g. momenta of molecules) and the macroscopic world at the global or systemic level (e.g. temperature). It is this link between micro and macro phenomena that was considered as one of the greatest triumphs of physics—statistical mechanics. In this framework it is possible to make extremely precise statements with respect to measurable systemic properties of systems composed of many particles which are described on a probabilistic basis.

Statistical mechanics makes it possible, in essence, to aggregate statistical information regarding individual components (microstates) in order to infer and understand the physical properties of the entire system (macrostates). It is essential to know how many microstates belong to a single macrostate. This number is called the *multiplicity*—the number of ways in which a macrostate can be generated by microstates. The logarithm of the multiplicity is called *entropy*, or (in this particular context) 'Boltzmann entropy'.

In physical systems in thermal equilibrium, it transpires that Boltzmann entropy is the same as the thermodynamic entropy,[1] introduced by Rudolf Clausius as a quantity that relates heat with temperature and the amount of work that can be obtained from steam engines. Statistical mechanics not only allows us to understand the thermodynamics of systems, but more generally provides us with a framework for understanding their phase diagrams. The knowledge of phase diagrams amounts to a full understanding of how systemic properties depend on their control variables. Phase diagrams are key to the management and control of complex systems.

6.1.1.2 Information theory

The second face of entropy appears in the context of simple (ergodic) information-producing sources. Here, entropy plays a twofold role. On the one hand it can be used to describe the length of the shortest possible codes that can encode a given stream of information in an unambiguous way. On the other hand it can be used to determine

---

[1] Thermodynamic entropy is obtained from Boltzmann entropy by a Legendre transformation, which relates energy states with probabilities.

whether a data stream can be encoded in such a way that even after it has been sent through a channel noisy enough to corrupt the data during transmission, the original data stream can be reconstructed in an error-free way from the received message. This is possible if the information production rate of a source is less than or equal to the so-called channel capacity.

To quantify the amount of information produced by a source, Claude Shannon introduced a mathematical measure. However, before publishing his results in a seminal 1948 paper [336], he asked John von Neumann what he should call it. The story goes that von Neumann suggested,

> You should call it entropy, for two reasons. In the first place your uncertainty function has been used in statistical mechanics under that name, so it already has a name. In the second place, and more important, no-one really knows what entropy really is, so in a debate you will always have the advantage [377].

The use of the term 'entropy' for an uncertainty measure of sources—the same term that Ludwig Boltzmann used to describe numbers of states in a system—is, quite understandably, a source of tremendous confusion. Counting numbers of microstates in systems is very different from the information production rates of sources.

6.1.1.3 *Maximum entropy principle*

The third area where the term entropy is used is in the context of statistical inference problems. Remarkably, under certain conditions, which we will study next, this entropy is equivalent to Boltzmann entropy. In problems of data analytics the question often arises as to what one might expect the most likely distribution function to be, given a specific data set and a statistical model. The most likely distribution is obtained by following a maximum likelihood approach, which maximizes an entropy functional. One compelling idea behind this approach is that it enables us to incorporate additional information, knowledge, or hypotheses about a system by using Lagrange multipliers. Such additional knowledge is typically given by the mean, the variance, or the higher moments of the distribution function that one wishes to infer. Additional knowledge or 'constraints' are typically obtained from some kind of experiment. Entropy maximization with constraints is often referred to as the *maximum entropy principle*. It was popularized by Edwin T. Jaynes [209]. The basic philosophy, of course, follows the earlier ideas of Boltzmann. The original contribution of Jaynes was to demonstrate that statistical mechanics can be formulated in a physics-free way. As a result, the concept became widely applicable to areas outside of physics.

6.1.1.4 *Trinity of entropy for simple systems: three-in-one*

For *simple systems* these three aspects of entropies are equivalent; namely, the formulas for Boltzmann entropy, the quantity that quantifies information, and the entropy in the maximum entropy principle. They are all identical.

The word *entropy* is one name for three very different concepts. To add to the confusion, in simple systems, the mathematical expressions for the three entropies turn out to be the same, $S = -k \sum_{i=1}^{W} p_i \log p_i$, where p_i denotes the probability[a] of (micro) state i and k is a constant. To avoid confusion, and before beginning to discuss entropy, we absolutely must specify which of the three concepts we are talking about. This entropy formula is sometimes called the Gibbs formula or Boltzmann–Gibbs–Shannon entropy. If an entropy formula can be written as a sum of functions $g(p_i)$ that depend on probabilities p_i only,

$$S = \sum_i g(p_i), \tag{6.1}$$

it is called a *trace form entropy*. For Boltzmann–Gibbs–Shannon entropy, $g(p_i) = -p_i \log p_i$.

---

[a] We will use p_i also for probabilities here. Whenever there is a possibility of confusion of probabilities with relative frequencies, we will use q_i for probabilities, and p_i for the relative frequencies.

What is a *simple system*? In the light of the three concepts of entropy, 'simple' means the following: a physical system is simple when there are no interactions or very weak interactions between its constituent elements and if there are no exogenous influences driving the system. An information-theoretic source is simple, if the symbols or states that it generates are produced independently and the source is ergodic. Simple statistical systems, for which the maximum entropy principle applies, involve statistical processes that basically follow multinomial statistics. We discuss these topics in more detail in Section 6.2.

6.1.1.5 *Entropy and equilibrium*

Physicists usually associate the term entropy with systems in *equilibrium*. This is because of the second law of thermodynamics, which states that the entropy of a closed physical system becomes maximal as it approaches equilibrium. For systems that are out-of-equilibrium, entropy ceases to be a function of the macroscopic physical control variables (macrostate), and its value starts to depend on the details of *how* the state was obtained. Entropy becomes path-dependent and thereby loses much of its practical usefulness. Some of its usefulness, however, may be retained in non-equilibrium situations but, to achieve this, we need to clearly distinguish between distinct notions of entropy.

We learn from information theory that sources must be *ergodic* for entropy to make sense as a measure of information. For non-ergodic systems, where information production rates are time-varying and non-stationary, the classic information-theoretic entropy concept loses its theoretic foundation and breaks down. Along the same lines, for the maximum entropy principle to hold, stochastic systems have to be composed of events that are sufficiently independent from one another, which is true for (multivalued) Bernoulli processes or multinomial statistics. As soon as processes become path- and history-dependent, the classical maximum entropy principle loses its theoretical support and becomes meaningless in the sense that, in general, its predictions will be wrong. In history-dependent stochastic systems, the probabilities that describe the system change as a function of the history of the events that occurred (were sampled) during the process. We discuss path- and history-dependent processes in more detail in Section 6.5.

Complex systems are inherently out-of-equilibrium, non-ergodic, or path- and history-dependent. So why do we need to discuss statistical mechanics and the 'entropy' of complex systems?

6.1.1.6   Entropy for complex systems—the challenges and promises

Complex systems exist in different phases. For different control-parameter settings, one and the same system might show radically different dynamical behaviour at the systemic level. The corresponding phase diagrams can be much more complicated than those for simple systems. As for simple systems, the knowledge of phase diagrams is key to understanding and controlling complex systems. In particular, the knowledge of critical parameter values where phase transitions occur are of paramount importance in terms of understanding systemic properties such as robustness, resilience, and efficiency. Phase diagrams are key to understanding systemic risk and collapse. They allow us to identify regions in parameter space where systems are expected to lose their ability to function and collapse. The statistical mechanics of complex systems will thus be essential to the understanding, management, and control of complex systems. The statistical mechanics of complex systems does not yet exist as a fully developed theory. It is, however, in the process of being established. Coming, eventually, to a fully consistent description of complex systems, from a statistical mechanics point of view, will have the following advantages:

- Classification of complex systems into universality or equivalence classes, Section 6.3.5.
- Statistical inference and prediction of path-dependent processes, Section 6.5.2.
- Understanding the origin and ubiquity of power laws in distribution functions, Sections 6.3.5, 6.5.2, and 6.6.2.
- Understanding the phase structure of specific complex systems.

As complex systems are typically driven out-of-equilibrium, non-ergodic, and both path- and history-dependent, it is not a straightforward task to arrive at a meaningful concept and at a definition of what the 'entropy' of a complex systems should be. For example, according to basic thermodynamic knowledge, irreversible thermodynamic processes are inherently path-dependent and, moreover, *thermodynamic entropy* ceases to exist as a function of observable macroscopic system parameters. This does not, however, imply that there is no *useful* notion of entropy at all. What we require from a notion of entropy at this point is to serve as a meaningful way of linking microscopic and macroscopic system properties. In the context of information theory, entropy must provide us with a useful measure of information production for non-ergodic and non-stationary sources. Finally, entropy must be able to predict distribution functions of path- and history-dependent stochastic processes.

To use entropy meaningfully in the context of complex systems, there are only two possible paths. The first is to partition a complex system into components that themselves are not complex, meaning that they are sufficiently close to equilibrium, to being independent, to being ergodic, as well as path-independent. If this is possible, which is, of course, usually not the case, one can safely apply classical entropy concepts. The highly non-trivial part is then to reassemble the results into a whole such that predictions about the systemic properties of the system remain valid. The second possibility is to deal with a system as it is and adequately modify the mathematical expressions for entropy such that the resulting predictions at the systemic level remain correct and are theoretically well-founded. In this chapter we follow only the second path.

The aim of this chapter is to show that the mathematical expressions for the three faces of entropy, which describe the thermodynamics, the information content, and the distribution functions of a system, are no longer identical, but are split into three distinct mathematical functionals. We will see how these expressions can be derived from general principles and applied successfully to specific path- and history-dependent processes. This chapter explores several new methodological directions for ultimately dealing with complex stochastic systems on the basis of a consistently generalized statistical mechanics.

6.2 Classical notions of entropy for simple systems

Whenever people discuss entropy, confusion seems to arise fairly quickly. This is mainly due to the need to—exactly—specify the framework within which the term entropy is used. As discussed, the same word is used for at least three very different concepts.

Concept 1 Entropy deals with the *thermodynamics* of physical objects. Entropy in thermodynamics is a quantity introduced by Rudolf Clausius as a heat- and work-related measure. It appears in thermodynamic equations, such as $dU = TdS - PdV$, where U denotes internal energy, T is temperature, P is pressure, V is volume, and S is entropy. Thermodynamic equations make it immediately clear that entropy in the context of physics has to be an *extensive* quantity. Extensive quantities are quantities that scale with system size. In the thermodynamic equation above, if one doubles the size of the system by adding an identical system to it, the internal energy and the volume obviously double, while temperature and pressure, which are the *intensive* system variables, do not change their values.

If entropy were not an extensive quantity and if one kept enlarging the system (going to the thermodynamic limit) the term TdS would vanish with respect to dU. This means that temperature would no longer feature in large systems, which is obviously nonsense. If U and V double but S does not, then this can only happen when $T = 0$, or if T were not an intensive variable, and if T were dependent on the scale of the system (which it is not). For thermodynamic systems, therefore, entropy must be an extensive variable. A second aspect of entropy was introduced by Boltzmann, who understood that the expression,

$$S_B = -k_B \sum_{i=1}^{W} p_i \log p_i, \tag{6.2}$$

is equivalent to the expression of Clausius's entropy.[2] Here W is the number of states, which are labelled i, the probability of finding state i is p_i, and k_B is a constant that is necessary to account for the correct dimensional units in the thermodynamic equations. Boltzmann, when considering gases, assumed that all states or configurations occurred with the same probability, $p_i = 1/W$, so that Equation (6.2) becomes,

$$S_B = k_B \log W, \tag{6.3}$$

which we refer to as Boltzmann entropy. This is simply the logarithm of the multiplicity of states that occur with the same probability. The fact that a macroscopic quantity like the entropy that appears in thermodynamic equations can be related to microscopic states (or configurations) was the starting point of statistical mechanics. Historically, the most fascinating aspect of this issue is that identification of the thermodynamic entropy with Boltzmann entropy made it possible to link macroscopic properties of a system (macrostate) with its microscopic causes (microstates) in a transparent and extremely simple way.

Concept 2 Entropy deals with quantifying *information production* by a source that emits sequences of symbols, that is, distinguishable states sent from a source to a receiver. Information production is tightly related to issues of coding information and to data compression. Whenever information is being quantified, the entropy functional appears naturally [336],

$$S_S = -k_S \sum_{i=1}^{W} p_i \log p_i, \tag{6.4}$$

which we refer to as *Shannon entropy*. Here, p_i are the relative frequencies of symbols from an alphabet of size W, which appear in the sequences emitted by a source.[3] Shannon entropy is the fundamental quantity in information theory. It tells us how efficient a particular stream of information can be coded and if the information source is an ergodic finite-state machine, which we will discuss next. The same form of entropy

---

[2] Boltzmann, in fact, never used the equation in this form; it was Planck who first used it in 1901 [309].

[3] We can set the constant $k_S = 1$. However, if we want to interpret S_S in units of binary bits per symbol, the constant needs to be chosen $k_S = 1/\log(2)$. For binary codes, each symbol of the alphabet is encoded by a uniquely identifiable string of numbers composed of 0 and 1.

keeps appearing in issues related to coding, data compression, and optimal prefix-free codes [239, 269]. Shannon asked if it were possible to code streams of information in such a way that they could be recovered in an error-free way after being sent through noisy channels. Noisy channels are channels that corrupt the coded signals during the process of transmission from sender to receiver. Depending on the capacity of a channel, a signal can be coded in such a way that it can be transmitted and then decoded with a vanishing error rate, as long as the entropy rate of the source is less than the capacity of the channel. This again was one of the outstanding and unexpected results of twentieth century science.

Concept 3 Entropy is used for *statistical inference*. Situations frequently arise where data are available in statistical form. In other words, information about the frequency of the occurrence of certain events is available. For example, whenever measurements are performed repeatedly with a finite resolution, such data appear. One way of representing statistical information is to use histograms. Statistical inference is a way of finding out how likely it is to observe a particular histogram; it allows us to compute the functional form of the *most likely* histogram.

Finding the most likely histogram is exactly what the *maximum entropy principle* is used for. For simple systems and processes the maximum entropy principle maximizes an entropy functional under the constraints that relative frequencies sum to one, $\sum_i p_i = 1$, and possibly under further constraints that are usually given in the form of moments of the distribution. The entropy functional needing to be maximized appears to be of the form [208],

$$S_{\mathcal{J}} = -\sum_{i=1}^{W} p_i \log p_i, \tag{6.5}$$

which we call the Jaynes maxent entropy. Here p_i are the relative frequencies of the occurrence of an outcome or a state i in a series of measurements, where each measurement can produce one of W possible outcomes. Jaynes was interested in a physics-free formulation of statistical mechanics, and, at the time, this earned him his share of negative polemics.

Obviously, the formula for entropy is degenerate for simple systems in the sense that for the three distinct entropy concepts (thermodynamics, entropy rate, and statistical inference), the same entropy functional, $-\sum_{i=1}^{W} p_i \log p_i$, appears.

The degeneracy of the entropy functional appears for simple systems only. Boltzmann, Shannon, and Jaynes were extremely clear about the conditions necessary for their respective entropy concepts to apply. We will see in this chapter that this degeneracy of entropy will disappear in the context of complex systems, that is, in systems that extend beyond the multinomial statistics of ideal gases and ergodic sources.

6.2.1 Entropy and physics

From classical physics we know that thermodynamic entropy must satisfy the following three thermodynamic principles.

1. For reversible (quasi-stationary) processes, the thermodynamic entropy $dS = dQ/T$ is an exact differential, meaning that the integral $\int_1^2 dS(\gamma)$ does not depend on the path in phasespace γ from 1 to 2, but only on the initial macrostate $\gamma(0)$ and the final state $\gamma(1)$. dQ is the heat added to or removed from the system, T is the temperature.

2. Entropy is additive, meaning that the entropy of a system composed of (identical) parts is the same as the sum of the entropies of the parts, $S_{AB} = S_A + S_B$.

3. Entropy in a closed system increases, $dS \geq 0$. If the macrostate of a closed system is given at a particular time, the most probable state of the system at any later time is one of higher or equal entropy. If the system is initially in equilibrium, the initial and final macrostates are the same and $dS = 0$.

Starting from these principles, we show that thermodynamic entropy, $dS = dQ/T$ and the entropy of statistical mechanics are indeed the same.[4]

The entropy of statistical mechanics in statistical equilibrium is defined as,

$$\sigma = \log \Delta\Gamma, \tag{6.6}$$

where $\Delta\Gamma$ is the phasespace volume that is accessible to the system, given a fixed energy.

We have to show that σ in Equation (6.6) indeed fulfils the three thermodynamic principles shown. First, σ has a definite value for a statistical system in equilibrium and therefore the change in entropy must be an exact differential.[5] $\Delta\Gamma$ can be interpreted as a measure of the uncertainty of our knowledge of the microstate a system actually occupies. In equilibrium this uncertainty is maximal. Once the accessible phasespace volume is known, the entropy is also known.

Second, σ is additive. Assume that we put together a system of N gas particles from two systems of N_1 and N_2 particles, respectively. Assume that those two systems have $\Delta\Gamma_1$ and $\Delta\Gamma_2$ states, respectively, then the number of states in the composed system is $\Delta\Gamma = \Delta\Gamma_1 \Delta\Gamma_2$, and additivity follows,

---

[4] We follow the classic development of Ch. Kittel [231].

[5] Note that a system that is out of equilibrium may be characterized by any set of microstates compatible with the constraints posed by the macrostate, for instance, the internal energy, U. In equilibrium, however, the system may be in any microstate compatible with the macrostate. This implies that $\Delta\Gamma_{\text{equilibrium}} \geq \Delta\Gamma$. Therefore, in a closed system with fixed macrostate values (e.g. $U = $ const.) it follows that $dS \geq 0$.

$$\sigma = \log \Delta\Gamma = \log \Delta\Gamma_1 \Delta\Gamma_2 = \sigma_1 + \sigma_2.$$

And third, to show the validity of the third principle, we assume that the equilibrium condition is given by the most probable condition. This allows us to exchange the average values of physical quantities for their values at the most likely configuration. It is much simpler to use the value at the most likely configuration than to compute the average value over the entire system; see the example below. If the equilibrium state is the most probable state, then the phasespace volume $\Delta\Gamma$ associated with the equilibrium properties is also a maximum. The entropy of a closed system therefore has its maximum when the system is in equilibrium. The tendency of entropy to increase must be interpreted in a statistical sense. This does not mean that entropy cannot decrease from time to time in individual trajectories of the system; it means that it will not, on average, decrease.

6.2.1.1 Example—average values and the most likely configuration

We show in the following example that it is often fully justified to replace the average values (which may be hard or impossible to compute for systems with many particles) with the value of the most likely configuration. We first compute the distribution function for a simple statistical system of N non-interacting spins sitting on a line. Each spin has a magnetic moment μ, and there is no external magnetic field. We want to compute the entropy of that system.

There are 2^N possible ways to arrange the spins so that the probability of finding a specific configuration c is $p_c = 1/2^N$. If we ask in how many configurations exactly n more spins point upwards than downwards, we can compute the distribution function of the magnetic moments. The magnetic moment of a system where $\frac{1}{2}(N+n)$ spins point upwards and $\frac{1}{2}(N-n)$ downwards is $m = n\mu$. Using elementary combinatorics, the number of configurations $W(n)$, where n spins more point upwards, is given by,

$$W(n) = \frac{N!}{\left[\frac{1}{2}(N+n)\right]! \left[\frac{1}{2}(N-n)\right]!}. \tag{6.7}$$

The probability of finding a net moment $m = n\mu$ is then,

$$p(m) = p(n\mu) = p_c W(n) = \left(\frac{1}{2}\right)^N \frac{N!}{\left[\frac{1}{2}(N+n)\right]! \left[\frac{1}{2}(N-n)\right]!}. \tag{6.8}$$

Repeated use of Stirling's formula, Section 8.2.1, and the formula for the logarithm, $\log(1 \pm \frac{n}{N}) = \pm\frac{n}{N} - \frac{n^2}{2N} + \cdots$ gives,

$$p(m) = \sqrt{\frac{2}{\pi N}} e^{-\frac{n^2}{2N}}. \tag{6.9}$$

Magnetization follows a Gaussian distribution with the mean and the maximum at zero. The most probable value coincides with the average value. Note that the distribution function becomes extremely peaked for large N, so that practically all of the distribution 'is concentrated' at the most likely position. We can see this very clearly in the following argument. Let us compute the entropy for the most likely macro condition $m = 0$ (zero magnetic moment). Starting from Equation (6.7), setting $n = 0$ and using Stirling's formula we get,

$$\sigma = \log W(0) = \log \frac{N!}{\left(\left[\frac{1}{2}(N)\right]!\right)^2} \sim N\log 2 - \frac{1}{2}\log \frac{\pi N}{2}. \tag{6.10}$$

Compare this result with the correct entropy that we get for *all* 2^N possible configurations, which is $\log 2^N = N\log 2$. For large N, the $\log N$ term in Equation (6.10) can be safely neglected, whereas $N\log 2$ cannot. This means that practically all the accessible phasespace has the same properties as the *most probable* part of phasespace. Note that this example can also be seen as a process: it is identical to a random walker that takes a sequence of N up or down steps with equal probability. Here, n is the net distance travelled from start to finish.

If the system size (e.g. number of particles) is large, the entropy of the thermodynamic system in equilibrium is insensitive to the conditions that characterize the system in the neighbourhood of the most probable condition. What matters is the value at the most likely condition.

6.2.1.2 *Composing systems—additivity and extensivity*

In classical statistical mechanics, the terms *additivity* and *extensivity* are often used interchangeably. They do, however, refer to different concepts, which must be properly distinguished. It is thus important to first clarify what it means to compose systems from subsystems.

For example, it is clear what it means to bring two containers, filled with ideal gas particles, into thermal contact; it is less clear, however, what it means if we combine two systems with strongly interacting elements. As these systems are brought together from initial isolation, interactions between the elements of both systems are forming. As new interactions form, the phasespace of the new system grows. While new states become accessible in the combined system, some states may become inaccessible. It might also become important *how* systems are brought into contact. This is the case if the way the interactions between elements of both systems are forming depends on the interface between the two systems. The way in which phasespace grows is strictly related to how systems are combined at a statistical level. As we will see in Section 6.4, the way systems are combined largely determines the notion of entropy as an *extensive* measure of phasespace volume.

In the following, assume systems with a finite number of states and denote with W_A and W_B the number of states in two initially separated systems A and B, respectively. The number of states accessible to the system, after we bring them together, is W_{AB}. Any quantity X that depends on the volume of phasespace (such as the statistical mechanics entropy from Equation (6.6)) is called *extensive* if the following relation holds,

$$X(W_{AB}) = X(W_A) + X(W_B) \qquad \text{(definition extensive)}. \qquad (6.11)$$

The quantity X is called *additive* if we have,

$$X(W_A W_B) = X(W_A) + X(W_B) \qquad \text{(definition additive)}. \qquad (6.12)$$

For systems in which the combined number of states equals the product of the individual number of states, $W_{AB} = W_A W_B$, the notions of additivity and extensivity are identical. Most systems that have been studied in classical statistical mechanics are of this kind. The phasespace of these systems grows exponentially in size (e.g. the number of particles N).

We distinguish additivity and extensivity. Additivity is defined as a property of a function (usually the log function), while extensivity is a statement about how systems behave (how their phasespace volumes change) when they are combined. For many complex systems we have that $W_{AB} < W_A W_B$, which means that phasespace grows subexponentially. In these cases it becomes essential to distinguish between additivity and extensivity.

6.2.1.3 Entropy and thermodynamic relations

We assumed that the condition of statistical equilibrium is given by the most probable condition of a closed system, which is the equivalent of saying that the entropy, σ, from Equation (6.6) is a maximum when the closed system is in equilibrium. The value of σ will depend on the internal energy, U, of the system, of the number of particles, N_i, of a given species, i, and on external variables, x_v, such as volume, magnetization, and so on. We write this as,

$$\sigma = \sigma(U, x_v, N_i). \qquad (6.13)$$

We can now discuss the condition of thermal equilibrium in a closed system where energy is conserved. Consider two subsystems, A and B, which can exchange energy but not particles or their containing volumes. Using the additivity property of entropy $\sigma = \sigma_A + \sigma_B$, in equilibrium we have,

$$d\sigma = d\sigma_A + d\sigma_B = 0 \qquad \text{or} \qquad d\sigma = \left(\frac{\partial \sigma_A}{\partial U_A}\right) dU_A + \left(\frac{\partial \sigma_B}{\partial U_B}\right) dU_B = 0. \qquad (6.14)$$

As total energy is conserved, $dU = dU_A + dU_B = 0$, the right part of Equation (6.14) becomes,

$$d\sigma = \left[\left(\frac{\partial \sigma_A}{\partial U_A}\right) - \left(\frac{\partial \sigma_B}{\partial U_B}\right)\right] dU_A = 0 \qquad \text{or} \qquad \frac{\partial \sigma_A}{\partial U_A} = \frac{\partial \sigma_B}{\partial U_B}, \qquad (6.15)$$

which allows us to define the *temperature* of system A,

$$\frac{\partial \sigma_A}{\partial U_A} = \frac{1}{\tau_A}. \tag{6.16}$$

In equilibrium, the temperatures of A and B are the same, $\tau_A = \tau_B$, which defines temperature as an equilibrium property that is related to entropy and energy. To relate this temperature to *absolute temperature* T in units of Kelvin, we have to assign physical units to it,

$$T = k_B \tau = \frac{1}{\beta}, \tag{6.17}$$

where k_B is called the Boltzmann constant, $1.38064852(79)\,10^{-23}$ Joule/Kelvin. β is called the *inverse temperature*.

Assume now that a system is in equilibrium and its entropy is $\sigma = \sigma(U, x_\nu, N_i)$. If we make reversible changes, such that the changed system is still in equilibrium, we have,

$$
\begin{aligned}
d\sigma &= \left(\frac{\partial \sigma}{\partial U}\right) dU + \sum_\nu \left(\frac{\partial \sigma}{\partial x_\nu}\right) dx_\nu + \sum_i \left(\frac{\partial \sigma}{\partial N_i}\right) dN_i \\
&= \frac{dU}{\tau} + \frac{1}{\tau} \sum_\nu X_\nu dx_\nu - \frac{1}{\tau} \sum_i \mu_i dN_i,
\end{aligned} \tag{6.18}
$$

where $X_\nu = \frac{\partial U}{\partial x_\nu}$ are called generalized forces and $\mu_i = -\frac{\partial U}{\partial N_i}$ are chemical potentials. We rewrite this result as,

$$dU = \tau d\sigma - \sum_\nu X_\nu dx_\nu + \sum_i \mu_i dN_i. \tag{6.19}$$

Let us consider a simple case where the number of particles are fixed, $dN_i = 0$, and the only external parameter is volume, $X_\mu = \Pi$. This leads us to,

$$dU = \tau d\sigma - \Pi dV. \tag{6.20}$$

The change in internal energy consists of two parts. The first, $\tau d\sigma$, reflects the energy change when the external parameters are kept constant. We identify $\delta Q = \tau d\sigma$ as the quantity of heat added to the system in a reversible process.[6] The second part is the energy change due to an external volume change. This is called mechanical work, δW. The work done must be $-pV$ and we identify the generalized force Π with pressure p. With these identifications we recover the first law of thermodynamics,

---

[6] δ is used here instead of d, as δQ it is not an exact differential, that is, Q is not a state function.

$$dU = \delta Q + \delta W, \qquad (6.21)$$

which is the statement of energy conservation. If we use practical thermodynamic units to make σ a thermodynamical quantity (like absolute temperature), we learn from $\delta Q = \tau\, d\sigma$, that a physical constant needs to be associated with it. It is clear that this constant must be the Boltzmann constant and,

$$S_{TD} = k_B \sigma. \qquad (6.22)$$

This is the relation between thermodynamic entropy, S_{TD}, and the entropy of statistical mechanics, σ.

Finally, note that the thermodynamic equation, Equation (6.20), dictates that σ is an *extensive* quantity. If it were not, and as the quantities U and V are extensive variables, imagine what would happen if the system were enlarged. In this scenario U and V would grow with system size, while σ would not. As the system size becomes large, the heat term in Equation (6.20) would effectively vanish. Large systems would not then contain a heat term, which is nonsense in physics. Therefore, σ must be extensive.

6.2.1.4 The Boltzmann distribution

In classical physics we encounter two fundamental notions of entropy. The entropy of Clausius, $dS_{TD} = dQ/T$, and the statistical mechanics entropy, $\sigma = \log \Delta\Gamma$. Boltzmann's triumph was to show that these are essentially identical, and by doing so, he linked macroscopic phenomena with details of microscopic configurations. A special case of the statistical mechanics entropy is Boltzmann entropy, $S_B = \log W$, where W is the number of discrete states accessible to the system. W is closely related to phases-pace volume. If the states do not appear with equal probabilities p_i, the Boltzmann–Gibbs–Shannon formula is appropriate, $S = -\sum_{i=1}^{W} p_i \log p_i$. This is sometimes also referred to as the Boltzmann single-particle entropy, or *reduced Boltzmann entropy*. If it is assumed that every state i is associated with an energy E_i, in physics, the distribution of p_i often follows an exponential called the *Boltzmann distribution*. If the average energy in the system is known to be U, the internal energy, one can obtain the Boltzmann distribution by maximizing the Boltzmann–Gibbs–Shannon entropy functional under this energy constraint. The constraint is guaranteed by using the Lagrangian multipliers α and β,

$$\Phi = -\sum_{i=1}^{W} p_i \log p_i - \alpha \sum_{i=1}^{W} (p_i - 1) - \beta \sum_{i=1}^{W} (p_i E_i - U). \qquad (6.23)$$

The first constraint ensures that the distribution is normalized; the second enforces that the average energy is U. To maximize Φ take derivatives with respect to p_i (for all $i = 1, \cdots, W$), and obtain $0 = -\log p_i - 1 - \alpha - \beta E_i$, which is simply the Boltzmann distribution,

$$p_i = e^{-1-\alpha} e^{-\beta E_i}. \qquad (6.24)$$

Here, α ensures the normalization of the distribution, $e^{1+\alpha} = \sum_i e^{-\beta E_i} = Z$; β is the inverse temperature, see Equation (6.17), and is given in physical units $[\beta] = \mathcal{J}^{-1}$; and Z is called the partition function. It is easy to verify (see Exercise 6.10) that the *Helmholtz free energy* $F = U - TS$ is given by,

$$F = -\frac{1}{\beta} \log Z. \tag{6.25}$$

To obtain the distribution function for a stochastic system we can often perform an effective two-step procedure:

- First, compute the derivative of the entropy, which is of trace form $S = \sum_i g(p_i)$, where $g(p_i) = -p_i \log p_i$; see Equation (6.1). In other words, we compute the derivatives $g'(p_i)$. We find that (up to a multiplicative factor) the derivative is simply the negative logarithm.

- Second, note that the derivative $g'(p_i)$ is a function of p_i. To find the distribution function p_i we have to invert function $g'(p_i)$, where we must take the constraint terms appropriately into account.

We will see that for complex systems we can follow exactly the same procedure: the derivative of the $g(p_i)$ is called a (negative) *generalized logarithm*. Its inverse function is the distribution function, if the constraint terms are properly taken into account.

6.2.1.5 Other entropies in physics

There are a vast number of entropy functionals that have been introduced in various contexts over the past five decades. Some of them were introduced in a classical context of complex systems. We will revisit some of them in Section 6.3.4. We finish this section by mentioning two entropies that play an important role in quantum mechanics.

von Neumann entropy It was introduced in quantum mechanics by von Neumann. In this straightforward generalization, the role of probabilities, p_i, in the Boltzmann–Gibbs–Shannon formula is played by the *density matrix*, which represents a quantum system in a mixed state. Mixed states are required to describe statistical ensembles of quantum systems or quantum systems in thermal equilibrium. States of quantum ensembles cannot be described by pure state vectors. For this the density operator is necessary. The density operator for a statistical mixture of pure states ψ_i is defined as $\rho = \sum_j p_j |\psi_j\rangle\langle\psi_j|$, and is a quantum version of a probability density function. The quantum version of the statistical physics entropy is then given by,

$$S_{vN} = -\mathrm{tr}(\rho \log \rho), \tag{6.26}$$

and is called the von Neumann entropy. Here, tr is the trace (sum over diagonal elements of the matrix), and ρ is the density operator, which (with respect to an orthogonal basis

ϕ_i) has the matrix elements $\rho_{ij} = \langle \phi_i | \rho | \phi_j \rangle$. Von Neumann entropy can be expressed in terms of the eigenvalues of the density matrix. As ρ is always a positive semi-definite operator, it has a spectral decomposition $\rho = \sum_i \lambda_i |\varphi_i\rangle \langle \varphi_i|$, where $|\varphi_i\rangle$ are orthonormal vectors, and the positive eigenvectors add up to one, $\sum \lambda_i = 1$. On this basis, the von Neumann entropy of a quantum system is given by,

$$S = -\text{tr}(\rho \log \rho) = -\sum_i \lambda_i \log \lambda_i. \qquad (6.27)$$

Rényi entropy   An entropy measure that appears in quantum information, computer science, and statistics is the Rényi entropy, which is sometimes also referred to as α-entropy,

$$S_\alpha = \frac{1}{1-\alpha} \log \left(\sum_{i=1}^{n} p_i^\alpha \right). \qquad (6.28)$$

All entropy functionals that we have encountered to date were of trace form; see Equation (6.1). Rényi entropy is not a trace form entropy. It is of the more complicated form, $F(\sum_i g(p_i))$, containing two functions F (here, the logarithm) and g (taking powers of α).

6.2.2   Entropy and information

The role of entropy in the context of information is related to the amount of uncertainty in the outcomes of stochastic processes. To make this connection clear, and to see how the notion of entropy is tied to the concept of uncertainty, we start with a simple example.

6.2.2.1   *A simple example*

Consider a system with W states A_i, with $i \in \{1, 2, \cdots, W\}$. Each of the states is associated with a respective probability, p_i. We represent the system in this scheme,

$$\begin{pmatrix} A_1 & A_2 & \cdots & A_W \\ p_1 & p_2 & \cdots & p_W \end{pmatrix}.$$

For example, a fair dice would be represented by the scheme,

$$\begin{pmatrix} A_1 & A_2 & A_3 & A_4 & A_5 & A_6 \\ \frac{1}{6} & \frac{1}{6} & \frac{1}{6} & \frac{1}{6} & \frac{1}{6} & \frac{1}{6} \end{pmatrix}.$$

If we consider two different schemes we can ask which of the two involves more uncertainty:

$$\begin{pmatrix} A_1 & A_2 \\ 0.5 & 0.5 \end{pmatrix} \qquad \text{or} \qquad \begin{pmatrix} A_1 & A_2 \\ 0.99 & 0.01 \end{pmatrix}.$$

Intuitively, the scheme on the left is more uncertain. For that scheme there is a 50:50 chance of A_1; in the scheme on the right it is almost certain that A_1 will be realized if we perform an 'experiment'.

Assume that we perform a sequence of observations on systems that follow these schemes without knowing their exact nature. An experiment generates a sequence of outcomes. With every new observation we learn something about the system and the scheme it follows. We obtain information about it. The information gained about a system is the same as the amount of uncertainty removed.[7]

> In loose terms, the relation between uncertainty and information reads: information obtained = uncertainty removed. Information-theoretic entropy is a measure that quantifies the average uncertainty in a system. It is the average 'surprise' that we get from observing or measuring a system.

Let us now use the information measure from Equation (6.4), to compute how much uncertainty is expected in a system by making successive observations or experiments. The formula $S_S(p_1,p_2,\ldots,p_W) = -\sum_{i=1}^{W} p_i \log p_i = \sum_{i=1}^{W} p_i \log 1/p_i$ can be interpreted as taking the average of the expression $\log 1/p_i$, which can be seen as the 'surprise factor' when observing the particular state i. If the probability p_i of observing state i is very low, and if we do observe it, the chances of surprise, $\log 1/p_i$, are high. Infrequently occurring outcomes contain more information than more frequently occurring ones. The entropy is the average of the surprise factor $I(p_i) = -\log p_i$, which is sometimes called the *information content* or *self-information* of an outcome i.

If we return to the two schemes here, using $S_S(p_1,p_2,\ldots,p_W) = -\sum_{i=1}^{W} p_i \log p_i$ for the scheme on the left, we get $S_S(p_1,p_2) = 1$, and for the scheme on the right we have $S_S(p_1,p_2) \sim 0.08$ (for base 2). For the left scheme we have a high average surprise factor, as both states are equiprobable or uniformly distributed. For the right scheme we expect state 1 to appear much more often and, whenever we observe a realization of state 1, we are not surprised. Whenever we observe state 2 we are surprised, but this does not happen often, so that on average the surprise is less likely for the right scheme and its entropy is lower.

6.2.2.2 Three reasons why Shannon entropy is a good measure of uncertainty

First, if we have complete certainty about the states in a system, the average surprise factor should be zero, $S_S = 0$. This is indeed the case if one single state i is realized with certainty, $p_i = 1$, and all the others never occur, $p_j = 0$. Shannon entropy reduces

---

[7] Note that if observations require effort or come at a cost, the implication is that work needs to be invested to reduce uncertainty. This is exactly what happens in driven complex systems that are powered by an energy potential. Energy flows can be (randomly) invested in reducing uncertainty. The currents drive the system away from its equilibrium states, which are the most common states. Structures can form.

to $S_S = -1 \log 1 = 0$, because of the mathematical facts that the limit $\lim_{x \to 0} x \log x = 0$ and $1 \log 1 = 0$.

Second, most uncertainty is found in systems where all states appear with the same probability. Given that there are W different states, this means $p_i = 1/W$. Note that when we talk about uncertainty, information gain, or entropy for the outcome of a given scheme or process, we talk about it *before* the experiment is done. After the experiment, once one particular state has been realized, all the others were not, and the entropy after the measurement is 0.

Third, we can use the measure S_S to understand what happens if we have more than one source of uncertainty. The simplest case is that of two independent systems, where the total entropy, the uncertainty or average surprise, is equal to the sum of the entropies of the independent systems. To see this, consider the two schemes,

$$\begin{pmatrix} A_1 & A_2 & \cdots & A_W \\ p_1 & p_2 & \cdots & p_W \end{pmatrix} \quad \text{and} \quad \begin{pmatrix} B_1 & B_2 & \cdots & B_N \\ q_1 & q_2 & \cdots & q_N \end{pmatrix}.$$

If the schemes are independent, the joint probability that state k is realized in scheme A and that state l is realized in scheme B is simply the product of the individual probabilities within their schemes, $p(k,l) = p_k q_l$. If we denote the Shannon entropy of the combined schemes A and B by $S_S(AB)$, we immediately see that,

$$S_S(AB) = -\sum_{k=1}^{W} \sum_{l=1}^{N} p(k,l) \log p(k,l) = -\sum_k \sum_l p_k q_l (\log p_k + \log q_l)$$
$$= S_S(A) + S_S(B). \tag{6.29}$$

Here, we used the fact that $\sum_{i=1}^{W} p_i = \sum_{j=1}^{N} q_j = 1$. We can also understand what happens to Shannon entropy if we look at coupled or correlated systems, where the output of one system is correlated with the outcome of another. Consider the two systems A and B and assume that whatever the output of system A is, it will determine the probability distribution for observing system B. In this case, we can write the joint probability of observing k from scheme A and l from B by, $p(k,l) = p_k p(l|k)$. Here, $p(l|k)$ is the *conditional probability* of observing state l, given that state k was observed in scheme A. We find,

$$S_S(AB) = -\sum_k \sum_l p_k p(l|k)(\log p_k + \log p(l|k))$$
$$= -\sum_k p_k \log p_k \sum_l p(l|k) - \sum_k p_k \sum_l p(l|k) \log p(l|k)$$
$$= S_S(A) + \sum_k p_k S_S^k(B) = S_S(A) + S_S(B|A). \tag{6.30}$$

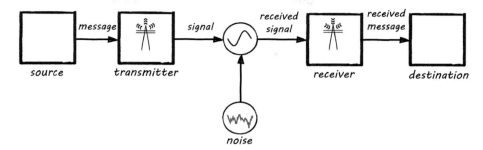

Figure 6.1 *Schematic diagram of a general communication system. After [336].*

The expression $S_S^k(B) = -\sum_l p(l|k) \log p(l|k)$ is called the *conditional entropy*. We denote the conditional entropy of B with respect to A by $S_S(B|A)$.

> The Shannon entropy of a combined system of correlated stochastic events is the entropy of one system plus the conditional entropy of the other. Note that the conditional Shannon entropy is always smaller than or equal to the unconditional, $S_S^k(B) \leq S_S(B)$. Intuitively, this means that the knowledge gained by knowing that A was in state k can only make our uncertainty about the outcome of B smaller— but never larger. In the worst case, knowledge about scheme A does not reduce our uncertainty about B. In this case, A and B are statistically independent. If conditional and unconditional entropies are the same, this indicates statistical independence.

These three arguments provide an intuitive explanation as to why Shannon entropy is a sensible measure of information. Note that S_S depends only on the probabilities of the states and nothing else. This dependence is also continuous, which is a property that a measure of information must have. In Section 6.2.2.10 we will turn the argument around: we start by specifying what properties a sensible measure for information must have. We formulate these properties as four axioms and then prove that the unique functional form of Shannon entropy can only be the Boltzmann–Gibbs–Shannon formula, $S_S = -\sum_i p_i \log p_i$.

6.2.2.3 Sources and information production

Encoding and sending messages. One intention of Shannon's 1948 paper [336] was to understand how to code messages so that they could be sent through noisy channels and then decode them in an error-free way. This is not a trivial problem. Transmitting a message through a noisy channel corrupts it and the message received will contain errors. How can the receiver ever be sure that the decoded message is finally 100% error-free? The schematic processes of coding, transmission, and decoding of a message is depicted in Shannon's original figure in Figure 6.1.

A source produces a signal. Before the signal is sent, it must be encoded in a form that can be sent through a channel, typically a physical device such as an electric cable or a mobile phone tower. For example, imagine that a source produces language in a sequence of words that are themselves composed of letters. If a given channel (cable, electromagnetic waves, mobile phone) can transmit bits, we will encode the language message in a binary form, that is, we will rewrite it in a sequence of zeros and ones. The encoded message is now passed through the channel—it is transmitted. Imagine that during the transmission it is damaged at a given position with a certain probability, p_{error}. It specifies the probability that any bit in the coded sequence has been (perhaps physically) flipped, lost, or erroneously added to the sequence. The receiver picks up a possibly corrupted signal and decodes it. In our example, the bit string is translated back to a sequence of letters, so that the message of the source can be understood in human language.

Decoding and error correcting During decoding, something must happen to correct the errors that were accumulated during the transmission process. Can the receiver know for sure that the message finally obtained is correct (error-free)? To give the receiver a chance to know 'how correct' the received message is, the simplest way of coding a message is to add redundancy. A simple way would be to send the message twice.[8] A receiver observing two identical messages can be pretty sure that the message was received correctly because the probability of the same error occurring twice is now very small, p_{error}^2. However, 'pretty sure' does not mean certainty.

Table 6.1 *How does error correction work in a coding-decoding scheme? Two coding schemes are shown, one without redundancy (top), and one where the strategy consists of sending the same message three times (bottom). In both cases, during transmission a bit is flipped in the noisy channel, causing an error. In the redundant case an error that occurred during transmission can be corrected during decoding if, for example, a majority rule is installed. At the place where the error occurs, two versions are correct and one contains the flipped bit. The two in the majority correct the minority. The decoded message is then corrected.*

Message	encoded	$\rightarrow$	received	decoded
no redundancy				
'Hi there'	100100111	$\rightarrow$	100100101	'Hi thor'
with redundancy				
	100100111	$\rightarrow$	100100111	
'Hi there'	100100111	$\rightarrow$	100100101	'Hi there'
	100100111	$\rightarrow$	100100111	

---

[8] Shannon thinks of a controller that can compare the sent and received messages and then tells the receiver where the errors occurred so that they can be corrected. The idea is that the amount of information for the error correction needing to be sent by the controller to the receiver can, in principle, be included in the message itself. Shannon provides us with a measure of the minimal redundancy we need to correct *all* the errors in a message [336]. This minimum may only be asymptotically achievable for infinitely long messages and may not be found in practice for actual coding strategies.

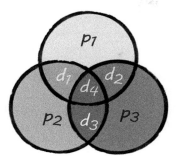

Figure 6.2 *Error-correcting coding with a so-called Hamming (7,4) code.*

In Table 6.1 we show two different coding schemes, one with and one without redundancy. The first coding scheme translates the message into bits and sends it through the channel. The second scheme translates the message into bits in the same way as before but sends it three times. Let us assume that during transmission one error occurs. In the first case, the decoded message contains the error. In the second encoding strategy, the error occurs in one of the three copies. Assuming that in the majority of cases (two) the signal is correct, we can correct the error. However, we are still not in a position to be certain that the signal received is 100% error-free. If we want more certainty that the received signal is correct, we have to add more redundancy, and thus reduce the *information transmission rate* through the channel.

Adding redundancy comes at a cost. To send more bits (due to increased redundancy) through the channel means that the transmission rate in *information per second* is reduced. If we send the message twice, the information transmission rate reduces by a factor of two. Information channels have a finite physical channel capacity—or a transmission rate—that is given in *bits per second* (not information per second). This rate is key to understanding how much information can be sent with practically no errors and the conditions under which errors become unavoidable.

There are better error-correcting coding strategies than sending messages multiple times. Many error-correction codes work along the lines of the so-called Hamming (7,4) code. Here coding works such that the binary bit string is first divided into substrings of length four. These four data bits we label $d1$, $d2$, $d3$, $d4$. Then three additional 'parity' bits, $p1$, $p2$, $p3$ are added and eventually seven bits are sent through the channel. The added redundancy is 7/4; hence the name given to the coding scheme. The three parity bits are added in the following way. The information carrying bits $d1$–$d4$ are written into the intersections of the three circles shown in Figure 6.2. The parity bits are then computed and written into the intersections of the circles. The parity bits are simply the binary sums of the data bits, $p1=d1+d2+d4$ (rem 2), $p2=d1+d3+d4$ (rem 2), and $p3=d2+d3+d4$ (rem 2). These seven bits $d1$ $d2$ $d3$ $d4$ $p1$ $p2$ $p3$ are then actually transmitted. The receiver writes the received bits in the same arrangement and checks if the parity bits are consistent with the data bits. If they are, no error has occurred. If there is a discrepancy, an error in a single data bit will always show up as a difference between two parity bits. In this way, a single error can immediately be detected and corrected. Note that if more than one error occurs, then the Hamming (7,4) code cannot correct them. The code is a so-called *single-bit-error-correcting code*.

6.2.2.4   Shannon's noisy channel coding theorem

It is intuitive to assume that for any given error rate p_{error}, no matter how cleverly you encode your signal, the receiver can never be 100% sure that the received and decoded message is indeed error-free. A tiny probability would still exist that the coding–decoding strategy could not correct *all* errors. The amazing result of Shannon's paper is that he proved that this intuition is wrong.

> Shannon proved that there are coding strategies that allow you to transmit a message practically error-free. The essence of the theorem is that this can only happen if the capacity of the channel is larger than the information production rate—the Shannon entropy—of the source.

In other words, for a noisy channel there is a threshold, called the channel capacity,[9] C, such that a message with entropy less than C can be coded in a way that it can be sent and decoded with an arbitrarily small error rate. If the entropy is larger than C, no such coding strategies exist and a non-vanishing error rate cannot be avoided.

Shannon demonstrates this fact with an existence theorem, but does not tell us how optimal error-correcting codes can be found. However, for any channel capacity $C > 0$, optimal encoding strategies (or codes) do exist. On average, they correct as many errors as are produced in the channel, and transmission in noisy channels can indeed become (asymptotically) error-free, if the information production rate of the source is small enough. To our knowledge, no such codes have ever been found in practice. Finding good (not optimal) encoding strategies has dominated information theory during the past decades. In Figure 6.3 we show the theoretical result of Shannon, namely, that there are strategies that allow for error-free transmission if the channel capacity C is larger than the information production rate of the source. We also see that for a given information production rate of the source $S_S > C$ there is a minimal error rate; one cannot find codes with a lower error rate.

6.2.2.5   Information production of a source

The *information production rate*, S_S of a source[10] is a characteristic property of the source. The simplest example is a random source, a 'machine' that produces symbols from a finite alphabet. For example, it produces the letters A, B, C, D, E with the respective chances or probabilities $q_A = 0.4$, $q_B = 0.1$, $q_C = 0.2$, $q_D = 0.2$, and $q_E = 0.1$. Machines that produce a finite number of states are called *finite-state machines*. The machine is depicted in Figure 6.4a. At every time step, it starts at its centre, takes one loop and,

---

[9] C is related but not identical to the transmission rate of the channel.

[10]   The information production rate is sometimes called *entropy production*.

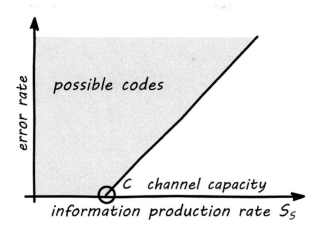

Figure 6.3 *Region where encoding strategies exist (shaded area). Every coding strategy represents one point in the plane. The information rate of the source is S_S. The more redundancy added in a strategy, the lower the error rate and the smaller the amount of information passed through the channel (in information per second). The amazing fact is that it is possible (in theory) to find coding strategies that transmit (infinitely long) messages at an error rate of exactly zero. For this to be possible, the information production rate of the source has to be smaller than the capacity of the channel (in bits per second), C.*

depending on which one it takes, produces one of the five letters and returns to its starting position at the center. If we 'run' this source (or scheme) and make an experiment, we obtain a sequence that might look something like this

$$x(N) = \text{AAACDCBDCEAADADACEDAEADCABEDADDCECAAAADA}$$

We denote the sequence of length N by $X(N) = (X_1, X_2, \cdots, X_{N-1}, X_N)$. Every element in the sequence x_i is from a symbol alphabet with $W = 5$ different symbols (letters), $x_i \in \{A, B, C, D, E\}$. From this experiment we can produce a histogram of how often the different letters have actually occurred. We find that $k_A = 16$, $k_B = 2$, $k_C = 7$, $k_D = 10$, and $k_E = 5$. As the machine has produced $N = 40$ outputs, the observed relative frequencies,[11] $p_i = k_i/N$, are $p_A = 0.4$, $p_B = 0.05$, $p_C = 0.175$, $p_D = 0.25$, and $p_E = 0.125$. If these relative frequencies approach the hardwired probabilities q_i, as the number of outputs becomes large, $N \to \infty$, the source is called *ergodic*. This machine is ergodic. The average rate of information produced by this source is, following Equation (6.4),

$$S_S(q_A, q_B, q_C, q_D, q_E) = -q_A \log q_A - q_B \log q_B - q_C \log q_C - q_D \log q_D - q_E \log q_E.$$
$$(6.31)$$

---

[11] Note that we now use q for probabilities and p for relative frequencies.

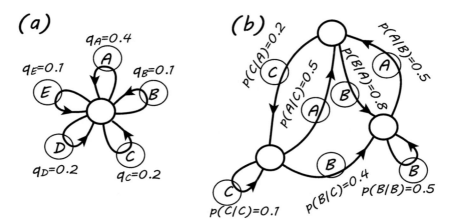

Figure 6.4 *(a) The finite-state machine produces a random sequence of letters from an alphabet of five letters. Each letter i appears with a fixed 'hardwired' probability q_i. (b) Example of a finite-state machine that produces letters $i = A, B,$ and C conditional on the letter that the machine produced in the previous time step. The transition probabilities from letter j to letter i are given by $p(i|j)$ and are again hardwired and do not change over time. Finite-state machines are Markov processes. After Shannon [336].*

A slightly more complicated example is a source that produces correlated random sequences of symbols. An example of such a finite-state machine is shown in Figure 6.4b. It produces three letters A, B, C with the respective probabilities $q_A = \frac{9}{27}, q_B = \frac{16}{27}$, $q_C = \frac{2}{27}$. However, successive letters are not drawn independently (see Exercise 6.11). What letter can be produced at the next time step depends on what letter has been produced at the previous time step. The conditional probabilities of the machine producing letter i, given that the preceding letter was letter j, are given by,

$$
\begin{array}{c|ccc}
p(i|j) & \multicolumn{3}{c}{i} \\
 & A & B & C \\
\hline
A & 0 & \frac{4}{5} & \frac{1}{5} \\
j \quad B & \frac{1}{2} & \frac{1}{2} & 0 \\
C & \frac{1}{2} & \frac{2}{5} & \frac{1}{10}
\end{array}
\tag{6.32}
$$

In this example, the probabilities of getting letter i are obtained by solving the set of equations $q_i = \sum p(i|j)q_j$. We call these probabilities the 'hardwired' or objective probabilities. As before, we can run an experiment with the machine to generate an actual sequence of letters. The machine might produce

$$x(N) = \text{ABBABABABABABABBBABBBBBBABABABABABBBACACABBABBB}$$

As before, we can now observe the relative frequencies p_A, p_B, and p_C. If the sequence is very long, $N \to \infty$, and if relative frequencies approach the objective hardwired frequencies of the machine q_A, q_B, and q_C, the source is ergodic. This machine is indeed ergodic, and its maximal information production rate is Shannon entropy,

$$S_S = -\sum_{i=1}^{3}\sum_{j=1}^{3} q_i p(j|i)\log p(j|i).$$ (6.33)

6.2.2.6 *Markov processes and ergodicity*

Before we continue the discussion about sources, we have to remember what a *Markov process* is, see Section 2.5.1.5, and we need to define *ergodic processes*.

A *Markov process* is a stochastic process in which the probability of reaching a possible state of the system at time $t+1$ depends only on the previous state where the system was found at time t. A Markov process has no memory of states visited before time t.

Finite-state machines that produce Markov processes can be seen as directed multi-graphs; see Section 4.2.3.4, where nodes represent the states α of the machine and the directed links $\alpha \to \beta$ represent 'letters' i of the alphabet that the machine can produce. Two different links between two nodes of a multigraph correspond to two distinct letters.

To define formally what we mean by an ergodic process [229], imagine a long sequence of N entries that was produced by a Markov process (a Markov sequence). The Markov process is fully specified by its objective probabilities q_i and conditional probabilities $p(i|j)$, just as we encountered them in the examples of sources. The relative frequency of visiting one of the possible states i in the system is $p_i = k_i/N$, where k_i is the number of visits to state i in the sequence.

A process $t \to X_t$ is called *ergodic*, if for all states $i \in \{1, 2, \cdots, W\}$ the expression holds

$$\lim_{N\to\infty} P\{|p_i - q_i| > \delta\} < \epsilon,$$ (6.34)

for any arbitrarily small values of $\epsilon > 0$ and $\delta > 0$. The notion of ergodicity is closely related to the law of large numbers and probabilistic convergence; see section 2.3.

This definition is read in the following way. For long sequences, the probability P that the difference between relative probabilities p_i and objective probabilities q_i is larger than δ, vanishes. No matter how small we make δ and ϵ, we can always find a value N' such that for all sequences of lengths $N > N'$, expression Equation (6.34) holds. In other words, as the Markov sequence becomes large, all relative frequencies p_i approach the objective frequencies q_i.

In practice, this also means that Markov processes with transition probabilities $p(i|j)$ that have a stationary state distribution q_i, which (if it exists) is a solution of the equation $q_i = \sum_{j=1}^{W} p(i|j)q_i$, are ergodic and that all relative frequencies $p_i = k_i/N$ (k_i being the number of times i was sampled in N trials) converge to the stationary solution, the marginal probabilities q_i. In other words, a Markov process is ergodic exactly if it has a (non-vanishing) stationary distribution function. Non-stationary processes are never ergodic. Processes that become trapped in subregions of their sample space are typically not ergodic.[12]

6.2.2.7   *Source properties*

As we will see in Section 6.2.2.10, Shannon entropy is a valid measure for entropy production only if the source fulfils the following properties.

> For Shannon entropy to be a sensible measure for describing the entropy production of a classic source, the source must be
>
> (i) Markov and
> (ii) ergodic.
>
> Ergodicity implies that the Markov process has non-vanishing, time-independent marginal probabilities, q. It is stationary in the sense that the objective probabilities are constant and must not change over time, $q \neq q(t)$.

We refer to a source with these properties as a *simple stochastic system*. More complicated sources, which is what many stochastic complex systems, in effect, are, can be non-ergodic, have memory, and have time-dependent marginal and conditional probabilities. They are path-dependent. Figure 6.5a shows an example of a finite-state machine that violates the stationarity requirement. The objective probabilities change over time. In Figure 6.5b a non-ergodic situation is depicted. No matter where we start the Markov process, after some time we will end up in the loop in the lower-right section of the machine. Once there, only the letters B will be produced. The process got stuck in a small region of its sample space. If we run an experiment, after a while the machine will only produce letters B.

$$x(N) = \text{CCCCCACCCBBBBBBBBBBBBBBBBBBBBBBBBBBB}\ldots$$

As the sequence gets longer the relative frequency in terms of observing letter A approaches $\lim_{N\to\infty} p_A = 0$, which is incompatible with the objective probabilities of producing the letter A, all of which are larger than zero. The process is therefore not ergodic. In other words, as not every state (node) of the machine is connected to any

---

[12] Such processes may still be considered ergodic if one restricts them to the subset of elements i of sample space where the relative frequencies are $p_i > 0$.

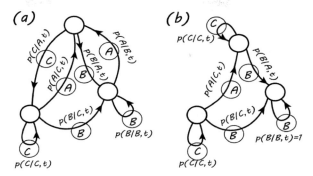

Figure 6.5 *(a) Example of a non-stationary finite-state machine that has time-dependent transition probabilities $p(i|j,t)$. (b) Example of a non-ergodic finite-state machine. After a few steps the machine is caught in a loop and always produces letters B with certainty. Letters A and C are no longer produced and their relative frequencies (which approach zero) do not match the objective (non-zero) probabilities. This makes the process non-ergodic. Many non-ergodic processes can be thought of as processes that get trapped in smaller and smaller regions of their phasespace.*

other node by a path that can be realized with non-zero probability, the process is not ergodic.

Example—What is the average information production of a Markov process? We are now in the position to compute the information production of a Markov process. Imagine a Markov process with W states $i \in \{1, 2, \cdots, W\}$. Suppose the transition probabilities between states are given by $p(i|k)$, and the objective probabilities are q_i. Now suppose that the system is in the particular initial state i. Using the definition of Shannon entropy from Equation (6.4), we obtain the information when the Markov process is moving from state i to any of the other states as $S_S^i = -\sum_{k=1}^{W} p(k|i) \log p(k|i)$, which we defined as the conditional entropy in Equation (6.30). The average of the conditional entropy across *all* initial states i is then simply given by,

$$S_S = \sum_i q_i S_S^i = -\sum_i \sum_k q_i p(k|i) \log p(k|i). \tag{6.35}$$

This is the average information production of a Markov process as it moves from one time step to the next.

6.2.2.8   *A note on information production and coding*

The key operational aspect of information theory is the coding and compression of messages. Codes are functions that convert strings composed from some alphabet (message alphabet) to strings from another alphabet (code alphabet), such as 0 and 1. The challenge is to represent message strings with the shortest possible codes. There is a wide variety of codes that differ in logic and performance. Codes should satisfy the property of being uniquely decodable, meaning that the code can be decoded such that no information is lost in the coding–decoding process. Codes that allow for a lossless compression are sometimes called prefix codes or Huffman codes. The better a lossless code, the shorter its code length will be. Shannon's (noiseless) source coding theorem establishes the limits for data compression and illustrates the practical meaning of Shannon entropy as the limit of compressibility of messages.

> Shannon's (noiseless) source coding theorem. Given N (large) i.i.d. random schemes, A, each with an entropy, $S(A)$, it is possible to compress them into $NS(A)$ bits with practically zero risk of information loss. If a lower number of bits is used for encoding, errors will almost certainly occur.

Shannon entropy is the fundamental limit to the compression of data from a given source. For lossless compression it is mathematically impossible to do better than Shannon entropy. For literature on the noisy coding theorem, see [257].

6.2.2.9   *Entropy rate of emitted sequences*

The concept of *entropy rate* (or information rate) is of central importance for the coding of symbol sequences $x(N)$. The entropy rate $S_{ER}(x(N))$ is the average length of the coded signal (usually in bits), given that the original sequence of symbols $x(N)$ is of length N, and also given that the code is uniquely decodable with no information being lost. Recall that all elements x_i in the sequence $x(N) = (x_1, x_2, \cdots, x_N)$ are elements from the message alphabet that consists of W different symbols. The *entropy rate* of the entire sequence (or sample) $x(N)$ is defined as the logarithm of the joint probability of sampling the specific sequence $x(N)$,

$$S_{ER}(x(N)) = -\frac{1}{N} \log p(x_1, x_2, \cdots, x_{N-1}, x_N). \tag{6.36}$$

For Markov processes we can verify that the entropy rate is merely the conditional entropy that we encountered previously in Equations (6.30) and (6.35). To see it explicitly, assume that the process is Markovian, so that we can write for the entropy rate,

$$S_{ER}(x(N)) = -\frac{1}{N} \log p(x_1, x_2, \cdots, x_{N-1}, x_N) = -\frac{1}{N} \log \prod_{n=1}^{N} p(x_n | x_{n-1})$$

$$= -\frac{1}{N} \sum_{n=1}^{N} \log p(x_n | x_{n-1}) = -\sum_{i=1}^{W} \sum_{j=1}^{W} p(i,j) \log p(i|j) \quad \text{(for} \quad N \to \infty\text{)}$$

$$= -\sum_{j=1}^{W} p_j \sum_{i=1}^{W} p(i|j) \log p(i|j), \tag{6.37}$$

where the transition rates from state j to i are given by the conditional probabilities $p(i|j)$, and $p(i,j)$ is the joint probability for i and j. We used the usual formula for the average,

$\frac{1}{N} \sum_{k=1}^{N} x(k) = \sum_{i=1}^{W} p_i x_i$. For independent Bernoulli processes, $p(i|j) = p_i$, and we get the familiar expression $S_{\text{ER}}(x(N)) = S_S = -\sum_i p_i \log p_i$.

To actually see how the entropy rate is related to the problem of coding, let us think of a source that produces messages consisting of sequences of letters from the message alphabet. The sequences $x(N) = (x_1, x_2, \cdots, x_N)$ are of length N. To transmit such a message we have to encode those letters in a form that is transmittable through a physical information channel. Assume that there are r symbols in the code alphabet. In electronic devices a conveniently transmittable format is a binary code alphabet with $r = 2$ letters (on and off). We now ask how many bits (in the code alphabet) per letter (in the message alphabet) are needed on average to transmit the message $x(N)$ through the channel. Short and efficient codes are obtained by assigning frequently occurring letters in the message alphabet to short binary code words, while infrequently observed letters can be represented by longer code words.

According to a mathematical theorem [239, 269], prefix-free codes, which are uniquely decodable codes exist if and only if the following condition holds,

$$\sum_{i=1}^{W} r^{-L_i} \leq 1. \tag{6.38}$$

Here, L_i is the length of the code word (in bits of an r-letter code alphabet) for letter i (of the message alphabet). For a binary code, $r = 2$, this means the following: if we know the probabilities q_i with which letters i are emitted from the source, we can find a prefix code with code words of length L_i, such that $-\log_2 q_i \leq L_i \leq 1 - \log_2 q_i$. This code is optimal. No shorter loss-free codes can be found. $L_i = \log_r 1/q_i$ is called the Shannon information content or *self-information* of letter i. The average length of the encoded messages now becomes,

$$\langle L \rangle = \lim_{N \to \infty} \frac{1}{N} \sum_{n=1}^{N} L_{x_n} \sim \lim_{N \to \infty} -\frac{1}{N} \sum_{n=1}^{N} \log_2 q_i = -\sum_{i=1}^{W} q_i \log_2 q_i = \frac{S_S(q)}{\log 2}, \tag{6.39}$$

or in other words, $S_S(q) \leq \langle L \rangle \log 2 \leq S_S(q) + \log 2$. This means that $S_S(q)$ is the lower bound for the *entropy rate* (in bits per letter) for i.i.d. processes for optimal prefix-free codes. If we transmit messages with fewer bits, they can no longer be uniquely decoded and the receiver loses information about the source. Longer loss-free codes can be found but are suboptimal in terms of transmission rates (in information per second).

For a more detailed mathematical discussion of issues concerning the coding of sequences we refer to the original works of Shannon, McMillan, and Breiman [68, 268, 336]. In particular, the Shannon–McMillan–Breiman theorem is worth studying in more detail.

6.2.2.10 *Shannon's axioms of information theory*

In his 1948 paper, Shannon specifies the requirements that any measure of information should fulfil. He derives these properties from his intuition that the 'information gained' about a source equals the 'uncertainty removed', for example, when performing experiments on random schemes that were discussed in Section 6.2.2. Shannon originally stated three requirements. A few years later A.I. Khinchin remarked:

> However, Shannon's treatment is not always sufficiently complete and mathematically correct so that, besides having to free the theory from practical details, in many instances I have amplified and changed both the statement of definitions and the statement of proofs of theorems [229].

Khinchin then added a fourth axiom. The three fundamental requirements of Shannon, in the completed version of Khinchin, are known as the *Shannon–Khinchin axioms* (SK). These axioms list the requirements needed for an entropy to be a reasonable measure, S, for the information or 'uncertainty' about a simple, finite probabilistic system. Here, by simple we mean stationary, Markovian, and ergodic. In particular, the Shannon–Khinchin axioms read:

- **Shannon–Khinchin's first axiom (SK1)** states that entropy $S(p_1, p_2, \cdots, p_W)$ is a continuous function of all its arguments p_i, and does not depend on anything other than the probabilities of the W states.

- **Shannon–Khinchin's second axiom (SK2)** states that, for a system with W states or outcomes, each of which occurs with probability $p_i \geq 0$, with $\sum_{i=1}^{W} p_i = 1$, the entropy $S(p_1, p_2, \cdots, p_W)$, must take its maximum for the equidistribution, meaning for $p_i = 1/W$, for all i. If all states are equally probable, entropy has its maximum.

- **Shannon–Khinchin's third axiom (SK3)** states that any entropy should remain the same when adding states to the system that never occur or occur with zero probability. Adding a state $W + 1$ where $p_{W+1} = 0$ does not change anything in the numerical value of entropy, $S(p_1, \cdots, p_W) = S(p_1, \cdots, p_W, 0)$. This is the axiom added by Khinchin.

- **Shannon–Khinchin's fourth axiom (SK4)** states the composition of two finite probabilistic ergodic systems (schemes) A and B. If the systems are independent of each other, entropy should be additive, meaning that the entropy of the combined system (AB) should be the sum of the individual systems, $S(AB) = S(A) + S(B)$. If the two systems are dependent on each other, the entropy of the combined system, that is, the information given by the realization of the two finite schemes A and B, $S(AB)$, is equal to the information gained by a realization of system A, $S(A)$, plus the expectation of the information gained by a realization of system B, conditional on the realization of system A, $S(AB) = S(A) + S(B|A)$. $S(B|A)$ is the conditional entropy that we encountered previously; see Equation (6.30). For obvious reasons, SK4 is also called the *composition axiom*.

With these axioms one can state the famous *uniqueness theorem* [229, 336].

Uniqueness theorem. Any functional $S(p)$ depending on the probability distribution $p = (p_1, \ldots, p_W)$ that is associated with a probabilistic system for which the four Shannon–Khinchin axioms hold, can only be of the form, $S(p) = -k \sum_{i=1}^{W} p_i \log p_i$, where $k > 0$ is a multiplicative constant. The proof is relatively simple and is well described in the original paper [336].

The Shannon–Khinchin axioms lead to a trace form entropy, meaning that entropy can be written in the form $S(p) = \sum_i g(p_i)$, where according to the uniqueness theorem, $g(p_i) = -p_i \log p_i$. We will see that for complex systems, where typically SK4 or other axioms are violated, g can be much more general. The first three Shannon–Khinchin axioms SK1-SK3 have the following implications for the functional form of g.

- SK1: the requirement that S depends continuously on p implies that g is a continuous function.
- SK2: the requirement that the entropy is maximal for the equidistribution $p_i = 1/W$ implies that g is a concave function.
- SK3: the requirement that adding a state $W + 1$ to a system that occurs with zero probability, $p_{W+1} = 0$, does not change the entropy, implies $g(0) = 0$.

The essence of the composition axiom SK4 is that it ensures the ergodicity of the system.[13] According to the composition axiom, the entropy of a system composed of subsystems A and B, equals the entropy of subsystem A plus the entropy of B, conditional on A. Note that the conditional entropy appearing in SK4 corresponds exactly to the information production of a Markov process, which we computed in Equation (6.35).

Classical information theory is the theory of ergodic sources. More precisely, it is the theory of finite-state Markov processes with stationary distribution functions. It can be, and has been, extended to non-stationary situations.

In the context of coding, the Shannon–Khinchin axioms identify four requirements for a functional S that measures the average amount of information (in units of bits) required to encode messages from sources generating sequences of elements (say, letters), x_i, with probabilities, p_i.

6.2.3   Entropy and statistical inference

In many situations the problem arises that, from given data, one would like to deduce the properties of the stochastic process that generated them. Solving this *inverse problem*

---

[13] More precisely, SK4 ensures the equivalence of all ergodic systems characterized by the same stationary distribution p. The exact way in which an ergodic system arrives at any particular state i with probability p_i is irrelevant. All possibilities relating to how an ergodic system can realize this state have the same information content.

is the essence of *statistical inference*. Statistical inference can be done in two ways: in a Bayesian, or a maximum-likelihood-based way.[14] Bayesian or ML-based inference methods are often used to test the quality of statistical models. If there is more than one possible model to describe the data, inference methods can be used to compare the quality of each model (hypothesis) with respect to alternative models. In this sense, inference methods can be used for model selection. If, for a given data set, the particular family of distribution functions is known, tailor-made ML estimators can sometimes be designed. In Section 3.4, we discuss ML estimators for the family of power law distribution functions. If additional information about a system, a source, or a data set, is available within a Bayesian or ML framework, this information can be incorporated into the design of the estimators. The *maximum entropy principle* is a particular framework for statistical inference, and such additional information can naturally be incorporated into it. The maximum entropy principle is deeply connected with the notion of entropy.

6.2.3.1 *The maximum entropy principle*

The maximum entropy principle is tightly related to the question of finding the *most likely* distribution function generated by a stochastic system. The technical idea behind the maximum entropy principle is to maximize a functional that is composed of an entropy part and a constraint part. The constraints, which are often given in terms of moments of the distribution function, allow us to include additional information or hypotheses about the system. They are simply added to the entropy part by using the method of Lagrangian multipliers.

Derivation of the maximum entropy principle   In the following derivation, we follow a simple argument that is very much in the spirit of Boltzmann (however, free of physics) and based purely on elementary combinatorial considerations. No information-theoretic concepts, such as information production or uncertainty are required. The argument came originally from Graham Wallis and was popularized by Edwin Thompson Jaynes [209]. Assume that we observe the outcome of a stochastic process and record the frequency of the outcomes (states) in a histogram. We are interested in the probability of observing a specific histogram, k.

Let us consider a simple example. Our stochastic process is a coin-tossing experiment (Bernoulli process). Assume we have one coin that is possibly biased and toss it N times. There are $W = 2$ possible states that we label with 0 for heads and 1 for tails. We get a sequence of events $x(N) = (x_1, x_2, \cdots, x_N)$, where $x_i \in \{0, 1\}$ stands for the outcome of one coin-tossing trial. The bias of the coin is q_0 for tossing heads and $q_1 = 1 - q_0$ for tails. For $q_0 = 1/2$ the coin is fair. The probabilities (q_0, q_1) for tossing 0 (heads) or 1 (tails) in a trial x_1 we refer to as the *bias* or the *constraints* of the system. The statistics of the possible outcomes of the coin-tossing sequences are visualized in the following binomial tree (Pascal triangle).

---

[14] Maximum-likelihood approaches are Bayesian approaches, even though some proponents of ML-based methods will deny this.

$$1q_0^0 q_1^0$$

$$1q_0^1 q_1^0 \qquad 1q_0^0 q_1^1$$

$$1q_0^2 q_1^0 \qquad 2q_0^1 q_1^1 \qquad 1q_0^0 q_1^2$$

$$1q_0^3 q_1^0 \qquad 3q_0^2 q_1^1 \qquad 3q_0^1 q_1^2 \qquad 1q_0^0 q_1^3$$

$$1q_0^4 q_1^0 \qquad 4q_0^3 q_1^1 \qquad 6q_0^2 q_1^2 \qquad 4q_0^1 q_1^3 \qquad 1q_0^0 q_1^4$$

Here the collection of all possible coin-tossing sequences of length $N = 4$ is shown. Let us denote the *histogram* of the sequence $x(N)$ by $k = (k_0, k_1)$, where k_0 is the number of heads (0) and k_1 the number of tails (1) that appeared in the sequence.

To obtain a particular sequence in the binomial tree here, one begins at the top ($N = 0$) and takes a step to the left every time heads appears and a step to the right for tails. After four trials the bottom line is reached, and it reads as follows: you get one sequence with four heads with the probability q_0^4. You get four possible sequences with three heads and one tails with probability $q_0^3 q_1$; six possible sequences with two heads and two tails with probability $q_0^2 q_1^2$, and so forth. The number of sequences leading to a given histogram is given by the binomial factors, $\binom{N}{n}$. They count the number of different ways in which n heads can be obtained in a sequence of length N. We refer to the combinatorial factor as the *multiplicity* M, as it tells us how many ways a particular histogram can be realized. The other terms describe the probabilities of observing any particular sequence with a histogram k. They depend on the bias of the coin alone.

The probability of finding a given histogram $k = (k_0, k_1)$ in the coin tossing experiment is the binomial distribution from Equation (2.64),

$$P(k|q) = \binom{N}{k_0} q_0^{k_0} q_1^{k_1} = \underbrace{\frac{N!}{k_0! k_1!}}_{\text{multiplicity}} \underbrace{q_0^{k_0} q_1^{k_1}}_{\text{probability}} = M(k)G(k|q), \tag{6.40}$$

where we used the explicit form of the binomial factor. The equation gives us the probability of finding a sample with the specific histogram k, given that the Bernoulli process is specified by the probabilities $q = (q_0, q_1)$. The probability $P(k|q)$ contains two parts, one being the multiplicity M that depends on the histogram k alone, and another part, $G(k,q) = q_0^{k_0} q_1^{k_1}$, that contains the bias probabilities q_i taken to the power of their appearance k_i. The first part is the multiplicity; the second contains the constraints. Exactly the same arguments hold for memoryless processes that have more than two states. If individual elements x_i are drawn from W different states, $x_n \in \{1, 2, \cdots, W\}$, the probability of finding the histogram $k = (k_1, k_2, \cdots, k_W)$, is given by the expression,

$$P(k|q) = \underbrace{\frac{N!}{k_1!k_2!\cdots k_W!}}_{\text{multiplicity}} \underbrace{q_1^{k_1} q_2^{k_2} \cdots q_W^{k_W}}_{\text{probability}} = M(k)\,G(k|q). \qquad (6.41)$$

Again, the probability $P(k|q)$ factorizes into the multiplicity, $M(k)$, that is independent of q, and a constraint part, $G(k|q)$, that contains the probabilities q_i. For W states the binomial factor becomes the *multinomial factor*, and the probabilities for sampling individual states $i \in \{1, \ldots, W\}$ are given by $q = (q_1, q_2, \cdots, q_W)$.

The aim is now to find the *most likely* histogram (or frequency distribution) k that maximizes the numerical value of $P(k|q)$. The histogram k^* that maximizes $P(k|q)$ is called the *maximum configuration*. As the logarithm is a monotonic function, the logarithm $\log P(k|q)$ will also be maximized by the histogram k^*. Taking the logarithm on both sides of Equation (6.41), we get,

$$\underbrace{\frac{1}{N}\log P(k|q)}_{-\text{relative entropy}} = \underbrace{\frac{1}{N}\log M(k)}_{\text{entropy } S} + \underbrace{\frac{1}{N}\log G(k|q)}_{-\text{cross-entropy}}. \qquad (6.42)$$

Here, we divided (scaled) both sides by the length of the sequence N (system size), that is, by the number of degrees of freedom of the system. To arrive at relative frequencies instead of the histogram, we can use $p = k/N$. Assuming that N is sufficiently large, p becomes the most likely distribution function.

The term on the left-hand side of Equation (6.42), $-\frac{1}{N}\log P(k = Np|q)$ is called *relative-entropy* or Kullback–Leibler divergence [243], $\frac{1}{N}\log P(Np|q) = -\sum_i p_i(\log p_i - \log q_i)$. As shown in Equation (6.43), we identify the first term on the right-hand side of Equation (6.42) with Shannon entropy $S_S(p)$, and the second term is the so-called *cross-entropy*, $-\frac{1}{N}\log G(pN|q) = -\sum_i p_i \log q_i$. With this we can rewrite Equation (6.42) as,

$$-\sum_i p_i \log \frac{p_i}{q_i} = -\sum_i p_i \log p_i + \sum_i p_i \log q_i, \qquad (6.43)$$

which states that the cross-entropy is equal to Shannon entropy plus relative entropy. The constraints in the maximum entropy principle are represented by the cross-entropy. At this point we make two observations:

Observation 1: factorization property. $P(k|q)$ factorizes into a term $M(k)$ (microstate multiplicity) that does not depend on q and a constraint term $G(k, q)$ (microstate probability) that contains all dependencies of q.

Observation 2: the logarithm of the multiplicity is simply the Boltzmann–Gibbs–Shannon entropy formula.

To make the second observation clear, we use $p_i = k_i/N$, and write,

$$
\begin{aligned}
S_{\text{MEP}}(p) &= \tfrac{1}{N}\log M(k) = \tfrac{1}{N}\log\left(\frac{N!}{k_1!k_2!\cdots k_W!}\right) \sim \tfrac{1}{N}\log\left(\frac{N^N}{k_1^{k_1}k_2^{k_2}\cdots k_W^{k_W}}\right) \\
&= \log N - \tfrac{1}{N}\sum_{i=1}^{W} k_i\log k_i \\
&= -\sum_{i=1}^{W} p_i\log p_i = S_{\text{BGS}}(p)
\end{aligned}
\tag{6.44}
$$

where we used Stirling's formula for the factorial; see Section 8.2.1.

> The logarithm of the multinomial multiplicity, divided by system size (degrees of freedom), is identical to the Boltzmann–Gibbs–Shannon formula, S_{BGS}.

Note that this is essentially equivalent to Boltzmann's thinking that we encountered in Equation (6.3), where Boltzmann entropy is the logarithm of the number of microstates the system can assume. The number of microstates corresponds directly to the multiplicity, that is, the degeneracy of microstates occurring with equal probability, $q_i = 1/W$. However, this reasoning does not need any recourse to physical systems—it is a pure result of elementary combinatorics. Also note that the Boltzmann–Gibbs–Shannon formula emerges only for processes that are multinomial in nature. In other words, systems must be composed of independent, indistinguishable components or must produce memoryless processes.

Let us take a look at the constraint term $G(k,q)$. For the sake of clarity, as will become clear below, we write the probabilities q_i in exponential form, $q_i = e^{-\alpha - \beta\epsilon_i}$. In physical systems, ϵ_i usually corresponds to the energy of state i and β is the inverse temperature. α is a normalization constant that ensures $\sum_i q_i = 1$. We see that,

$$
\frac{1}{N}\log G(k,q) = -\frac{1}{N}\sum_{i=1}^{W} k_i(\alpha + \beta\epsilon) = -\alpha\sum_{i=1}^{W} p_i - \beta\sum_{i=1}^{W} p_i\epsilon_i.
\tag{6.45}
$$

Using this result, the log-probability of finding the histogram in Equation (6.42) becomes,

$$
\frac{1}{N}\log P(k|q) = \underbrace{-\sum_{i=1}^{W} p_i\log p_i}_{\text{max ent entropy}} \underbrace{-\,\alpha\sum_{i=1}^{W} p_i - \beta\sum_{i=1}^{W} p_i\epsilon_i}_{\text{constraints}}.
\tag{6.46}
$$

All that remains is to find the maximum. To do this we maximize the equation by taking derivatives with respect to k_i, and setting the result to zero,

$$0 = \frac{\partial}{\partial k_i} \log P(k|q) = \frac{\partial}{\partial p_i} \left(-\sum_{i=1}^{W} p_i \log p_i - \alpha \sum_{i=1}^{W} p_i - \beta \sum_{i=1}^{W} p_i \epsilon_i \right). \qquad (6.47)$$

The result of this procedure is a set of equations that determine the most likely occurrence frequencies p_i^*, or the most likely histogram $k_i^* = N p_i^*$. Equation (6.47) is called the *maximum entropy principle*. The entropy that appears in Equation (6.47) is called the *Jaynes maxent entropy*, or S_{MEP}. The constants α and β play the role of Lagrangian multipliers in the optimization problem. α ensures the normalization of p_i^* and β fixes the first moment; here it is the average energy. It now becomes clear why we wrote the probabilities q in exponential form.

If the probability of finding a histogram of a stochastic process factorizes into a multiplicity M and a term G that contains the constraints, then a maximum entropy principle exists. This means that one can infer the most likely distribution function about a stochastic system by maximizing the entropy functional $-\sum_i p_i \log p_i$, under the constraints that are added with Lagrangian multipliers. For example, if higher moments of the distribution function are known, these can be added as additional constraint terms using Lagrangian multipliers.

In Section 6.5 we discuss that factorizations into multiplicity and constraint terms also exist for systems that are more complicated than those based on multinomial statistics. This will lead us to generalized entropy functionals for complex systems.

6.2.4 Limits of the classical entropy concept

We discussed three different approaches to statistical systems and encountered the concept of entropy in three very different contexts. In statistical mechanics we encountered *Boltzmann entropy*, S_B, as a quantity that links the number of microscopic states in a system to a quantity that allows us to compute macroscopic properties of simple (non-interacting) physical systems, like the ideal gas. The macroscopic equations are the thermodynamic equations that relate observable parameters, like pressure, volume, or temperature. We found that entropy, when used in thermodynamic equations, must be an extensive quantity. We discussed the conceptual difference between a system being extensive and a mathematical formula being additive. For simple systems the notions of extensive and additive effectively coincide. For complex systems the difference will become relevant.

In the context of information theory, we encountered *Shannon entropy*, S_S which allows us to quantify the rate of information production of a stationary and ergodic source without memory. We have understood how the same quantity is used to understand the fundamental limits of the compressibility of messages in terms of optimal coding. We have seen how the concept of the entropy rate is naturally related to the conditional entropy for Markov processes, and how Boltzmann–Gibbs–Shannon entropy is related to Bernoulli processes. We have discussed the existence of coding schemes that allow us to transmit messages through noisy channels in an error-free fashion, as long as the information production rate is less than the channel capacity.

In the context of statistical inference, we discussed a method of finding the most likely distribution function of a stochastic system if we know the constraining facts of the system, such as constraints on its states or moments of the distribution function. We encountered the maximum entropy, S_{MEP}, as the logarithm of the multiplicity of the number of sequences that lead to the same frequency distribution or histogram. The derivation of the maximum entropy principle can be done as a purely combinatorial exercise.

All three approaches to entropy of statistical systems lead to the degenerate functional form of the Boltzmann–Gibbs–Shannon entropy,

$$S_B = S_S = S_{\text{MEP}} = -\sum_i p_i \log p_i. \tag{6.48}$$

In the derivation of the maximum entropy principle, we understood the origin of this degeneracy.

The degeneracy of entropy for simple systems, $S_B = S_S = S_{\text{MEP}}$, is rooted in the simple combinatorial fact that the logarithm of the multinomial factor is nothing but the entropy functional, $\log N!/(k_1! k_2! \cdots k_W!) \sim -N \sum_{i=1}^{W} p_i \log p_i$. This relation is exact if systems are large $N \to \infty$. In other words, any stochastic system that is based on multinomial statistics has an entropy of this form, regardless of the context used. As the underlying statistics behind non-interacting physical systems and behind ergodic sources are multinomial, they are (multistate) Bernoulli processes, and all approaches become the same. We will learn that for complex systems this degeneracy of entropic forms no longer exists.

6.3   Entropy for complex systems

In this section, we will develop a consistent framework for an entropy that is meaningful for (at least several) complex systems. We will discuss the fact that complex systems violate ergodicity. For many statistical complex systems, regardless of whether they are physical or non-physical, the assumptions of the Shannon–Khinchin axioms SK1–SK3 hold. What is different for complex systems is that the fourth Shannon–Khinchin axiom, the composition axiom, is explicitly violated. This means that whenever complex systems are brought together, when they grow, or when they are split, it matters *how* they are combined or separated. This is due to interactions that are building up between the elements of the subsystems as they merge or separate. We will understand the importance of knowing how the phasespace volume of systems changes as a function of system size.

This knowledge will enable us to classify a large class of non-ergodic statistical complex systems.

The violation of ergodicity has far-reaching consequences for the statistical description and treatment of systems. We will discuss possibilities for addressing the loss of ergodicity and how this plays out in the areas of physical systems, information theory, and statistical inference. The purpose of this undertaking is to:

- classify statistical complex systems,
- understand the origin of fat-tailed and power law distribution functions that dominate the statistics,
- understand macroscopic properties based on the nature of phasespace,
- analytically handle and predict statistics of history- or path-dependent processes, and to
- develop a maxent framework for history- or path-dependent processes.

We start with the Shannon–Khinchin axioms and analyse what happens if the fourth axiom is dropped. We then discuss the implications for systems if we require them to be extensive. Finally, we deal with statistical inference in situations where processes are no longer multinomial but become history- or path-dependent.

We will see explicitly that the degeneracy of the three main entropy concepts will break down for complex systems in general. One single system will be described by three different entropy functionals depending on the aspect we are looking at: the thermodynamical, the information-theoretic, or the maxent entropy aspect. This will be discussed in detail for two simple path-dependent processes: the Pólya urn process and the sample space reducing process. We will find that, for complex systems, in general, $S_B \neq S_S \neq S_{\mathrm{MEP}}$.

6.3.1   Complex systems violate ergodicity

In previous chapters, we have argued that complex systems are systems made up of many elements or components that strongly interact with each other. States of these elements change as a function of these interactions, which generally happen on networks; and the interaction networks change as a function of the states of the elements. We learned that such co-evolving systems are effectively changing their boundary conditions, as the system evolves. Moreover, many complex systems are open systems, driven, and generally out-of-equilibrium. As a result of these characteristics:

Complex systems are usually non-Markovian and intrinsically non-ergodic. Complex systems are typically

- evolutionary,
- path- and history-dependent,

- have long memory,
- have long-range and co-evolving interactions.

Depending on the context, these features generally make systems non-ergodic and multinomial statistics inapplicable. They cannot be interpreted as finite-state machines. They are also frequently not extensive, and the fourth Shannon–Khinchin axiom that ensures ergodicity of a system is generally violated.

6.3.1.1 *Violation of the composition axiom*

Complex systems generally violate the fourth Shannon–Khinchin axiom SK4, which states how systems must be composed in such a way that they can be described with classic entropy concepts. As noted in Section 6.2.2, this axiom ensures that the entropy of a system composed of two subsystems A and B, must be equal to the entropy of system A plus the entropy of system B conditional on A, $S(AB) = S(A) + S(B|A)$. If systems A and B are independent, $S(B|A) = S(B)$. The uniqueness theorem states that if Shannon–Khinchin axioms 1–4 hold, the only form entropy can take is a trace form, $S_S = \sum_{i=1}^{W} g(p_i)$, and $g(x) = -x \log x$ (up to a multiplicative constant).

As the fourth Shannon–Khinchin axiom SK4 ensures that systems and processes are ergodic, we cannot assume its validity for complex systems that are typically non-ergodic.

For a notion of entropy that can be used for complex systems, SK4 needs to be replaced by more general composition rules or has to be given up altogether. The philosophy that we will follow here is to assume that SK4 does not hold in the context of complex systems,[15] and to explore the consequences for the uniqueness theorem. We do, however, assume that for many systems, the first three axioms remain perfectly valid. If SK4 is given up, then obviously, the expression for entropy must become more general than the Boltzmann–Gibbs–Shannon formula, $-\sum_{i=1}^{W} p_i \log p_i$. In Section 6.3.8 we will also encounter systems that violate more axioms than SK4.

---

[15] In the context of thermodynamics, the violation of the composition axiom has an important and clear intuitive meaning. SK4 states that any possibility of realizing any one of W states i with fixed probabilities p_i is equivalent in terms of information content. In equilibrium a system has infinite time to sample all states, no matter *how* the system moves through phasespace. In driven systems this changes. The faster a system is driven, the less time it has to explore the vicinity of its present location in phasespace. As a consequence, not all possible ways of realizing states i remain equivalent and SK4 will no longer hold. Think, for example, of decision-making under time pressure. If a decision can be made without time constraints, we can consider all options and arrive at the optimal decision once all options have been ranked according to their expected payoffs. If a decision is forced by a time limit, one can no longer explore all options and can only choose the best option from those that could be explored within the time limit. How we make decisions now depends on how we explore our options, and again SK4 is violated.

6.3.2   Shannon–Khinchin axioms for complex systems

We now consider those complex systems for which the following axioms hold.

Axioms for complex systems that violate SK4.

- SK1 holds: entropy $S(p_1, p_2, \cdots, p_W)$ is a continuous function of all its arguments p_i. It does not depend on any other variables.
- SK2 holds: entropy $S(p_1, p_2, \cdots, p_W)$ assumes a maximum value for the case, where all W states are equally probable (equidistribution), meaning that $p_i = 1/W$.
- SK3 holds: adding a state $W + 1$ to the system that never occurs, $p_{W+1} = 0$, does not change the entropy, $S(p_1, p_2, \cdots, p_W) = S(p_1, p_2, \cdots, p_W, 0)$.
- The entropy is a trace form entropy.[a]

[a] Here we assume that the resulting entropy is of trace form, $S_g = \sum_i^W g(p_i)$, which is not the most general assumption. This assumption is not essential but it simplifies the mathematical analysis. In Section 6.5.2 we will see that trace form entropies appear naturally under very general conditions.

6.3.3   Entropy for complex systems

We are now able to establish a generalization of Shannon's uniqueness theorem for complex systems for which the SK4 axiom is violated but SK1, SK2, and SK3 remain valid. By doing this we will see that the resulting entropy, $S_{c,d}$, is a generalization of Shannon entropy, S_S. We will see in Section 6.4 how this generalized entropy $S_{c,d}$ can be interpreted as an extensive quantity that can be used to describe non-extensive complex systems (that violate SK4). An extensive entropy, as we argued before, is fundamental for thermodynamic equations to make sense. We show how the growth of phasespace determines the specific form of the generalized entropy. In Section 6.5 we will see that exactly the same generalized entropy $S_{c,d}$ appears naturally when we generalize the maximum entropy principle to non-multinomial processes. As for simple systems, we have to clarify exactly in which context we use the term entropy. We now generalize the uniqueness theorem.

Complex systems classification theorem. All systems for which the axioms SK1, SK2, and SK3 hold can be

1. classified uniquely by two numbers, c and d,
2. represented by a two-parameter entropy,

$$S_{c,d}(p) = \frac{r}{c} A^{-d} e^A \left[\sum_{i=1}^{W} \Gamma(1 + d, A - c \log p_i) - r p_i \right], \tag{6.49}$$

where $\Gamma(a,b) = \int_b^\infty dt\, t^{a-1} e^{-t}$ is the incomplete Gamma function, see Equation (8.14), r is a constant composed of c and d, $A = \frac{cdr}{1-(1-c)r}$, and e is the base of the logarithm. The constant $r > 0$ can be chosen freely, but, depending on the values of d, must satisfy,

$$
\begin{array}{ll}
d > 0: & r < \frac{1}{1-c}, \\
d = 0: & r = \frac{1}{1-c}, \\
d < 0: & r > \frac{1}{1-c}.
\end{array}
\tag{6.50}
$$

The entropy of Equation (6.49) is called (c,d)-entropy, complex systems entropy, or simply $S_{c,d}$.

The proof of this theorem is interesting in itself and we follow [175]. To show that the entropy for complex systems is indeed given by Equation (6.49), we remember from Section 6.2.2 that the statement 'axioms SK1–SK3 hold', for $S_g(p) = \sum_{i=1}^{W} g(p_i)$, is equivalent to saying that the function $g(x)$ is a continuous, concave function with $g(0) = 0$. We now look at the state of maximal entropy, which following SK2, is obtained for equidistribution, $p_i = 1/W$. For this case we get,

$$
S_g(W) = \sum_{i=1}^{W} g\left(\frac{1}{W}\right) = W g\left(\frac{1}{W}\right).
\tag{6.51}
$$

We now examine how the maximal entropy value behaves under simple scaling transformations. In particular, we ask, what happens if we change the number of states from W states to λW states? How does entropy change? For systems with many states W we find the generic scaling law,

$$
\lim_{W \to \infty} \frac{S_g(W\lambda)}{S_g(W)} = \lim_{W \to \infty} \lambda \frac{g\left(\frac{1}{\lambda W}\right)}{g\left(\frac{1}{W}\right)} = \lambda^{1-c}.
\tag{6.52}
$$

To see that this statement is correct, let us define $f(z) = \lim_{x \to 0} \frac{g(zx)}{g(x)}$ with $(0 < z < 1)$. We can easily prove that for systems satisfying SK1, SK2, and SK3, $f(z) = z^c$, for $0 < c \leq 1$. Note that in this definition zx corresponds to $1/(\lambda W)$.

Theorem 1 *Let g be a continuous, concave function on $[0,1]$ with $g(0) = 0$ (which is guaranteed by SK1–SK3) and let $f(z) = \lim_{x \to 0^+} g(zx)/g(x)$ be continuous, then f is of the form $f(z) = z^c$, with $c \in (0,1]$.*

Proof *The proof is obtained by inserting a term $g(bx)$ in the denominator and the numerator,*

$$
f(ab) = \lim_{x \to 0} \frac{g(abx)}{g(x)} = \lim_{x \to 0} \frac{g(abx)}{g(bx)} \frac{g(bx)}{g(x)} = f(a)f(b).
\tag{6.53}
$$

In Section 3.2 we learn that the only solution f to the scaling equation $f(ab) = f(a)f(b)$ is that f is a power law. All pathological solutions are excluded by the requirement that f is continuous. Therefore $f(ab) = f(a)f(b)$ implies that $f(z) = z^c$ is the only possible solution of this equation. Further, as $g(0) = 0$, also $\lim_{x \to 0} g(0x)/g(x) = 0$, and it follows that $f(0) = 0$. This necessarily implies that $c > 0$. $f(z) = z^c$ also has to be concave, as $g(zx)/g(x)$ is concave in z for arbitrarily small, fixed $x > 0$. Therefore, $c \leq 1$. Note that $c > 1$ explicitly violates SK2, and $c \leq 0$ violates SK3.

There is a second generic scaling law for entropy. Let us substitute λ in Equation (6.52) by $\lambda \to W^a$ and define the quantity, $\frac{S(W^{1+a})}{S(W)W^{a(1-c)}}$. We find that the expression does not yield 1, as one could naively expect, but,

$$\lim_{W \to \infty} \frac{S(W^{1+a})}{S(W)W^{a(1-c)}} = (1+a)^d, \tag{6.54}$$

for large W. To show that this statement is true, we proceed as before and define the quantity,

$$h_c(a) = \lim_{W \to \infty} \frac{S(W^{1+a})}{S(W)} W^{a(c-1)} = \lim_{x \to 0} \frac{g(x^{1+a})}{x^{ac}g(x)}, \tag{6.55}$$

where $x = 1/W$. With this we formulate a second theorem.

Theorem 2 *Assume that SK1, SK2, and SK3 hold. Let g be as before and assume that $h_c(a)$ is given by Equation (6.55), then $h_c(a) = (1+a)^d$, where d is a real constant.*

Proof *We compute $h_c(a)$ by inserting terms (as before) that cancel each other to obtain a factorization of h_c,*

$$h_c(a) \quad = \lim_{x \to 0} \frac{g(x^{a+1})}{x^{ac}g(x)} = \frac{g\left((x^b)^{\left(\frac{a+1}{b}-1\right)+1}\right)}{(x^b)^{\left(\frac{a+1}{b}-1\right)c}g(x^b)} \frac{g(x^b)}{x^{(b-1)c}g(x)} = h_c\left(\frac{a+1}{b}-1\right) h_c(b-1), \tag{6.56}$$

for some constant, b. By a transformation of variables, $a = bb' - 1$, one obtains $h_c(bb' - 1) = h_c(b-1)h_c(b'-1)$. Setting $H(x) = h_c(x-1)$ one again gets $H(bb') = H(b)H(b')$, and $H(x) = x^d$ for some constant d, and consequently $h_c(a) = (1+a)^d$. For the case that $c = 1$, the concavity of g implies $d \geq 0$; for $c < 1$, d can also have negative values.

Note that, in principle, $h_c(a)$ depends on c and a. However, it turns out that h_c does not depend on c. $h_c(a)$ is independent of the first scaling property. Therefore, the scaling law of $h_c(a)$ can be seen as an orthogonal correction term to the first scaling law captured by the exponent c. Indeed, one can go to successively higher corrections, as was pointed out in [366].

There are two scaling laws for all statistical systems for which SK1, SK2, and SK3 hold,

$$\lim_{W \to \infty} \frac{S_g(W\lambda)}{S_g(W)} = \lambda^{1-c} \qquad \text{with} \qquad 0 \le c < 1$$

$$\lim_{W \to \infty} \frac{S(W^{1+a})}{S(W)W^{a(1-c)}} = (1+a)^d \quad \text{with} \quad d \text{ real-valued.} \qquad (6.57)$$

All such systems can be uniquely classified by two scaling exponents, c and d, which characterize the change of entropy by adding or removing states to the system. Systems that have the same exponents are said to be in the same (c,d)-equivalence class.[a]

[a] Scaling exponents can also be worked out for entropies that are not of trace form, as we will see in Section 6.3.5.

To complete the argument of why Equation (6.49) is indeed the entropy for complex systems that violate SK4, we now have to ask which functions S_g are compatible with the two scaling laws in Equations (6.57). The function in Equation (6.49) solves both scaling requirements Equations (6.57), for all possible combinations of values for c and d. $S_{c,d}$ depends on the unique scaling exponents c and d that fully characterize the scaling behaviour of the entropy. Different stochastic systems will, in general, have different scaling exponents for their entropy.

We mentioned that r in Equation (6.49) can be chosen relatively freely. For $d > 0$, a practical choice for r is,

$$r = \frac{1}{1 - c + cd}, \qquad (6.58)$$

for which the constant $A = 1$, and Equation (6.49) simplifies to,

$$S_{c,d}(p) = \frac{e}{c(1-c+cd)} \sum_{i=1}^{W} \Gamma(1+d, 1 - c\log p_i) - \frac{1}{1-c+cd}. \qquad (6.59)$$

For $d < 0$, a convenient choice is $r = \exp(-d)/(1-c)$. Every choice of r provides us with an equivalent representative of the equivalence class (c,d), that is, r has no effect on the the properties of the entropy in the limit of large W. The choice of r, however, encodes finite-size characteristics of the system.

It is important to note that $S_{c,d}$ in Equation (6.49) is not unique, and that there are other functions that also solve the two scaling relations in Equations (6.57). Perhaps the simplest possibility is,

$$S_{c,d} \sim \sum_{i=1}^{W} p_i^c \left(\log \frac{1}{p_i} \right)^d.$$ (6.60)

This function gives us the correct asymptotic properties but does not work for *all* possible values of c and d. In particular, using Equation (6.60) leads to problems with the other axioms, concavity in particular, which occur in the important parameter region $d = 0$.

6.3.4   Special cases

Starting from the $S_{c,d}$ entropy given in Equation (6.49), we can now discuss a few important special cases.

6.3.4.1   *Classical entropy of Boltzmann, Gibbs, Shannon, and Jaynes*

$c = 1$, $d = 1$. For $d = 1$ the incomplete Gamma function is $\Gamma(d+1, x) = \int_x^\infty t e^{-t} dt = e^{-x}(x+1)$. Using this result in Equation (6.49) we obtain the classical entropy functional as it appears in the Boltzmann entropy, the Shannon entropy, or in the classical maximum entropy principle; we recover the Boltzmann–Gibbs–Shannon formula,

$$S_{1,1} = \sum_i g_{1,1}(p_i) = -\sum_i p_i \log p_i + 1.$$ (6.61)

In Figure 6.6 we show the position of the classical entropy functional in the parameter space of c (y-axis) and d (x-axis). The regions where SK2 and SK3 are violated are shaded in dark grey. The classical entropy functional appears at the point $c = 1$ and $d = 1$. The irrelevant constant[16] 1 in Equation (6.61) appears as a consequence of the relaxation of SK4. It does not affect any properties of the entropy and leaves the validity of SK1, SK2, and SK3 untouched.

6.3.4.2   *Tsallis entropy*

$c = q$, $d = 0$. In this case, the incomplete Gamma function becomes $\Gamma(1, x) = e^{-x}$. Using this in Equation (6.49) we get,

$$S_{q,0} = \sum_i g_{q,0}(p_i) = \frac{1}{q} \left(\frac{1 - \sum_i p_i^q}{q-1} + 1 \right).$$ (6.62)

This special case is widely known as Tsallis entropy, up to a factor. In Figure 6.6 we show the region in the (c, d)-parameter space where Tsallis entropy appears along the line characterized by $d = 0$ and $c = q$. For values of c larger than one or below zero, SK2 and SK3 are violated.

---

[16] Note that $g_{1,1}(p_i) = -\int_0^{p_i} dx \log x = -p_i(\log p_i - 1)$. However, by a scaling transformation of g one can obtain $(1/B)g_{1,1}(Bp_i) = -p_i \log p_i$, where B is the base of the logarithm.

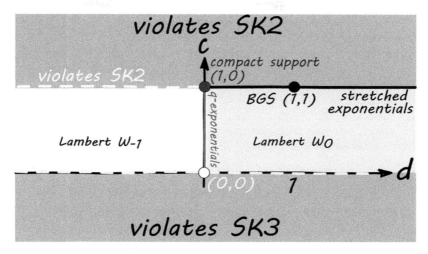

Figure 6.6 *Overview of the locations of various entropies and their associated distribution functions in the plane spanned by the scaling exponents c and d. Entropies that appear in the same location are asymptotically equivalent; they belong to the same equivalence class. Boltzmann–Gibbs–Shannon entropy corresponds to the point (1,1), Tsallis entropy to the line (c,0), where $0 < c \leq 1$. The entropy for stretched exponentials is located at the line (1,d), where $d > 0$.*

Tsallis entropy plays a role in the statistical mechanics of complex systems, mainly because it is tightly related to power laws in distribution functions. It has found literally hundreds of applications in science and statistics [380]. A detailed overview of Tsallis entropy and its applications is found in [379]. It was originally introduced in the contexts of classification processes [184] and multifractals [378]. The simplest way of motivating Tsallis entropy is to use the well-known representation of the log function, $\log x = \lim_{q \to 1} \frac{x^{1-q}-1}{1-q}$, and then 'forget' about taking the limit. One can then define the *generalized logarithm* as $\log_q x = \frac{x^{1-q}-1}{1-q}$, often simply called q-logarithm.[17] If one replaces the logarithm in the Boltzmann–Gibbs–Shannon formula by the q-logarithm, $\log_q$, one arrives at the definition of Tsallis entropy,

$$S_q = \sum_i p_i \log_q \frac{1}{p_i} = \sum_i p_i \frac{p_i^{-(1-q)} - 1}{1 - q} = \frac{1 - \sum_i p_i^q}{q - 1}. \tag{6.63}$$

This definition has led to some criticism of Tsallis entropy as an ad hoc concept that allows one to introduce a fitting parameter, q, which can be used to fit power laws but does not provide a deeper understanding of the origin of power laws. This is unjustified, however, as Tsallis entropy, as a special case of complex systems entropy,

---

[17] Do not confuse the q with the base of the logarithm in the notation of the q-logarithm, $\log_q x$.

Equation (6.49), is deeply connected to statistical systems where SK1, SK2, and SK3 hold and SK4 is violated.[18]

The importance of Tsallis entropy arises from the fact that, if it is maximized under constraints, as in Equation (6.23), where the Boltzmann–Gibbs–Shannon formula is replaced by Tsallis entropy of Equation (6.63), it yields power law distribution functions. These distribution functions are q-exponentials that we encountered and discussed in Section 2.4.1.4. Note that in Equation (6.62) we again obtain an additive, irrelevant constant 1 and an overall multiplicative factor[19] that is not present in the original definition of Tsallis entropy of Equation (6.63).

6.3.4.3   *Anteonodo–Plastino entropy*

$c = 1$, $d > 0$. For these parameters it is not possible to reduce the incomplete Gamma function to simpler functions. By setting $c = 1$ in Equation (6.49) we get,

$$S_{1,d>0} = \sum_i g_{1,d}(p_i) = \frac{e}{d} \sum_i \Gamma(1+d, 1-\log p_i) - \frac{1}{d}. \qquad (6.64)$$

This entropy has been introduced as an entropy that is tightly related to stretched exponential functions [12] and is referred to as Anteonodo–Plastino entropy. If Equation (6.64) is maximized under constraints as in Equation (6.23), the corresponding distribution functions are stretched exponentials; see Section 2.4.2.2. In Figure 6.6 we show the region where Anteonodo–Plastino entropy is realized in parameter space, namely $c = 1$ and $d > 0$. For the special case of $d = 2$ we can further simplify the Gamma function and get the expression,

$$S_{1,2} = 2\left(1 - \sum_i p_i \log p_i\right) + \frac{1}{2} \sum_i p_i (\log p_i)^2. \qquad (6.65)$$

This is an example where the entropy is a superposition of the Boltzmann–Gibbs–Shannon formula and the alternative expression for the (c, d)-entropy we encountered in Equation (6.60). The asymptotic behaviour is dominated by the second term.

6.3.5   Classification of complex systems based on their entropy

During the past five decades a number of entropy functionals have been proposed. Most of them were introduced in a specific context, such as special relativity, quantum information, or for pure mathematical reasons. Most of them are of trace form, with the exceptions of Rényi entropy and Landsberg–Vedral entropy. The scaling exponents c

---

[18] Note that for the special case of Tsallis entropy, a unique composition rule SK4* can be engineered so that if SK1, SK2, SK3, and SK4* are valid, Tsallis entropy emerges uniquely [1].

[19] Applying a simple scaling transformation $g_{q,0}(p_i) = \mu^{-1} g_{q,0}(\lambda p_i)$ would solve the problem. Here, $\lambda = q^{1/(q-1)}$ and $\mu = c\lambda^{-c}$. Tsallis entropy is $S_q(p) = \mu S_{q,0}(\lambda p)$.

and d can be computed for any given entropy by using Equations (6.57). We show this explicitly for two examples.

- Given the Boltzmann–Gibbs–Shannon formula, $S = \sum_{i=1}^{W} g(p_i)$, with $g(x) = -x\log(x)$, we find that $f(z) = z$, where we use Equation (6.52). This implies that $c = 1$. From Equation (6.52) it follows that $h_1(a) = 1 + a$, which implies $d = 1$. Therefore, the classical entropy formula $S = -\sum_i p_i \log p_i$ belongs to the equivalence class $(c, d) = (1, 1)$.
- Given Tsallis entropy, $S_q = \sum_{i=1}^{W} g_q(p_i)$, with $g_q(x) = x\log_q \frac{1}{x}$, with some simple algebra we obtain $f(z) = z^q$, which means $c = q$, and $h_q(a) = 1$, that is, $d = 0$. Therefore, Tsallis entropy belongs to the equivalence class $(c, d) = (q, 0)$.

Table 6.2 *Classification of some entropies introduced in various contexts during the past five decades in terms of their (c, d)-equivalence class. They can all be characterized uniquely by the two scaling exponents c and d (given that $W \to \infty$). Except for Rényi entropy and Landsberg–Vedral entropy, which are not of trace form, all listed entropies are special cases of $S_{c,d}$ entropy given in Equation (6.49). By using Equations (6.57) it can be easily verified that $S_{q>1}$, S_b and S_E are asymptotically identical. So are $S_{q<1}, S_A, S_{SM}$, and S_κ, as well as S_η and S_γ. Here we used $r = (1 - c + cd)^{-1}$.*

	Entropy	c	d	Reference
Boltzmann–Gibbs–Shannon	$S_{BGS} = \sum_i p_i \log(1/p_i)$	1	1	
Rényi	$S_\alpha = \frac{1}{1-\alpha} \log \sum_i p_i^\alpha$	1	1	[321]
Tsallis 1	$S_{q<1} = \frac{1 - \sum_i p_i^q}{q-1}$ $(q < 1)$	$q < 1$	0	[378]
Kaniadakis	$S_\kappa = -\sum_i p_i (p_i^\kappa - p_i^{-\kappa})/(2\kappa)$ $(0 < \kappa \leq 1)$	$1 - \kappa$	0	[216]
Abe	$S_{A,q} = -\sum_i (p_i^q - p_i^{1/q})/(q - 1/q)$	$0 < q \leq 1$	0	[38]
Sharma–Mittal	$S_{SM,r,\kappa} = -\sum_i p_i^r (p_i^\kappa - p_i^{-\kappa})/(2\kappa)$	$r - \kappa$	0	[38]
Landsberg–Vedral	$S_{LV,q} = \left(\left(\sum_i p_i^q \right)^{-1} - 1 \right)/(q-1)$	$2 - q$	0	[38]
Tsallis 2	$S_{q>1} = \left(1 - \sum_i p_i^q \right)(q - 1)$ $(q > 1)$	1	0	[378]
Curado	$S_b = \sum_i (1 - e^{-bp_i}) + e^{-b} - 1$ $(b > 0)$	1	0	[98]
Exponential	$S_E = \sum_i p_i \left(1 - e^{\frac{p_i - 1}{p_i}} \right)$	1	0	[383]
Anteonodo–Plastino	$S_\eta = \sum_i \left(\Gamma\left(\frac{\eta+1}{\eta}, -\log p_i \right) - p_i \Gamma\left(\frac{\eta+1}{\eta} \right) \right)$ $(\eta > 0)$	1	$1/\eta$	[12]
Ubriaco	$S_\gamma = \sum_i p_i \log^{1/\gamma}(1/p_i)$	1	$1/\gamma$	[386]
Shafee	$S_\beta = \sum_i p_i^\beta \log(1/p_i)$	β	1	[335]
Hanel–Thurner 1	$S_{c,d} = r \sum_i \left(\frac{e}{c} \Gamma(d+1, 1 - c\log p_i) - p_i \right)$	c	d	[175]
Hanel–Thurner 2	$S_{c,d} = \sum_i p_i^c \log^d(1/p_i)$	c	$d \neq 0$ Appendix [366]

We collect the scaling exponents c and d for a number of entropies in Table 6.2. It is immediately clear that many of these entropies have the same scaling exponents c and d, even though their defining formulas look quite different. This means that these entropies are essentially identical for large systems ($W \to \infty$), as we will see in detail in Section 6.3.5. Entropies with the same exponents belong to the same (c, d)-equivalence class. All of the examples, except the Rényi entropy and the Landsberg–Vedral entropy,[20] are special cases of the (c, d)-entropy that we derived from the axioms in Equation (6.49). In the table it is referred to as Hanel–Thurner 1. Hanel–Thurner 2 is the case discussed in Equation (6.60).

6.3.5.1 Equivalence classes—when are two entropies equivalent?

In Table 6.2 we find that a number of seemingly different entropies show the same set of scaling exponents, which means that they are asymptotically identical. A simple way of seeing if two trace form entropies $S_A = \sum_i g_A(p_i)$ and $S_B = \sum_i g_B(p_i)$ are equivalent is to check if their asymptotic ratio is a constant. In particular, if $\lim_{x\to 0^+} g_A(x)/g_B(x) = \phi$ is a positive finite constant, then the entropies S_A and S_B are equivalent. This is easy to see,

$$\lim_{x\to 0^+} \frac{g_A(zx)}{g_A(x)} = \frac{g_A(zx)}{g_B(zx)} \frac{g_B(x)}{g_A(x)} \frac{g_B(zx)}{g_B(x)} = \phi \frac{1}{\phi} \frac{g_B(zx)}{g_B(x)} = \lim_{x\to 0^+} \frac{g_B(zx)}{g_B(x)}, \qquad (6.66)$$

and S_A and S_B have the same exponent c. By an analogous argument the same result can be obtained for the exponent d that appears in the second asymptotic property in Equations (6.57). A consequence of this argument is that, if one entropy is a scaled version of the other, $g_B(x) = \mu g_A(\lambda x)$, then g_B and g_A are equivalent. Here, μ and λ are some constants.

> The comprehensive classification of entropies into (c, d)-equivalence classes brings order to the zoo of entropies. The classification is not restricted to trace form entropies.

6.3.5.2 Non-trace form entropies—a note on Rényi entropy

So far we have restricted our analysis to trace form entropies. This is by no means necessary, as we can demonstrate for the Rényi entropy [321], which is sometimes also referred to as α-entropy. It is defined as,

$$S_\alpha = \frac{1}{1-\alpha} \log \left(\sum_{i=1}^{W} p_i^\alpha \right), \qquad (6.67)$$

where $\alpha \geq 0$ and $\alpha \neq 1$. Clearly, Rényi entropy is not a trace form entropy, $S(p) = \sum_i g(p_i)$, but is of the form $S = G(\sum_{i=1}^{W} g(p_i))$, where G is the logarithm, $G(x) =$

---

[20] Remember that Rényi and the Landsberg–Vedral entropy are not trace form entropies. They can nevertheless be classified in terms of the scaling exponents c and d.

$\log(x)/(1-\alpha)$, and g the power function, $g(x) = x^{\alpha}$. One can nevertheless compute the scaling properties of these entropic forms in exactly the same way as for trace form entropies. Here we get,

$$\lim_{W \to \infty} \frac{S_{\alpha}(\lambda W)}{S_{\alpha}(W)} = \lim_{s \to \infty} \frac{G(\lambda f_g(\lambda^{-1})s)}{G(s)}, \qquad (6.68)$$

where $f_g(z) = \lim_{x \to 0} g(zx)/g(x)$. The expression $f_G(s) = \lim_{s \to \infty} G(sy)/G(y)$, provides the starting point of the analysis, which then becomes slightly more involved. In particular, for Rényi entropy with $G(x)$ and $g(x)$ specified, we obtain after some algebra that $(c,d) = (1,1)$, which is the same as for the Boltzmann–Gibbs–Shannon entropy [175]. This finding is not surprising because Rényi entropy is purely additive, which means that the composition rule SK4 is, $S_{\alpha}(AB) = S_{\alpha}(A) + S_{\alpha}(B)$. In other words, subsystems A and B are statistically independent.

6.3.6 Distribution functions from the complex systems entropy

We have seen in Sections 6.2.1 and 6.2.3 that we can obtain distribution functions of a process by maximizing the corresponding entropy under certain constraints, see Equation (6.23). We also learned that the distribution function associated with the Boltzmann–Gibbs–Shannon formula and its particular constraints is the exponential or Boltzmann distribution. What happens if we replace the entropy $-\sum_i p_i \log p_i$ in Equation (6.23) by the (c,d)-entropy given in Equation (6.49)? We follow exactly the same two-step process that we discussed in Section 6.2.1.4. Before we do that, however, we introduce the concept of the *generalized logarithm*.

For any given trace form entropy $S = \sum_i g(p_i)$, the generalized logarithm associated with that entropy is the negative derivative of the function $g(p_i)$,

$$\Lambda(p_i) = -\frac{d}{dp_i} g(p_i). \qquad (6.69)$$

For generalized logarithms we require[a] that $\Lambda(1) = 0$ and $\Lambda'(1) = 1$. Its inverse function $\mathcal{E} = \Lambda^{-1}$ determines the distribution function, $p(x) = \mathcal{E}(-x)$, if the constraint terms are taken properly into account in the variable x; see [173–175].

[a] To guarantee these requirements, $\Lambda(1) = 0$ and $\Lambda'(1) = 1$, scaling transformations of the kind $g(x) \to ag(bx)$ can be used. Note that these scaling transformations do not change the (c,d)-equivalence class of the entropy. See Exercise 6.13.

The generalized logarithm for the (c,d)-entropy can easily be computed by using Equation (6.49). To differentiate the incomplete Gamma function, note that $\frac{d}{dx} \int_{a(x)}^{b(x)} dt f(t) = b'(x) f(b(x)) - a'(x) f(a(x))$, where a' and b' are the first derivatives of the functions a and b. From $\Lambda_{c,d,r}(x) = -dg_{c,d,r}(x)/dx$ we get,

$$\Lambda_{c,d,r}(x) = r\left[1 - x^{c-1}\left(1 - \frac{1 - (1-c)r}{dr}\log(x)\right)^d\right].$$ (6.70)

Finally, the generalized exponential function, the inverse function of Λ, can be found, for example, by using tables of special functions, and is given by,

$$p_{c,d,r}(x) = \mathcal{E}_{c,d,r}(-x) = e^{-\frac{d}{1-c}\left[W_k\left(b(1+\frac{x}{r})^{\frac{1}{d}}\right) - W_k(b)\right]},$$ (6.71)

where the generalized exponential function $\mathcal{E}_{c,d,r}$ can be obtained by inverting the generalized logarithm of Equation (6.70). Clearly, the distribution function depends on the scaling exponents c and d. Here, $W_k(x)$ is the Lambert-W function, which cannot be written in closed form; see Section 8.1.4. It is defined as the solution to the equation $x = W(x)\exp(W(x))$. $W_k(x)$ is the kth branch of the Lambert function that has only two *real* solutions $W_k(x)$, the branch $k = 0$ and the branch $k = -1$. Branch $k = 0$ covers the classes for $d \geq 0$, and branch $k = -1$ those for $d < 0$. Here, b is a constant and is defined as $b = ze^z$ with $z = \frac{r(1-c)}{1-r(1-c)}$. Note that for $d > 0$ the formulas simplify by using the convenient choice $r = 1/(1 - c + cd)$. If we write (c,d) instead of (c,d,r), then we will always imply this choice of the parameter r.

In Figure 6.6 we show the different regions of c and d and the associated entropies and distribution functions. Depending on the sign of d, we have to use the first branch of the Lambert-W function $k = -1$ for $d < 0$ (white region), and the zero'th branch for $d > 0$ (light-grey region). Depending on the specific values of c and d, there are a few special cases that can be worked out by using the properties of the Lambert-W function.

- **Boltzmann distribution** Setting $c = 1$ and $d = 1$ in Equation (6.71) and using the fact[21] that $\lim_{z\to 0}\frac{1}{z}W(z(1-x)) = 1 - x$, we get immediately that,

$$p_{(1,1)}(x) = e^{-x-1+1} = e^{-x}.$$ (6.72)

- **Stretched exponential distribution** All cases associated with $(c,d) = (1,d)$ for $d > 0$ are associated with stretched exponential distributions. To see this, remember that $d > 0$ requires the branch $k = 0$ of the Lambert-W function. Performing a Taylor expansion for $W_0(x) \sim x - x^2 + \ldots$, for small values of x, the limit $c \to 1$ of Equation (6.71) turns out to be a stretched exponential,

$$p_{1,d,r}(x) = e^{-dr\left((1+\frac{x}{r})^{\frac{1}{d}}-1\right)}.$$ (6.73)

Note that the limit $d \to 1$ can be performed by simply setting $d = 1$.

---

[21] For example, Wolfram alpha can be used to verify this fact.

- **Power laws (q-exponentials)** The case $0 < c \le 1$ and $d = 0$ yields q-exponentials, or Tsallis distributions, $p_{c,0}(x)$, which show asymptotic power laws. We start from the (c,d)-logarithm of Equation (6.70). We then set $r = 1/(1 - c + cd)$ and note that, with this choice, the factor $\frac{1-r(1-c)}{dr} = c$ and therefore $\Lambda_{c,d}(y) = r\left[1 - y^{c-1}(1 - c\log y)^d\right]$. We then take the limit $d \to 0$, the parameter r becomes $r = 1/(1 - c)$, and we obtain $\Lambda_{c,0}(y) = (y^{c-1} - 1)/(c - 1) = \log_{2-c}(y)$. Inverting $\Lambda_{c,0}$ gives a q-exponential $\mathcal{E}_{c,0}$ and $p_{c,0}(x) = \mathcal{E}_{c,0}(-x)$ becomes,

$$p_{c,0}(x) = \exp_{2-c}(-x) = (1 + (1 - c)x)^{\frac{1}{c-1}}. \tag{6.74}$$

- **Generalized exponential distribution** Cases for all other values of c and d. The distribution function for all other combinations of the scaling exponents c and d are given by the Lambert-W exponentials provided in Equation (6.75). We encountered them in Equation (2.59).

All statistical systems for which the axioms SK1, SK2, and SK3 hold, and that are characterized by a trace form entropy have distribution functions that are Lambert-W exponentials,

$$p_{c,d}(x) \sim e^{-\frac{d}{1-c}W_k\left(\frac{(1-c)r}{1-(1-c)r}\exp\left(\frac{(1-c)r}{1-(1-c)r}\right)(1+\frac{x}{r})^{\frac{1}{d}}\right)} \qquad \begin{array}{ll} k = -1 & \text{for} \quad d < 1 \\ k = 0 & \text{for} \quad d \ge 0 \end{array} \tag{6.75}$$

These include all important special cases, such as the Boltzmann distribution, power laws (q-exponentials), and stretched exponentials. There are essentially no other possibilities.

What do generalized exponentials or Lambert-W exponentials look like? Lambert-W exponentials ($d \ne 0$) are very similar to q-exponentials ($d = 0$). We show this in Figure 6.7. The tails of Lambert-W exponentials for various values of c and d are very similar to the power law tails of q-exponentials; see Figure 6.7a. The tail exponents of the q-exponentials are uniquely determined by c. For the Lambert-W exponentials, c also dominates the tail; however, d can be seen as a small correction to the dominating power law decay. In Figure 6.7b we see that this correction in the Lambert-W exponentials for $d < 0$ results in a slight bending of the tail to the left (positive second derivative), while for $d > 0$, Figure 6.7c, the second derivative is always negative. The figure shows more than ten orders of magnitude on the x-axis. Note how small the correction or influence of d is. For many practical reasons and, in particular, for many experimental data, it will be practically impossible to distinguish Lambert-W exponentials from q-exponentials.

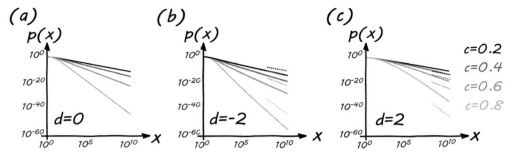

Figure 6.7 *Comparison of (a) q-exponentials (asymptotic power laws) and Lambert-W exponentials with (b) d = −2 and (c) d = 2. The tails of Lambert-W exponentials are extremely similar to the power law tails of the q-exponentials. The plot shows more than ten orders of magnitude on the x axis. For many experimental data, it will be hard—if not impossible—to distinguish between the two types of functions. Note, however, that the Lambert-exponentials for (b) d < 0 show a slight bending to the left (positive second derivative), while for (c) d > 0 the second derivative is always negative. This is also true for q-exponentials.*

6.3.7 Consequences for entropy when giving up ergodicity

In a nutshell, this section dealt with the consequences of giving up the composition axiom or the fourth Shannon–Khinchin axiom SK4, which ensures ergodicity of stochastic systems. We learned that, when giving up SK4, the entropy becomes a more general function than the classical Boltzmann–Gibbs–Shannon formula, $-\sum_i p_i \log p_i$. The entropy becomes characterized solely by the way its maximum changes with the number of available states. We found that there are two generic scaling laws that govern this change. For systems with many states, the corresponding scaling exponents c and d uniquely determine the general form of entropy, which turns out to be (c,d)-entropy, $S_{c,d}$. This can be considered as the entropy for complex systems that violate ergodicity but otherwise fulfil SK1, SK2, and SK3. We showed that these scaling laws also exist for entropies that are not of trace form, such as Rényi entropy.

We have learned that relaxing ergodicity has the following consequences:

- It becomes possible to classify different complex systems by the scaling behaviour of their generalized entropy. Systems can be classified into (c,d)-equivalence classes that are characterized by two scaling exponents.

- (c,d)-entropy naturally explains the origin of a large number of entropies that have been suggested in the past decades. Many of them are identical for systems with a large number of states. Systems that belong to the same (c,d)-equivalence class show identical statistical behaviour, in particular, they have identical tails in their distribution functions.

- (c,d)-entropy naturally explains the existence of a wide spectrum of distribution functions that occur in complex stochastic systems, such as power laws and stretched exponentials. The Boltzmann distribution (exponential distribution) is also included as a special case.

- (c,d)-entropy naturally explains the ubiquity of power laws. In particular, it explains how Tsallis entropy is rooted in the Shannon–Khinchin axioms. The general form of distribution functions are Lambert-W exponentials, which are extremely similar to q-exponentials (asymptotic power laws). It might be hard to distinguish Lambert-W exponentials from q-exponentials in experimental data. Thus, Lambert-W exponentials can be easily mistaken for power laws.

- (c,d)-entropy will allow us to understand the statistics of processes with memory or path-dependence; see Section 6.5.

6.3.8   Systems that violate more than the composition axiom

We have discussed the consequences arising from violating Shannon–Khinchin axiom SK4. Interestingly, there are systems and processes that violate more than one axiom [168]. These include the entropy associated with the maximum entropy for Pólya urn processes and the sample space reducing process; see Section 3.3.5. We will discuss these processes and their entropies in Sections 6.6.1 and 6.6.2.

6.4   Entropy and phasespace for physical complex systems

We mentioned earlier in this chapter that entropy in many physical statistical systems scales extensively with system size and that it would be hard to make sense of thermodynamical quantities (macrostate variables) and their relations if this were not so. Obviously, in a thermodynamic equation such as $dU = TdS$, the macro variables U and S have to scale in the same way if the intensive variable T is kept constant. For non-physical systems or processes, where there is no need for thermodynamic equations, entropy does not necessarily have to be extensive.

We now show that if entropy is extensive, the way phasespace volume $W(N)$ depends on system size N uniquely determines the functional form of the entropy of a classical system. We give a concise criterion when this entropy is of the classical Boltzmann–Gibbs–Shannon form and when it becomes (c,d)-entropy. We show that extensive (c,d)-entropy only exists if a dynamically or statistically relevant fraction of degrees of freedom vanishes in the thermodynamic limit. This means that, as such systems become large, the bulk of the degrees of freedom is 'frozen' and statistically inactive. We will see that systems governed by generalized (c,d)-entropies are therefore systems whose phasespace volume effectively collapses to a low-dimensional 'surface'. Examples of such processes include self-reinforcing processes that follow a winner-takes-all dynamics.

6.4.1 Requirement of extensivity

There are two types of macro variables. *Intensive* variables, such as temperature or pressure, are scale-independent, meaning that they characterize a system independently of its scale. And there are *extensive* variables, which do depend on scale (compare Section 6.2.1.3). A quantity is called extensive if, when the system is enlarged by a factor, the quantity scales by the same factor. For example, internal energy is an extensive variable. If the volume of a room is doubled (e.g. by opening a door to another room) the number of particles is doubled and, assuming they have the same energy on average (i.e. if the intensive variables are kept constant), the internal energy doubles. A variable is called intensive if, when the system is enlarged, the variable does not change, given that everything else remains the same. If the volume of a room is doubled by opening the door to the adjacent room which is of equal temperature, temperature does not double. In Section 6.2.1 we discussed the difference between the notion of *extensivity* in the context of entropy, $S(W_{AB}) = S(W_A) + S(W_B)$, and *additivity*, $S(W_A W_B) = S(W_A) + S(W_B)$. Only for systems with independent components do both notions coincide.

6.4.1.1 Imposing the requirement of extensivity

Consider a system with N elements. The number of possible states (microstates) in the system depends on the number of elements, N. The function $W(N)$ captures this dependence. $W(N)$ is system-specific and will be a different function for different systems. In the following, we only consider systems, where $W(N)$ is a monotonically increasing function. This excludes situations where adding an element reduces the degrees of freedom of the system. The function $W(N)$ tells us how many states in phasespace are effectively occupied. Following this logic, extensivity means that entropy is proportional to N, $S(W(N)) = \varphi N$, where φ is a positive constant. From SK2 it follows that for trace form entropies S_g the maximum is found for the uniform distribution $p_i = 1/W$ for all states i, and we can write $S_g(W) = \sum_{i=1}^{W} g(p_i) = Wg(1/W)$. The extensivity condition then implies that,

$$\frac{d}{dN} S_g(W(N)) = W(N)' \left(g\left(\frac{1}{W}\right) - \frac{1}{W} g'\left(\frac{1}{W}\right) \right) = \varphi. \tag{6.76}$$

This equation allows us to link the function $W(N)$, which describes how phasespace grows with system size N, to the asymptotic properties of g, and in particular to c and d.

6.4.2 Phasespace volume and entropy

If we know the phasespace volume $W(N)$ as a function of system size N, we can directly compute the generalized entropy and its (c, d)-equivalence class.

6.4.2.1 How phasespace determines entropy

To find the derivative g' in Equation (6.76), we make use of the generic scaling relation from Equations (6.57), which applies to all systems that satisfy the first three SK axioms

SK1, SK2, and SK3. As before, we use $x = 1/W$. Taking the derivative with respect to λ of the expression, $\lim_{x \to 0} \frac{g(\lambda x)}{g(x)} = \lambda^c$, we get $\lim_{x \to 0} g'(\lambda x) = \lim_{x \to 0} \frac{c}{x} \lambda^{c-1} g(x)$. For $\lambda = 1$ and large W, the derivative of g is $g'(1/W) = c \, W \, g(1/W)$. Using this expression in Equation (6.76) gives $W(N)'(1 - c)g(1/W) = \varphi$ and the extensivity condition, $S_g(W(N)) = Wg(1/W) = \varphi N$, finally leads us to the first scaling exponent c,

$$\frac{1}{1 - c} = \lim_{N \to \infty} N \frac{W'}{W}. \tag{6.77}$$

We use the shorthand notation $W' = \frac{d}{dN} W(N)$ and $W = W(N)$.

To compute d we use the second scaling relation in Equations (6.57), $(1 + a)^d = \lim_{x \to 0} \frac{g(x^{1+a})}{x^{ac} g(x)}$. Taking the derivative with respect to a on both sides, setting $a \to 0$ and using the extensivity condition, we get the second exponent d that determines the (c, d)-equivalence class,

$$d = \lim_{N \to \infty} \log W \left(\frac{1}{N} \frac{W}{W'} + c - 1 \right). \tag{6.78}$$

We learn that once the dependence of phasespace volume on the system size is known, $W(N)$, and if extensivity is required, the scaling exponents of entropy, (c, d), are uniquely specified by Equations (6.77) and (6.78). These determine the corresponding (c, d)-equivalence class and the (c, d)-entropy that is given by Equation (6.49).

6.4.2.2 Simple examples

With these formulas we can compute several simple examples.

- **Exponential phasespace growth,** $W(N) = e^N$. Equation (6.77) gives $1/(1 - c) = N$, which means that for large N, $c = 1$. From Equation (6.78) we obtain $d = N\frac{1}{N} = 1$. Exponential phasespace growth is therefore associated with the classical entropy functional given by $(c, d) = (1, 1)$. Systems with an exponentially growing phasespace are of the Boltzmann, Shannon, or Jaynes type.

- **Power law phasespace growth,** $W(N) = N^b$. From Equation (6.77) we get $1/(1 - c) = NbN^{-1} = b$; from Equation (6.78) we obtain $d = b\log N \left(\frac{1}{N} \frac{N}{b} + 1 - \frac{1}{b} - 1 \right) = 0$. Power law phasespace growth is associated with $(c, d) = (1 - \frac{1}{b}, 0)$. Associated systems are of Tsallis-type and are dominated by power law tails in their distribution functions.

- **Stretched exponential phasespace growth,** $W(N) = \exp(\lambda N^\gamma)$. Equation (6.77) gives $1/(1 - c) = \lambda \gamma N^{\gamma - 1}$, which means $c = 1$. From Equation (6.78) we get $d = \lambda N^\gamma \left(\frac{1}{N} \frac{1}{\lambda \gamma N^{\gamma - 1}} \right) = \frac{1}{\gamma}$. Stretched exponential phasespace growth is related to Anteonodo-Plastino entropy, $(c, d) = (1, \frac{1}{\gamma})$, whose associated distribution functions are stretched exponentials.

6.4.2.3 A formula for phasespace volume as a function of system size

We can further compute how the phasespace volume of a system depends on system size if the scaling exponents c and d are known. To derive the corresponding formula we first take the derivative with respect to N on both sides of the relation that we obtained in Section 6.4.2.1, $g(1/W) = \frac{\varphi N}{W}$ and get,

$$-g'\left(\frac{1}{W}\right) = \frac{W^2}{W'}\frac{\varphi}{W}\left(1 - N\frac{W'}{W}\right) = \varphi\left(\frac{W}{W'} - N\right) = -\varphi Nc = -\Lambda\left(\frac{1}{W}\right), \quad (6.79)$$

where we made use of the property $g'(x) = -\Lambda(x)$ and used Equation (6.77). Note that we assume N to be large. Remember that the inverse of the generalized logarithm Λ is $\mathcal{E} = \Lambda^{-1}$, which represents the functional form of the distribution function. If we now invert $\Lambda(1/W) = \varphi Nc$ we get,

$$W(N) = \frac{1}{\mathcal{E}_{c,d}(-\varphi cN)} \sim \exp\left[\frac{d}{1-c}\mathcal{W}_k\left(b\left(\frac{\varphi cN}{r}\right)^{\frac{1}{d}}\right)\right], \quad (6.80)$$

where we made use of Equation (6.71), and where $\mathcal{W}_k$ is the Lambert-W_k function and $b = ze^z$ with $z = (1-c)r/(1-(1-c)r)$. We find that the phasespace volume grows like a Lambert-W exponential in N. Note that the Lambert-W exponential form implies that, for most complex systems where $(c,d) \neq (1,1)$, the number of states $W(N)$ grows subexponentially with N.

6.4.2.4 Subexponential phasespace growth and strongly constrained processes

Subexponential phasespace growth often results from strong constraints in the system. A simple example may help to illustrate this point. Imagine a simple coin-tossing experiment. We start with N fair coins and throw all of them in sequence. Clearly, the phasespace grows exponentially with system size, that is, the number of coins N.

Number of coins	States	Number of states
1 coin	↑, ↓	$W(1) = 2$
2 coins	↑↑, ↑↓, ↓↑, ↓↓	$W(2) = 4$
N coins		$W(N) = 2^N$

Now imagine the same process with one modification. We generate a sequence of N ↑ and ↓ symbols with equal probability just as before, but now we impose a constraint that enforces the following rule: after the first appearance of ↑, all subsequent symbols must be ↑.

Number of coins	States	Number of states
1 coin	↑, ↓	$W(1) = 2$
2 coins	↑↑, ↓↑, ↓↓	$W(2) = 3$
3 coins	↑↑↑, ↓↑↑, ↓↓↑, ↓↓↓	$W(3) = 4$
N coins		$W(N) = N + 1$

We immediately see that this constraint changes the phasespace growth to become subexponential; in this case, we have linear growth in N.[22]

> The growth of phasespace volume determines extensive entropy, and vice versa.

6.4.3 Some examples

To illustrate the previous theoretical results, we present several examples of systems whose phasespace volumes do not increase exponentially with system size. Systems with subexponentially growing phasespace volumes are associated with extensive entropies that are not of Boltzmann, Shannon, or Jaynes form, but are realizations of (c, d)-entropy.

6.4.3.1 Ageing random walks

Consider a one-dimensional random walk of the following kind. The random walker can make up or down moves. The walker starts at step $t = 1$. Every time the random walker takes an up–down decision, it walks for $[t^\beta]_+$ times steps in the chosen direction. t is the total number of steps the walker has taken up to that point. $0 \leq \beta < 1$ is a positive number, and the bracket $[.]_+$ means rounding to the next highest integer. In other words, if the walker has taken t steps in the process to date, and makes the decision 'up', it will continue walking for another $[t^\beta]_+ - 1$ steps in the up direction. It comes to a rest at step $t + [t^\beta]_+$, where the next up–down decision is made. For example, fix $\beta = 0.5$. At step $t = 1$ the random walker decides to go up, $\omega(1) = 1$, where $\omega(t)$ marks the direction the walker goes at every time step N, for up $\omega(t) = 1$, for down $\omega(t) = -1$. It has to go up $[t^{0.5}]_+ = 1$ steps. At $t = 2$ it decides to go down, $\omega(2) = -1$. It now has to continue to go down for $[2^{0.5}]_+ = 2$ time steps, that is, $\omega(3) = \omega(2) = -1$. At $t = 4$ it can decide again, and so on. After t steps the walker is at position $x_\beta(t) = \sum_{n=1}^{t} \omega(n)$, see Figure 6.8a. This random walk[23] is a process in which the number of independent decisions per time step decreases as the process 'ages'. The walker becomes more and more constrained over time and becomes increasingly persistent in its direction. Let us denote by $k(t)$ the number of random decisions within the entire walk up to step t. Clearly, k grows like $t^{1-\beta}$,

---

[22] Note that we do no longer sample the entire phasespace of 2^N states but are trapped in a small region. Compare with the situation in Figure 6.5b.

[23] For experts in random walks, it may be interesting to know that the continuum limit of this process is well-defined. By defining (a generalized) Wiener measure by $dW(t) = \lim_{dt \to 0} dt^{1-\frac{\beta}{2}} \omega([t/dt]_+)$ it is possible to show that for $W(t) = \int_0^t dW(\tau)$ one gets $\langle W(t)^2 \rangle \sim t^{2-\beta}$. As $2 - \beta \geq 1$ these processes are super-diffusive.

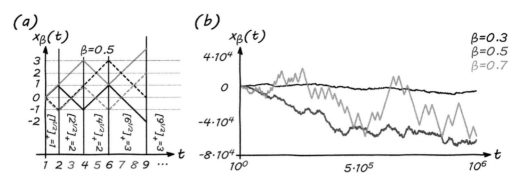

Figure 6.8 *(a) Four realizations of an 'ageing' random walk with $\beta = 0.5$. All possible paths run along the lines of the underlying grid. The walker does not take a decision at every time step but once a direction is decided at time step t, it is followed for the next t^β time steps. The random walker takes fewer and fewer decisions per time step as the process unfolds. (b) Three realizations with $\beta = 0.3$ (black), 0.5 (dark-grey), and 0.7 (light-grey). As the fraction of steps, where the walkers can freely decide to go up or down, vanishes in the limit $t \to \infty$, eventually the walkers will almost never change their direction, irrespective of the value β. The walks become increasingly persistent.*

and the number of possible sequences grows like a stretched exponential, $W(t) \sim 2^{t^{1-\beta}}$. Consequently, the associated extensive entropy is of the (c,d)-equivalence class $\left(1, \frac{1}{1-\beta}\right)$. Realizations for several values of β are shown in Figure 6.8b.

6.4.3.2   A social network model: join-a-club spin system

Imagine a situation where a person wants to join a club made up of opinionated members, all with a binary opinion on some matter, yes/no, black/white, or up/down. To join the club is easy; the only rule is that, on joining, you are required to make friends with a given fraction a of all the $N(t)$ club members that exist at time t. The situation is depicted schematically in Figure 6.9. As the network grows, the network becomes more and more connected, the average degree increases, but its *connectancy*, $\eta = 2L/N(N-1)$ remains constant; see Section 4.3.1. L is the number of links in the network.

Once members have joined, they hear their friends' opinions: up or down. To minimize conflict, at every time step every member changes her opinion according to the majority of her friends. To allow for the possibility that some people will also have an opinion different from those around them, each person is equipped with a little 'free will'— or 'energy'—ϵ that can be used to maintain an opposite opinion. This dynamics can be phrased as a spin model on a growing constant-connectancy network. To this end, consider an Ising-type spin model (spins are up or down) on a network with N nodes that carry the spin and L links. Whenever two nodes are linked, a spin–spin interaction is possible. Two (anti-) parallel spins contribute $\mathcal{J}^+$ ($\mathcal{J}^-$) to the total system energy E. Let $\epsilon = E/N$ denote the energy density per node, and let $\mu = \mathcal{J}^+$ be the energy cost for maintaining a link. $n^{+(-)}$ is the number of up (down) spins. We assume that the network

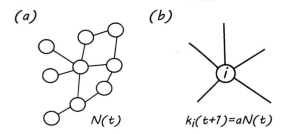

Figure 6.9 *(a) A club has N(t) members at time t. (b) As a new member joins at the next time step, it has to make friends with aN(t) members. The new member is the node labelled i; its aN(t) open links must be linked randomly to the existing members. As the network grows, new members must make more friends than those that joined earlier. The resulting network is a* constant-connectancy *network.*

connectivity (degree) $k = \eta N/2$ is large enough for mean field approximations to be reasonably accurate. The total energy $E = N\epsilon$ of the network can be estimated by,

$$E = -\frac{L\left[\left(n^+(n^+ - 1) + n^-(n^- - 1)\right)\mathcal{J}^+ + 2n^+n^-\mathcal{J}^-\right]}{N(N-1)} + \mu L = \eta n^+(N - n^+)\Delta\mathcal{J}, \quad (6.81)$$

where $\Delta\mathcal{J} = \mathcal{J}^+ - \mathcal{J}^-$. Solving this quadratic equation for n^+, one finds,

$$n^+ = \frac{N}{2}\left(1 - \sqrt{1 - \frac{2\epsilon}{k\Delta\mathcal{J}}}\right) \sim \frac{\epsilon}{\eta\Delta\mathcal{J}}. \quad (6.82)$$

The right-hand side is an approximate solution for n^+ if $2\epsilon/k\Delta\mathcal{J}$ is small. It is clear that fixing the energy ϵ per node determines n^+. The number of states W corresponds to the number of ways n^+ up spins can be distributed over N nodes, which is $W = \binom{N}{n^+}$. We are now ready to combine the spin system with the dynamics of the growing network. To increase the size of the network N in an extensive way, ϵ has to be kept fixed such that the total energy $E = N\epsilon$ of the spin system is extensive.

Constant-connectancy growth   Let us assume that the network grows as described before, that is, $\eta = $ const. during the the growth process. In this case, the degree of the network is $k = \eta N/2$ and $n^+ \sim \epsilon/\eta\Delta\mathcal{J} = $ const. For the phasespace dependence on N, it follows that,

$$W(N) = \binom{N}{n^+} \sim \left(\frac{N}{n^+}\right)^{n^+}\exp(-n^+) \sim N^{n^+}, \quad (6.83)$$

where we used Stirling's formula for the binomial factor; see Section 8.2.1. From Equations (6.77) and (6.78) we immediately obtain the (c,d)-equivalence class $(1 - \frac{1}{n^+}, 0)$. This means that constant-connectancy networks require a (c,d)-entropy that turns out to be the Tsallis entropy with $c = 1 - \frac{1}{n^+}$.

Constant-connectivity growth   If we assume instead that the network grows in such a way that each new member is required to make m friends, so that the average *degree k* in the network remains constant over time, then we get from Equation (6.82) that $n^+ = aN$, with $0 < a < 1$ constant. For the phasespace growth, Stirling's approximation yields,

$$W = \binom{N}{aN} \sim b^N \quad \text{with} \quad b = a^{-a}(1-a)^{a-1} > 1. \tag{6.84}$$

From Equation (6.77) and (6.78) we find $c = 1$ and $d = 1$. The extensive entropy is $(1, 1)$-entropy. In this case, Boltzmann–Gibbs–Shannon entropy governs the growing system.

> In this example, we learn that the way in which a network grows dynamically can determine the extensive entropy of a spin system operating on that network.

6.4.3.3 Black hole entropy

In thermodynamics the number of particles usually scales with volume if pressure or temperature are kept constant. The following example shows what happens if phasespace growth is not exponential in volume but exponential in the surface. Black holes provide an example of this kind. Black holes are objects in space whose gravitational fields are so strong that no matter, light, or signal can escape from them. The boundary of a black hole is given by the Schwarzschild radius, or the event horizon. Outside observers can obtain no information about the events inside the event horizon. However, dynamics of matter, gases, and particles are going on inside black holes, and there is a well-defined statistical mechanics that describes it. The thermodynamics of gravitational systems is highly non-trivial. For example, it can lead to negative specific heat [42, 185, 361], meaning that the more energy is added to the system, the colder it becomes. Whatever the thermodynamics within black holes may be, the information about it cannot pass the event horizon. A black hole is specified by very few macro variables such as its mass, electric (and possibly magnetic monopole) charge, and its angular momentum. Many internal microstates lead up to these macro variables, and as we learned in Section 6.2, thermodynamic entropy quantifies this multiplicity. One may be tempted to associate a thermodynamic entropy with a black hole [40, 41]. The entropy that was proposed by Beckenstein and Hawking is proportional to the area of the event horizon,

$$S_{BH} = \frac{A}{4L_P^2}, \tag{6.85}$$

where A is the surface of the event horizon and the Planck length is $L_P = Gh/(2\pi c^3)$, with G the gravitational constant, h the Planck constant, and c the speed of light.

We can now ask what the *extensive entropy* is for black holes. If we assume that the event horizon radius of a spherical black hole is R, its volume and surface are $V = \frac{4\pi}{3}R^3$ and

$A = 4\pi R^2$, respectively. If we further assume that the particle number $N \sim V = \frac{4\pi}{3} R^3$ scales with the volume V, which is usually fine for non-gravitational situations, and the number of observable states in a black hole scales exponentially with the surface, then we have,

$$W(N) = e^A = e^{4\pi R^2} = e^{4\pi \left(\frac{3V}{4\pi}\right)^{2/3}} \sim e^{\lambda N^{2/3}}. \tag{6.86}$$

We immediately identify this phasespace growth as a stretched exponential and know from Equations (6.77) and (6.78) that $c = 1$, and $d = 3/2$. As a consequence, the (c,d)-equivalence class is $(1, 3/2)$ and the extensive entropy must be $S_{1,3/2}$. The same argument was used in [382]. One should keep in mind that this is a hypothetical example. Its physical content is questionable. The entropy in Equation (6.85) has not, to date, been demonstrated to be correct; nor does the associated extensive entropy, $S_{1,3/2}$, tell us how to perform experiments on combining black holes. However, the example does make it very clear that if the number of (observable) elements does not scale like the volume of the system, but instead like a subvolume or surface of lower dimension, phasespace grows in a stretched exponential manner.

6.4.4 What does non-exponential phasespace growth imply?

In previous examples, we experienced a recurring theme: (c,d)-entropy becomes the extensive entropy of a system if the system becomes more constrained over time (the rate of free decisions in the random walk, or the majority opinion formation rule in the growing constant-connectancy network), or if the number of states in a system does not scale with the volume of the system, as might happen with black holes.

We summarize the implications of non-exponential phasespace growth.

- We learned that the phasespace volume as a function of system size determines the scaling exponents (c,d) of the corresponding extensive entropy. This is given by $S_{c,d}$.

- All systems with exponential phasespace growth are of Boltzmann, Shannon, or Jaynes type.

- For subexponential phasespace growth, the phasespace volume, when compared to the exponential case, becomes vanishingly small, $\lim_{N\to\infty} W(N)/e^N = 0$. The phasespace collapses to a point or a fractal of measure zero.

- Thus, for many subexponential systems, the bulk of the statistically relevant degrees of freedom is frozen. Most elements are statistically inactive. Only a small fraction of components are free (unconstrained enough) to 'run' the statistical system. The majority of components are constrained to such a degree that they play no further role in the statistics of the system.

continued

- Extensive entropy describes phasespace growth but (except for the Boltzmann-Gibbs-Shannon entropy) usually does not predict the correct distribution function when used in the maximum entropy principle.
- We conjecture the applicability of extensive entropies in physical situations, where surface effects play an important role, that is, where the effective degrees of freedom are concentrated on surfaces. This could be the case for self-organized critical systems, spin systems with dense meta-structures, such as spin-domains, vortices, instantons, caging, and so on, and in anomalous diffusion.

Anomalous diffusion is a phenomenon where not all of three-dimensional space is available for the diffusion of particles, but where surface effects from the porous material (think of a sponge or sandstone) play a role. Anomalous diffusion is described by the anomalous diffusion equation, which is a non-linear Fokker–Planck equation of the type,

$$\frac{\partial}{\partial t} p^{\mu}(x,t) = -\frac{\partial}{\partial x}\left[F(x)p^{\mu}(x,t)\right] + D\left(\frac{\partial}{\partial x}\right)^2 p^{\nu}(x,t),$$ (6.87)

where $F = k_1 - k_2 x$ is the external force. μ, ν, and k_2 are positive real numbers. It is interesting to note that this equation is solved by a q-exponential function; see [380]. The porous media equation is important for describing phenomena such as diffusion of proteins in cells or crude oil in sandstone.

6.5   Maximum entropy principle for complex systems

In Section 6.2.3, in particular in Equation (6.42), we learned that for multinomial processes with prior probabilities q, the logarithm of the probability $P(k|q)$ of finding the distribution function k, factorizes into two parts: the logarithm of the multiplicity, which is independent of q and that we identify with the Jaynes *max entropy functional*, and a constraint term called *cross-entropy*. In short,

$$\underbrace{\frac{1}{N}\log P(k|q)}_{-\text{relative entropy}} = \underbrace{\frac{1}{N}\log M(k)}_{\text{entropy } S} + \underbrace{\frac{1}{N}\log G(k|q)}_{\text{cross-entropy}}.$$ (6.88)

The term $-\frac{1}{N}\log P(k|q)$ we call *relative entropy* or *divergence*. We have seen that this factorization is the basis of the maximum entropy principle that we derived in Equation (6.47). We will now show that under certain circumstances, such a factorization is also possible for non-multinomial processes, in particular, for simple path- or history-dependent processes. We will demonstrate in an example that it is possible to start from a set of dynamical update equations of a stochastic path-dependent process and derive the corresponding max entropy functional for it. Given this entropy, we can then predict the distribution function of this process. We follow the original derivation of [178].

6.5.1 Path-dependent processes and multivariate distributions

For multinomial processes the temporal order of events is irrelevant. In other words, the probability $p(x(N))$ of finding a sequence of length N, $x(N) = (x_1, x_2, \cdots, x_N)$, is invariant under permutations of its elements x_n. What are the consequences for path-dependent systems, in which this symmetry is broken? Broken invariance means that different sequences $x(N)$ and $x'(N)$ with the same histogram, $k(x) = k(x')$ will, in general, have different probabilities of occurring, $p(x) \neq p(x')$. As an immediate consequence, $P(k|q)$ no longer factorizes into a multiplicity term and the probability of sequences x with histogram k. As soon as this symmetry is broken, things become more complicated. This is because the notion of multiplicity, which uniquely quantifies the degeneracy of microstates with respect to the macrostate (here, this is the histogram), may no longer exist. As a consequence, Boltzmann entropy, which is the logarithm of multiplicity, will not exist. In general, we are now forced to abandon the statistics of events and focus on the statistics of paths. It is therefore surprising that for several simple path-dependent processes, a factorization of $P(k|q)$ into a multiplicity term and constraint terms is still possible. If this can be accomplished, a maximum entropy principle for path-dependent processes does exist. We will see that the entropy functional will not then be of the Boltzmann–Gibbs–Shannon type, but is the (c, d)-entropy.

6.5.2 When does a maximum entropy principle exist for path-dependent processes?

As before, we make use of the Shannon–Khinchin axioms introduced in Section 6.2.2.10. The first axiom SK1 states that entropy depends only on the probabilities p_i. Multiplicity depends on the histogram $k = pN$, and must not depend on other parameters. To get rid of the N-dependence in the multiplicity, we must introduce a scaling factor $\phi(N)$ that removes the remaining N-dependence from entropy (logarithm of multiplicity), such that SK1 is fulfilled.[24] If a factorization $P(k|q) = \hat{M}(k)\hat{G}(k|q)$ exists, then a generalized maximum entropy principle also exists with,

$$\underbrace{\frac{1}{\phi(N)} \log P(k|q)}_{-\text{gen. relative entropy}} = \underbrace{\frac{1}{\phi(N)} \log \hat{M}(k)}_{\text{generalized entropy}} + \underbrace{\frac{1}{\phi(N)} \log \hat{G}(k|q)}_{\text{constraints}}. \qquad (6.89)$$

---

[24] SK1 ensures that the factorization of $P(k|q) = M(k)G(k|q)$ into a q-independent multiplicity $M(k)$, and a q-dependent constraint term $G(k|q)$ is not arbitrary.

6.5.2.1 *Towards a generalized maximum entropy principle*

The following arguments are technical. The reader who is not so mathematically inclined can skip this section and continue with the summary. We proceed as before in Section 6.2.3. To obtain the most likely distribution function we try to find the most likely histogram, meaning histogram k^* that maximizes $P(k|q)$ for a given N. We denote the generalized relative entropy by,

$$D(p|q) = -\frac{1}{\phi(N)}\log P(k|q) \qquad \text{where} \qquad p = k/N. \tag{6.90}$$

The maximal distribution $p^* = k^*/N$ therefore minimizes $D(p|q)$, and is obtained by solving,

$$\frac{\partial}{\partial p_i}\left(D(p|q) - \alpha\left(\sum_{j=1}^{W} p_i - 1\right)\right) = 0, \tag{6.91}$$

for all $i = 1, 2, \cdots, W$. α is the Lagrange multiplier that ensures the normalization of the probability distribution p. To compute the derivative, let us define e_i as a W-dimensional vector, whose ith component is 1, and all the others are 0. With e_i and the definition of the derivative we have,

$$\frac{\partial}{\partial p_i}D(p|q) \sim \frac{N}{\phi(N)}\frac{\log P(k_1,\ldots,k_i-1,k_{i+1},\ldots,k_W|q) - \log P(k_1,\ldots,k_W|q)}{1}$$
$$= \frac{N}{\phi(N)}\log\frac{P(k-e_i|q)}{P(k|q)} = \frac{N}{\phi(N)}V_i(k|q), \tag{6.92}$$

where N comes from the change from p to k and $V_i(k|q) = \log P(k-e_i|q) - \log P(k|q)$. Obviously, $V_i(k|q)$ is the ith component of a W-dimensional vector $V(k|q)$. The idea behind the factorization is to find a q-independent basis where the first coordinate of V does not depend on q. If such a basis exists we can write the generalized relative entropy in two terms, namely, where we identify entropy with the first component and cross-entropy, which contains the constraints of the process, with the remaining components. The problem of factorizing $P(k|q)$ reduces to finding an appropriate basis for V.

Let $b_{ji}(k)$ be the ith component of the jth basis vector for any given histogram, k, so that we can write,

$$V_i(k|q) = \sum_{j=1}^{W} c_j(k|q)b_{ji}(k), \tag{6.93}$$

where $c_j(k|q)$ are the coordinates of V in the (not necessarily orthonormal) basis $b_{ji}(k)$. As can be easily verified, not all bases are compatible with SK1-SK3 (see condition (i) in the following theorem). For reasons that become clear below, we choose the Ansatz,

$$b_{ji}(k) = \frac{\kappa_{ji}}{\gamma_T(N, k_i)} \log \frac{M_{u,T}(k - e_i)u(N)}{M_{u,T}(k)}, \tag{6.94}$$

where the functions $M_{u,T}(k)$ are so-called *deformed multinomial* factors, which are based on *deformed factorials*. For their respective definitions, see Section 8.1.8 and 8.1.7. $M_{u,T}$ depends on two functions, u and T. κ_{ji} are appropriately chosen constants, and $\gamma_T(N, r) = N [T(r/N) - T((r - 1)/N)]$ is a factor that depends on function T. With the basis in Equation (6.94) we can write,

$$\frac{P(k - e_i|q)}{P(k|q)} = \prod_{j=1}^{W} \left(\frac{M_{u,T}(k - e_i)u(N)}{M_{u,T}(k)} \right)^{\frac{c_j(k|q)}{\gamma_T(N, k_i)} \kappa_{ji}} = \prod_{j=1}^{W} u \left(NT \left(\frac{k_i}{N} \right) \right)^{c_j(k|q)\kappa_{ji}}. \tag{6.95}$$

Note that this can be done for *any* process that produces sequences of length N, $x(N) = (x_1, x_2, \cdots, x_N)$, and x_n takes values from $x_n \in \{1, \ldots, W\}$. We can now formulate the following.

Theorem *Consider processes that produce sequences $x(N) = (x_1, x_2, \cdots, x_N)$, with $x_n \in \{1, \cdots, W\}$ and that can be parametrized by variables q. The process produces histograms k with probability $P(k|q)$. Let N be large and $k^*(N|q)$ be the histogram that maximizes $P(k|q)$. Assume that a basis of the form given in Equation (6.94) can be found, for which two conditions hold:*

(i) $\kappa_{1i} = 1$, for all $i = 1, \ldots, W$

(ii) for fixed values of N and q, the coordinate $c_1(k|q)$ of $V(k|q)$ in this basis, as defined in Equation (6.93), becomes a non-zero constant at $k^*(N|q)$.[25]

Under these conditions $P(k|q)$ factorizes, $P(k|q) = M_{u,T}(k)G_{u,T}(k|q)$, with,

$$\frac{G_{u,T}(k - e_i|q)}{G_{u,T}(k|q)} = \prod_{j=2}^{W} u \left(NT \left(\frac{k_i}{N} \right) \right)^{c_j(k|q)\kappa_{ji}}. \tag{6.96}$$

Moreover, there is a maximum entropy principle with generalized entropy $S(p) = \frac{1}{\phi(N)} \log M_{u,T}(k)$, for some scaling function $\phi(N)$. The factors $u(.)^{c_j(k|q)\kappa_{ji}}$ in Equation (6.96) represent the constraint terms in the maximum entropy principle.[26] The solution of the maximum entropy principle is given by $p^* = k^*/N$.

---

[25] Condition (ii) means that the first derivatives of $c_1(k|q)$ vanish at $k = k^*$ under the constraint, $\sum k_i = N$.

[26] Typically, processes have relevant and irrelevant dimensions depending on whether components of the process pointing in the direction of a particular basis vector die out by themselves or not. Only relevant dimensions need to be (experimentally) constrained. These define the macro variables needed to describe the macrostate of the system. In this direct sense, the relevant dimensionality of a system can be measured, that is, in numbers of the required independent macrostate variables.

The meaning of the theorem is that the existence of a maximum entropy principle can be seen as a geometric property of a given process. This reduces the problem to finding a convenient basis that does not violate axioms SK1-SK3. The latter is guaranteed by the theorem. A convenient basis is given in Equation (6.94).

Condition (ii) of the theorem guarantees the existence of primitive integrals $M_{u,T}(k)$ and $G_{u,T}(k|q)$. If condition (i) is violated, the first basis vector b_{1i} of Equation (6.94) introduces a functional in p that will generally violate SK2. Conditions (i) and (ii) together determine $S(p)$ up to a multiplicative constant c_1, which can be absorbed in a normalization constant. $G_{u,T}$ may be difficult to construct in practice. However, for solving the maximum entropy principle, it is not necessary to know $G_{u,T}$; it is sufficient to know the derivatives of the logarithm. These are obtained by taking the logarithm of Equation (6.96).

6.5.2.2   *If a factorization exists—what then?*

We have seen that for some path-dependent systems (those compatible with the conditions of the theorem) a factorization for the general case of non-multinomial processes does indeed exist, if we replace $\hat{M}$ and $\hat{G}$ in Equation (6.89) by the deformed multinomial $M_{u,T}$, and a corresponding $G_{u,T}$, respectively. The latter is given in Equation (6.96). In analogy to Equation (6.42), the factorization reads,

$$\underbrace{\frac{1}{\phi(N)}\log P(k|q)}_{-\text{generalized rel. ent.}} = \underbrace{\frac{1}{\phi(N)}\log M_{u,T}(k)}_{\text{generalized ent. } S(p)} + \underbrace{\frac{1}{\phi(N)}\log G_{u,T}(k|q)}_{-\text{generalized cross-ent.}}. \qquad (6.97)$$

$\phi(N)$ has to be chosen such that for large N the *generalized relative entropy*, $-\frac{1}{\phi(N)}\log P(k|q)$ neither becomes 0, nor diverges for large N. $S(p) = \frac{1}{\phi(N)}\log M_{u,T}(k)$ is the *generalized entropy*, and $C(p|q) = -\frac{1}{\phi(N)}\log G_{u,T}(k|q)$ is the *generalized cross-entropy*.

> In complete analogy with the multinomial case, the generalized cross-entropy equals the generalized entropy plus the generalized relative entropy.

6.5.2.3   *Deriving the maximum entropy for path-dependent processes*

We can now compute the generalized entropy from Equation (6.97) by using the deformed multinomials and their definition from Section 8.1.8,

$$S(p) = \lim_{N\to\infty} \frac{1}{\phi(N)}\log M_{u,T}(k)$$

$$= \frac{1}{\phi(N)}\left[\sum_{n=1}^{N}\log u(n) - \sum_{i=1}^{W}\sum_{n=1}^{NT(k_i/N)}\log u(n)\right]$$

$$= \sum_{n=1}^{N} \frac{1}{N} \frac{N \log u(n)}{\phi(N)} - \sum_{i=1}^{W} \sum_{n=1}^{NT(p_i)} \frac{1}{N} \frac{N \log u(n)}{\phi(N)}$$

$$= \int_{0}^{1} dy \frac{N \log u(Ny)}{\phi(N)} - \sum_{i=1}^{W} \int_{0}^{T(p_i)} dy \frac{N \log u(Ny)}{\phi(N)}$$

$$= -\sum_{i=1}^{W} \int_{0}^{p_i} dz\, T'(z) \frac{N \log u(NT(z))}{\phi(N)} + \int_{0}^{1} dz\, T'(z) \frac{N \log u(NT(z))}{\phi(N)}, \qquad (6.98)$$

Here, $T'(z)$ is the derivative with respect to z. We replaced the sums over n by integrals, which is justified for large N. We immediately find that the resulting generalized entropy is of trace form $\sum_{i=1}^{W} g(p_i)$. Recall the definition of the generalized logarithm from Equation (6.69) as $\Lambda = -\frac{d}{dx} g(x)$, and that transformations of the form $S(p) \rightarrow \mu S(\lambda p) - \delta$ for $\mu, \lambda > 0$ do not change the (c, d)-equivalence class of S, which was explicitly shown in [173, 177, 365]. This then allows us to identify the generalized logarithm Λ by comparing the last line of Equation (6.98) with trace form entropies $S(p) = -\sum_{i=1}^{W} \int_{0}^{p_i} dz \Lambda(z)$. We find that,

$$\Lambda(z) = \mu T'(z) \frac{N}{\phi(N)} \log u(NT(z)) - \delta, \qquad (6.99)$$

where $\mu > 0$ and δ are constants that can be determined by using the properties of the generalized logarithm, $\Lambda(1) = 0$ and $\Lambda'(1) = 1$. Equation (6.99) is a differential equation that can be solved by a separation Ansatz. Taking derivatives of Equation (6.99), first with respect to z, and then with respect to N, one solves the equation by separation of variables with a separation constant ν. Setting $\delta = \log \lambda$ one gets,

$$\Lambda(z) = \frac{T'(z) T(z)^{\nu} - T'(1)}{T''(1) + \nu T'(1)^2}$$

$$u(N) = \lambda^{(N^{\nu})}$$

$$\phi(N) = N^{1+\nu}. \qquad (6.100)$$

For any appropriate choice of T and ν, see Section 8.1.8, one obtains the (c, d)-entropies [175, 176].

For the simplest possible example, let us specify the function T to be the identity, $T(z) = z$, and $\lambda > 1$. We obtain

$$S(p) = \frac{1 - \sum_{i=1}^{W} p_i^c}{c - 1} \qquad \text{where} \qquad c = 1 + \nu, \qquad (6.101)$$

which is of (c,d)-universality class $(c,0)$, or Tsallis entropy.[27] Other choices of T will in general lead to other (c,d)-entropies with other values for c and d.

6.5.2.4   *The most likely distribution function for path-dependent processes*

The expressions $\phi(N)$ and $u(x)$ from Equation (6.100) can now be used in Equations (6.91) and (6.95) to finally obtain the most likely distribution from the maximum relative entropy,

$$p_i^* = T^{-1}\left(\left[\frac{\log \lambda}{\alpha} \sum_{j=1}^{W} c_j(Np^*|q)\kappa_{ji}\right]^{-\frac{1}{\nu}}\right), \tag{6.102}$$

which must be solved self-consistently, as p^* also appears in c_j on the right-hand side; α is the Lagrange multiplier. Here T^{-1} is the inverse function of T. If only the first two basis vectors are relevant (the generalized entropy and one constraint term), we get distributions of the form,

$$p_i^* = T^{-1}\left(\left[1 + \nu(\hat{\alpha} + \hat{\beta}\epsilon_i)\right]^{-\frac{1}{\nu}}\right), \tag{6.103}$$

with $\hat{\alpha} = \frac{1}{\nu}(\frac{\log \lambda}{\alpha} c_1 - 1)$, $\hat{\beta} = \frac{\log \lambda}{\alpha \nu} c_2(Np^*|q)$. In a polynomial basis, specified by $\kappa_{ji} = (i-1)^{j-1}$, the equally spaced 'energy levels' are given by $\epsilon_i = (i-1)$. Note that $c_1 = 1$, and $c_2(p^*N|q)$ depend on bias terms. Again, for the simplest choice of $T(x) = x$, Equation (6.103) becomes a q-exponential, with the tail asymptotically approaching a power law.

6.5.3   Example—maximum entropy principle for path-dependent random walks

We now show the existence of stochastic path-dependent, out-of-equilibrium processes, whose time-dependent distribution functions can be predicted by the maximum entropy principle. Consider a W-dimensional random process producing sequences x that increase in length at every time step N. At time N the sequence is $x(N) = (x_1, x_2, \cdots, x_N)$, with every element taking values from $x_i \in \{1, 2, \cdots, W\}$. The probability $p(i|k,q)$ of producing $x_{N+1} = i$ depends on the histogram k of the sequence $(x_1, x_2, \cdots, x_N)$, and the prior probabilities (or biases) q only. The probability of finding a given histogram, $P(k|q)$ can be defined recursively,

$$P(k|q) = \sum_{i=1}^{W} p(i|k - e_i, q)P(k - e_i|q), \tag{6.104}$$

where e_i is a W-dimensional vector, whose component i is one, and all other entries are zero.

---

[27] In fact, all power law functions $T(x) = x^k$, with $k > 0$, lead to Tsallis entropy.

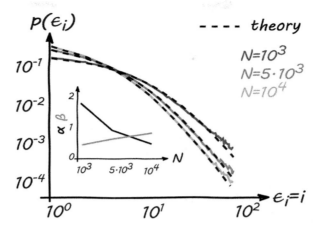

Figure 6.10 *Distribution functions of a path-dependent W-dimensional random process, based on the deformed factorial $N!_u$ with $u(r) = (\lambda^{(r^v)} - 1)/(\lambda - 1)$. Numerical results of the distributions of the process that follows Equation (6.105), are shown for $W = 50$ states i, $\lambda = 1.1$, and $v = 0.25$ ($c = 1.25$) for various lengths N (solid lines). Distributions follow the theoretical result (dashed lines) from Equation (6.103). Dashed lines show $p_i = (1 - (1 - c)(\alpha + \beta\epsilon_i))^{1/(1-c)}$, with $\epsilon_i = i - 1$. The inset shows the change of β with iterations N.*

For a concrete example,[28] let us think of a process with the specific transition probability,

$$p(i|k,q) = \frac{q}{Z(k)} \prod_{j=i+1}^{W} f(k_j) \quad \text{with} \quad f(y) = \lambda^{y^v}, \tag{6.105}$$

where $Z(k)$ is the normalization constant, and $\lambda > 0$. $p(i|k,q)$ is the probability of sampling state i in the next step, given the initial prior probabilities $q = (q_1, \cdots, q_W)$ and the fact that the process up to this point in time has produced the histogram, k. Let us choose the simplest prior distribution $q_i = 1/W$. The process defined in Equation (6.105) becomes more persistent as the sequence gets longer. In the beginning, all states are equiprobable. With every realization of state i in the process, all states $j < i$ become even more probable—the process is self-reinforcing. In this sense, the process shares similarities with the Pólya urn process that we discuss in Section 6.6.1. One can now find adequate basis vectors $b_{ji}(k)$ using the deformed multinomials $M_{u,T}(k)$ based on $u(y) = \lambda^{(y^v)}$ and $T(y) = y$. We also use a polynomial basis, $\kappa_{ji} = (i - 1)^{j-1}$. From Equation (6.103) we finally obtain $p_i^* = [1 - (1 - c)(\alpha + \beta\epsilon_i)]^{1/(1-c)}$. In Figure 6.10 the distribution functions p_i, obtained from numerical simulations of sequences with $W = 50$ states, are shown for sequence lengths $N = 1000$, 5000, and $10\,000$ (solid lines). The distributions exactly follow the theoretical result (dashed lines), p^*. While the power exponent $-\frac{1}{v}$ does not change with N, β increases with N (inset), which indicates that the process becomes more persistent as it evolves; it 'ages'.

---

[28] This is a process invented for the sake of the example. It might not be of any practical relevance.

In summary, we pursued the following line of reasoning. We first specified the dynamics of a path-dependent process. From its update rules, we followed a recipe to compute an entropy functional. Maximization of this functional yields the prediction for the distribution function at any time in the future. The prediction is in excellent agreement with numerical simulations of the process.

We explored the general conditions under which a maximum entropy principle for non-multinomial systems exists. In particular, we focused on processes that are non-Markovian and path-dependent. We formulated a theorem stating that a maximum entropy principle exists for a range of path-dependent processes, where the probability of sampling the next state depends on the histogram k of the states sampled up to that point, $p(i|k,q)$. As a consequence of the first Shannon–Khinchin axiom SK1, for such processes, a factorization into a generalized multiplicity and a probability term does, in fact, exist. The generalized multiplicity takes the form of deformed multinomial factors $M_{u,T}$. The generalized entropy is simply the logarithm of the deformed multinomial, $S(p) = \frac{1}{\phi(N)} \log M_{u,T}(k)$. Entropy is the logarithm of multiplicity.

Some path-dependent processes can be understood with statistics based on deformed multinomials. The structure of deformed multinomial factors implies two important facts:

- The maximum entropy is always a (c,d)-entropy. In the simplest case ($T(x) = x$), we find the (c,d)-equivalence class $(c,0)$, and $S_{c,0}$, that is, Tsallis entropy emerges naturally.
- Deformed multinomials always lead to trace form entropies.

In Section 6.3.2 we required entropies to be of trace form. This requirement is not necessary for systems and processes that are governed by deformed multinomial statistics.

6.6   The three faces of entropy revisited

We conclude this chapter with an extended example. In Section 6.2.4 we showed that for simple systems or processes, entropy is degenerate in the sense that thermodynamic extensive entropy, information-theoretic entropy, and the entropy-based inference method of the maximum entropy principle, all lead to the same Boltzmann–Gibbs–Shannon functional,

$$S_B = S_S = S_{\text{MEP}} = -\sum_i p_i \log p_i. \tag{6.106}$$

We understood that the origin of this degeneracy is the fact that the processes underlying such systems are all based on multinomial statistics of multistate Bernoulli processes. In the following example, we will demonstrate for two explicitly path-dependent processes that the degeneracy of entropy, which is present in simple systems, vanishes. For these processes we will find distinct expressions for the extensive, information-theoretic, and statistical inference-based entropy. We consider two processes, the Pólya urn process, and the sample space reducing (SSR) process, which we encounter in Section 3.3.5. Both processes are models of i processes that are frequently found in complex systems. The Pólya urn process is a process that is tightly related to history-dependent, preferential, self-reinforcing selection processes. SSR processes are associated with systems that become more constrained over time. The number of states that can be reached by the system reduces as the process evolves in time. They possibly provide the simplest model of driven dissipative systems, and offer a particularly simple framework for understanding the ubiquity of power laws in history- or path-dependent processes.

6.6.1   The three entropies of the Pólya urn process

The following section is mathematically challenging. However, it is not necessary to understand the mathematical details to grasp the basic message. The reader may jump from here to the main results in Section 6.7.

6.6.1.1   The Pólya urn process

Pólya urns are models for understanding self-reinforcing processes. A Pólya urn process is based on an urn that is initially filled with balls of different colours [112, 314]. Every colour is seen as a state; if there are ten colours in the urn, we have $W = 10$ states. The essence of a Pólya urn process is that every time a ball of a given colour is drawn from the urn, it is returned into the urn and an additional δ balls of the same colour are added; see Figure 6.11. These processes are self-reinforcing processes. By drawing a certain colour, because of the subsequent influx of δ new balls of the same colour, the prior probability of drawing this colour in the future is increased. In that sense, the Pólya urn process is path-dependent; it shows the *rich-get-richer* and the *winner-takes-all* (WTA) dynamics. In the limit of no reinforcement, $\delta \to 0$, we obtain a simple multinomial urn process, that is, drawing with replacement.

Pólya urns are related to the beta-binomial distribution, Dirichlet processes, Chinese restaurant problem, and several models of population genetics. Their mathematical properties are well studied in the literature [211, 394]. Pólya urns have been used in a wide variety of practical applications, including response-adaptive clinical trials [374], tissue growth models [51], institutional development [96], computer data structures [20], resistance to reform in EU politics [150], ageing of alleles and Ewens's sampling formula [106, 197], image segmentation and labelling [26], and the emergence of novelties in evolutionary scenarios [9, 376].

A Pólya urn is initially filled with a_i balls of colour $i = \{1, \cdots, W\}$. One draws the first ball of colour $x_1 = i$ with probability $q_i = a_i/A$, where $A = \sum_{i=1}^{W} a_i$ is the number of all balls initially contained in the urn. If we draw a ball of colour i we replace it and add another δ balls of the same colour to the urn. As a consequence, the probability of drawing another colour (state) i after drawing N times is given by,

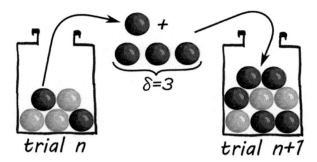

Figure 6.11 *Schematic illustration of a Pólya urn process. When a ball of a given colour is drawn, it is replaced by $1 + \delta$ balls of the same colour. Then the next ball is drawn and the process is repeated for N iterations. Here, $\delta = 3$ and we have two states, $W = 2$, grey and black. This process creates a self-reinforcing, history-dependent dynamics. The underlying statistics is of a non-multinomial nature.*

$$p(i|k(N),\theta) = \frac{a_i + k_i \delta}{A + N\delta} = \frac{q_i + k_i \gamma}{1 + N\gamma}, \qquad (6.107)$$

where $k(N)$ is the histogram of the colours drawn up to step N; $\gamma = \delta/A$ is the reinforcement parameter; $\theta = (q, \gamma)$ is the set of parameters that characterize the process; and q_i are the prior probabilities. If $\gamma = 0$, the Pólya urn process is simply *drawing with replacement*, in which case it is equivalent to a Bernoulli process. If $\gamma > 0$, then the probability of drawing colour i in the $(N + 1)$th sample depends on the *history* of samples x in terms of histograms $k(N)$.

Pólya urn processes exhibit a cross-over behaviour between dynamics that asymptotically behave like Bernoulli processes (weak reinforcement) and WTA dynamics (strong reinforcement). For intermediate reinforcement strength γ, the way the system behaves depends on the details of the random events that happen early on in the process—how sequences of samples, $x(N) = (x_1, \cdots, x_N)$ behave for large N depends on samples, x_n, that were drawn at times n much smaller than N. The 'young' Pólya urn process 'decides' if it equilibrates and becomes a Bernoulli process, or if it enters a WTA scenario.

Pólya urns operate on the edge of WTA dynamics. How can we understand this statement? If we measure the histogram $k(N_1)$ of the Pólya urn process after N_1 steps, then we may continue the process by thinking of it as a modified Pólya urn in a new initial condition, where we redefine $a_i(N_1) = a_i + k_i(N_1)\delta$ and $A(N_1) = A + \delta N_1$. For this modified Pólya urn process, the parameters have been modified from $\theta = (q, \gamma)$ to $\theta' = (q', \gamma')$ with,

$$\begin{aligned} q_i' &= \frac{q_i + \gamma k_i(N_1)}{1 + \gamma N_1}\gamma \\ \gamma' &= \frac{\gamma}{1 + N_1 \gamma}. \end{aligned} \qquad (6.108)$$

As a consequence, the effective reinforcement reduces, $\gamma' < \gamma$, and $\gamma'(N_1) \to 0$ as $N_1 \to \infty$. The distribution q' becomes modified by the random history of the process $x(N_1)$, and the effective reinforcement parameter γ' decreases over time. Whether a particular realization of a process (specified by θ) enters the WTA dynamics depends on whether the modified Pólya urn (specified by θ') enters the WTA dynamics or not. This depends on the paths that $x(N_1)$ took within the first N_1 steps. If, in those first steps, one of the states i can acquire most of the weight, the process will enter WTA dynamics, that is, i will be sampled almost always in the future.[29]

We can now discuss the *three entropies* of Pólya urn processes. In the following, we focus only on WTA scenarios that are associated with strong reinforcement. For weak reinforcement we asymptotically face the situation of a multinomial process, where the three entropies become degenerate (Boltzmann–Gibbs–Shannon formula) and Section 6.2.4 applies.

6.6.1.2   The information production entropy of the Pólya urn process

Before we compute the information production rate for the Pólya urn process, we have to clarify what the entropy rate for path-dependent systems means. In Equation (6.36) we defined the entropy rate of a sequence $x(N)$ as the logarithm of the joint probability of sampling the specific sequence $x(N) = (x_1, x_2, \cdots, x_N)$,

$$S_{\mathrm{ER}}(x(N)) = -\frac{1}{N} \log p(x_1, x_2, \cdots, x_{N-1}, x_N). \tag{6.109}$$

Remember that this is the shortest code (in bits per symbol), given that our alphabet consists of messages of length N. In Section 6.2.2 we discussed Shannon's question of how many bits per letter are needed on average to transmit Markovian messages through possibly noisy information channels.

In path- or history-dependent processes, each x_n may depend on earlier events, x_m, with $m < n$. This allows us to write the probability of sampling the sequence $p(x(N))$ as,

$$p(x(N)) = \prod_{n=1}^{N} p(x_n | x(n-1)). \tag{6.110}$$

If we denote the initial sequence as the empty set by $x(0) = \emptyset$, and the initial distribution by $q_i = p(i|\emptyset)$, we obtain for the entropy rate,

$$S_{\mathrm{ER}}(x(N)) = -\frac{1}{N} \sum_{n=1}^{N} \log p(x_n | x(n-1)). \tag{6.111}$$

[29] Non-linear Pólya urns, where the effective reinforcement decays more slowly as time progresses, almost certainly enter WTA dynamics.

We are now ready to compute the entropy rate for the Pólya urn process. In WTA scenarios, the relative frequency p_j of observing the winner j approaches 1, that is, p concentrates on the winner state by essentially sampling only that state. The process may run for several steps (N_0) and after those it practically only samples the winner j, while all other 'loser states', $i \neq j$, are never drawn. Without knowing the exact initial distribution of the loser states, we can assume that they all have equal probabilities, and using Equation (6.107) we get,

$$p_i(N) \sim \frac{1 - p_j(N)}{W - 1} = \frac{1 - q_j}{(W - 1)(1 + \gamma N)}, \tag{6.112}$$

where q_j is the probability of initially sampling the winner j. $p_i(N)$ means the probability of sampling i at time step N. Therefore, the probability of drawing the winner again at the Nth trial becomes,

$$p_j(N) = 1 - \sum_{i \neq j}^{W-1} p_i(N) \sim 1 - \frac{1 - q_j}{1 + \gamma N}. \tag{6.113}$$

If we draw the winner j repeatedly in a sequence x for N times in a row, to obtain the probability of x we have to multiply all the probabilities $p_j(n) = p(x_n|x(n-1))$ for $n = 1, \cdots, N$, and compute the information production $S_{\mathrm{ER}}(x) = -\frac{1}{N} \log \prod_{n=1}^{N} p_j(n) = -\frac{1}{N} \sum_{n=1}^{N} \log p_j(n)$. Finally, we arrive at,

$$S_{\mathrm{ER}}(x) = -\frac{1}{N} \sum_{n=1}^{N} \log p(x_n|x(n-1)) = -\frac{1}{N} \sum_{n=1}^{N} \log\left(1 - \frac{1 - q_j}{1 + \gamma n}\right) \sim \frac{1}{N} \frac{1 - q_i}{\gamma} \log N, \tag{6.114}$$

where, in the last step, we use that $\log(1 - x) \sim 1/x$ and $\sum_{n=1}^{N} \frac{1}{n} \sim \log N$. The entropy production per sample of a Pólya urn process asymptotically approaches zero as $S_{\mathrm{ER}} \sim \mathcal{O}(\log(N)/N)$. Intuitively, this is exactly what one would expect, namely, the surprise factor becomes very low. The total entropy production, that is, the number of bits required to encode the entire sequence grows logarithmically, $N S_{\mathrm{ER}} \sim \log N$.

6.6.1.3   *The extensive entropy of the Pólya urn process*

If a process does not visit all possible W states in its phasespace but effectively occupies only a subset of $\bar{W}$, what does this mean? It could mean that the probability of only those $\bar{W}$ states is larger than zero. However, this view is too restrictive in many circumstances, and it is wise to consider another interpretation. Remember that if $p_i = 1/\bar{W}$ on $\bar{W}$ states, then sampling independently from this distribution yields $S_S(p) = \log \bar{W}$. We may now think of an arbitrary distribution p on W states (not concentrated on $\bar{W}$ states) with the property that $\log \bar{W} = S_S(p)$ and think of p as being *effectively concentrated* on only $\bar{W}$ states. As a consequence, we can compute $\bar{W}$ of a sequence $x(N)$ sampled from a Pólya urn in WTA dynamics through $\bar{W}(n) \sim \exp(S_S(p(n)))$, where $p(n)$ is the distribution of balls in the urn after $n-1$ steps. Using p from Equation (6.113) and (6.112) in the Shannon entropy S_S we get,

$$\bar{W}_{\text{Pólya}}(N) \sim (1+\gamma N)^{\frac{1-q_j}{\gamma}}, \qquad (6.115)$$

where we again use the approximations $\log(1+x) \sim x$ and $\sum_{n=1}^{N} 1/n \sim \log N$. From Equation (6.77) it follows that $1/(1-c) = (1-q_j)/\gamma$ and $d=0$. Therefore, the (c,d)-equivalence class of the Pólya urn process is $(1 - \frac{\gamma}{1-q_j}, 0)$, and the entropy is given by,

$$S_{\text{EXT}} = S_{c,0} \qquad \text{with} \qquad c = 1 - \frac{\gamma}{1-q_j}. \qquad (6.116)$$

Note that c becomes negative for sufficiently large γ, which means that the axiom SK3 can be violated by Pólya urns that experience WTA dynamics. If SK3 is violated, the predictions of Equation (6.77) and (6.78) may no longer be accurate. However, it can be shown that $S_{\text{EXT}} = S_{c,0}$ also holds for violated SK3; see [363].

6.6.1.4   *The maximum entropy of the Pólya urn process*

For Pólya urn processes, the probability of observing a sequence $x(N)$ is the same as the probability of observing any other sequence $x'(N)$ if the respective histograms are the same, $k(N) = k'(N)$. As a consequence, the probability $P(k|\theta) = M(k)G(k|\theta)$ factorizes into the multiplicity $M(k)$, which is given by the multinomial factor and the sequence probability, $G(k|\theta)$. One might conclude that the number of degrees of freedom therefore scales like $\phi = N$ and that the Pólya maximum entropy is Jaynes maxent entropy, S_J, plus generalized cross-entropy terms. If γ is small, this is indeed the case, as the urn essentially behaves like a Bernoulli process. If γ is sufficiently large, the Pólya urn is likely to enter the WTA dynamics if one state is selected repeatedly at the very beginning of the process and gains an advantage over all the other states.

For a sufficiently large γ, one finds that $G(k|\theta)$ can be written as $G(k|\theta) = \tilde{M}(k)\tilde{G}(k|\theta)/M(k)$, so that $MG = \tilde{M}\tilde{G}$. For details, see [168]. This means that the probability for the histogram $P = \tilde{M}\tilde{G}$, no longer depends on the multinomial factor M at all.[30] For $\gamma > 0$ the expression $\log\tilde{M}$ scales very differently than multinomial multiplicities do. With $\phi = 1$ the maxent functional, $S_{\text{MEP}} = \frac{1}{\phi}\log\tilde{M}$ becomes a well-defined *generalized* entropy, and $S_{\text{cross}} = -\frac{1}{\phi}\log\tilde{G}$ a *generalized* cross-entropy functional. In [168] it is shown in detail that,

$$\begin{aligned} S_{\text{MEP}} &\sim -\sum_{i=1}^{W} \log(p_i + 1/N) \\ S_{\text{cross}} &\sim -\frac{1}{\gamma}\sum_{i=1}^{W} q_i \log(p_i + 1/N). \end{aligned} \qquad (6.117)$$

---

[30] Comparing with Section 6.5.2, note that the factorization of the histogram probability P into $\tilde{M}$ and $\tilde{G}$ corresponds to a different choice of basis.

The numerical values for the WTA dynamics are,

$$S_{\text{MEP}} \sim (W-1)\log N + \text{const.}$$
$$S_{\text{cross}} \sim \frac{1-q_j}{\gamma}\log N + \text{const.}$$

$$(6.118)$$

The intrinsic instability of self-reinforcing processes makes it hard to predict the distribution function $p = (p_1, \ldots, p_W)$ with maximum entropy. However, quite remarkably, ensembles of Pólya urn processes show stable frequency *rank* distributions. If we want to predict the relative frequencies of states that are rank-ordered,[31] the corresponding rank distribution $\tilde{p} = (\tilde{p}_1, \cdots, \tilde{p}_W)$ can indeed be predicted with high accuracy. For further details, see the original literature [168].

6.6.2   The three entropies of sample space reducing processes

We discussed SSR processes in detail in Section 3.3.5. These are processes whose sample space reduces as they evolve in time. They provide a natural way of explaining the origin and the ubiquity of power laws in complex systems, and Zipf's law in particular [93]. SSR processes are typically irreversible, dissipative processes that are driven between sources and sinks. Complicated, driven, dissipative processes and self-organized critical systems can sometimes be decomposed into simpler SSR processes. Examples of SSR processes include diffusion and search processes on networks [90], fragmentation processes, sentence formation in linguistics [367], and cascading processes [92].

6.6.2.1   The SSR process

SSR processes can be viewed as processes in which the currently occupied state determines the sample space for the next step. If the system is in state i, it can sample states from a sample space Ω_i that depends on i. Ω_i contains all the possible states that can be reached from state i. Often, sample spaces are nested along the process, meaning that $\Omega_i \subset \Omega_j$, whenever state i occurs later in the process than j. As the process evolves, the sample space successively becomes smaller, hence the name 'sample space reducing'. Eventually, an SSR process ends in a sink state, $i = 1$ (Ω_1 is the empty set; there are no more states to reach) and stops. The dynamics of such systems is irreversible and non-ergodic. To obtain a statistics of the process, the process has to be restarted when it stops in the sink state. Restarting resets the sample space from Ω_1 to Ω_W, and the SSR process becomes quasi-ergodic, however, without detailed balance.[32] No detailed balance means that there is a current running through the system that leads to a stationary frequency distribution p, which is a solution to equation $p_i = \sum_{j=1}^{W} p(i|j)p_j$.

A simple way of depicting an SSR process is by a ball that bounces downwards on a staircase with random distances. It never jumps upwards. Every stair represents a state i.

---

[31] The largest frequency has rank $r = 1$, the second largest rank $r = 2$, and so on.
[32] The detailed balance condition is $p(i|j)p_j = p(j|i)p_i$. It does not hold for SSR processes, as it has a direction.

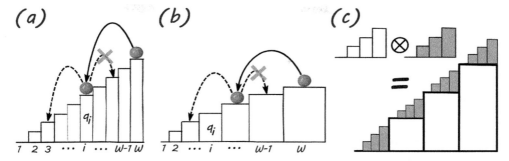

Figure 6.12 *(a) Schematic view of an SSR process. A ball bounces downwards with random jump distances but never upwards. Once it reaches state 1, the process is restarted at the top. After running the process many times, the visiting probabilities of states i approach Zipf's law, $p_i = i^{-1}$. (b) An SSR process with non-uniform prior probabilities q_i is shown. For a large class of prior probabilities, the visiting distributions still follow exactly Zipf's law. (c) Visualization of a possible combination of two SSR processes with the Cartesian product.*

State $i = 1$ corresponds to the bottom of the staircase, $i = W$ to the top; see Figure 6.12a. Obviously, successive sample spaces are nested, as the ball goes down.

A ball at step i can sample all steps j that are lower than itself $j < i$. If the steps carry prior probabilities $q = (q_1, \cdots, q_W)$—which can be intuitively thought of as the widths of the steps, Figure 6.12b—the process will visit state $j < i$ with probability q_j/Q_{i-1}, where $Q_i = \sum_{s=1}^{i} q_s$ is the cumulative distribution of q up to state i. Remarkably, regardless of q, SSR processes still follow Zipf's law in the visiting distributions, $p_i = i^{-1}$, with some rare exceptions; see Section 3.3.5 or [90]. By restarting the process whenever it reaches the sink state $i = 1$, one forces the process to become quasi-ergodic, meaning that a stationary distribution p exists, despite the fact that the SSR process itself is irreversible. By allowing the process to jump to any position with a given frequency (driving rate) $1 - \lambda$, the visiting distributions remain exact power laws, $p_i = i^{-\lambda}$. In the following, we discuss the three entropies of this 'staircase' process.

6.6.2.2   The entropy production rate of the SSR process

As SSR processes are Markov processes, the probability of sampling state x_n depends only on the previous sample x_{n-1}. We can therefore use Equation (6.37) to compute the probability of sampling the entire process, and obtain the entropy rate,

$$S_{\text{ER}}(x(N)) = -\frac{1}{N} \sum_{n=1}^{N} \log p(x_n | x_{n-1}) \sim -\sum_{i=1}^{W} \sum_{j=1}^{W} p_i p(j|i) \log p(j|i). \qquad (6.119)$$

The entropy production rate of the SSR process is simply the average *conditional entropy*, which is exactly the result we found for Markov processes. For uniform priors, $q_i = 1/W$, one obtains the numerical value of the SSR entropy production,

$$S_{\text{ER}} = p_1 \log W + p_1 \sum_{i=2}^{W} \frac{1}{i} \log i \sim \frac{1}{2} \log W + 1 + \mathcal{O}(1/\log(W)), \tag{6.120}$$

where we replaced sums $\sum_{i=a}^{b} f(i)$ by integrals $\int_{a-1/2}^{b+1/2} dx f(x)$, and assumed $p_i = p_1/i$ and $p(j|i) = 1/(i-1)$ for $j < i$. Note that $p_1 \sim 1/\log W$. The 1 in Equation (6.120) arises from the restarting procedure.

6.6.2.3 *The extensive entropy of the SSR process*

One can quantify how the phasespace volume of an SSR process grows by the number of decisions that the process takes along its path. For a Bernoulli process with W states, this means a choice from W possible successor states at *every* time step. After N samples, the process has realized one specific path out of the W^N possibilities. How does the effective number of decisions grow for SSR processes?

In Section 6.6.1 we discuss what we mean by the *effective number of states*, $\bar{W}$, that a process effectively visits. If an SSR process is in state i, it can only visit states in Ω_i, which means that it may sample from $W(i) = i - 1$ states. Restarting means that whenever we reach state $i = 1$, we can sample from all $W_1 = W$ states. A realization of a path $x(N)$ is one possibility out of,

$$\hat{W}(x(N)) = \prod_{n=1}^{N} W(x_n). \tag{6.121}$$

The average *effective* amount of choice has to be $\bar{W}^N = \hat{W}(x(N))$, and one gets,

$$\prod_{n=1}^{N} W(x_n) = \prod_{i=1}^{W} W(i)^{k_i} \sim \prod_{i=1}^{W} W_i^{p_i N} = \bar{W}^N, \tag{6.122}$$

where in the first step we use that i has been sampled k_i out of N times and in the second we set $k_i = p_i N$, where p_i are the relative frequencies of elements i in the sequence $x(N)$. From this it follows that $\bar{W} = W^{p_1} \prod_{i=2}^{W} (i-1)^{p_i}$. Using that the distribution function of the SSR process is Zipf's law, $p_i = p_1/i$, where $1/p_1 = \sum_{i=1}^{W} 1/i \sim \log W$, the average amount of choice per step involved in sampling a typical SSR sequence $x(N)$, is given by the numerical value,

$$S_{\text{EXT}} = \log \bar{W} = \frac{1}{2} \log W + 1 + \mathcal{O}(1/\log W). \tag{6.123}$$

The 1 comes again from restarting the process. This result implies that $\bar{W} \sim e\sqrt{W}$.

The definition of extensivity is tightly related to the way systems are composed. An SSR process A with W_A states can be combined with another SSR process B with W_B states to form a combined process AB by substituting an entire copy of B for each state A;

see Figure 6.12c. In this case, for the effective number of states in the combined system, we get,

$$\bar{W}_{AB} = e\sqrt{W_A W_B} = \frac{1}{e}\bar{W}_A \bar{W}_B. \tag{6.124}$$

If process A is composed N times with itself in the same manner, we obtain $\bar{W}_{A^N} = e(\bar{W}_A/e)^N$. In other words, the quasi-ergodic SSR process has an exponentially growing phasespace and the extensive entropy is the Boltzmann–Gibbs–Shannon formula,

$$S_{\text{EXT}}(p) = S_S(p) = -\sum_{i=1}^{W} p_i \log p_i. \tag{6.125}$$

This is an amazing result. For other ways of combining SSR processes, which may very well exist, S_{EXT} is not necessarily given by the Boltzmann–Gibbs–Shannon entropy.

6.6.2.4   *The max entropy of the SSR process*

To complete the example we have to derive the maximum entropy principle for the SSR process, which involves some combinatorial work. Assume that the SSR process (with restarting) X has the histogram $k = (k_1, \ldots, k_W)$. We have to determine its probability $P(k|q) = M(k)G(k|q)$ after N observations and then find the maximum configuration k^* that maximizes $P(k|q)$. To compute M we first decompose any sampled sequence $x(N) = (x_1, \cdots, x_N)$ into shorter sequences x^r, such that $x(N) = (x^1(N_1), x^2(N_2) \cdots x^R(N_R))$ is a concatenation of such shorter sequences. Note that $\sum_{r=1}^{R} N_r = N$. Every sequence $x^r(N_r)$ is a sample of executing X until it stops in the sink state. We refer to x^r as one 'run' of X. This means that any run $x^r(N_r) = (x_1^r, x_2^r, \cdots x_{N_r}^r)$ is a monotonously decreasing sequence of states, $x_n^r > x_{n+1}^r$, that ends at state $x_{N_r}^r = 1$, where the run stops and the process is restarted. As every run ends in the sink state $i = 1$, the number of runs is equal to the number of times state 1 is sampled, meaning that $R = k_1$. Arranging $x(N)$ in a table with W columns and k_1 rows allows us to determine the probability G and the multiplicity M of a sequence x. We denote a state that is visited in a run by a star $*$, and a state that is not visited by a minus sign $-$.

$r \times i$	W	$W-1$	$W-2$	$\cdots$	2	1
1	$*$	$-$	$-$	$\cdots$	$*$	$*$
2	$-$	$*$	$*$	$\cdots$	$-$	$*$
3	$*$	$-$	$*$	$\cdots$	$-$	$*$
$\vdots$	$\vdots$	$\vdots$	$\vdots$	$\vdots$	$\vdots$	$\vdots$
$R-2$	$-$	$*$	$*$	$\cdots$	$-$	$*$
$R-1$	$-$	$*$	$-$	$\cdots$	$*$	$*$
R	$-$	$-$	$*$	$\cdots$	$-$	$*$
	k_W	k_{W-1}	k_{W-2}	$\cdots$	k_2	k_1

$$\tag{6.126}$$

From this scheme we can directly read off the number M of sequences $x(N)$ that have the same histogram k. Note that column i is $k_1 = R$ entries long and contains k_i items. One can therefore produce all those sequences $x(N)$ by rearranging k_i visits to state i in a column of k_1 possible positions. Each column $i > 1$ therefore contributes to M with a binomial factor $\binom{k_1}{k_i} = k_1!/k_i!/(k_1 - k_i)!$. As a consequence one finds $M(k) = \prod_{i=2}^{W} \binom{k_1}{k_i}$, and the reduced maximum entropy $\frac{1}{N} \log M$ is found,

$$S_{\text{MEP}} = -\sum_{i=2}^{W} \left[p_i \log \frac{p_i}{p_1} + (p_1 - p_i) \log \left(1 - \frac{p_i}{p_1} \right) \right]. \tag{6.127}$$

Its numerical value is

$$S_{\text{MEP}} = p_1 \sum_{i=2}^{W} \left(1 - \frac{1}{i} \right) \log \left(1 - \frac{1}{i} \right) + p_1 \sum_{i=2}^{W} \frac{1}{i} \log i \sim \frac{1}{2} \log W + 1 + \mathcal{O}(1/\log W). \tag{6.128}$$

Similarly, one can determine the probability of sampling a particular sequence $x(N)$. Each visit to a state $i > 1$ in the sequence $x(N)$ contributes to the probability of the next visit to a state $j < i$ with a factor $1/Q_{i-1}$, where $Q_i = \sum_{n=1}^{i} q_n$. Only if $i = 1$ do we fail to obtain a renormalization factor, as the process restarts and all states i are valid targets with probability q_i. It thus follows that $G(k|q, N) = \prod_{i=1}^{W} q_i^{k_i} \prod_{j=2}^{W} Q_{j-1}^{-k_j}$, and the cross-entropy is,

$$S_{\text{cross}}(p|q) = -\sum_{i=1}^{W} p_i \log q_i + \sum_{i=2}^{W} p_i \log Q_{i-1}. \tag{6.129}$$

Note that the maxent entropy of SSR processes in Equation (6.128) is not of trace form, as the state $i = 1$ remains entangled with every other state j. This is obviously a consequence of the restarting procedure.[33] The SSR process is an example of a process that violates more than one SK axiom.

For sufficiently large W, the numerical values for the information production S_{ER}, for the maxent entropy S_{MEP}, and for the generalized cross-entropy S_{cross}, are all the same,

$$S_{\text{ER}} \sim S_{\text{MEP}} \sim S_{\text{cross}} \sim \frac{1}{2} \log W + \mathcal{O}(1 + 1/\log W). \tag{6.130}$$

Much of what is true for Markov processes remains true for SSR processes. SSR processes are 'made' Markovian by restarting the process whenever it stops in the sink

---

[33] Note that in the general context of driven systems, this means that we look jointly at two processes, the reloading process (driving process) and the relaxation process. The interaction of both processes will in general entangle states and maxent entropies will no longer be of trace form.

Table 6.3 *Information-theoretic entropy rate S_{ER}, extensive entropy S_{EXT}, and the maxent entropy, S_{MEP} for the Pólya urn and sample space reducing processes. The expressions are generally valid for a large number of iterations N and many states W.*

Process	S_{ER}	S_{EXT}	S_{MEP}
Pólya (WTA)	$\frac{1-q_j}{\gamma}\frac{1}{N}\log N$	$S_{1-\frac{\gamma}{1-q_j},0}$	$-\sum_i \log p_i$
SSR	$1+\frac{1}{2}\log W$	$S_{1,1}=-\sum_i p_i \log p_i$	$-\sum_{i=2}^{W}\left[p_i\log\frac{p_i}{p_1}+(p_1-p_i)\log\left(1-\frac{p_i}{p_1}\right)\right]$

state $i = 1$. Comparing Equation (6.130) with the information production of Bernoulli processes, which yields $\log W$, note that SSR processes only need half the information to encode a message of a given length.

> One important lesson we learn from SSR processes is that driven and dissipative systems show enhanced compressibility. For systems in equilibrium, the number of paths $x(N)$ that share the same histogram k, is given by $\binom{N}{k}$. This is the maximal number of paths that exist under system-specific constraints. Driven systems, which are not in equilibrium, also have to fulfil these constraints. However, as the set of equilibrium states is maximal, the set of non-equilibrium states is smaller. In other words, a non-equilibrium state is 'better localized' in sample space than equilibrium processes. In this sense stationary non-equilibrium states can have more 'structure'. We recall that additional information can only reduce entropy. The existence of these reduced subsets of phasespace that can be identified with structures, is the reason why sequences emitted from non-equilibrium systems can be compressed better.

In conclusion, we have shown in the last section that for path-dependent processes, the entropy formula is no longer degenerate for the three concepts of entropy: the extensive entropy, the information-theoretic entropy rate, and the maxent entropy. We have shown explicitly for the Pólya urn process and for the SSR process that the corresponding three entropy formulas can be computed and that they are indeed different. The results are summarized in Table 6.3. The Pólya urn process is an example of a self-reinforcing process that, with strong reinforcement, shows WTA dynamics. The SSR process is a process that describes systems that increasingly self-constrain their dynamics as they evolve. Both, self-reinforcement and sample space reduction are key features of countless complex phenomena in nature and man-made systems. It might be possible for real processes to be decomposed into simple self-reinforcing and/or SSR processes of the kind that we presented here. If this is possible one can, in principle, also deduce and understand the statistical properties of such systems in a much deeper way than is currently the case.

6.7 Summary

In this chapter, we reviewed the three most popular notions of entropy. The first notion was introduced in physics as a quantity to understand thermodynamics. This entropy is an extensive quantity and can be used to relate microscopic and macroscopic (global or systemic) variables of a system. The second notion is information-theoretic entropy and was used to quantify the amount of information that systems produce as they evolve. The third notion of entropy is used as a statistical method for inferring statistical distribution functions from data. We discuss how the mathematical expressions associated with these three concepts of entropy change when passing from simple to complex systems. We demonstrate in several concrete examples how these concepts of entropy still make perfect sense in the context of complex systems, which are generally out-of-equilibrium, non-ergodic, and do not follow simple multinomial statistics.

The main purpose of the exercise performed in this chapter was to unambiguously show that for any process that cannot be based on Bernoulli processes or traced back to them, it needs to be exactly specified which of the three concepts of entropy we are talking about. In general, all three 'entropies' have to be computed to characterize a particular complex system. To naively use the Boltzmann–Gibbs–Shannon formula, $-\sum_i p_i \log p_i$ as a one-size-fits-all approach—as is often done in the mainstream literature in physics, information theory, and maxent applications—is, in many cases, doomed to fail and leads to confusion and nonsense. In particular, we have shown the following:

- For simple systems the concepts of extensive entropy, information-theoretic entropy and the entropy that appears in the maximum entropy principle, all lead to the same entropy functional, $S = -\sum_i p_i \log p_i$. The entropy concepts are degenerate. The essence of simple systems and processes resides in the fact that they are all built on multinomial statistics of Bernoulli processes. We have shown that Bernoulli processes (independent processes) generically lead to the same entropy, regardless of which concept is used. Mathematically, for systems that fulfil the four Shannon–Khinchin axioms with respect to scaling, information production, and maximum configuration, the only possible entropy functional is, $S = -\sum_i p_i \log p_i$.

- Complex systems often are not ergodic and are driven. They thus violate the fourth Shannon–Khinchin axiom. The generic entropy formula for systems that fulfil the first three Shannon–Khinchin axioms is $S_{c,d} \sim \sum_{i=1}^{W} \Gamma(1+d, 1 - c \log p_i)$, where c and d are scaling exponents that are specific to the system. Practically all entropies suggested over the past half century are special cases of this form, including the Boltzmann–Gibbs–Shannon entropy ($S_{1,1}$), Tsallis entropy ($S_{c,0}$), and entropies related to stretched exponentials ($S_{1,\nu}$).

- The scaling exponents c and d can be used to classify all statistical complex systems that are compatible with the first three Shannon–Khinchin axioms, SK1–SK3, into (c, d)-equivalence classes.

- We learned how the phasespace of a system is related to its extensive entropy. This can be used to understand a number of problems, such as anomalous diffusion or ageing spin systems, where the number of constraints grows with system size.

- The maximum entropy principle can naturally be extended to non-multinomial systems. For systems with self-reinforcing winner-takes-all dynamics, $S_{c,d}$ appears as the natural entropy formula. For relatively simple path-dependent systems, the methodology developed allows us to start from update equations of a path-dependent system and compute its maxent entropy. Maximization under constraints provides us with the distribution functions of (possibly non-stationary) systems.

- There are complex systems that violate more than one Shannon–Khinchin axiom. We showed that it is still possible to compute the three entropies for such systems. We use Pólya urns and sample space reducing processes to exemplify how one can compute entropy formulas for the three notions of entropy. The Pólya urn process is an example of a self-reinforcing history-dependent process; the sample space reducing process is an example of a driven system that typically produces power law distribution functions. These processes are relatively simple when compared to stochastic processes that occur in actual non-egodic, evolutionary, complex adaptive systems, which themselves are often self-reinforcing, path-dependent, and frequently composed of multiple interacting subprocesses. We demonstrate, using these concrete examples, that the degeneracy of the three entropy concepts is, in general, explicitly broken for complex systems.

- The three entropy concepts capture information about distinct properties of the underlying system. It remains to be seen if, and to what extent, complex adaptive systems can be classified into families that share the same triplet of entropies (S_{EXT}, S_{ER} and S_{MEP}).

6.8 Problems

Problem 6.1 Show that the conditional Shannon entropy in Equation (6.30) is always less than or equal to the unconditional entropy, $S_S^k(B) \leq S_S(B)$.

Problem 6.2 Show that the probabilities of producing the letters A, B, C in the finite-state machine shown in Figure 6.4, $q_A = 9/27, q_B = 16/27, q_C = 2/27$, are compatible with the conditional probabilities given in Equation (6.32). Use $p_i = \sum_j p(i|j) p(j)$.

Problem 6.3 Consider a W-dimensional random walker. At every time step the walker makes one step of unit length in the positive or negative direction of the ith dimension. Assume that the prior probabilities are $q_i = 1/W$. Use the maximum entropy principle from Equation (6.47) to compute the distribution function that describes the position of the walker after N time steps.

Problem 6.4 Show that if the volume of an ideal gas composed of N particles is scaled by a factor λ, the increase in entropy is $\Delta\sigma = N\log\lambda$. Hint: express the ratio of the volumes before and after the scaling in terms of a ratio of their phasespace volumes.

Problem 6.5 Similar to Equation (6.23), show that if you use information that constrains the second moment of the distribution to $\sum_i p_i E_i^2 = K$, the maximization of the functional Φ gives a Gaussian distribution.

Problem 6.6 Show that both Equations (6.49) and (6.60) solve the two scaling relations in Equations (6.57).

Problem 6.7 Show that if Tsallis entropy is maximized with two Lagrangian parameters, α and β, in other words if

$$\Phi = -\sum_{i=1}^{W} \frac{1 - \sum_i p_i^c}{c-1} - \alpha\sum_{i=1}^{W}(p_i - 1) - \beta\sum_{i=1}^{W}(p_i E_i - U),$$

is maximized, the resulting distribution function p_i is a q-exponential. What is the value of q?

Problem 6.8 Compute c and d for the Kaniadakis entropy given in Table 6.2, $S_\kappa = \sum_i p_i(p_i^\kappa - p_i^{-\kappa})/(-2\kappa)$. Discuss its distribution function.

Problem 6.9 (a) Show that the solution to the differential equation $\frac{d}{dt}x(t) = \alpha x(t)$ is $x(t) = x_0 e^{\alpha t}$, (the exponential function). (b) Show that the solution to the equation $\frac{d}{dt}x(t) = x^q(t)$ is $x(t) = [1 + (1-q)t]^{\frac{1}{1-q}}$, (the q-exponential). (c) Show that the solution to the delayed differential equation $\frac{d}{dt}x(t) = \alpha x(t-\tau)$ is $x(t) = e^{\frac{1}{\tau}W(\alpha\tau)t}$, (the Lambert-W exponential). Here $W(x)$ is the Lambert-W function, for which $W(x)e^{W(x)} = x$ holds; see Section 8.1.4. Use the Ansatz, $x(t) = \exp(\frac{1}{\tau}f(\alpha\tau)t)$, with f some function.

Problem 6.10 Compute the Helmholtz free energy, $F = U - TS$, for the distribution function $p_i = \frac{1}{Z}e^{-\beta E_i}$. Hint: use $S = -\sum_i p_i \log p_i$ and remember that the internal energy is the average energy, $U = \langle E_i\rangle$.

Problem 6.11 Take the finite-state machine given in Figure 6.4b. The process produces a Markov process with transition probabilities $p(i|j)$ from letter j to i. How could you compute the asymptotic relative frequencies of the letters of the alphabet emitted by the machine? Implement a code on the computer (for instance, using Matlab or Octave) to simulate the finite-state machine and compare the experimental results with the theoretical results.

Problem 6.12 Suppose you are given a trace form (c, d, r)-entropy, $S_{c,d,r}(p) = \sum_{i=1}^{W} g(p_i)$, from Equation (6.49). Compute the scale transformation $\mu S_{c,d}(\lambda x, r) + \delta x$ with $\lambda > 0$, μ, and δx being constants. Determine how μ, δx, and the constant r' have to depend on $\lambda > 0$, c, d, and r for equation $\mu S_{c,d,r}(\lambda x) + \delta x = S_{c,d,r'}(x)$ to be an exact identity. Hint: compute the left-hand side $\mu S_{c,d}(\lambda x, r) + \delta x$ and the right-hand side $S_{c,d,r'}(x)$ of the equation and then compare terms.

Problem 6.13 Show that the scaling transformations of the kind $g(x) \to ag(bx)$ do not change the (c, d)-equivalence class of a trace form entropy.

Problem 6.14 Show that Equation (6.120) is indeed correct.

7

The Future of the Science of Complex Systems?

When one considers the science of complex systems, the classical concepts immediately come to mind: algorithmic complexity, genetic algorithms and networks, fitness landscapes, econophysics, allometric scaling, network theory, self-organized criticality, and percolation. All these concepts and directions are about two decades old, or older. Since then, much progress has been made on elucidating and elaborating many of the details of these topics, and the research community has grown from a few dozen to the now hundreds who participate at conferences on complex systems. But what is new? What were the achievements of the past two decades? Or can it be that the science of complex systems was a product of the 1980s that has run out of new ideas and concepts, and is only capable of working out details? If so, is complexity science obsolete and will it be replaced by something new? What are the next steps forward? When we discussed these questions with W. Brian Arthur he offered the following picture:

> The state of complex systems is like a large army that is ready to strike. It is positioned in the countryside and confronts the other army that represents all the unsolved problems in complexity. But our army is stuck. It does not move, it stands still. What is needed is that some sally out at different points, so that the others will follow to attack the massive problems ahead.

The intention of this book was exactly this. To help prepare the way for new progress in a number of directions where we are currently stuck. In particular, we tried to make progress in the following directions:

- Clarifying the origin of scaling laws, in particular for driven non-equilibrium systems.
- Deriving the statistics of driven systems on the basis of understanding their driving and relaxation processes, and their relations.
- Categorizing probabilistic complex systems. Once we know which universality class a particular system belongs to, we know how it behaves statistically, how to identify its relevant parameters, and where its transition and breaking points might be.

Introduction to the Theory of Complex Systems. Stefan Thurner, Rudolf Hanel, and Peter Klimek,
Oxford University Press (2018). © Stefan Thurner, Rudolf Hanel, and Peter Klimek.
DOI: 10.1093/oso/9780198821939.001.0001

- Meaningful generalization of statistical mechanics, and information theory so that they finally become useful for complex systems.
- Unifying the many different approaches to evolution and co-evolution into a single mathematical framework.
- Developing mathematical formalisms for co-evolutionary dynamics of states and interactions.

We believe that pushing these directions will finally help to develop a more coherent theory of complex systems than we currently have. And perhaps more importantly, it will help to finally establish a coherent framework for quantifying, monitoring, and managing systemic risk, resilience, robustness, and efficiency. This might help us see what we can understand in evolution, innovation and creativity, and what we cannot understand. The emerging framework will have to be radically combined with big data sets, otherwise it will not be viable and might not survive the century.

We do not believe that the science of complex systems is obsolete. Never before has it been more urgent to understand complex systems and to be able to manage them. Never before have we been so well prepared for that job. We are entering an era where practically everything that moves on this planet, as well as everything that does not, is monitored by sensors. Quite conceivably, we will eventually record almost everything that happens—we will literally copy the planet into a virtual data world. At the same time we are producing the computational means to handle those data. What is missing are the mathematical tools, concepts, and algorithms to *make sense* of those data in a quantitative and predictive manner. Only then will the management of complex systems be possible.

Will artificial intelligence and machine learning do the job of complexity science? We do not believe that. So far, artificial intelligence and machine learning are fantastic methods for recognizing and learning patterns. But they are not yet making sense of those patterns, and are not linking them with possible underlying mechanisms and causes at various scales. They recognize and use patterns to play chess and games like 'Go', and to translate English into Chinese, but as yet, they do not perform the science of co-evolving systems.

What might happen is that complexity science will not be called 'complexity science' in the future. However, in our opinion the use of mathematics and computer and data science, in combination with a disciplinary understanding of systems, will continue to be a fascinating challenge until we understand what can be learned about complexity and what can not. The bottleneck in our opinion is the current state of the available mathematical framework and concepts. With progress in data availability and computing power, the lack of appropriate algorithms and of ways to create true understanding from data will become increasingly clear. Problems associated with complexity will keep following us for a while. Practically all systems evolve depending on their context and the context changes with the evolution of the systems. We see more and more of this in fantastically comprehensive data sets. However, our human mind is still 'bad' at understanding co-evolutionary dynamics—we need to keep sharpening our tools.

ST Vienna January 2018

8

Special Functions and Approximations

8.1 Special functions

Here we collect a few special functions that occur throughout the text.

8.1.1 Heaviside step function

The Heaviside step function $\Theta(x)$ is a discontinuous function that is zero for $x < 0$, and one for $x \geq 0$,

$$\Theta(x) = \begin{cases} 0 & x < 0 \\ 1 & x \geq 0 \end{cases}.$$ (8.1)

8.1.2 Dirac delta function

The Dirac delta function $\delta(x)$ is not really a function. In simple terms, it is an object that is zero everywhere, except at $x = 0$, where it is 'infinitely' large. It is something like a function that is 'infinitely narrow' and has an 'infinitely large' spike at $x = 0$,

$$\delta(x) = \begin{cases} \infty & x = 0 \\ 0 & x \neq 0 \end{cases}.$$ (8.2)

More precisely, the Dirac delta function $\delta(x)$ is the limit of a sequence of functions. At its simplest, it can be seen as the limit of a Gaussian function that becomes infinitely narrow in the limit,

$$\delta(x) = \lim_{a \to 0} \frac{1}{|a| \sqrt{\pi}} e^{-(x/a)^2}.$$ (8.3)

This is not the only way to define the Dirac delta function. It can also be seen as a box function centred around $x = 0$, with width $1/n$ and height n in the limit,

Introduction to the Theory of Complex Systems. Stefan Thurner, Rudolf Hanel, and Peter Klimek,
Oxford University Press (2018). © Stefan Thurner, Rudolf Hanel, and Peter Klimek.
DOI: 10.1093/oso/9780198821939.001.0001

$$\delta(x) = \lim_{n \to \infty} n \left[\Theta \left(x - \frac{1}{2n} \right) - \Theta \left(x + \frac{1}{2n} \right) \right]. \tag{8.4}$$

Here, $\theta(x)$ is the Heaviside step function, Equation (8.1). There are many more ways to define it. For all definitions, the following two properties hold: the integral over the Dirac delta is one,

$$\int_{-\infty}^{\infty} \delta(t) = 1, \tag{8.5}$$

and

$$\int_{-\infty}^{\infty} f(x)\delta(x-y) \, dx = f(y). \tag{8.6}$$

The latter property is an often-used practical relation. Note that the Heaviside step function can be written as,

$$\theta(x) = \int_{-\infty}^{x} \delta(y) \, dy. \tag{8.7}$$

8.1.3   Kronecker delta

The Kronecker delta δ_{ij} is defined as,

$$\delta_{ij} = \begin{cases} 1 & i = j \\ 0 & i \neq j \end{cases}, \tag{8.8}$$

where i and j are integers. The following property is often useful in computations,

$$\sum_i a_i \delta_{ik} = a_k. \tag{8.9}$$

The Kronecker delta can be seen as a discrete version of the Dirac delta. It can also be interpreted as a matrix with entries $\delta_{ii} = 1$ in the diagonal, and zero everywhere else.

8.1.4   The Lambert-W function

The Lambert-W function is a special function that cannot be expressed in closed form. It is defined as the solution to the equation,

$$x = \mathcal{W}_k(x) e^{\mathcal{W}_k(x)}. \tag{8.10}$$

In general, the solutions x can be complex numbers. Here, we are only interested in real-valued solutions. The problem then arises that the Lambert-W function is not

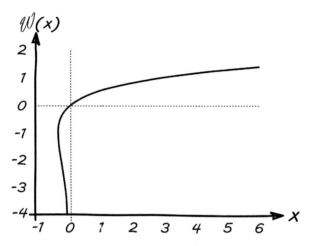

Figure 8.1 *The real solutions of the Lambert-W function. Obviously, this is not a single-valued function. For this reason it is split into two parts, one where $W \geq -1$, which we call $W_0(x)$, and another where $W \leq -1$ that is called $W_{-1}(x)$.*

single-valued; see Figure 8.1. This can be solved by decomposing the function into a part where $W \geq -1$, in which it is defined by a single-valued function $W_0(x)$, with $W_0(0) = 0$ and $W_0(-1/e) = -1$. For $W \leq -1$ we find another single-valued function called $W_{-1}(x)$. It decreases from $W_{-1}(-1/e) = -1$ to $W_{-1}(0) = -\infty$. All other solutions yield complex numbers.

A remarkable feature of the Lambert-W function is that it is the inverse function of $x\log x$, meaning that $[W(x)]^{-1} = x\log x$. It is also noteworthy that the Lambert-W function arises naturally in the context of delayed differential equations of the form,

$$\frac{d}{dt}x(t) = ax(t - \tau). \tag{8.11}$$

The solution is the Lambert-W exponential distribution, $x(t) = e^{\frac{1}{\tau}W(a\tau)t}$.

8.1.5   Gamma function

The Gamma function is defined as the integral over a power function times an exponential,

$$\Gamma(x) = \int_0^\infty t^{x-1}e^{-t}dt. \tag{8.12}$$

For integer values of $x = n$ the Gamma function is,

$$\Gamma(n) = (n - 1)!, \tag{8.13}$$

where the factorial is defined as $n! = n \cdot n - 1 \cdots 2 \cdot 1$. The Gamma function can be seen as an extension of the factorial to real and complex numbers.

8.1.6 Incomplete Gamma function

The incomplete Gamma function differs from the Gamma function only in terms of range of integration, and is given by,

$$\Gamma(x,b) = \int_b^\infty t^{x-1} e^{-t} dt. \tag{8.14}$$

8.1.7 Deformed factorial

Deformed factorials are generalizations of the factorial $N! = \prod_{n=1}^N n$, and are defined as,

$$N!_u = \prod_{n=1}^N u(n), \tag{8.15}$$

where $u(n)$ $(n = 0, 1, 2, \cdots)$ are positive, monotonically increasing functions on the natural numbers [50, 202, 313].

8.1.8 Deformed multinomial

Deformed multinomials are based on deformed factorials with a specific choice of the function u. We define deformed multinomials as,

$$M_{u,T}(k) = \frac{N!_u}{\prod_i \left\lfloor NT\left(\frac{k_i}{N}\right)\right\rfloor!_u}, \tag{8.16}$$

where T is a continuous, monotonically increasing function, with $T(0) = 0$, and $T(1) = 1$. The symbol $\lfloor x \rfloor$ defines the largest integer that is less than x.

8.1.9 Generalized logarithm

The generalized logarithm is defined as,

$$\Lambda(x) = -\frac{d}{dx} g(x), \tag{8.17}$$

where $g(x)$ is the function that appears in the trace form entropy, $S = \sum_i g(p_i)$. If the first three Shannon–Khinchin axioms SK1, SK2, and SK3 hold, $g(p_i)$ is continuous, concave, and obeys $g(0) = 0$, respectively. Generalized logarithms have the same two properties as the logarithm at $x = 1$, $\Lambda(1) = 0$ and the derivative $\Lambda'(1) = 1$.

8.1.10   Pearson correlation coefficient

The Pearson correlation coefficient is a measure of the linear correlation between two (random) variables X and Y; see [302]. It is defined as,

$$\rho_{XY} = \frac{\langle (X - \mu_X)(Y - \mu_Y) \rangle}{\sigma_X \sigma_Y}, \tag{8.18}$$

where $\langle . \rangle$ is the expectation value, and μ and σ are the mean and standard deviation of the variables, respectively. The Pearson correlation coefficient takes values from -1 (anti-correlated) to 1 (perfectly correlated). A value of $\rho_{XY} = 0$ means that the variables are not correlated.

The definition of the *sample correlation coefficient*, which is used for concrete data with N samples, is given by,

$$\rho_{xy} = \frac{\sum_{i=1}^{N}(x_i - \bar{x})(y_i - \bar{y})}{\sqrt{\sum_{i=1}^{N}(x_i - \bar{x})^2}\sqrt{\sum_{i=1}^{N}(y_i - \bar{y})^2}}, \tag{8.19}$$

where we denote the sample mean by $\bar{x} = \frac{1}{N}\sum_{i=1}^{N} x_i$. The Pearson correlation coefficient is invariant (does not change) under the affine transformation of the variables, $X \rightarrow a + bX$ and $Y \rightarrow c + dY$, with $a, b, c,$ and d constants.

8.1.11   Chi-squared distribution

If one has a Gaussian distributed random variable X, and if there are N realizations of it, $x_1, x_2, \cdots, x_N$, then the sum of their squared values,

$$q = \sum_{i=1}^{N} x_i^2, \tag{8.20}$$

is distributed as a chi-squared distribution. One says that there are N 'degrees of freedom'. The probability density function $p_{\chi^2}(x, N)$ of the chi-squared distribution is,

$$p_{\chi^2}(x|N) = \begin{cases} \dfrac{x^{\frac{N}{2}-1}e^{-\frac{x}{2}}}{2^{\frac{N}{2}}\Gamma\left(\frac{N}{2}\right)} & x > 0 \\ 0 & x \le 0 \end{cases}, \tag{8.21}$$

where N is the number of degrees of freedom. The cumulative chi-squared distribution is given by,

$$P(X < x|k) = \frac{\gamma(\frac{k}{2}, \frac{x}{2})}{\Gamma(\frac{k}{2})}, \tag{8.22}$$

where $\gamma(a, x) = \int_0^x t^{a-1} e^{-t} dt$ is the *lower* incomplete Gamma function. The chi-squared distribution is seen in many test statistics, where squares of random numbers appear.

8.2 Approximations

8.2.1 Stirling's formula

Stirling's formula, or Stirling's approximation, is a way of computing the factorial $n!$ for large values of n. It is given by,

$$\ln n! = n \ln n - n + O(\ln n). \tag{8.23}$$

This is immediately visible when noting that,

$$\ln n! = \ln 1 + \ln 2 + \cdots + \ln n = \sum_{k=1}^{n} \ln k \sim \int_1^n \ln x \, dx = n \ln n - n + 1 \sim n \ln n - n. \tag{8.24}$$

The approximation that takes the second order correction into account yields a somewhat better approximation,

$$n! \sim \sqrt{2\pi} \, n^{n+\frac{1}{2}} e^{-n}. \tag{8.25}$$

8.2.2 Expressing the exponential function as a power

One of the definitions of the exponential function is through the limit,

$$e^x = \lim_{n \to \infty} \left(1 + \frac{x}{n}\right)^n. \tag{8.26}$$

For large n, an extremely useful approximation can be obtained by using,

$$e^x \sim \left(1 + \frac{x}{n}\right)^n, \tag{8.27}$$

which approximates the exponential function by a power law!

8.3   Problems

Problem 8.1 Show that the following scaling relation for the Dirac delta is true, $\int_{-\infty}^{\infty} \delta(ax)\,dx = 1/|a|$.

Problem 8.2 By using the integral definition of the Gamma function and integration by parts, show that $\Gamma(n+1) = n\Gamma(n)$. Then show that $\Gamma(n)/\Gamma(n+k) \sim n^{-k}$ for large n.

Problem 8.3 Show that $n! = \int_{0}^{\infty} t^n e^{-t}\,dt$. Use the method of induction and start to show that for $n = 0$ the formula is correct. Then assume that it is correct for n and show that it is correct for $n+1$. You might want to use integration by parts.

Problem 8.4 Assume that $g(x) = -x\log x + x$. What is the generalized logarithm?

References

[1] Abe, S. Axioms and uniqueness theorem for Tsallis entropy. *Physics Letters A 271* (2000), 74–9.

[2] Abe, S., and Thurner, S. Hierarchical and mixing properties of static complex networks emerging from fluctuating classical random graphs. *International Journal of Modern Physics C 17*, 9 (2006), 1303–11.

[3] Abrahams, E., Anderson, P., Licciardello, D., and Ramakrishnan, T. Scaling theory of localization: absence of quantum diffusion in two dimensions. *Physical Review Letters 42* (1979), 673–6.

[4] Abramson, G. Ecological model of extinctions. *Physical Review E 55*, 1 (1997), 785–8.

[5] Acemoglu, D., Ozdaglar, A., and Tahbaz-Salehi, A. Systemic risk and stability in financial networks. *American Economic Review 105*, 2 (2015), 564–608.

[6] Adamic, L. A., and Huberman, B. A. Zipf's law and the internet. *Glottometrics 3* (2002), 143–50.

[7] Albert, R., and Barabási, A.-L. Statistical mechanics of complex networks. *Reviews of Modern Physics 74*, 1 (2002), 47–97.

[8] Albert, R., Jeong, H., and Barabási, A.-L. Internet: diameter of the world-wide web. *Nature 401* (1999), 130–1.

[9] Alexander, J., Skyrms, B., and Zabell, S. Inventing new signals. *Dynamic Games and Applications 2* (2012), 129–45.

[10] Amaral, L. A. N., and Meyer, M. Environmental changes, coextinction, and patterns in the fossil record. *Physical Review Letters 82*, 3 (1999), 652–5.

[11] Anderson, R. M., and May, R. M. *Infectious Diseases of Humans: Dynamics and Control*, Oxford University Press, 1992.

[12] Anteneodo, C., and Plastino, A. Maximum entropy approach to stretched exponential probability distributions. *Journal of Physics A 32* (1999), 1089–97.

[13] Arthur, W. B. *Increasing Returns and Path Dependence in the Economy*. University of Michigan Press, Ann Arbor, 1994.

[14] Arthur, W. B. *The Nature of Technology: What It Is and How It Evolves*. Free Press, Simon and Schuster, 2009.

[15] Arthur, W. B., and Polak, W. The evolution of technology within a simple computer model. *Complexity 11*, 5 (2006), 23–31.

[16] Atkinson, Q. D., and Gray, R. D. Curious parallels and curious connections, phylogenetic thinking in biology and historical linguistics. *Systematic Biology 54*, 4 (2005), 513–26.

[17] Auerbach, F. Das Gesetz der Bevölkerungskonzentration. *Petermanns Geographische Mitteilungen 59* (1913), 74–6.

[18] Axelrod, R. The dissemination of culture: a model with local convergence and global polarization. *Journal of Conflict Resolution 41*, 2 (1997), 203–26.

[19] Axtell, R. L. Zipf distribution of US firm sizes. *Science 293* (2001), 1818–20.

[20] Bagchi, A., and Pal, A. Asymptotic normality in the generalised Pólya–Eggenberger urn model, with an application to computer data structures. *SIAM Journal on Algebraic Discrete Methods 6* (1985), 394–405.

[21] Bagley, R. J., and Farmer, J. D. Spontaneous emergence of a metabolism. *Artificial Life II 10* (1992), 93–141.

[22] Bagley, R. J., Farmer, J. D., and Fontana, W. Evolution of a metabolism. *Artificial Life II 10* (1992), 141–58.

[23] Bak, P., and Sneppen, K. Punctuated equilibrium and criticality in a simple model of evolution. *Physical Review Letters 71*, 24 (1993), 4083–6.

[24] Bak, P., Tang, C., and Wiesenfeld, K. Self-organized criticality: an explanation of $1/f$ noise. *Physical Review Letters 59*, 4 (1987), 381–4.

[25] Balassa, B. The purchasing-power parity doctrine: a reappraisal. *Journal of Political Economy 72*, 6 (1964), 584–96.

[26] Banerjee, A., Burlina, P., and Alajaji, F. Image segmentation and labeling using the Pólya urn model. *IEEE Transactions Image Processing 8* (1999), 1243–53.

[27] Barabási, A.-L. *Network Science*. Cambridge University Press, 2016.

[28] Barabási, A.-L., and Albert, R. Emergence of scaling in random networks. *Science 286* (1999), 509–12.

[29] Barabási, A.-L., and Oltvai, Z. N. Network biology: understanding the cell's functional organization. *Nature Reviews Genetics 5*, 2 (2004), 101–13.

[30] Barnes, E. R. An algorithm for partitioning the nodes of a graph. *SIAM Journal on Algebraic Discrete Methods 3*, 4 (1982), 541–50.

[31] Barouch, D. H., O'Brien, K. L., Simmons, N. L., King, S. L., Abbink, P., Maxfield, L. F., Sun, Y.-H., La Porte, A., Riggs, A. M., Lynch, D. M., Clark, S. L., Backus, K., Perry, J. R., Seaman, M. S., Carville, A., Mansfield, K. G., Szinger, J. J., Fischer, W., Muldoon, M., and Korber, B. Mosaic HIV-1 vaccines expand the breadth and depth of cellular immune responses in rhesus monkeys. *Nature Medicine 16* (2010), 319–23.

[32] Barton, N. H., and Charlesworth, B. Why sex and recombination? *Science 281* (1998), 1986–90.

[33] Barunik, J., Aste, T., Di Matteo, T., and Liu, R. Understanding the source of multifractality in financial markets. *Physica A 391* (2012), 4234–51.

[34] Battiston, F., Nicosia, V., and Latora, V. Structural measures for multiplex networks. *Physical Review E 89* (2014), 032804.

[35] Battiston, S., Puliga, M., Kaushik, R., Tasca, P., and Caldarelli, G. Debtrank: too central to fail? Financial networks, the Fed and systemic risk. *Scientific Reports 2* (2012), 514.

[36] Batty, M. *Cities and Complexity: Understanding Cities with Cellular Automata, Agent-Based Models, and Fractals*. MIT Press, 2007.

[37] Bavelas, A. Communication patterns in task-oriented groups. *Journal of the Acoustical Society of America 22*, 6 (1950), 725–30.

[38] Beck, C. Generalised information and entropy measures in physics. *Contemporary Physics 50* (2009), 495–510.

[39] Beck, C., and Cohen, E. G. Superstatistics. *Physica A 322* (2003), 267–75.

[40] Beckenstein, J. Black holes and the second law. *Lettere al Nuovo Cimento 4* (1972), 737–40.

[41] Beckenstein, J. Black holes and entropy. *Physical Review D 7* (1973), 2333–46.

[42] Beckenstein, J. Generalized second law of thermodynamics in black-hole physics. *Physical Review D 9* (1974), 3292–300.

[43] Benjamini, Y. Discovering the false discovery rate. *Journal of the Royal Statistical Society: Series B 72*, 4 (2010), 405–16.

[44] Benjamini, Y., and Hochberg, Y. Controlling the false discovery rate: a practical and powerful approach to multiple testing. *Journal of the Royal Statistical Society: Series B 57*, 1 (1995), 289–300.

[45] Benkö, G., Flamm, C., and Stadler, P. F. A graph-based toy model of chemistry. *Journal of Chemical Information and Computer Sciences 43*, 4 (2003), 1085–93.

[46] Benz, A. O. Living reviews in solar physics. *Flare Observations 5* (2008), 1–64.

[47] Berge, C. *Graphs and Hypergraphs*. North-Holland Publishing Company, Amsterdam, 1973.

[48] Berry, A. C. The accuracy of the Gaussian approximation to the sum of independent variates. *Transactions of the American Mathematical Society 49* (1995), 122–36.

[49] Bettencourt, L., Lobo, J., Helbing, D., Kühnert, C., and West, G. Growth, innovation, scaling, and the pace of life in cities. *Proceedings of the National Academy of Sciences USA 104*, 17 (2007), 7301–6.

[50] Bhargava, M. The factorial function and generalizations. *American Mathematical Monthly 107* (2000), 783–99.

[51] Binder, B., and Landman, K. Tissue growth and the Pólya distribution. *Australasian Journal of Engineering Education 15* (2009), 35–42.

[52] Bland, J. M., and Altman, D. G. Multiple significance tests: the Bonferroni method. *British Medical Journal 310*, 6973 (1995), 170.

[53] Blondel, V. D., Guillaume, J.-L., Lambiotte, R., and Lefebvre, E. Fast unfolding of communities in large networks. *Journal of Statistical Mechanics 2008*, 10 (2008), P10008.

[54] Boccaletti, S., Bianconi, G., Criado, R., Del Genio, C. I., Gómez-Gardenes, J., Romance, M., Sendina-Nadal, I., Wang, Z., and Zanin, M. The structure and dynamics of multilayer networks. *Physics Reports 544*, 1 (2014), 1–122.

[55] Bollobás, B. *Modern Graph Theory*, Vol. 184. Springer, 1998.

[56] Bollobás, B. Random graphs. In *Modern Graph Theory*. Springer, 1998, pp. 215–52.

[57] Bollobás, B., and Riordan, O. The diameter of a scale-free random graph. *Combinatorica 24*, 1 (2004), 5–34.

[58] Bollobás, B., and Riordan, O. *Percolation*. Cambridge University Press, 2006.

[59] Boltzmann, L. Über die mechanische Bedeutung des zweiten Hauptsatzes der Wärmetheorie. *Sitzungsberichte der Kaiserlichen Akademie der Wissenschaften, Wien 53* (1866), 195–220.

[60] Boltzmann, L. Über das Wärmegleichgewicht zwischen mehratomigen Gasmolekülen. *Sitzungsberichte der Kaiserlichen Akademie der Wissenschaften, Wien 63* (1871), 397–418.

[61] Bonacich, P. Power and centrality: a family of measures. *American Journal of Sociology 92*, 5 (1987), 1170–82.

[62] Bondy, J. A., and Murty, U. S. R. *Graph Theory with Applications*, Vol. 290. Macmillan, London, 1976.

[63] Borgatti, S. P. Centrality and network flow. *Social Networks 27*, 1 (2005), 55–71.

[64] Bornholdt, S., and Ebel, H. World wide web scaling exponent from Simon's 1955 model. *Physical Review E 64*, 3 (2001), 035104.

[65] Boss, M., Elsinger, H., Summer, M., and Thurner, S. The network topology of the interbank market. *Quantitative Finance 4* (2005), 677–84.

[66] Bouchaud, J.-P., and Potters, M. *Theory of Financial Risk and Derivative Pricing*. Cambridge University Press, 2003.

[67] Brandes, U., Delling, D., Gaertler, M., Görke, R., Hoefer, M., Nikoloski, Z., and Wagner, D. *On Modularity NP Completeness and Beyond*. Universität Karlsruhe Fakultät für Informatik, 2006.

[68] Breiman, L. The individual ergodic theorem of information theory. *Annals of Mathematical Statistics 28* (1957), 809–911.

[69] Briggs, K., and Beck, C. Modelling train delays with q-exponential functions. *Physica A 378* (2007), 498–504.

[70] Brock, D., and Moore, G. *Understanding Moore's Law: Four Decades of Innovation*. Chemical Heritage Foundation, 2006.

[71] Brown, J., and West, G. *Scaling in Biology*. Oxford University Press, 2000.

[72] Browning, T. R. Applying the design structure matrix to system decomposition and integration problems: a review and new directions. *IEEE Transactions on Engineering Management 48*, 3 (2001), 292–306.

[73] Caldarelli, G., and Chessa, A. *Data Science and Complex Networks: Real Case Studies with Python*. Oxford University Press, 2016.

[74] Castellano, C., Fortunato, S., and Loreto, V. Statistical physics of social dynamics. *Reviews of Modern Physics 81*, 2 (2009), 591–646.

[75] Centola, D. The spread of behavior in an online social network experiment. *Science 329* (2010), 1194–7.

[76] Cerf, V., Dalal, Y., and Sunshine, C. Specification of internet transmission control protocol. *Internet History by Ronda Hauben: From the ARPANET to the Internet. TCP Digest (UUCP)* (1974).

[77] Chaitin, G. J. On the length of programs for computing finite binary sequences: statistical considerations. *Journal of the ACM 16* (1969), 145–59.

[78] Chakrabarti, B. K., Chakraborti, A., and Chatterjee, A. *Econophysics and Sociophysics: Trends and Perspectives*. John Wiley & Sons, 2007.

[79] Chmiel, A., Klimek, P., and Thurner, S. Spreading of diseases through comorbidity networks across life and gender. *New Journal of Physics 16*, 11 (2014), 115013.

[80] Christakis, N. A., and Fowler, J. H. *Connected: The Surprising Power of Our Social Networks and How They Shape Our Lives*. Little, Brown, 2009.

[81] Christensen, K., and Moloney, N. R. *Complexity and Criticality*. World Scientific, 2005.

[82] Chu, H.-Y., Sprouffske, K., and Wagner, A. The role of recombination in evolutionary adaptation of *Escherichia coli* to a novel nutrient. *Journal of Evolutionary Biology 30*, 9 (2017), 1692–711.

[83] Chung, F. R. *Spectral Graph Theory*, Vol. 92. American Mathematical Society, 1997.

[84] Clark, A. G., Markow, T. A., Kaufman, T. C., Kellis, M., Gelbart, W., Iyer, V. N., et al. Evolution of genes and genomes on the *Drosophila* phylogeny. *Nature 450* (2007), 203–18.

[85] Clauset, A., Shalizi, C. R., and Newman, M. E. J. Power-law distributions in empirical data. *SIAM Review 51*, 4 (2009), 661–703.

[86] Clauset, A., and Young, M. Scale invariance in global terrorism. *arXiv:physics/0502014* (2005).

[87] Cohen, R., ben Avraham, D., and Havlin, S. Percolation critical exponents in scale-free networks. *Physical Review E 66* (2002), 036113.

[88] Colizza, V., Barrat, A., Barthelemy, M., and Vespignani, A. The role of the airline transportation network in the prediction and predictability of global epidemics. *Proceedings of the National Academy of Sciences USA 103* (2005), 2015–20.

[89] Conrad, B., and Mitzenmacher, M. Power laws for monkeys typing randomly: the case of unequal probabilities. *IEEE Transactions on Information Theory 50*, 7 (2004), 1403–14.

[90] Corominas-Murtra, B., Hanel, R., and Thurner, S. Extreme robustness of scaling in sample space reducing processes explains Zipf's law in diffusion on directed networks. *New Journal of Physics 18* (2016), 093010.

[91] Corominas-Murtra, B., Hanel, R., Zavojanni, L., and Thurner, S. How driving rates determine the statistics of driven non-equilibrium systems with stationary distributions. *Scientific Reports 8*, (2018), 10837.

[92] Corominas-Murtra, B., Hanel, R., and Thurner, S. Sample space reducing cascading processes produce the full spectrum of scaling exponents. *Scientific Reports 7* (2017), 11223.

[93] Corominas-Murtra, B., Hanel, R., and Thurner, S. Understanding scaling through history-dependent processes with collapsing sample space. *Proceedings of the National Academy of Sciences USA 112* (2015), 5348–53.

[94] Cristelli, M., Tacchella, A., and Pietronero, L. The heterogeneous dynamics of economic complexity. *PLoS ONE 10*, 2 (2015), e0117174.

[95] Croft, W. Evolutionary linguistics. *Annual Review of Anthropology 37* (2008), 219–234.

[96] Crouch, C., and Farrell, H. Breaking the path of institutional development? Alternatives to the new determinism. *Rationality and Society 16* (2004), 5–43.

[97] Crow, J. F., and Kimura, M. *An Introduction to Population Genetics Theory*. Harper and Row, New York, 1970.

[98] Curado, E., and Nobre, F. On the stability of analytic entropic forms. *Physica A 335* (2004), 94–106.

[99] Darwin, C. *The Origin of Species by Means of Natural Selection, or, the Preservation of Favored Races in the Struggle for Life*, Vol. 1. International Science Library, The Werner Company, 1896.

[100] Davidsen, J., Ebel, H., and Bornholdt, S. Emergence of a small world from local interactions: modeling acquaintance networks. *Physical Review Letters 88*, 12 (2002), 128701.

[101] Dawkins, R. *The God Delusion*. Random House, 2016.

[102] De Las Rivas, J., and Fontanillo, C. Protein–protein interactions essentials: key concepts to building and analyzing interactome networks. *PLoS Computational Biology 6* (2010), e1000807.

[103] Dehmer, M., Emmert-Streib, F., and Pickl, S. *Computational Network Theory: Theoretical Foundations and Applications*, Vol. 5. John Wiley & Sons, 2015.

[104] Dobson, I., Carreras, B. A., Lynch, V. E., and Newman, D. E. Complex systems analysis of series of blackouts: cascading failure, critical points, and self-organization. *Chaos 17* (2007), 026103.

[105] Dodds, P. S., Muhamad, R., and Watts, D. J. An experimental study of search in global social networks. *Science 301* (2003), 827–9.

[106] Donnelly, P. Partition structures, Pólya urns, the Ewens sampling formula, and the ages of alleles. *Theoretical Population Biology 30* (1986), 271–88.

[107] Dosi, G., and Nelson, R. R. An introduction to evolutionary theories in economics. *Journal of Evolutionary Economics 4*, 3 (1994), 153–72.

[108] Duch, J., and Arenas, A. Community detection in complex networks using extremal optimization. *Physical Review E 72*, 2 (2005), 027104.

[109] Durkheim, E. *The Division of Labour in Society*, translated from the French edition of 1893 by WD Halls, 1984.

[110] Eagle, N., Macy, M., and Claxton, R. Network diversity and economic development. *Science 328* (2010), 1029–31.

[111] Edwards, S. F., and Anderson, P. W. Theory of spin glasses. *Journal of Physics F 5*, 5 (1975), 965.

[112] Eggenberger, F., and Pólya, G. Über die Statistik verketteter Vorgänge. *Zeitschrift für Angewandte Mathematik und Mechanik 1* (1923), 279–89.

[113] Eigen, M. Selforganization of matter and the evolution of biological macromolecules. *Naturwissenschaften 58*, 10 (1971), 465–523.

[114] Eigen, M., and Schuster, P. The hypercycle. *Naturwissenschaften 65*, 1 (1978), 7–41.

[115] Einstein, A. Über die von der molekularkinetischen Theorie der Wärme geforderte Bewegung von in ruhenden Flüssigkeiten suspendierten Teilchen. *Annalen der Physik 322*, 8 (1905), 549–60.

[116] Engineering and Technology History Wikipedia. Milestones: Fleming valve, 1904, 2004. [Online; accessed 7 March 2017].

[117] Erdős, P., and Rényi, A. On random graphs. i. *Publicationes Mathematicae 6* (1959), 290–7.

[118] Erdős, P., and Rényi, A. On the evolution of random graphs. *Publication of the Mathematical Institute of the Hungarian Academy of Sciences 5*, 1 (1960), 17–60.

[119] Erwin, D. H., and Valentine, J. W. *The Cambrian Explosion: The Construction of Animal Biodiversity*. Roberts, 2012.

[120] Esseen, C.-G. A moment inequality with an application to the central limit theorem. *Skand. Aktuarietidskr. 39* (1956), 160–70.

[121] Euler, L. Solutio problematis ad geometriam situs pertinentis. *Commentarii Academiae Scientiarum Petropolitanae 8* (1741), 128–40.

[122] Evans, D. J., and Searles, D. J. The fluctuation theorem. *Advances in Physics 51*, 7 (2002), 1529–85.

[123] Fagiolo, G. Clustering in complex directed networks. *Physical Review E 76*, 2 (2007), 026107.

[124] Farmer, J. D. Market force, ecology, and evolution. *Industrial and Corporate Change 11*, 5 (2002), 895–953.

[125] Farmer, J. D. Physicists attempt to scale the ivory towers of finance. *Computing in Science and Engineering 1* (1999), 26–39.

[126] Fermi, E. On the origin of the cosmic radiation. *Physical Review 75* (1949), 1169–74.

[127] Fernández-Gracia, J., Suchecki, K., Ramasco, J. J., San Miguel, M., and Eguíluz, V. M. Is the voter model a model for voters? *Physical Review Letters 112*, 15 (2014), 158701.

[128] Fick, A. Über Diffusion. *Annalen der Physik 170*, 1 (1855), 59–86.

[129] Fisher, R. A. Applications of student's distribution. *Metron 5* (1925), 90–104.

[130] Flyvbjerg, H., Sneppen, K., and Bak, P. Mean field theory for a simple model of evolution. *Physical Review Letters 71* (1993), 4087–90.

[131] Fontana, W. Algorithmic chemistry: a model for functional self-organization. *Artificial Life II* (1991), 159–202.

[132] Fontana, W., Wagner, G., and Buss, L. W. Beyond digital naturalism. *Artificial Life 1*, 1–2 (1993), 211–27.

[133] Fortunato, S. Community detection in graphs. *Physics Reports 486*, 3 (2010), 75–174.

[134] Freeman, L. C. A set of measures of centrality based on betweenness. *Sociometry* (1977), 35–41.

[135] Friedman, D. Evolutionary economics goes mainstream: a review of the theory of learning in games. *Journal of Evolutionary Economics 8*, 4 (1998), 423–32.

[136] Frobenius, F. G. Über Matrizen aus nicht negativen Elementen. *Sitzungsberichte der Preussischen Akademie der Wissenschaften, Berlin* (1912), 456–77.

[137] Gabaix, X. Zipf's law for cities: an explanation. *Quarterly Journal of Economics 114* (1999), 739–67.

[138] Gai, P., and Kapadia, S. Contagion in financial networks. *Proceedings of the Royal Society of London A 466*, 2120 (2010), 2401–23.

[139] Galam, S. Minority opinion spreading in random geometry. *European Physical Journal B 25*, 4 (2002), 403–6.

[140] Galam, S. Contrarian deterministic effects on opinion dynamics: the hung elections scenario. *Physica A 333* (2004), 453–60.

[141] Galam, S. Sociophysics: a review of Galam models. *International Journal of Modern Physics C 19*, 3 (2008), 409–40.

[142] Galilei, G. *Discorsi e dimostrazioni matematiche intorno a due nuove scienze attenenti alla mecanica & movimenti locali.* Leida, Appresso Gli Elsevirii, 1638.

[143] Gardiner, C. W. *Handbook of Stochastic Methods (2nd ed.).* Springer, Berlin, 1985.

[144] Gardner, C. Measuring value in healthcare. *arXiv:0806.2397* (2008).

[145] Garlaschelli, D., and Loffredo, M. I. Patterns of link reciprocity in directed networks. *Physical Review Letters 93*, 26 (2004), 268701.

[146] Gell-Mann, M. What is complexity? *Complexity 1* (1995), 16–19.

[147] Gell-Mann, M., and Lloyd, S. Information measures, effective complexity, and total information. *Complexity 2* (1996), 44–52.

[148] Gell-Mann, M., and Low, F. Quantum electrodynamics at small distances. *Physical Review 95* (1954), 1300–12.

[149] Georgeot, G., and Giraud, O. The game of go as a complex network. *Europhysics Letters 97*, 6 (2012), 68002.

[150] Geppert, T. EU Agrar- und Regionalpolitik: Wie vergangene Entscheidungen zukünftige Entwicklungen beeinflussen – Pfadabhängigkeit und die Reformfähigkeit von Politikfeldern. *PhD Thesis, University of Bamberg Press,* (2012).

[151] Girko, V. L. Circular law. *Theory of Probability and its Applications 29*, 4 (1985), 694–706.

[152] Girvan, M., and Newman, M. E. J. Community structure in social and biological networks. *Proceedings of the National Academy of Sciences USA 99*, 12 (2002), 7821–6.

[153] Gopikrishnan, P., Plerou, V., Amaral, L. A. N., Meyer, M., and Stanley, H. E. Scaling of the distributions of fluctuations of financial market indices. *Physical Review E 60* (1999), 5305–16.

[154] Gould, S. J. *The Structure of Evolutionary Theory.* Harvard University Press, 2002.

[155] Gower, J. C., and Ross, G. J. Minimum spanning trees and single linkage cluster analysis. *Applied Statistics* (1969), 54–64.

[156] Granovetter, M. S. The strength of weak ties. *American Journal of Sociology 78*, 6 (1973), 1360–80.

[157] Gross, T., D'Lima, C. J. D., and Blasius, B. Epidemic dynamics on an adaptive network. *Physical Review Letters 96* (2006), 208701.

[158] Gross, T., and Sayama, H. *Adaptive Networks, Theory, Models and Applications.* Springer, 2009.

[159] Gubin, S. P., Koksharov, Y. A., Khomutov, G., and Yurkov, G. Y. Magnetic nanoparticles: preparation, structure and properties. *Russian Chemical Reviews 74*, 6 (2005), 489–520.

[160] Guimera, R., Mossa, S., Turtschi, A., and Amaral, L. A. N. The worldwide air transportation network: anomalous centrality, community structure, and cities' global roles. *Proceedings of the National Academy of Sciences USA 102*, 22 (2005), 7794–9.

[161] Gumbel, E. J. The return period of flood flows. *Annals of Mathematical Statistics 12* (1941), 163–90.

[162] Gutenberg, B., and Richter, C. Frequency of earthquakes in California. *Bulletin of the Seismological Society of America 34* (1944), 185–8.

[163] Hackett, A. *70 Years of Best Sellers, 1895–1965.* R. R. Bowker Company, New York, 1967.

[164] Hadeler, K. Stable polymorphisms in a selection model with mutation. *SIAM Journal on Applied Mathematics 41*, 1 (1981), 1–7.

[165] Haldane, A. G., and May, R. M. Systemic risk in banking ecosystems. *Nature 469* (2011), 351–5.

[166] Hallam, A., and Wignall, P. B. *Mass Extinctions and their Aftermath.* Oxford University Press, 1997.

[167] Hanel, R., Corominas-Murtra, B., Liu, B., and Thurner, S. Fitting power-laws in empirical data with estimators that work for all exponents. *PLoS ONE 12* (2017), e0170920.

[168] Hanel, R., Corominas-Murtra, B., and Thurner, S. Understanding frequency distributions of path-dependent processes with non-multinomial maximum entropy approaches. *New Journal of Physics 19* (2017), 033008.

[169] Hanel, R., Kauffman, S. A., and Thurner, S. Phase transition in random catalytic networks. *Physical Review E 72*, 3 (2005), 036117.

[170] Hanel, R., Kauffman, S. A., and Thurner, S. Towards a physics of evolution: critical diversity dynamics at the edges of collapse and bursts of diversification. *Physical Review E 76*, 3 (2007), 036110.

[171] Hanel, R., Pöchacker, M., Schölling, M., and Thurner, S. A self-organized model for cell-differentiation based on variations of molecular decay rates. *PLoS ONE 7*, 5 (2012), e36679.

[172] Hanel, R., Pöchacker, M., and Thurner, S. Living on the edge of chaos: minimally nonlinear models of genetic regulatory dynamics. *Philosophical Transactions of the Royal Society of London A 368*, 1933 (2010), 5583–96.

[173] Hanel, R., and Thurner, S. Generalized Boltzmann factors and the maximum entropy principle: entropies for complex systems. *Physica A 380* (2007), 109–14.

[174] Hanel, R., and Thurner, S. Generalized-generalized entropies and limit distributions. *Brazilian Journal of Physics 39* (2009), 413–16.

[175] Hanel, R., and Thurner, S. A comprehensive classification of complex statistical systems and an ab-initio derivation of their entropy and distribution functions. *Europhysics Letters 93* (2011), 20006.

[176] Hanel, R., and Thurner, S. When do generalized entropies apply? How phase space volume determines entropy. *Europhysics Letters 96* (2011), 50003.

[177] Hanel, R., Thurner, S., and Gell-Mann, M. Generalized entropies and the transformation group of superstatistics. *Proceedings of the National Academy of Sciences USA 108* (2011), 6390–4.

[178] Hanel, R., Thurner, S., and Gell-Mann, M. How multiplicity determines entropy and the derivation of the maximum entropy principle for complex systems. *Proceedings of the National Academy of Sciences USA 111* (2014), 6905–10.

[179] Harary, F., and Palmer, E. M. *Graphical Enumeration*. Elsevier, 2014.

[180] Hartfield, M., and Keightley, P. D. Current hypotheses for the evolution of sex and recombination. *Integrative Zoology 7*, 2 (2012), 192–209.

[181] Hartl, D. L., and Clark, A. G. *Principles of Population Genetics*, Vol. 116. Sinauer Associates, Sunderland, 1997.

[182] Harvey, P. H., and Pagel, M. D. *The Comparative Method in Evolutionary Biology*, Vol. 239. Oxford University Press, 1991.

[183] Hausmann, R., Hidalgo, C. A., Bustos, S., Coscia, M., Simoes, A., and Yildirim, M. A. *The Atlas of Economic Complexity: Mapping Paths to Prosperity*. MIT Press, 2014.

[184] Havrda, J., and Charvat, F. Quantification method of classification processes. Concept of structural α-entropy. *Kybernetika 3* (1967), 30–5.

[185] Hawking, S. Black hole explosions? *Nature 248* (1974), 30–1.

[186] Hebb, D. O. *The Organization of Behavior*. John Wiley & Sons, 1949.

[187] Helbing, D., Farkas, I., and Vicsek, T. Simulating dynamical features of escape panic. *Nature 407* (2000), 487–91.

[188] Hey, T., Tansley, S., Tolle, K. M. *The Fourth Paradigm: Data-Intensive Scientific Discovery*, Vol. 1. Microsoft Research, Redmond, 2009.

[189] Hidalgo, C. A., and Hausmann, R. The building blocks of economic complexity. *Proceedings of the National Academy of Sciences USA 106*, 26 (2009), 10570–5.

[190] Hidalgo, C. A., Klinger, B., Barabási, A.-L., and Hausmann, R. The product space conditions the development of nations. *Science 317* (2007), 482–7.

[191] Hill, W. G., and Robertson, A. The effect of linkage on limits to artificial selection. *Genetics Research 8*, 3 (1966), 269–4.

[192] Hofbauer, J., and Sigmund, K. *Evolutionary Games and Population Dynamics*. Cambridge University Press, 1998.

[193] Holland, H. D. *The Chemical Evolution of the Atmosphere and Oceans*. Princeton University Press, 1984.

[194] Holland, J. *Adaptation in Natural and Artificial Systems: An Introductory Analysis with Applications to Biology, Control and Artificial Intelligence*. University of Michigan, 1975.

[195] Holland, P. W., and Leinhardt, S. Transitivity in structural models of small groups. *Comparative Group Studies 2*, 2 (1971), 107–24.

[196] Holley, R. A., and Liggett, T. M. Ergodic theorems for weakly interacting infinite systems and the voter model. *Annals of Probability* (1975), 643–63.

[197] Hoppe, F. Pólya-like urns and the Ewens' sampling formula. *Journal of Mathematical Biololgy 20* (1984), 91–4.

[198] Horstmeyer, L., Kuehn, C., and Thurner, S. Network topology near criticality in adaptive epidemics. *arXiv:1805.09358* (2018).

[199] Husak, G. J., Michaelsen, J., and Funk, C. Use of the gamma distribution to represent monthly rainfall in Africa for drought monitoring applications. *International Journal of Climatology 27*, 7 (2007), 935–44.

[200] Ideker, T., Galitski, T., and Hood, L. A new approach to decoding life: systems biology. *Annual Review of Genomics and Human Genetics 2*, 1 (2001), 343–72.

[201] Sweeting, M. J., Sutton, A. J., and Lambert, P. C. What to add to nothing? Use and avoidance of continuity corrections in meta-analysis of sparse data. *Statistics in Medicine 23*, 9 (2004), 1351–75.

[202] Jackson, F. On *q*-definite integrals. *Quarterly Journal of Pure and Applied Mathematics 41* (1910), 193–203.

[203] Jackson, M. O. A survey of network formation models: stability and efficiency. *Group Formation in Economics: Networks, Clubs, and Coalitions* (2005), 11–49.

[204] Jackson, M. O. *Social and Economic Networks*. Princeton University Press, 2010.

[205] Jain, S., and Krishna, S. Autocatalytic sets and the growth of complexity in an evolutionary model. *Physical Review Letters 81*, 25 (1998), 5684–7.

[206] Jain, S., and Krishna, S. A model for the emergence of cooperation, interdependence, and structure in evolving networks. *Proceedings of the National Academy of Sciences USA 98*, 2 (2001), 543–7.

[207] Jarque, C. M., and Bera, A. K. A test for normality of observations and regression residuals. *International Statistical Review* (1987), 163–72.

[208] Jaynes, E. Information theory and statistical mechanics. *Physical Review 106* (1957), 620–30.

[209] Jaynes, E. *Probability Theory: The Logic of Science*. Cambridge University Press, 2003.

[210] Jensen, T. R., and Toft, B. *Graph Coloring Problems*, Vol. 39. John Wiley & Sons, 2011.

[211] Johnson, N., and Kotz, S. *Urn Models and Their Application: An Approach to Modern Discrete Probability Theory*. John Wiley & Sons, New York, 1977.

[212] Johnson, N., Spagat, M., Restrepo, J., Bohorquez, J., Suarez, N., Restrepo, E., and Zarama, R. From old wars to new wars and global terrorism. *arXiv: 0506213* (2005).

[213] Johnson, S. C. Hierarchical clustering schemes. *Psychometrika 32*, 3 (1967), 241–54.

[214] Joyce, G. F. The antiquity of RNA-based evolution. *Nature 418* (2002), 214–21.

[215] Kadanoff, L. Scaling laws for Ising models near T_c. *Physics 2* (1966), 263–72.

[216] Kaniadakis, G. Statistical mechanics in the context of special relativity. *Physical Review E 66* (2002), 056125.

[217] Karger, D. R. Global min-cuts in RNC, and other ramifications of a simple min-cut algorithm. In *Proceedings of the Fourth Annual ACM-SIAM Symposium on Discrete Algorithms* (1993), Vol. 93, pp. 21–30.

[218] Katz, L. A new status index derived from sociometric analysis. *Psychometrika 18*, 1 (1953), 39–43.

[219] Kauffman, S. A. *At Home in the Universe: The Search for the Laws of Self-Organization and Complexity*. Oxford University Press, 1996.

[220] Kauffman, S. A., and Levin, S. Towards a general theory of adaptive walks on rugged landscapes. *Journal of Theoretical Biology 128*, 1 (1987), 11–45.

[221] Kauffman, S. A. Metabolic stability and epigenesis in randomly constructed genetic nets. *Journal of Theoretical Biology 22*, 3 (1969), 437–67.

[222] Kauffman, S. A. *Origin of Order: Self-Organization and Selection in Evolution*. Oxford University Press, 1993.

[223] Kauffman, S. A. *Investigations*. Oxford University Press, 2000.

[224] Kauffman, S. A., and Johnsen, S. Coevolution to the edge of chaos: coupled fitness landscapes, poised states, and coevolutionary avalanches. *Journal of Theoretical Biology 149*, 4 (1991), 467–505.

[225] Kauffman, S. A., and Weinberger, E. D. The NK model of rugged fitness landscapes and its application to maturation of the immune response. *Journal of Theoretical Biology 141*, 2 (1989), 211–45.

[226] Keeler, J., and Farmer, J. D. Robust space-time intermittency and 1/f noise. *Physica D 23*, 5 (1986), 413–35.

[227] Kelarev, A., and Quinn, S. A combinatorial property and power graphs of groups. *Contributions to General Algebra 12*, 58 (2000), 3–6.

[228] Kermack, W. O., and McKendrick, A. G. A contribution to the mathematical theory of epidemics. *Proceedings of the Royal Society of London A 115* (1927), 700–21.

[229] Khinchin, A. *Mathematical Foundations of Information Theory*. Dover Publlications, New York, 1957.

[230] Kitano, H. Systems biology: a brief overview. *Science 295* (2002), 1662–4.

[231] Kittel, C. *Elementary Statistical Physics*. Dover Publications, New York, 1958.

[232] Kivelä, M., Arenas, A., Barthelemy, M., Gleeson, J. P., Moreno, Y., and Porter, M. A. Multilayer networks. *Journal of Complex Networks 2*, 3 (2014), 203–71.

[233] Klimek, P., Hausmann, R., and Thurner, S. Empirical confirmation of creative destruction from world trade data. *PLoS ONE 7*, 6 (2012), e38924.

[234] Klimek, P., Lambiotte, R., and Thurner, S. Opinion formation in laggard societies. *Europhysics Letters 82* (2008), 28008.

[235] Klimek, P., Thurner, S., and Hanel, R. Evolutionary dynamics from a variational principle. *Physical Review E 82*, 1 (2010), 011901.

[236] Kolmogorov, A. N. Three approaches to the quantitative definition of information. *Problems in Information Transmission 1* (1965), 1–7.

[237] Kolmogorov, A. N. *Foundations of the Theory of Probability*. Chelsea, New York, 1950.

[238] Kossinets, G., Kleinberg, J., and Watts, D. J. The structure of information pathways in a social communication network. In *Proceedings of the 14th ACM SIGKDD International Conference on Knowledge Discovery and Data Mining* (2008), ACM, pp. 435–43.

[239] Kraft, L. *A Device For Quantizing, Grouping, and Coding Amplitude Modulated Pulses*. MS Thesis, Electrical Engineering Department, MIT, Cambridge, 1949.

[240] Kramer, E. M., and Lobkovsky, A. E. Universal power law in the noise from a crumpled elastic sheet. *Physical Review E 53* (1996), 1465–9.

[241] Krapivsky, P. L., and Ben-Naim, E. Scaling and multiscaling in models of fragmentation. *Physical Review E 50* (1994), 3502–7.

[242] Krapivsky, P. L., and Redner, S. Dynamics of majority rule in two-state interacting spin systems. *Physical Review Letters 90*, 23 (2003), 238701.

[243] Kullback, S., and Leibler, R. On information and sufficiency. *Annals of Mathematical Statistics 22* (1951), 79–86.

[244] Kumer, D. A., and Das, K. P. Modeling extreme hurricane damage using the generalized Pareto distribution. *American Journal of Mathematical and Management Sciences 35*, 1 (2016), 55–66.

[245] Lane, N. *Life Ascending: The Ten Great Inventions of Evolution*. Profile Books, 2010.

[246] Langton, C. G. Computation at the edge of chaos: phase transitions and emergent computation. *Physica D 42* (1990), 12–37.

[247] Lasi, H., Fettke, P., Kemper, H.-G., Feld, T., and Hoffmann, M. Industry 4.0. *Business and Information Systems Engineering 6*, 4 (2014), 239.

[248] LeBaron, B. Building the Santa Fe artificial stock market. *Working Paper, Brandeis University* (2002).

[249] Leduc, M., and Thurner, S. Incentivizing resilience in financial networks. *Journal of Economic Dynamics and Control 82* (2017), 44–66.

[250] Lenski, R. E., Ofria, C., Pennock, R. T., and Adami, C. The evolutionary origin of complex features. *Nature 423* (2003), 139–44.

[251] Leskovec, J., and Horvitz, E. Planetary-scale views on a large instant-messaging network. In *Proceedings of the 17th International Conference on the World Wide Web* (2008), ACM, pp. 915–24.

[252] Liggett, T. M. Coexistence in threshold voter models. *Annals of Probability* (1994), 764–802.

[253] Liggett, T. M. *Interacting Particle Systems*, Vol. 276. Springer, 2012.

[254] Lindeberg, J. W. Eine neue Herleitung des Exponentialgesetzes in der Wahrscheinlichkeitsrechnung. *Mathematische Zeitschrift 15* (1922), 211–25.

[255] Lloyd, S. Measures of complexity: a non-exhaustive list. *Control Systems Magazine, IEEE 21* (2001), 7–8.

[256] Lotka, A. The frequency distribution of scientific productivity. *Journal of the Washington Academy Sciences 16* (1926), 317–23.

[257] MacKay, D. *Information Theory, Inference, and Learning Algorithms*. Cambridge University Press, 2003.

[258] Malamud, B. D., Morein, G., and Turcotte, D. L. Forest fires: an example of self-organized critical behavior. *Science 281* (1998), 1840–2.

[259] Mandelbrot, B. *An Informational Theory of the Statistical Structure of Language*, Vol. 83. Butterworth, 1953.

[260] Mann, H. B., and Whitney, D. R. On a test of whether one of two random variables is stochastically larger than the other. *Annals of Mathematical Statistics* (1947), 50–60.

[261] Mantegna, R. N., and Stanley, H. E. *An Introduction to Econophysics: Correlations and Complexity in Finance*. Cambridge University Press, 1999.

[262] Mantegna, R. N. Hierarchical structure in financial markets. *European Physical Journal B 11*, 1 (1999), 193–7.

[263] Martin, W., Baross, J., Kelley, D., and Russell, M. J. Hydrothermal vents and the origin of life. *Nature Reviews Microbiology* 6, 11 (2008), 805–14.

[264] Matsumaru, N., di Fenizio, P. S., Centler, F., and Dittrich, P. On the evolution of chemical organizations. In *Proceedings of the 7th German Workshop of Artificial Life* (2006), pp. 135–46.

[265] Matthew, M. *Matthew 25:29 in Luther, M.* Hans Lufft, Wittenberg, 1545.

[266] May, R. M., and Lloyd, A. L. Infection dynamics on scale-free networks. *Physical Review E* 64, 6 (2001), 066112.

[267] McCord, K. R. *Managing the Integration Problem in Concurrent Engineering.* PhD thesis, MIT, 1993.

[268] McMillan, B. The basic theorems of information theory. *Annals of Mathematical Statistics* 24 (1953), 196–219.

[269] McMillan, B. Two inequalities implied by unique decipherability. *IEEE Transactions Information Theory* 2 (1956), 115–16.

[270] McNerney, J., Farmer, J. D., Redner, S., and Trancik, J. E. Role of design complexity in technology improvement. *Proceedings of the National Academy of Sciences* 108, 22 (2011), 9008–13.

[271] Mehta, M. L. *Random Matrices,* Vol. 142. Academic Press, 2004.

[272] Meyer-Ortmanns, H., and Thurner, S. *Principles of Evolution: From the Planck Epoch to Complex Multicellular Life.* Springer, 2011.

[273] Mézard, M., Parisi, G., and Virasoro, M. *Spin Glass Theory and Beyond: An Introduction to the Replica Method and its Applications,* Vol. 9. World Scientific Publishing, 1987.

[274] Miller, G. A., and Chomsky, N. Finitary models of language users. In *Handbook of Mathematical Psychology* (Ed. D. Luce), John Wiley & Sons, 1963.

[275] Minkoff, E. *Evolutionary Biology.* Addison Wesley, 1983.

[276] Mitzenmacher, M. A brief history of generative models for power law and lognormal distributions. *Internet Mathematics* 226–251 (2003).

[277] Montemurro, M. A. Beyond the Zipf-Mandelbrot law in quantitative linguistics. *Physica A* 300 (2001), 567–78.

[278] Moore, G. E. Cramming more components onto integrated circuits, reprinted from electronics. *IEEE Solid-State Circuits Society Newsletter* 38, 8 (1965), 114–17.

[279] Moreno, Y., Pastor-Satorras, R., and Vespignani, A. Epidemic outbreaks in complex heterogeneous networks. *European Physical Journal B* 26, 4 (2002), 521–9.

[280] Myhrvold, N. Moore's law corollary: Pixel power. *New York Times* 7 (2006).

[281] Nagy, B., Farmer, J. D., Bui, Q. M., and Trancik, J. E. Statistical basis for predicting technological progress. *PLoS ONE* 8, 2 (2013), 1–7.

[282] Naudts, J. *Generalised Thermostatistics. Springer,* 2011.

[283] Nelson, R. R., and Winter, S. G. *An Evolutionary Theory of Economic Change.* Harvard University Press, 2009.

[284] Nevzorov, V. *Records.* Translations of a Mathematical Monograph, American Mathematical Society, 2001.

[285] Newman, M. E. J. Spread of epidemic disease on networks. *Physical Review E* 66, 1 (2002), 016128.

[286] Newman, M. E. J. Fast algorithm for detecting community structure in networks. *Physical Review E* 69, 6 (2004), 066133.

[287] Newman, M. E. J. Finding community structure in networks using the eigenvectors of matrices. *Physical Review E* 74, 3 (2006), 036104.

[288] Newman, M. E. J., and Girvan, M. Finding and evaluating community structure in networks. *Physical Review E* 69, 2 (2004), 026113.

[289] Newman, M. E. J., and Palmer, R. G. *Modeling Extinction*. Oxford University Press, 2003.

[290] Newman, M. E. J. Power laws, Pareto distributions and Zipf's law. *Contemporary Physics 46*, 5 (2005), 323–51.

[291] Newman, M. E. J. *Networks: An Introduction*. Oxford University Press, 2010.

[292] Noh, J. D., and Rieger, H. Random walks on complex networks. *Physical Review Letters 92*, 11 (2004), 118701.

[293] Nolan, J. *Stable Distributions: Models for Heavy Tailed Data*. Birkhauser, New York, 2003.

[294] Ogburn, W. F. *Social Change with Respect to Culture and Original Nature*. BW Huebsch, Incorporated, 1922.

[295] Opsahl, T., Agneessens, F., and Skvoretz, J. Node centrality in weighted networks: generalizing degree and shortest paths. *Social Networks 32*, 3 (2010), 245–51.

[296] Otto, S. P., and Lenormand, T. Resolving the paradox of sex and recombination. *Nature Reviews Genetics 3*, 4 (2002), 252–61.

[297] Page, K. M., and Nowak, M. A. Unifying evolutionary dynamics. *Journal of Theoretical Biology 219*, 1 (2002), 93–8.

[298] Page, L., Brin, S., Motwani, R., and Winograd, T. The pagerank citation ranking: bringing order to the web. Techical report, Stanford InfoLab, 1999.

[299] Pareto, V. *Cours d'Économie Politique*. New edition by G.-H. Bousquet and G. Busino, Librairie Droz, Geneva, 1964.

[300] Parisi, G. Complex systems: a physicist's viewpoint. *Physica A 263*, 1–4 (1999), 557–64.

[301] Pastor-Satorras, R., and Vespignani, A. Epidemic spreading in scale-free networks. *Physical Review Letters 86*, 14 (2001), 3200–3.

[302] Pastor-Satorras, R., and Vespignani, A. Immunization of complex networks. *Physical Review E 65*, 3 (2002), 036104.

[303] Pearson, K. Note on regression and inheritance in the case of two parents. *Proceedings of the Royal Society of London 58* (1895), 240–2.

[304] Pearson, K. On the criterion that a given system of deviations from the probable in the case of a correlated system of variables is such that it can be reasonably supposed to have arisen from random sampling. *The London, Edinburgh, and Dublin Philosophical Magazine and Journal of Science 50*, 302 (1900), 157–75.

[305] Pearson, K. *On the Theory of Contingency and its Relation to Association and Normal Correlation*. Cambridge University Press, 1904.

[306] Perkins, T. J., Foxall, E., Glass, L., and Edwards, R. A scaling law for random walks on networks. *Nature Communications 5* (2014), 5121.

[307] Perron, O. Zur Theorie der Matrizen. *Mathematische Annalen 64*, 2 (1907), 248–63.

[308] Peters, O., Hertlein, C., and Christensen, K. A complexity view of rainfall. *Physical Review Letters 88* (2001), 018701.

[309] Planck, M. On the law of distribution of energy in the normal spectrum. *Annalen der Physik 4* (1901), 553–63.

[310] Poledna, S., Molina-Borboa, J., van der Leij, M., Martinez-Jaramillo, S., and Thurner, S. Multilayer network nature of systemic risk in financial networks and its implications for the costs of financial crises. *Journal of Financial Stability 20* (2015), 70–81.

[311] Poledna, S., and Thurner, S. Elimination of systemic risk in financial networks by means of a systemic risk transaction tax. *Quantitative Finance 16* (2016), 1599–613.

[312] Poledna, S., and Thurner, S. Basel III capital surcharges for G-SIBs are far less effective in managing systemic risk in comparison to network-based, systemic risk-dependent financial transaction taxes. *Journal of Economic Dynamics and Control 77* (2017), 230–46.

[313] Pólya, G. Über ganzwertige Polynome in algebraischen Zahlkörpern. *Journal für Reine und Angewandte Mathathematik 149* (1919), 97–116.

[314] Pólya, G. Sur quelques points de la théorie des probabilités. *Annales de l'Institute Henri Poincare 1* (1930), 117–161.

[315] Powner, M. W., Gerland, B., and Sutherland, J. D. Synthesis of activated pyrimidine ribonucleotides in prebiotically plausible conditions. *Nature 459* (2009), 239–42.

[316] Press, W., and Dyson, F. Iterated prisoner's dilemma contains strategies that dominate any evolutionary opponent. *Proceedings of the National Academy of Sciences USA 109* (2012), 10409–13.

[317] Raup, D. M., and Sepkoski, J. J. Mass extinctions in the marine fossil record. *Science 215* (1982), 1501–3.

[318] Redner, S. *A Guide to First-Passage Processes.* Cambridge University Press, 2001.

[319] Reed, M., and Simon, B. *Methods of Modern Mathematical Physics. Functional Analysis.* Academic Press, 1980.

[320] Reed, W. J., and Hughes, B. D. From gene families and genera to incomes and internet file sizes: why power laws are so common in nature. *Physical Review E 66* (2002), 067103.

[321] Rényi, A. On measures of entropy and information. In *Proceedings of the Fourth Berkeley Symposium on Mathematical Statistics and Probability, Vol 1: Contributions to the Theory of Statistics.* The University of California Press, 1961.

[322] Rodriguez-Iturbe, I., Ijjasz-Vasquez, E. J., Bras, R. L., and Tarboton, D. G. Power law distributions of discharge mass and energy in river basins. *Water Resources Research 28*, 4 (1992), 1089–93.

[323] Rokach, L., and Maimon, O. Clustering methods. In *Data Mining and Knowledge Discovery Handbook.* Springer, 2005, pp. 321–52.

[324] Royer, L., Reimann, M., Andreopoulos, B., and Schroeder, M. Unraveling protein networks with power graph analysis. *PLoS Computational Biology 4*, 7 (2008), e1000108.

[325] Sahal, D. A theory of progress functions. *AIIE Transactions 11*, 1 (1979), 23–9.

[326] Saichev, A., and Sornette, D. Power law distribution of seismic rates. *Tectonophysics 431* (2007), 7–13.

[327] Samuelson, P. *Economics.* Tata McGraw Hill, 2010.

[328] Schelling, T. C. Dynamic models of segregation. *Journal of Mathematical Sociology 1*, 2 (1971), 143–86.

[329] Schumpeter, J. A. *Theorie der Wirtschaftlichen Entwicklung: eine Untersuchung über Unternehmergewinn, Kapital, Kredit, Zins und den Konjunkturzyklus.* Duncker and Humblot, 1911.

[330] Schumpeter, J. A. *Business Cycles*, Vol. 1. McGraw-Hill, New York, 1939.

[331] Schweitzer, F., Fagiolo, G., Sornette, D., Vega-Redondo, F., Vespignani, A., and White, D. R. Economic networks: the new challenges. *Science 325* (2009), 422–5.

[332] Semiconductor Industry Association. Global semiconductor sales top $335 billion in 2015, 2016. [Online; accessed 7-March-2017].

[333] Sepkoski, J. J. Ten years in the library: new data confirm paleontological patterns. *Paleobiology 19* (1993), 43–51.

[334] Serrano, M. Á., Boguná, M., and Vespignani, A. Extracting the multiscale backbone of complex weighted networks. *Proceedings of the National Academy of Sciences USA 106*, 16 (2009), 6483–8.

[335] Shafee, F. Lambert function and a new non-extensive form of entropy. *IMA Journal of Applied Mathematics 72* (2007), 785–800.

[336] Shannon, C. A mathematical theory of communication. *Bell System Technical Journal 27* (1948), 379–423, 623–56.

[337] Shields, R. Cultural topology: the seven bridges of Königsberg, 1736. *Theory, Culture and Society 29*, 4-5 (2012), 43–57.

[338] Silverberg, G., and Verspagen, B. The size distribution of innovations revisited: an application of extreme value statistics to citation and value measures of patent significance. *Journal of Econometrics 139*, 2 (2007), 318–39.

[339] Simmel, G. *Sociologie. Untersuchungen über die Formen der Vergesellschaftung*. Duncker und Humblot, 1908.

[340] Simon, H. A. On a class of skew distribution functions. *Biometrika 42*, 3/4 (1955), 425–40.

[341] Smith, E., and Morowitz, H. J. *The Origin of Nature and Life on Earth*. Cambridge University Press, 2016.

[342] Soffer, S. N., and Vázquez, A. Network clustering coefficient without degree-correlation biases. *Physical Review E 71*, 5 (2005), 057101.

[343] Solé, R. V., and Manrubia, S. C. Extinction and self-organized criticality in a model of large-scale evolution. *Physical Review E 54*, 1 (1996), R42.

[344] Solomonoff, R. J. The mechanization of linguistic learning. In *Proceedings of the Second International Congress on Cybernetics, Namur, Belgium, 1958* (1960), pp. 180–93.

[345] Sood, V., and Redner, S. Voter model on heterogeneous graphs. *Physical Review Letters 94*, 17 (2005), 178701.

[346] Sornette, D. *Critical Phenomena in Natural Sciences*. Springer, 2006.

[347] Sornette, D., and Zajdenweber, D. The economic return of research: the Pareto law and its implications. *European Physical Journal B 8* (1999), 653–64.

[348] Spearman, C. The proof and measurement of association between two things. *American Journal of Psychology 15*, 1 (1904), 72–101.

[349] Stadler, P. F., Fontana, W., and Miller, J. H. Random catalytic reaction networks. *Physica D: Nonlinear Phenomena 63*, 3-4 (1993), 378–92.

[350] Stanley, H. E. *Introduction to Phase Transitions and Critical Phenomena*. Oxford University Press, 1971.

[351] Stanley, H. E. Scaling, universality, and renormalization: three pillars of modern critical phenomena. *Reviews of Modern Physics 71*, 2 (1999), S358.

[352] Steels, L. The synthetic modeling of language origins. *Evolution of Communication 1*, 1 (1997), 1–34.

[353] Steels, L., and Vogt, P. Grounding adaptive language games in robotic agents. In *Proceedings of the Fourth European Conference on Artificial Life* (1997).

[354] Stein, C. A bond for the error in the normal approximation to the distribution of a sum of dependent random variables. In *Proceedings of the Sixth Berkley Symposium on Mathematical Statistics and Probability 2* (1972), pp. 583–602.

[355] Stigler, S. M. *The History of Statistics: The Measurement of Uncertainty Before 1900*. Harvard University Press, 1986.

[356] Stokic, D., Hanel, R., and Thurner, S. Inflation of the edge of chaos in a simple model of gene interaction networks. *Physical Review E 77* (2008), 061917.

[357] Suchecki, K., Eguíluz, V. M., and San Miguel, M. Voter model dynamics in complex networks: role of dimensionality, disorder, and degree distribution. *Physical Review E 72*, 3 (2005), 036132.

[358] Sumbaly, R., Kreps, J., and Shah, S. The big data ecosystem at linkedin. In *Proceedings of the 2013 ACM SIGMOD International Conference on Management of Data* (2013), ACM, pp. 1125–34.

[359] Sznajd-Weron, K., and Sznajd, J. Opinion evolution in closed community. *International Journal of Modern Physics C 11*, 6 (2000), 1157–65.

[360] Tao, T. *Topics in Random Matrix Theory*, Vol. 132. American Mathematical Society, Providence, 2012.

[361] Thirring, W. Systems with negative specific heat. *Zeitschrift für Physik 235* (1970), 339–52.

[362] Thompson, P. Learning by doing. *Handbook of the Economics of Innovation 1*. Elsevier, 2010 429–76.

[363] Thurner, S., Corominas-Murtra, B., and Hanel, R. Three faces of entropy for complex systems: information, thermodynamics, and the maximum entropy principle. *Physical Review E 96* (2017), 032124.

[364] Thurner, S., Feurstein, M., and Teich, M. Multiresolution wavelet analysis of heartbeat intervals discriminates healthy patients from those with cardiac pathology. *Physical Review Letters 80* (1998), 1544–7.

[365] Thurner, S., and Hanel, R. Entropies for complex systems: Generalized-generalized entropies. In *Complexity, Metastability and Nonextensivity, AIP 965* (2007), pp. 68–75.

[366] Thurner, S., and Hanel, R. What do generalized entropies look like? An axiomatic approach for complex, non-ergodic systems. In *Recent Advances in Generalized Information Measures and Statistics, Bentham Science eBook;* see also appendix in *arXiv:1104.2070* (2013).

[367] Thurner, S., Hanel, R., Liu, B., and Corominas-Murtra, B. Understanding Zipf's law of word frequencies through sample-space collapse in sentence formation. *Journal of the Royal Society Interface 12* (2016), 20150330.

[368] Thurner, S., Klimek, P., and Hanel, R. Schumpeterian economic dynamics as a quantifiable model of evolution. *New Journal of Physics 12*, 7 (2010), 075029.

[369] Thurner, S., and Poledna, S. Debtrank-transparency: controlling systemic risk in financial networks. *Scientific Reports 3* (2013), 1888.

[370] Thurner, S., Szell, M., and Sinatra, R. Emergence of good conduct, scaling and Zipf laws in human behavioral sequences in an online world. *PLoS ONE 7*, 1 (2012), e29796.

[371] Thurner, S., and Tsallis, C. Nonextensive aspects of self-organized, scale-free, gas-like networks. *Europhysics Letters 72* (2005), 197–203.

[372] Thurner, S., Windischberger, C., Moser, E., Walla, P., and Barth, M. Scaling laws and persistence in human brain activity. *Physica A 326* (2003), 511–21.

[373] Thurston, R. H. *A History of the Growth of the Steam-Engine*. D. Appleton, 1883.

[374] Tolusso, D., and Wang, X. Interval estimation for response adaptive clinical trials. *Computational Statistics and Data Analysis 55* (2011), 725–30.

[375] Travers, J., and Milgram, S. The small world problem. *Psychology Today 1* (1967), 61–7.

[376] Tria, F., Loreto, V., Servedio, V., and Strogatz, S. The dynamics of correlated novelties. *Scientific Reports 4* (2014), 5890.

[377] Tribus, M., and McIrvine, E. C. Energy and information. *Scientific American 224* (1971), 179–88.

[378] Tsallis, C. Possible generalization of Boltzmann–Gibbs statistics. *Journal of Statistical Physics 52* (1988), 479–87.

[379] Tsallis, C. *Introduction to Nonextensive Statistical Mechanics: Approaching a Complex World.* Springer, 2009.

[380] Tsallis, C. Bibliography of works on Tsallis entropy. *http://tsallis.cat.cbpf.br/TEMUCO.pdf* (2017).

[381] Tsallis, C., and Bukman, D. Anomalous diffusion in the presence of external forces: exact time-dependent solutions and their thermostatistical basis. *Physical Review E 54* (1996), R2197.

[382] Tsallis, C., and Cirto, L. Black hole thermodynamical entropy. *European Physical Journal C 73* (2013), 2487.

[383] Tsekouras, G., and Tsallis, C. Generalized entropy arising from a distribution of q indices. *Physical Review E 71* (2005), 046144.

[384] Turcotte, D. L. *Fractals and Chaos in Geology and Geophysics*. Cambridge University Press, 1997.

[385] Turcotte, D. L. Self-organized criticality. *Reports on Progress in Physics 62* (1999), 1377–429.

[386] Ubriaco, M. Entropies based on factional calculus. *Physics Letters A 373* (2009), 2516–19.

[387] Ugander, J., Karrer, B., Backstrom, L., and Marlow, C. The anatomy of the facebook social graph. *arXiv:1111.4503* (2011).

[388] United States Census Bureau, 2017. [Online; accessed 2 January 2018].

[389] Vannimenus, J., and Toulouse, G. Theory of the frustration effect. ii. Ising spins on a square lattice. *Journal of Physics C 10*, 18 (1977), L537.

[390] Vázquez, A. Growing network with local rules: preferential attachment, clustering hierarchy, and degree correlations. *Physical Review E 67*, 5 (2003), 056104.

[391] Vazquez, F., Eguíluz, V. M., and San Miguel, M. Generic absorbing transition in coevolution dynamics. *Physical Review Letters 100*, 10 (2008), 108702.

[392] Viksnins, G. *Economic Systems in Historical Perspective*. Kendall Hunt, 1997.

[393] von Neumann, J., and Morgenstern, O. *Theory of Games and Economic Behavior*. Princeton University Press, 1944.

[394] Wallstrom, T. The equalization probability of Pólya urn. *American Mathematical Monthly 119* (2012), 516–18.

[395] Wang, R., Nisoli, C., Freitas, R., Li, J., McConville, W., Cooley, B., Lund, M., Samarth, N., Leighton, C., Crespi, V. H., and Schiffer, P. Artificial 'spin ice' in a geometrically frustrated lattice of nanoscale ferromagnetic islands. *Nature 439*, (2006), 303–06.

[396] Wang, Y., Chakrabarti, D., Wang, C., and Faloutsos, C. Epidemic spreading in real networks: an eigenvalue viewpoint. In *22nd International Symposium on Reliable Distributed Systems* (2003), IEEE, pp. 25–34.

[397] Ward, P. D., Botha, J., Buick, R., De Kock, M. O., Erwin, D. H., Garrison, G. H., Kirschvink, J. L., and Smith, R. Abrupt and gradual extinction among late permian land vertebrates in the Karoo basin, South Africa. *Science 307* (2005), 709–14.

[398] Wasserman, S., and Faust, K. *Social Network Analysis: Methods and Applications*, Vol. 8. Cambridge University Press, 1994.

[399] Watts, D. J. *Small Worlds: The Dynamics of Networks between Order and Randomness*. Princeton University Press, 1999.

[400] Watts, D. J. A simple model of global cascades on random networks. *Proceedings of the National Academy of Sciences USA 99*, 9 (2002), 5766–71.

[401] Watts, D. J. *Six Degrees: The Science of a Connected Age*. WW Norton and Company, 2004.

[402] Watts, D. J., and Strogatz, S. H. Collective dynamics of 'small-world' networks. *Nature 393* (1998), 440–2.

[403] Weibull, W. A statistical distribution function of wide applicability. *Journal of Applied Mechanics 18* (1951), 293–7.

[404] Weinberger, E. D., and Fassberg, A. NP completeness of Kauffman's N-K model, a tuneably rugged fitness landscape. *Santa Fe Institute Working Paper: 1996-02-003* (1996).

[405] Weisstein, E. Fundamental forces, 2007. [Online; accessed 22-January-2018].

[406] West, D. B. *Introduction to Graph Theory*, Vol. 2. Prentice Hall Upper Saddle River, 2001.

[407] West, G., Brown, J., and Enquist, B. A general model for the origin of allometric scaling laws in biology. *Science 276* (1997), 122–6.

[408] Wigner, E. P. Characteristic vectors of bordered matrices with infinite dimensions. In *The Collected Works of Eugene Paul Wigner*. Springer, 1993, pp. 524–40.

[409] Willis, J. C. *Age and Area*. Cambridge University Press, 1922.

[410] Wilson, K. The renormalization group: critical phenomena and the Kondo problem. *Reviews of Modern Physics 47*, 4 (1975), 773–840.

[411] Wong, C.-Y., and Wilk, G. Tsallis fits to PT spectra for pp collisions at the LHC. *Acta Physica Polonica B 43*, 11 (2012), 2047.

[412] Wright, S. The roles of mutation, inbreeding, crossbreeding, and selection in evolution. In *Proceedings of the Sixth International Congress on Genetics 1* (1932), pp. 355–66.

[413] Wright, T. P. Factors affecting the cost of airplanes. *Journal of the Aeronautical Sciences 3*, 4 (1936), 122–8.

[414] Yang, Z., Algesheimer, R., and Tessone, C. J. A comparative analysis of community detection algorithms on artificial networks. *Scientific Reports 6* (2016), 30750.

[415] Yates, F. Contingency tables involving small numbers and the χ^2 test. *Supplement to the Journal of the Royal Statistical Society 1*, 2 (1934), 217–35.

[416] Yule, U. G. A mathematical theory of evolution, based on the conclusions of Dr. J. C. Willis, F.R.S. *Philosophical Transactions of the Royal Society B 213* (1925), 402–10.

[417] Zhang, B., and Horvath, S. A general framework for weighted gene co-expression network analysis. *Statistical Applications in Genetics and Molecular Biology 4* (2005), 17.

[418] Zhou, D., Huang, J., and Schölkopf, B. Learning with hypergraphs: clustering, classification, and embedding. In *Advances in Neural Information Processing Systems* (2007), pp. 1601–8.

[419] Zipf, G. K. *The Psychobiology of Language*. Houghton-Mifflin, 1935.

[420] Zipf, G. K. *Human Behavior and the Principle of Least Effort*. Addison Wesley, 1949.

Index